I0071954

Exploring Robotics and Robotic Systems

Exploring Robotics and Robotic Systems

Edited by **Rowland Wilson**

NYRESEARCH
PRESS

New York

Published by NY Research Press,
23 West, 55th Street, Suite 816,
New York, NY 10019, USA
www.nyresearchpress.com

Exploring Robotics and Robotic Systems
Edited by Rowland Wilson

© 2016 NY Research Press

International Standard Book Number: 978-1-63238-471-3 (Hardback)

This book contains information obtained from authentic and highly regarded sources. Copyright for all individual chapters remain with the respective authors as indicated. All chapters are published with permission under the Creative Commons Attribution License or equivalent. A wide variety of references are listed. Permission and sources are indicated; for detailed attributions, please refer to the permissions page and list of contributors. Reasonable efforts have been made to publish reliable data and information, but the authors, editors and publisher cannot assume any responsibility for the validity of all materials or the consequences of their use.

The publisher's policy is to use permanent paper from mills that operate a sustainable forestry policy. Furthermore, the publisher ensures that the text paper and cover boards used have met acceptable environmental accreditation standards.

Trademark Notice: Registered trademark of products or corporate names are used only for explanation and identification without intent to infringe.

Printed in the United States of America.

Contents

Permissions

List of Contributors

Preface

It is often said that books are a boon to mankind. They document every progress and pass on the knowledge from one generation to the other. They play a crucial role in our lives. Thus I was both excited and nervous while editing this book. I was pleased by the thought of being able to make a mark but I was also nervous to do it right because the future of students depends upon it. Hence, I took a few months to research further into the discipline, revise my knowledge and also explore some more aspects. Post this process, I began with the editing of this book.

Robotics is an umbrella discipline which brings together several different engineering domains such as mechanical engineering, electrical engineering and computer engineering, as well as computer systems for their control, sensory feedback, and information processing. Robotics is a dynamic field, as technological advances continue; building new robots serves diverse practical purposes, both domestically and commercially. Assembly robots, welding robots, combat robots are some of the popular categories of robots in the current times. This book provides a holistic overview of robotics by discussing its components, behavior, cognition and applications. It brings forth contributions of experts and scientists which will provide innovative insights into this field. Students, researchers, experts, engineers and all associated with robotics will benefit alike from this book.

I thank my publisher with all my heart for considering me worthy of this unparalleled opportunity and for showing unwavering faith in my skills. I would also like to thank the editorial team who worked closely with me at every step and contributed immensely towards the successful completion of this book. Last but not the least, I wish to thank my friends and colleagues for their support.

Editor

Virtualized traffic at metropolitan scales

David Wilkie [1], Jason Sewall [2], Weizi Li [1] and Ming C. Lin [1]*

[1] *Department of Computer Science, University of North Carolina at Chapel Hill, Chapel Hill, NC, USA,* [2] *Parallel Computing Laboratory, Intel Corporation, Santa Clara, CA, USA*

Few phenomena are more ubiquitous than traffic in urban scenes, and few are more significant economically, socially, or environmentally. Many virtual-reality applications and systems, including virtual globes and immersive multi-player worlds that are often set in a large-scale modern or futuristic setting, feature traffic systems. Virtual-reality models can also aid in addressing the challenges of real-world traffic – the ever-present gridlock and congestion in cities worldwide: traffic engineers and planners can diagnose system instabilities and explore control strategies in virtual worlds reconstructed from available sensor data. To create these VR systems with traffic mimicking real-world conditions, road network models need to be created and represented. Traffic needs to be realistically and efficiently simulated. To analyze real-world scenarios, the traffic conditions need to be estimated and reconstructed. To create virtual scenarios, such as simulated cities, traffic needs to be intelligently and efficiently routed. These applications all require research advances in road network capture and modeling, intelligent traffic routing and simulation, and traffic state estimation and reconstruction. New systems need to be designed that combine these components with visual and analytical infrastructure. In this paper, we present some state-of-the-art approaches for these areas as well as our vision for unified virtual-reality traffic systems that combine and integrate them to achieve virtualized traffic.

Keywords: virtualized traffic, road networks, traffic simulation, traffic routing, urban scenes

Edited by:
Anthony Steed,
University College London, UK

Reviewed by:
Guillaume Moreau,
Ecole Centrale de Nantes, France
Yiorgos L. Chrysanthou,
University of Cyprus, Cyprus

*Correspondence:
Ming C. Lin,
Department of Computer Science,
University of North Carolina at
Chapel Hill, 201 S. Columbia Street,
Chapel Hill, NC 27599, USA
lin@cs.unc.edu

1. Introduction

In both the physical world and many virtual cities, traffic is a ubiquitous phenomenon. Modeling these systems realistically and efficiently is an ongoing challenge that can lead to more immersive and more engaging virtual urban environments, as well as enable visual analysis and control to significantly improve real-world outcomes.

Typical urban environments are rife with traffic, which has significant implications both economically and socially around the world. Systematic failings are regular events; gridlock and traffic jams cost 2.9 billion gallons of wasted fuel and cost over 121 billion dollars every year in the U.S. alone, as discussed in Schrank et al. (2012). A major drive to solve these challenges entails using the road network more efficiently via *Intelligent Traffic Systems*, combining sensing, communications, and traffic controls in a managed system. Virtual-reality models have a key role to play in (1) visualizing real-world conditions, (2) visualizing potential alternative control strategies, (3) communicating conditions and proposals to the public, and (4) providing analysts a real-world context and interface.

As self-driving and other cars become more prevalent, a shared model of the road network could significantly improve their performance. Fleets of autonomous vehicles roving the road network can build a map, including up-to-date details about road blockages and closures. Such a model can be used by the vehicles to plan more effectively than would be possible using only real-time sensor data or static road network maps. Rendered views of the model can let controllers and analysts validate the maps.

For the virtual world, research efforts have focused on creating, representing, and visualizing "digital urbanscapes" [e.g., Pausch et al. (1992), Cremer et al. (1997), and Donikian et al. (1999)]. These efforts have yielded significant results on a number of fronts, including satellite and aerial imagery melded into virtual globes, the creation of 3D models of cities, real-time rendered worlds, and new procedural modeling techniques. Google Earth and Microsoft Virtual Earth, which feature imagery of most cities and 3D models, are both products of this success. The creation of 3D city models from sensor data has been a focus area as has work on streaming imagery and geometry for real-time rendering of massive models on everything from high-end workstations to mobile devices.

Despite the impressive progress that has been made, these efforts focused only on static models of the world, ignoring dynamic elements such as traffic. The evolution of flight and driving simulators from single-user programs to vast virtual worlds magnifies the demand for a higher level of detail and for dynamic elements. The addition of traffic to these worlds, especially driven by real-time sensor data, would considerably enhance the realism and immersion of these worlds. For other applications, such as video games set in modern cities, city planning programs, and traffic simulators, the need for traffic is even more clear. Further, these traffic systems need to be realistic and need to behave intelligently, often while operating within a tight computational budget. There are existing systems for traffic simulation, [e.g., MITSIM (2011) and SUMO (2009)] and ongoing active research on new methodologies [see Delling et al. (2009) for details]. There has also been work on incorporating realistic traffic in virtual environments, such as Plumert et al. (2004), Thomson et al. (2005), and Wang et al. (2005), but these have focused on small scenarios: single roads or intersections, predominantly.

These applications demand advancements in modeling and simulating *virtualized traffic*, i.e., *the creation of virtual traffic flows directly from real-world data*. In this paper, we present a vision based on the synthesis of our recent work to construct virtualized traffic at metropolitan scales for VR applications. First, for the creation and representation of virtual road networks, we review related approaches and suggest a direction forward for virtual models suitable for the above applications, as discussed above. Our approach builds off of approaches to create simulation-enabled representations from GIS road network models, which are now publicly available in vast databases. This step involves building formal models atop of the noisy, human-authored data before creating a geometric representation. These models can also serve as a prior model to be fused with traffic sensor data. Second, to simulate virtualized traffic, we present approaches for creating hybrid traffic simulators that allow for level-of-detail simulations, enabling low-cost simulation of background traffic that can be switched to high-fidelity traffic simulation when necessary (see Section 2). Third, to estimate and communicate traffic conditions, we present a method that filters traffic sensor data, discrete in both space and time, to recreate a continuous estimate of traffic

conditions that can be used to create visualizations or predict future traffic patterns, as described in Section 3. Fourth, to enable low-cost intelligent routing for city-scale navigation, we present an approach that manages traffic flows via "*participatory traffic routing*," in which the plans for individual vehicles are aggregated to coordinate their routes, as presented in Section 4. Using this approach, simulated, virtualized vehicles can react intelligently to the addition of roads or to anomalous conditions. Finally, in Section 5, we discuss our approaches and the directions for future work they point toward.

2. Road Networks

Road networks are an integral part of the traffic system. These networks can be models of real-world networks, captured using a variety of techniques, or can be artificial networks, generated to represent fictitious locales in virtual worlds. In either case, these networks are geometrically and topologically complex, making manual creation difficult for large-scale models.

There have been a variety of efforts directed at the problem of representing, designing, and acquiring road networks. Thomas and Donikian (2000) presented an early and influential approach to modeling human environments including streets, buildings, and free spaces. The model is used to create behavioral animations of pedestrians and vehicles in highly detailed scenarios. Willemsen et al. (2006) discuss "ribbon networks" for modeling the paths of agents in a road network, such that kinematic behavior is preserved. Wang et al. (2002) propose curves parameterized by their arc-length for motion-control and animation suitable for description of vehicle motion. Garcia-Dorado et al. (2014) take the novel approach of letting desired traffic behavior guide the creation of road networks and cityscapes. While these approaches do a good job of representing roads, they often do not address the challenges of creating large-scale models, especially from noisy data, and some road network models would not be able to support efficient simulation due to their geometric representations.

2.1. Levels of Detail
Road networks can be represented by differing levels of detail, from polygonal graphs to meshes and parameterized curves. These representations are suitable for different applications and have different costs associated for creation and use.

2.1.1. Directed Graph
A road network can be represented very simply as a directed graph for which edges, E, represent segments of roadway, and the vertices, V, represent intersections between roads and terminus points. Each edge can have attributes such as length, number of lanes, and speed limit. Representations such as this are suitable for online map display and for applications that do not require a high level of detail, such as simple routing operations.

However, a network such as this would not be suited for any visual analysis, animation, or highly detailed simulation, as the geometric details of the road are lost. Some problems would also be difficult to handle with this representation, such as determining which roads correspond to a sequence of global-positioning system (GPS) coordinates.

2.1.2. Polyline Road Networks

To address these shortcomings, road networks must be embedded in space; typically a 2D planar (or a height-field) is used. In this representation, the vertices, V^i, i.e., intersections, have coordinates indicating their positions, and each edge e, i.e., each road segment, has additional vertices, V_e^g, that approximate the geometry of the road. The edges and vertices can have additional attributes, A, as above.

This is known as the *polyline road network* representation (Safra et al., 2006), and it is the most common for road networks and is used for nearly all geographic-information system (GIS) applications, including in online mapping and routing applications, and are widely available in public data sources.

This representation also has its shortcomings. For one, lanes are not individually represented. This limits the level of detail at which simulations can be done and limits the level of analysis. Second, the polyline (or polygonal chain) geometry can be a source of significant error – real-world cars "cut" corners, following curvilinear paths derived from their kinematics (LaValle, 2006). This is exacerbated at intersections, at which simulated cars could take unrealistic turns.

This leads into the third shortcoming, that intersections are modeled as abstract, dimensionless junctions of edges. In reality, they have a geometric extent and can encompass complicated topological relationships. A consequence of this representation is that complex intersections either must be modeled using multiple vertices or their geometric and topological details must be sacrificed: modeling the intersections with multiple vertices means that vertices no longer solely represent the topological relationships between roads, but instead also represent the geometric extent and internal connections of the intersection.

2.1.3. Lanes, Geometry, and Intersections

To address these issues, additional details can be added to the representation. First, individual lanes can be represented. These lanes need geometric representations, which can be distinct or can reference a shared road segment geometry. Second, intersections also need a more complex representation: their geometric extents and topological relations need to be captured. This includes not only a bounding geometry but also *internal lanes* that connect the incoming and outgoing lanes. Third, a more detailed representation can be used for the various geometries, such as using a higher number of vertices or using a representation that can model curves, such as a spline. Finally, a model is needed for each intersection that defines the behavior of the intersection, whether the intersection uses a traffic light logic, stop sign, or some other rules of the road.

2.2. Road Network Creation

The complexity of road network geometry and topology is such that the manipulation and creation of their representations is computationally costly. Several approaches show promise for the automated creation of road network models, using GIS data, sensors, or procedural generation, depending on the desired application. We consider here only those methods for creating reconstructions or models of real-world road networks.

2.2.1. Creation from Sensing

Numerous methods for creating GIS road geometry from satellite imagery have been proposed, many of which are described in surveys including Park et al. (2002), Mena (2003), and Fortier et al. (1999). At this point, these methods are limited to creating relatively low detail networks due to the limited resolution of satellite imagery. Similar methods using aerial imagery from aircraft and UAVs have the potential to create more detailed road maps, along with 3D reconstructions of terrain and buildings.

Rich road network models can also be created by car-mounted sensors, which are becoming increasingly common as cars with auto-pilot, adaptive cruise-control, and self-driving functions enter the market. These sensors range from simple GPS devices to laser and radar sensors capable of mapping the geometry of the roads. Fleets of vehicles such as this have the potential, in the future, to use their sensors not only for real-time navigation but also to collaboratively create full 3D models of road networks. From the point clouds and reconstructed meshes created by these methods, logical road structures, as described above, need to be extracted in order to be used for traffic simulation and animation.

2.2.2. Creation from GIS Data

Digital representations of real-world road networks are commonly available in the form of GIS *polyline road networks*, but many applications, such as microscopic traffic simulation or high-quality animation, require a higher level of detail and accuracy. Some higher quality data sets already exist, but are usually limited to relatively small scale scenarios. Other techniques focus on extrapolating a detailed road network from the 2D GIS data. The SUMO traffic simulation system (Krajzewicz et al., 2012), can now generate networks from GIS input. A recent work (Wang et al., 2014) demonstrated an approach to creating 3D geometry from GIS inputs.

In Wilkie et al. (2012), we presented work on extrapolating plausible, high detail road network models suitable for microscopic simulation and animation from publicly available geographic-information system (GIS) data, i.e., polyline roads. The resulting model is composed of both (1) a graph that captures the topological relationships of the lanes that make up the road network and (2) a geometric representation that captures the shape of the roads as an optimal sequence of arcs and line segments. The model has the following features:

- It has sufficient data for high detail traffic simulation, visualization, and analysis, created automatically.
- The geometric road representation is C^1 continuous.
- Geometric operations such as calculating the distance between cars and the position along the road can be computed efficiently.

Our approach creates a *polyline road network* as described above. While each GIS data set has some notion of a road (or "way"), the usage is not necessarily consistent nor is the definition necessarily useful for further processing. We enforce a formal definition for a road on the data, easing filtering steps and enabling the creation of more complex road network features. We then create geometric representations for the roads and their intersections,

FIGURE 1 | A selection of road networks generated by our approach. **(A)** A large-scale highway scene. **(B)** A highway overpass created by our method for a 3D application. **(C)** A road network created by our approach and overlaid on a satellite image. **(D)** Ramps connecting to a divided highway. Wilkie et al. (2012). Image © IEEE.

including for details such as highway ramps. Some example road networks generated with this approach can be seen in **Figure 1**.

In addition, we plan to release the code for creating a 3D road network directly from GIS data as described in this section at: http://gamma.cs.unc.edu/RoadLib.

3. Traffic Simulation

Research on traffic simulation began in 1950s and is still an active field. Helbing (2001) serves as a good entry point to the field, although there have been many developments since. Doniec et al. (2008) apply strong behavioral models to agent-based simulation, as does Lu et al. (2014b). Shen and Jin (2012) discuss urban agent-based simulation with a focus on smooth lane changes, and recent work by Lu et al. (2014a) and focuses on preventing agent collisions in rural environments.

An essential component of traffic systems is all the vehicles – the cars, trucks, motorcycles, etc. The state of the vehicles is time-varying (except at very severe traffic jams), and a representation is needed for both the vehicles that are on the roads at a particular time and for the boundary conditions of the system.

At any moment of time, we can refer to the *state* of the traffic system, which is the specification of all time-varying values, such as how many cars are on the road; how fast the cars are moving; and how many cars are entering the system. The representation of the state is dependent on how the dynamics of the traffic system are modeled. There are two primary approaches. First, every vehicle can be represented. This is the most direct approach: for every car in the real, there is a virtual car in the model with a known position along a lane, a velocity, and perhaps other parameters. This type of representation is referred to as *microscopic* in the literature or *agent-based* if the car models encapsulate some

decision making ability. Second, the vehicles can be represented as average quantities over some spatial discretization, i.e., a lane can be divided into cells, and each of these cells can contain a density value and velocity value that represent the average of these statistics in the real-world. This type of representation is called *macroscopic*. Each state representation has its associated dynamics definition, i.e., how the cars actually move. Like the state representation, there are two broad categories for these models – microscopic and macroscopic. For the former, individual vehicles are simulated, typically by calculating an acceleration from the state of the vehicle and the vehicle ahead of it along the lane. For the latter, differential equations define the evolution of the density and velocity fields over time. In either case, the model of the dynamics can be embodied in a simulation, which, if given initial and boundary conditions, can model the evolution of the traffic dynamics over time.

Real-world road networks span whole continents. Working with networks, this large would be impractical, and so some region of interest must be extracted. Cars pass into and out of this region, and boundary modeling can be used to control the edge behavior of the system. "Internal boundaries," such as large parking lots, can be similarly represented.

3.1. Animation and Immersion with Continuum Simulation

We have shown in Sewall et al. (2010) that it is possible to create realistic animations and immersive simulations using continuum (or macroscopic) simulations. This requires a continuum simulation model that is well defined for all road network features, including changes in speed limit, intersections, highway ramps, etc. In our work, we used the state-of-the-art *Aw-Rascle-Zhang* model of traffic dynamics (Aw and Rascle, 2000; Zhang, 2002) on

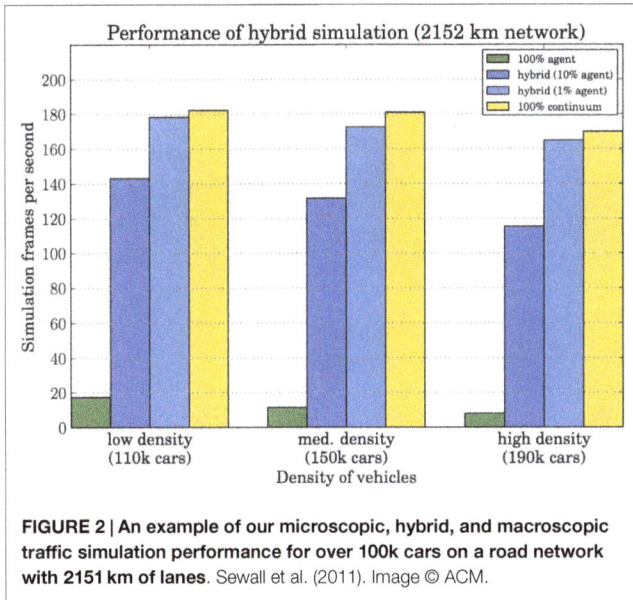

FIGURE 2 | An example of our microscopic, hybrid, and macroscopic traffic simulation performance for over 100k cars on a road network with 2151 km of lanes. Sewall et al. (2011). Image © ACM.

TABLE 1 | Sequence comparison results (in percentile) along the NGSIM 101 freeway.

Metric	Agent-based	Continuum	Hybrid
Flux LCSS	0.934	0.541	0.820
Flux EDR	0.951	0.685	0.861

The rows show that our hybrid technique, in contrast to agent-based, microscopic traffic simulation, gives only a modest drop (in the mid to upper 80 percentile) in comparison scores with the real-world data.

a fully featured road network as described above. This simulation is visualized using *carticles*, microscopic cars that are coupled with the macroscopic simulation. These cars match the traffic simulation state determined by the continuum simulator as well as determine when certain events occur within the simulation, such as lane changes. With this approach, we can easily simulate many thousands of cars at faster than real time (Sewall et al., 2010).

3.2. Level-of-Detail Simulation

Though the above work adds individual vehicles to the continuum traffic simulation, there are scenarios in which a more detailed, microscopic simulation is needed. If a user is interacting with the traffic, for example, heterogeneous, agent-like vehicles would create a more realistic and immersive experience. Given the computational cost of microscopic simulation, however, it is typically not feasible to simulate a large area using this method. To address this challenge, we proposed an approach in Sewall et al. (2011) in which subsets of the road network can be simulated using a microscopic simulator while the surrounding region can be simulated using the computationally cheaper macroscopic simulator. We are also able to *interactively* switch between simulation strategies for a particular region, allowing levels of detail based on, for example, viewing distance – all in real time. An example of our method's performance can be seen in **Figure 2**.

In addition, we have validated our simulation results by modifying string-distance metrics, such as LCSS and EDR, to compare agent-based, continuum, and level-of-detail simulation using NGSIM data. Our level-of-detail simulation has been able to achieve comparable accuracy while maintaining overall excellent performance, as shown in **Table 1**.

3.3. Other Issues

Above, we discussed our prior work on level-of-detail traffic simulation. However, further work is possible here as microscopic simulation still may not provide sufficient detail for some applications. Remember that in these simulation formulations, the space

in which the vehicles can move is restricted to a single dimension, the centerline of the lane. Real-world vehicles do not adhere to this constraint, and there are many applications that are only possible if this is relaxed: reconstructions and visualizations of anomalous traffic conditions, dangerous intersections, or accidents would require deeper models of vehicle movement. More immersive interactive traffic also requires these agents as real-world traffic does not neatly align itself to the lane. However, removing this constraint creates additional difficulties: allowing 2D movement of the cars requires providing a motion planning strategy that accounts for their non-holonomic kinematics, such as LaValle and Kuffner (2001), which is computationally costly.

4. Traffic Reconstruction

As discussed above, one area of research has focused on the creation and representation of "digital urbanscapes," which are digital representations of cities and, more generally, geographic areas. Commercial systems are now available that feature such models, obtained by using reconstruction techniques and aerial photography or other sensor measurements. However, these models are static: they lack dynamic elements, such as virtual people or traffic, which would increase their realism and user immersion.

Creating a digital representation of traffic that corresponds to real-world conditions is called "virtualized traffic" and was first introduced in van den Berg et al. (2009). To create virtualized traffic representations, in Wilkie et al. (2013), we proposed an efficient approach to creating animations that match observations, which could originate either from real-world traffic sensors or from a procedural system to allow the creation of controlled traffic animations. An overview of our approach is shown in **Figure 3**. We can see that there are two primary stages: data analysis and visualization. In the data analysis stage, the traffic state is estimated via a filter and a macroscopic traffic simulation model. In the visualization stage, a reconstruction of the traffic is created by an agent-based traffic simulator that is controlled via boundary conditions and a target velocity field. The created traffic can be displayed in 3D VR settings and is consistent with the traffic sensors and/or control points. An example of our reconstruction method's accuracy can be seen in **Figure 4**.

4.1. State Estimation

To create a traffic state estimate, the first part of our approach, we use sensor measurements (or control points), a macroscopic traffic simulator, and an Ensemble Kalman Smoother (EnKS), which was first introduced in Evensen (2003). This approach is summarized as follows. In **Figure 5**, we can see that each lane is discretized into cells, which hold continuum quantities. We create

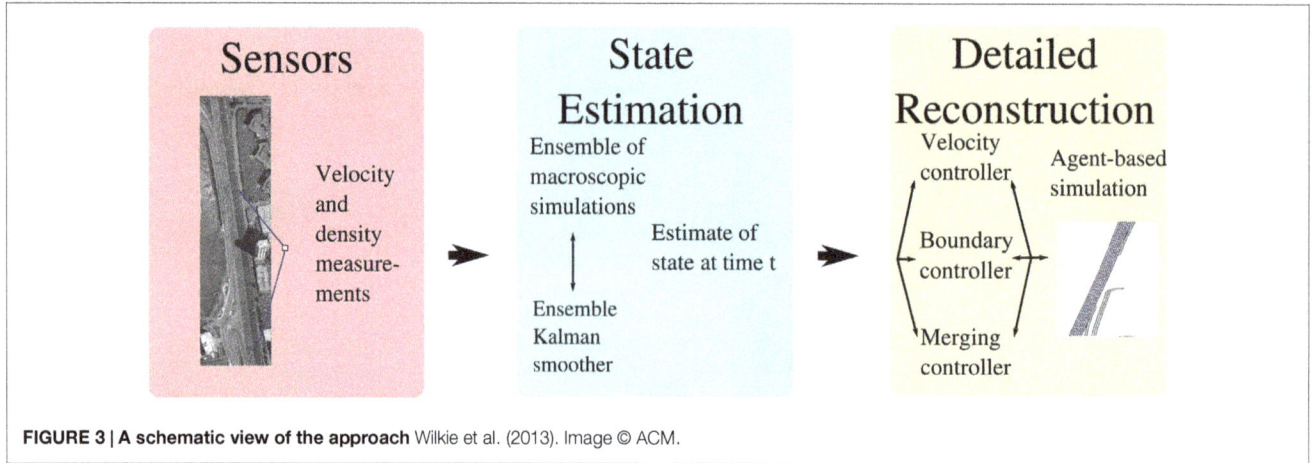

FIGURE 3 | **A schematic view of the approach** Wilkie et al. (2013). Image © ACM.

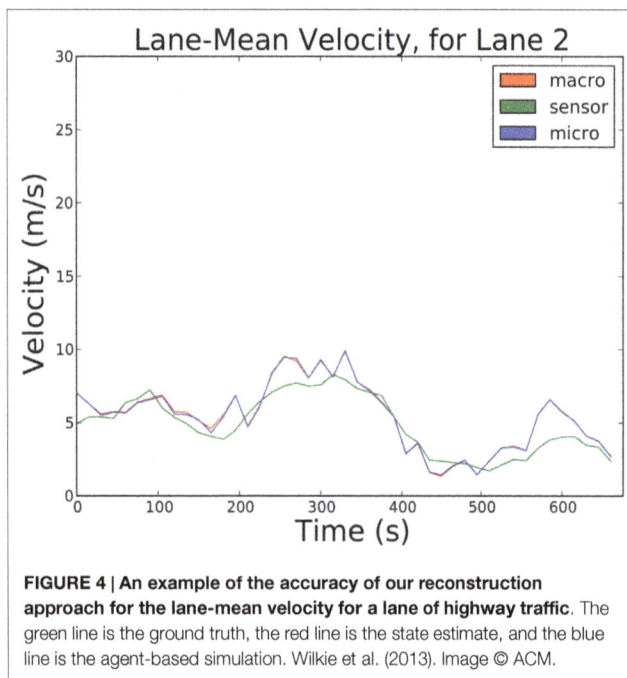

FIGURE 4 | **An example of the accuracy of our reconstruction approach for the lane-mean velocity for a lane of highway traffic**. The green line is the ground truth, the red line is the state estimate, and the blue line is the agent-based simulation. Wilkie et al. (2013). Image © ACM.

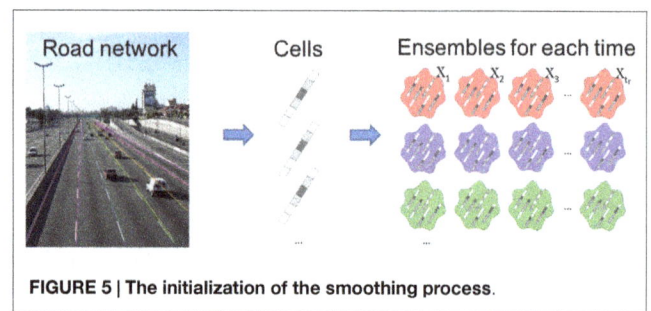

FIGURE 5 | **The initialization of the smoothing process**.

to place the vehicles. The vehicles are simulated using a simplified microscopic traffic simulator. Each car is first advected according to the target velocity field. Additionally, each vehicle has a velocity scaling factor, drawn from a bounded normal distribution, to model the variance in preferred driving speeds. The cars are also subject to a simplified leader-follower relationship. Other details, such as merging and boundary conditions, are also determined by the underlying state estimate. An example of reconstructed highway traffic, recreated at interactive rates, can be seen in **Figure 7**.

multiple estimates of the traffic for each lane, which is called the ensemble. Then, we update each ensemble member to maximize the likelihood of the given observations, which again come from either sensor measurements or specified control points. As seen in **Figure 6**, this update is a combination of simulation, to evolve the ensemble member forward in time, and statistics, to align the member with the observations using a calculated Kalman Gain matrix. Finally, we use the mean of the ensemble of states for each timestep as the estimate of the traffic state.

4.2. Reconstruction Animation

Once a state has been estimated for each timestep, we create a traffic animation. To do this, we first create cars for each lane in a way that matches the target density field. This is done by separating the cars by a distance calculated from the target density. If a higher variance in the separation distances is desired, additional noise can be added to the distances or a Poisson process can be used

4.3. Other Issues

The method described above created detailed animations of a single highway. For larger scale reconstructions with tighter computational budgets, an alternative approach is needed to creating the estimate of the traffic state. One such interpolation method for offline data is the *adaptive smoothing method*, described in Treiber and Kesting (2013), which uses a superposition of two convolution kernels that account for traffic flows. It remains to be seen how well approaches such as this work in the context of visual reconstruction and whether observable details in the traffic are lost.

Another potential area for further research is fitting full microscopic simulation models to the traffic data. Our work creates vehicles using relatively simple agents that follow the estimated velocity field. Full microscopic simulations would require fitting behavioral parameters based on the data. However, a larger dataset is required to avoid the solution being underdetermined.

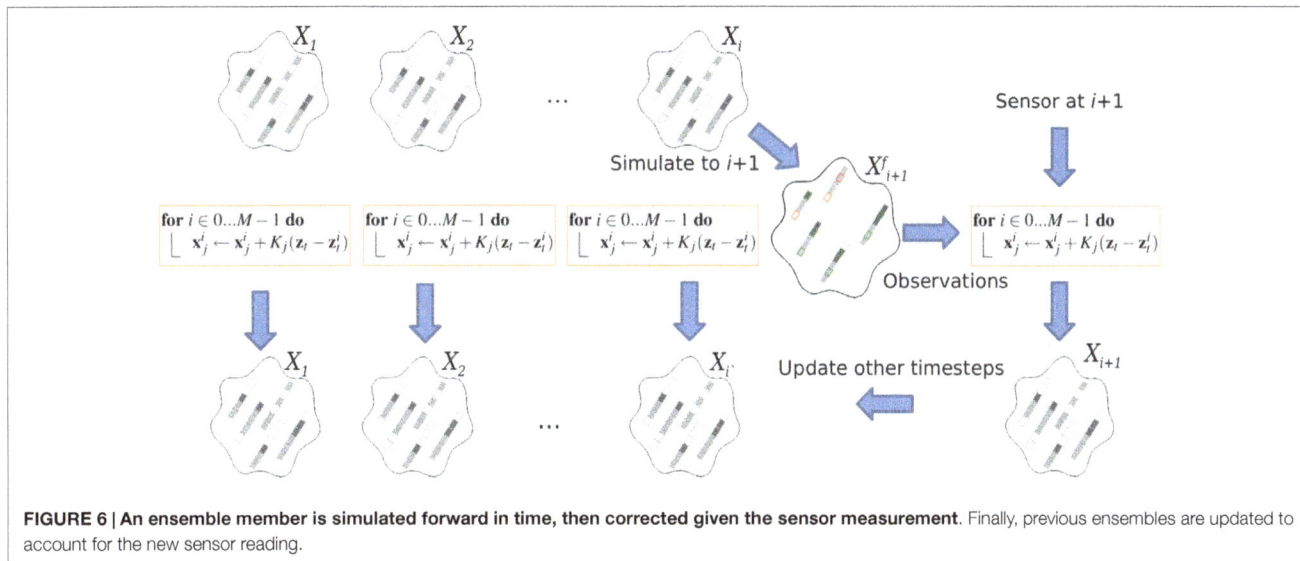

FIGURE 6 | An ensemble member is simulated forward in time, then corrected given the sensor measurement. Finally, previous ensembles are updated to account for the new sensor reading.

FIGURE 7 | On the bottom half of the divided highway is virtualized traffic matching real-world conditions as measured by loop-detector sensors. On the top half of the highway is virtualized traffic matching user-specified control points. The traffic is animated atop a GIS-derived road network. Wilkie et al. (2013). Image © ACM.

5. Vehicle Routing

The problem of routing a car through a road network can be abstractly described as the problem of finding a route through a graph, preferably a route that optimizes some desired criteria, which is solvable using Djikstra's algorithm or an A* algorithm. However, in their basic form, these approaches do not yield good performance for large road networks. Optimizations and alternative approaches that take advantage of the structure of the road network can yield computational speedups on the order of 10^5 and 10^6, as described in Delling et al. (2009). These approaches build hierarchies of the road network and pre-compute distance tables and shortest path regions.

In the context of virtual reality, these optimization efforts are likely less necessary for most applications as the scale of the road networks is often smaller: few applications need continent-sized road networks. However, in virtual-reality applications, the challenge is that every vehicle must be routed, and they need to be routed in a way that does not artificially create congestion. If every car is routed naively, the "shortest" routes will be assigned more traffic than is optimal, creating congestion patterns in some

areas while leaving other parts of the road network underutilized. This can be especially frustrating in city simulation scenarios where the drivers are expected to behave intelligently to user road network designs.

5.1. Self-Aware and Participatory Route Planning

To address this challenge, we proposed in Wilkie et al. (2011) a method that can route thousands of vehicles in a manner that minimizes congestion and travel time. By utilizing a method for estimating each vehicle's effect on the traffic density field, our method allows the route planner to take other vehicles into account without any explicit cooperation.

Our method work considers the cars' own planned routes as a source of information. The method routes cars individually, and then uses the planned routes as estimates for the cars' trajectories in the near future. These trajectories, in aggregate, form an estimate of the future traffic pattern. This method is composed of (a) a *route planner* that computes paths for cars through a time-varying density field defined on the road network and (b) an *update* that modifies the density field according to the calculated route.

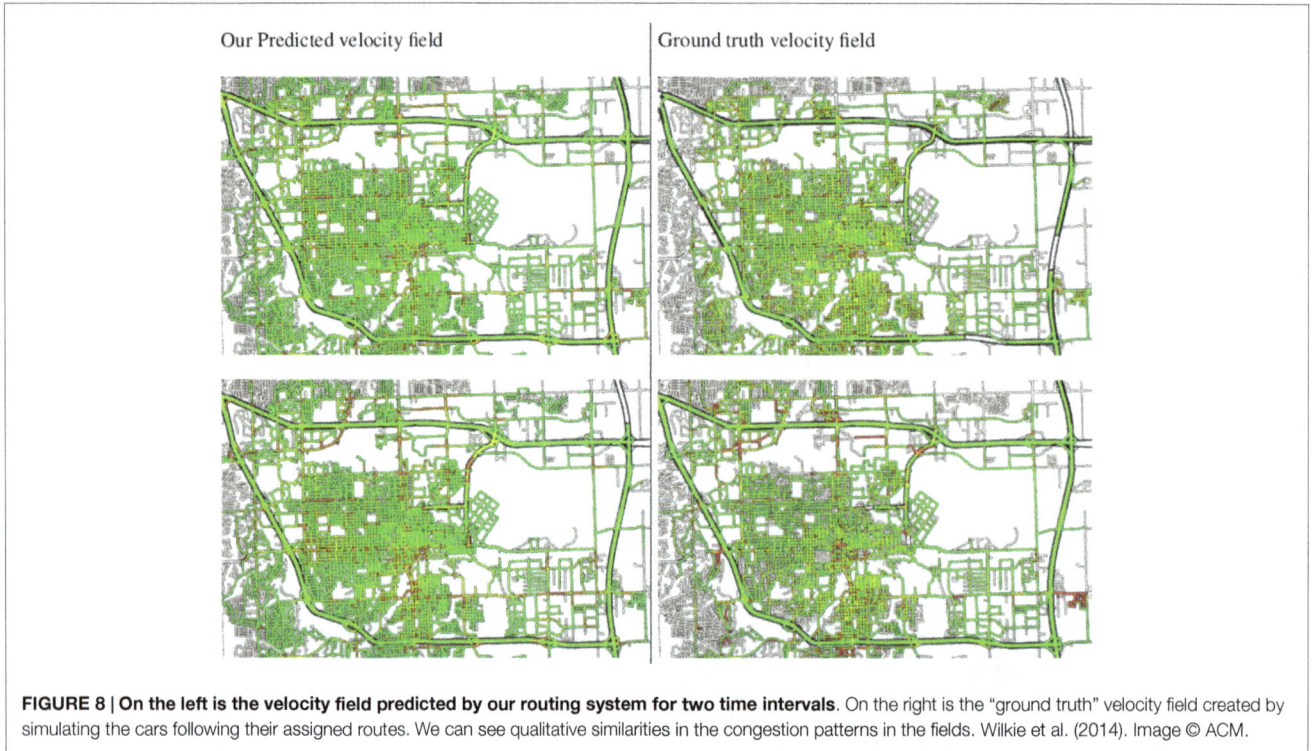

FIGURE 8 | On the left is the velocity field predicted by our routing system for two time intervals. On the right is the "ground truth" velocity field created by simulating the cars following their assigned routes. We can see qualitative similarities in the congestion patterns in the fields. Wilkie et al. (2014). Image © ACM.

The *route planner* makes use of a stochastic A*-search algorithm through a time-dependent density field. This field is composed of Gaussian distributions, discrete in time and space, defined over the road network graph, G. For each road explored, the cost of traversing the road is the estimated travel time

$$\tau_e(t) = \frac{\ell_e}{f_e(\rho_e(t))},$$

which is the length of the road ℓ_e divided by the estimated velocity, $f_e(\rho_e(t))$, which is a function of the estimated density. The function uses values for the maximum density and maximum velocity to determine the current velocity, which can be estimated using a number of models, as discussed in Greenshields et al. (1935) and Work et al. (2010).

Once a route has been planned, that route is considered an estimate for where that car will go in the future. For each car routed, the method adds a marginal amount of density to the road network along the planned path. For each edge of the route, the travel time estimates are used to calculate the probability that the car is on that edge during each time interval, $q_e(t)$. This car is then added to the density field for that road segment, taking into account the length of the road,

$$\rho_e(t) = \rho_e(t) + q_e(t)/\ell_e$$

The above approach uses a relatively simple model of traffic, however, and ignores certain phenomena, including congestion and delay caused by traffic lights and the spread of congestion and grid lock through a road network. In Wilkie et al. (2014), we presented an extension that models these phenomena in order to include their costs in the routing calculation. With these models,

our approach is able to predict the evolution of the traffic conditions in the future, allowing vehicles to take those conditions into account when routing and behave more intelligently. In **Figure 8**, for example, we can see the velocity field predicted by our router as well as the velocity field that occurred when the routes were simulated by a state-of-the-art microscopic traffic simulator.

5.2. Other Issues
Our approach can create routes for vehicles that simulate intelligent routing behavior and realistic traffic patterns. However, it is still relatively computationally costly. Years of research has gone into optimizing planning for static road networks, but it is not clear which if any of these approaches can be used in tandem with our approach.

6. Conclusion

Virtual environments, including training simulations, virtual globes, games set in cities, city simulations, traffic controller systems, and others, can all benefit from realistic, visual, efficient, and detailed traffic systems. Creating these systems, however, poses a number of ongoing research challenges. We have surveyed our approaches to creating virtual road networks and to simulating, reconstructing, and routing virtualized traffic using real-world data. We believe that these methods are among the first steps toward creating efficient virtualized traffic systems for realistic, metropolitan-scale virtual environments.

Much work remains to be done to fully achieve this vision, however. In each of the areas we discussed, there are promising future directions for research. For road network generation, current accuracy is limited to the accuracy and availability of the road network data and challenges remain in realistic procedural road network generation. For microscopic (or agent-based) traffic simulation, more detailed simulation models can be developed, such as models that allow more realistic movement within and between lanes, and better driver behavior models for intersections and routing are needed to enable more realistic traffic simulation and avoid deadlock situations in dense scenarios. For virtualized traffic reconstruction from data, further research is needed to improve the efficiency of the approaches and extend them to handle GPS data signals and complex road networks. For traffic routing, algorithmic improvements are needed to decrease the compute cost, especially in the context of interactive applications with a large number of vehicles.

Acknowledgments

This research is supported in part by National Science Foundation, Award #IIS-1247456.

References

Aw, A., and Rascle, M. (2000). Resurrection of "second order" models of traffic flow. *SIAM J. Appl. Math.* 60, 916–938. doi:10.1137/S0036139997332099

Cremer, J., Kearney, J., and Willemsen, P. (1997). Directable behavior models for virtual driving scenarios. *Trans. Soc. Comput. Simul. Int.* 14, 87–96.

Delling, D., Sanders, P., Schultes, D., and Wagner, D. (2009). "Engineering route planning algorithms," in *Algorithmics of Large and Complex Networks* (Berlin: Springer), 117–139.

Doniec, A., Mandiau, R., Piechowiak, S., and Espié, S. (2008). A behavioral multi-agent model for road traffic simulation. *Eng. Appl. Artif. Intell.* 21, 1443–1454. doi:10.1016/j.engappai.2008.04.002

Donikian, S., Moreau, G., and Thomas, G. (1999). Multimodal driving simulation in realistic urban environments. *Prog. Syst. Robot Anal. Cont. Des. (LNCIS)* 243, 321–332. doi:10.1007/BFb0110555

Evensen, G. (2003). The ensemble kalman filter: theoretical formulation and practical implementation. *Ocean Dyn.* 53, 343–367. doi:10.1007/s10236-003-0036-9

Fortier, A., Ziou, D., Armenakis, C., and Wang, S. (1999). *Survey of Work on Road Extraction in Aerial and Satellite Images.* Technical Report, 241: Center for Topographic Information Geomatics.

Garcia-Dorado, I., Aliaga, D. G., and Ukkusuri, S. V. (2014). "Designing large-scale interactive traffic animations for urban modeling," in *Computer Graphics Forum*, Vol. 33 (Wiley Online Library), 411–420.

Greenshields, B., Bibbins, J. R., Channing, W. S., and Miller, H. H. (1935). "A study of traffic capacity," in *Highway Research Board Proceedings* (Washington, DC: Highway Research Board), Vol. 14, 448–477.

Helbing, D. (2001). Traffic and related self-driven many-particle systems. *Rev. Mod. Phys.* 73, 1067–1141. doi:10.1371/journal.pone.0094351

Krajzewicz, D., Erdmann, J., Behrisch, M., and Bieker, L. (2012). Recent development and applications of SUMO – Simulation of Urban MObility. *Int. J. Adv. Syst. Meas.* 5, 128–138.

LaValle, S. M. (2006). *Planning Algorithms.* New York, NY: Cambridge University Press.

LaValle, S. M., and Kuffner, J. J. (2001). Randomized kinodynamic planning. *Int. J. Rob. Res.* 20, 378–400. doi:10.1177/02783640122067453

Lu, X., Chen, W., Xu, M., Wang, Z., Deng, Z., and Ye, Y. (2014a). Aa-fvdm: an accident-avoidance full velocity difference model for animating realistic street-level traffic in rural scenes. *Comput. Animat. Virtual Worlds* 25, 83–97. doi:10.1002/cav.1540

Lu, X., Wang, Z., Xu, M., Chen, W., and Deng, Z. (2014b). A personality model for animating heterogeneous traffic behaviors. *Comput. Animat. Virtual Worlds* 25, 363–373. doi:10.1002/cav.1575

Mena, J. (2003). State of the art on automatic road extraction for gis update: a novel classification. *Pattern Recognit. Lett.* 24, 3037–3058. doi:10.1016/S0167-8655(03)00164-8

MITSIM. (2011). *MITSIM.* MIT Intelligent Transportation Systems. Available at: http://mitsim.sourceforge.net/

Park, J., Saleh, R., and Yeu, Y. (2002). "Comprehensive survey of extraction techniques of linear features from remote sensing imagery for updating road spatial databases," in *ASPRS-ACSM Annual Conference and FIG XXII Congress.*

Pausch, R., Crea, T., and Conway, M. (1992). A literature survey for virtual environments – military flight simulator visual systems and simulator sickness. *Presence Teleoperators Virtual Environ.* 1, 344–363.

Plumert, J. M., Kearney, J. K., and Cremer, J. F. (2004). Children's perception of gap affordances: bicycling across traffic-filled intersections in an immersive virtual environment. *Child Dev.* 75, 1243–1253. doi:10.1111/j.1467-8624.2004.00736.x

Safra, E., Kanza, Y., Sagiv, Y., and Doytsher, Y. (2006). "Efficient integration of road maps," in *Proceedings of the 14th Annual ACM International Symposium on Advances in Geographic Information Systems, GIS '06* (New York, NY: ACM), 59–66.

Schrank, D., Eisele, B., and Lomax, T. (2012). *Ttis 2012 Urban Mobility Report.* (Texas: Texas A&M, College Station).

Sewall, J., Wilkie, D., and Lin, M. C. (2011). "Interactive hybrid simulation of large-scale traffic," in *ACM Transactions on Graphics (TOG)*, Vol. 30, (New York, NY: ACM), 135.

Sewall, J., Wilkie, D., Merrell, P., and Lin, M. (2010). "Continuum traffic simulation," in *Proceedings of Eurographics 2010, Computer Graphics Forum*, Vol. 29 (Blackwell Publishing), 439–448.

Shen, J., and Jin, X. (2012). Detailed traffic animation for urban road networks. *Graph. Models* 74, 265–282. doi:10.1016/j.gmod.2012.04.002

SUMO. (2009). *SUMO – Simulation of Urban MObility.* Available at: http://sourceforge.net/projects/sumo/

Thomas, G., and Donikian, S. (2000). Modelling virtual cities dedicated to behavioural animation. *CGF* 19, 71–80. doi:10.1111/1467-8659.00399

Thomson, J. A., Tolmie, A. K., Foot, H. C., Whelan, K. M., Sarvary, P., and Morrison, S. (2005). Influence of virtual reality training on the roadside crossing judgments of child pedestrians. *J. Exp. Psychol. Appl.* 11, 175. doi:10.1037/1076-898X.11.3.175

Treiber, M., and Kesting, A. (2013). "Traffic flow dynamics," in *Traffic Flow Dynamics: Data, Models and Simulation, ISBN 978-3-642-32459-8*, Vol. 2013 (Berlin: Springer-Verlag), 1.

van den Berg, J., Sewall, J., Lin, M., and Manocha, D. (2009). "Virtualized traffic: reconstructing traffic flows from discrete spatio-temporal data," in *Virtual Reality Conference, 2009. VR 2009. IEEE* (IEEE), 183–190.

Wang, H., Kearney, J., and Atkinson, K. (2002). "Arc-length parameterized spline curves for real-time simulation," in *Proc. 5th International Conference on Curves and Surfaces* (Brentwood, TN: CiteSeer), 387–396.

Wang, H., Kearney, J., Cremer, J., and Willemsen, P. (2005). "Steering behaviors for autonomous vehicles in virtual environments," in *Proc. IEEE Virtual Reality Conf*, 155–162.

Wang, J., Lawson, G., and Shen, Y. (2014). Automatic high-fidelity 3d road network modeling based on 2d gis data. *Adv. Eng. Software* 76, 86–98. doi:10.1016/j.advengsoft.2014.06.005

Wilkie, D., Baykal, C., and Lin, M. C. (2014). "Participatory route planning," in *Proc. of SIGSPATIAL, 2014.* ACM.

Wilkie, D., Sewall, J., and Lin, M. (2012). Transforming gis data into functional road models for large-scale traffic simulation. *IEEE Trans. Vis. Comput. Graph* 18, 890–901. doi:10.1109/TVCG.2011.116

Wilkie, D., Sewall, J., and Lin, M. (2013). Flow reconstruction for data-driven traffic animation. *ACM Trans. Graph. (TOG)* 32, 89. doi:10.1145/2461912.2462021

Wilkie, D., van den Berg, J. P., Lin, M. C., and Manocha, D. (2011). "Self-aware traffic route planning," in *AAAI.*

Willemsen, P., Kearney, J., and Wang, H. (2006). Ribbon networks for modeling navigable paths of autonomous agents in virtual environments. *IEEE Trans. Vis. Comput. Graph* 12, 331–342. doi:10.1109/TVCG.2006.53

Work, D., Blandin, S., Tossavainen, O., Piccoli, B., and Bayen, A. (2010). A traffic model for velocity data assimilation. *Appl. Math. Res. Express* 2010, 1–35. doi:10.1093/amrx/abq002

Zhang, H. (2002). A non-equilibrium traffic model devoid of gas-like behavior. *Trans. Res. B Methodol.* 36, 275–290. doi:10.1016/S0191-2615(00)00050-3

Conflict of Interest Statement: The authors declare that the research was conducted in the absence of any commercial or financial relationships that could be construed as a potential conflict of interest.

2

Utilization of human-like pelvic rotation for running robot

Takuya Otani[1,2]*, Kenji Hashimoto[3,4], Masaaki Yahara[5], Shunsuke Miyamae[5], Takaya Isomichi[5], Shintaro Hanawa[6], Masanori Sakaguchi[6,7], Yasuo Kawakami[6], Hun-ok Lim[4,8] and Atsuo Takanishi[4,9]

[1] *Graduate School of Advanced Science and Engineering, Waseda University, Tokyo, Japan,* [2] *Japan Society for the Promotion of Science, Tokyo, Japan,* [3] *Waseda Institute for Advanced Study, Tokyo, Japan,* [4] *Humanoid Robotics Institute (HRI), Waseda University, Tokyo, Japan,* [5] *Graduate School of Creative Science and Engineering, Waseda University, Tokyo, Japan,* [6] *Faculty of Sport Sciences, Waseda University, Tokyo, Japan,* [7] *Faculty of Kinesiology, University of Calgary, Calgary, AB, Canada,* [8] *Faculty of Engineering, Kanagawa University, Yokohama, Japan,* [9] *Department of Modern Mechanical Engineering, Waseda University, Tokyo, Japan*

Edited by:

Giuseppe Carbone,
University of Cassino and
South Latium, Italy

Reviewed by:

Ye Zhao,
The University of Texas at Austin, USA
Fernando Gomez-Bravo,
Huelva University, Spain
Mingfeng Wang,
University of Cassino and
South Latium, Italy

*Correspondence:

Takuya Otani,
Graduate School of Advanced
Science and Engineering, Waseda
University, #41-304, 17 Kikui-cho,
Shinjuku-ku, Tokyo 162-0044, Japan
t-otani@takanishi.mech.
waseda.ac.jp

The spring loaded inverted pendulum is used to model human running. It is based on a characteristic feature of human running, in which the linear-spring-like motion of the standing leg is produced by the joint stiffness of the knee and ankle. Although this model is widely used in robotics, it does not include human-like pelvic motion. In this study, we show that the pelvis actually contributes to the increase in jumping force and absorption of landing impact. On the basis of this finding, we propose a new model, spring loaded inverted pendulum with pelvis, to improve running in humanoid robots. The model is composed of a body mass, a pelvis, and leg springs, and, it can control its springs while running by use of pelvic movement in the frontal plane. To achieve running motions, we developed a running control system that includes a pelvic oscillation controller to attain control over jumping power and a landing placement controller to adjust the running speed. We also developed a new running robot by using the SLIP[2] model and performed hopping and running experiments to evaluate the model. The developed robot could accomplish hopping motions only by pelvic movement. The results also established that the difference between the pelvic rotational phase and the oscillation phase of the vertical mass displacement affects the jumping force. In addition, the robot demonstrated the ability to run with a foot placement controller depending on the reference running speed.

Keywords: humanoid, human motion analysis, running, pelvis, joint elasticity

Introduction

To realize human motions, researchers performed motion capture experiments or simulations. In motion capture experiments, researchers collect and analyze data, and on the basis of the results, verify or refute their hypothesis to clarify the mechanisms of human motion. However, for ethical reasons, motions that pose a risk of injury for human subjects cannot be tested, despite the possibility of improving those motions by coordinated training (World Medical Association, 1964). Some investigators have used simulations to evaluate their models or hypotheses of human motion. However, simulations have problems such as long processing time or model error (Rose and Gamble, 2005). To resolve these problems, we have sought to conduct research on human motion and sport science by using a biped humanoid robot that can mimic human motions. When such a humanoid robot realizes human-like motion or uses an instrument as does a human, its human-like motions can be compared with those that do not mimic human motions. By using the humanoid robot, we can measure various data as joint's angle, angular velocity, torque, robot's attitude, and so on. For example, the risk of injury can be evaluated according to the joint's torque. This comparison should be useful for verifying the characteristics of human motions and testing instruments, which in turn should improve the performance characteristics of both humans and robots.

We previously developed a biped humanoid robot named WABIAN-2R to mimic human motions and their underlying mechanisms (Ogura et al., 2006). WABIAN-2R, which is equipped with a 2-degree-of-freedom (DOF: roll, yaw) pelvis, shows a human-like gait with a stretched knee (Hashimoto et al., 2013, 2014). However, this robot is limited to walking. In human running, the joints of the standing leg require about 1000 W (Dalleau et al., 1998), which is much higher than the power of the actuator in WABIAN-2R.

In previous studies, running has been realized in a number of robots. Raibert et al. developed running robots with a linear spring leg (Raibert, 1986). Hyon et al. developed a biologically inspired biped robot based on a model of a dog's leg (Hyon et al., 2003). These robots are not human-like. However, a few recent studies have shown that biped humanoid robots can run. For example, ASIMO can run at a speed of 2.5 m/s (Honda Motor Company Ltd.; Takenaka et al., 2011). Furthermore, Toyota's biped humanoid robot can also run, using a zero moment point (ZMP)-based running control system (Tajima et al., 2009). The athlete robot developed by Niiyama et al. has a human-like musculoskeletal system built to achieve dynamic motions such as running (Niiyama et al., 2012). The biped robot MABEL, which has leg elasticity that originates from a leaf spring, can run the fastest among all presently available biped robots, attaining a speed of 3 m/s with axial constraints on the Y-axis (Grizzle et al., 2009). However, none of these robots mimic human running characteristics.

A number of important characteristics of human running have been identified by researchers in the field of biomechanics and sports science, including head stabilization (Pozzo et al., 1990), moment compensation using the upper body and arms (Collins et al., 2009), and leg stiffness (Gunther and Blickhan, 2002). However, a potentially important but as yet unreported characteristic is that a human's pelvic movement in the frontal plane increases takeoff forces and absorbs landing impacts. In this paper, we provide evidence for this characteristic by an analysis of human motion. We then propose a running model that combines the traditional spring loaded inverted pendulum (SLIP) model (Chapman and Caldwell, 1983; Blickhan, 1989; McMahon and Cheng, 1990) and a pelvis. The proposed model, called spring loaded inverted pendulum with pelvis (SLIP^2), is composed of a body mass, a pelvis, and leg springs. We then describe a running control system composed of a pelvic oscillation controller to attain control over jumping power, and a foot placement controller to adjust the running speed, and a running robot that uses the SLIP^2 model, with which we successfully conducted a hopping and running experiment.

The remainder of the paper is organized as follows. In Section "Introduction," we analyze human pelvic movement and describe the running control method, which includes control of pelvic movement and foot placement. In Section "Materials and Methods," we describe the design of the running robot used in this study. In Section "Results," we present and discuss experimental results. Finally, in Section "Conclusions and Future Work," we present the conclusions and future work.

Materials and Methods

Pelvic Movement Analysis

To identify characteristics of human running motion that could be useful for designing a robot, we conducted a series of motion capture experiments with human subjects. A motion capture system with eight infrared cameras (Motion Analysis Corp., Santa Rosa, CA, USA) was used to determine three-dimensional marker positions at 240 Hz. The spherical retro-reflective markers were attached on the skin of the pelvis, thigh, shank, and rearfoot. Ground reaction forces were simultaneously collected at 2400 Hz using a force plate (Bertec Corp., Columbus, OH, USA) for the identification of the stance phase. We asked seven subjects (gender: male, height: 1724 ± 36 mm, weight: 64 ± 6 kg) to perform regular running motion at 3.5 m/s and collected five sets of data for each subject. None of the participants had any muscular, neurological, or joint disorders that could affect their performance. The study was approved by the Office of Research Ethics at Waseda University and written informed consent was obtained from all participants. These subjects were given several practice trials to ensure that they could land with a natural running style on the force plate. When the running speed was within 5% of the target speed and the subject could land on the force plate, data collection began.

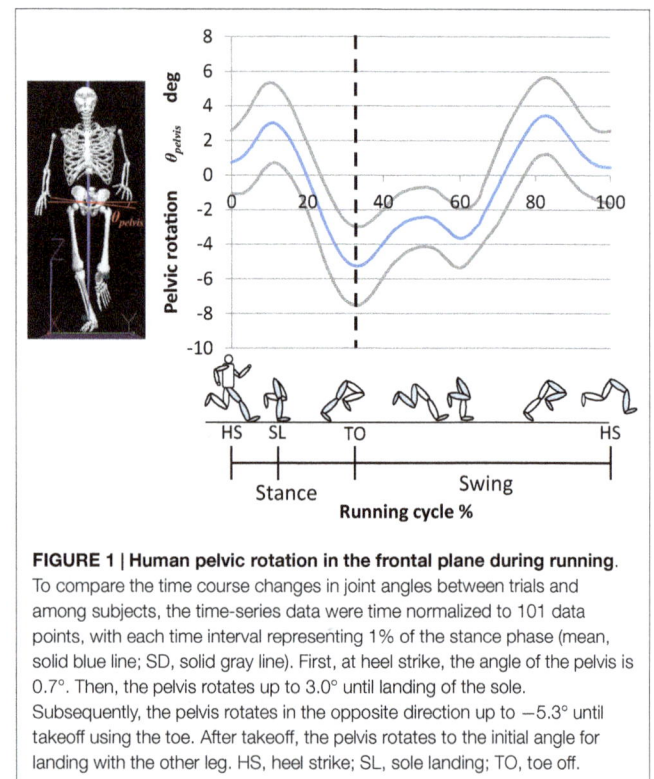

FIGURE 1 | Human pelvic rotation in the frontal plane during running.
To compare the time course changes in joint angles between trials and among subjects, the time-series data were time normalized to 101 data points, with each time interval representing 1% of the stance phase (mean, solid blue line; SD, solid gray line). First, at heel strike, the angle of the pelvis is 0.7°. Then, the pelvis rotates up to 3.0° until landing of the sole. Subsequently, the pelvis rotates in the opposite direction up to −5.3° until takeoff using the toe. After takeoff, the pelvis rotates to the initial angle for landing with the other leg. HS, heel strike; SL, sole landing; TO, toe off.

Figure 1 shows the pelvic movement in the frontal plane during running. The pelvis levels off at an angle of 0°. The measurements show that the angle of the pelvis is 0.7° at heel strike, at which point the idling leg is lowered. The pelvis then rotates up to 3.0° until landing of the sole. Subsequently, the pelvis rotates in the opposite direction up to −5.3° until takeoff using the toe. After takeoff, the pelvis rotates to the initial angle for landing with the other leg, i.e., it rotates to lower the idling leg and then rotates in the opposite direction to raise it.

The pelvic movement during the stance phase resembles a sine wave. Usually, the vertical movement of the center of mass during human running is modeled as a SLIP. We calculated the characteristic oscillation period of the vertical movement of the center of mass from data on the ground reaction force and vertical movement of the center of mass, as well as that of the pelvic movement, by measuring the period of peak-to-peak movement. The characteristic oscillation period of the vertical movement of the center of mass was 0.29 ± 0.03 s, and that of the pelvic movement was 0.30 ± 0.03 s. Given the similarity of these values, pelvic movement will affect movement of the center of mass effectively. This result suggests that the pelvic movement in the frontal plane, which acts as sine wave, increases the takeoff force and absorbs the landing impact. The pelvic movement at landing absorbs the landing impact, and the subsequent movement increases the takeoff force.

SLIP² Model

In previous studies, human running was modeled by a SLIP (McMahon and Cheng, 1990), whereas human walking was modeled by an inverted pendulum (Kajita and Espiau, 2008). The SLIP model is composed of a body mass and a spring leg and is based on the linear relationship between the ground reaction force and the vertical displacement of the body during running (Chapman and Caldwell, 1983; Blickhan, 1989; McMahon and Cheng, 1990). Moreover, the knee and ankle joints of the standing leg act like torsion springs that give rise to leg stiffness (Gunther and Blickhan, 2002), which along with joint stiffness vary with the running speed (Farley and González, 1996). In this model, movement in the flight phase is modeled as a parabolic motion of a mass point. With this model, which describes human running in a simple and straightforward way, the motion of the human leg in the stance phase of running stores energy and releases it like a spring. However, the SLIP model differs from the actual human body because it does not take the pelvis into account.

On the basis of the human motion analysis presented in the previous section, we propose a new model, SLIP² (**Figure 2**), which is composed of an upper body, a pelvis, and spring legs. During stance, human motion is modeled by SLIP², whereas during flight, it is modeled as a parabola. According to the SLIP² model, running is realized by two component processes: attaining sufficient jumping power, and controlling the running speed. To realize running, we implemented a pelvic oscillation controller for attaining jumping power and a foot placement controller for controlling the running speed.

Pelvic Oscillation Control

The pelvic oscillation control method is used for storing energy. In this method, the pelvis is controlled by using the natural frequency

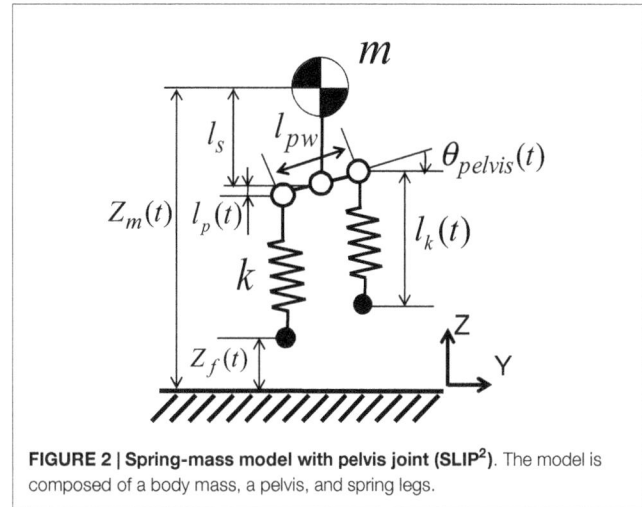

FIGURE 2 | Spring-mass model with pelvis joint (SLIP²). The model is composed of a body mass, a pelvis, and spring legs.

calculated from the mass weight and leg stiffness in the stance phase, and controlled to move it to the landing joint angles of pelvis in the flight phase, with the objective of attaining sufficient jumping power. The hip roll axis is controlled to be aligned with the pelvis roll axis. In the stance phase, the model, which has elastic elements, can store and release the energy by resonance. The pelvic movement is modeled as a linear displacement, and the equation of motion in the vertical direction in the stance phase is given by:

$$m\ddot{z}_m(t) + k(z_m(t) - l_s - l_p(t) - l_k(0)) - mg = 0$$

where m is the weight of the body mass; $z_m(t)$, the vertical displacement of the mass; l_s, the distance between the body mass and the pelvis; $l_p(t)$, the vertical displacement of the pelvis caused by its rotation; $l_k(t)$, the length of the leg spring; k, the leg stiffness; g, the gravitational acceleration ($= 9.8$ m/s²); and t, the time of the stance phase.

Based on the human motion analysis, the pelvic movement in the stance phase is expressed as follows:

$$\theta_{\text{pelvis}}(t_{\text{stance}}) = A \sin(\omega t_{\text{stance}} + \varphi)$$

where A is the pelvic rotation amplitude; ω, the natural frequency; t_{stance}, the time of the stance phase; and ϕ, the phase difference between the mass vertical movement and the pelvic movement.

To use the resonance effectively, the phase difference between the mass vertical movement and the pelvic movement in the stance phase is important. To analyze the effect of the pelvic movement, we calculated the mass vertical movement of the SLIP² model. At first, the SLIP² model fell from the mass height; 0.9 m. After landing, the SLIP² model moved its pelvis according to the above equation.

We performed on the three conditions that: the pelvic rotation amplitude was 0°, the phase difference was 0 rad or π rad for verifying the influence of the phase difference. In this calculation, we set the mass 55 kg, the leg stiffness 16 kN/m, and the pelvic rotation amplitude 0° or 5°. **Figure 3** shows the mass height of the body in this calculation. We plotted the results until the mass height was at a peak after jumping. When the phase difference was 0 rad, the mass height became higher than the fall height. On

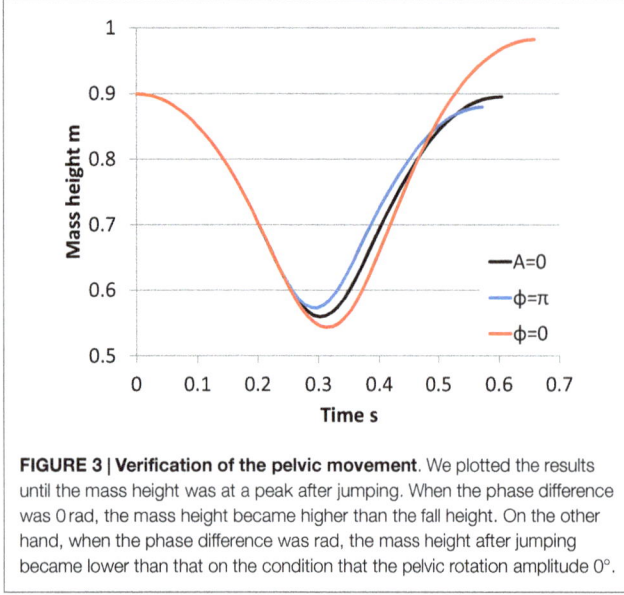

FIGURE 3 | Verification of the pelvic movement. We plotted the results until the mass height was at a peak after jumping. When the phase difference was 0 rad, the mass height became higher than the fall height. On the other hand, when the phase difference was rad, the mass height after jumping became lower than that on the condition that the pelvic rotation amplitude 0°.

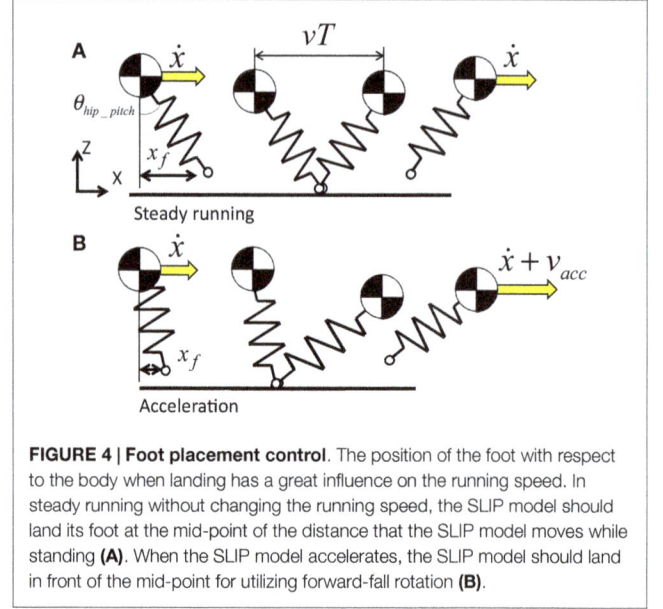

FIGURE 4 | Foot placement control. The position of the foot with respect to the body when landing has a great influence on the running speed. In steady running without changing the running speed, the SLIP model should land its foot at the mid-point of the distance that the SLIP model moves while standing **(A)**. When the SLIP model accelerates, the SLIP model should land in front of the mid-point for utilizing forward-fall rotation **(B)**.

the other hand, when the phase difference was π rad, the mass height after jumping became lower than that on the condition that the pelvic rotation amplitude 0°. Therefore, when the phase of the mass movement and that of the pelvic movement are close, the SLIP^2 can attain the high jumping power. When the phases are not close, however, the SLIP^2 cannot jump because the pelvic movement impedes the mass vertical movement. It indicates that the pelvic movement influences on the mass vertical movement, and the phase difference is important for the effectiveness of the attaining jumping power using the pelvic movement.

To utilize the phase difference between the mass movement and the pelvic movement effectively, the SLIP^2 model must land with the defined joint angle of the pelvis and hip roll at landing. When the joint angles at landing are different from the defined angles, the phase of the mass movement in the vertical direction while standing differs from that of the pelvic movement. The defined angle can be calculated by the difference between the phases to change the effectiveness of the pelvic movement for attaining jumping power. To solve this problem, we estimated the next landing time from the velocity of the mass in the vertical direction on takeoff, and moved the pelvis to the angle $\theta_{\text{pelvis\_ini}}$ to start pelvic oscillation from the same angle as in the previous running cycle. Because the movement of the mass traces a parabola in the flight phase, the next landing time T_{landing} is given by:

$$T_{\text{landing}} = \frac{2V_z}{g}$$

where V_z is the velocity of the mass in the vertical direction at takeoff.

The pelvic movement in the flight phase is expressed as follows:

$$\theta_{\text{pelvis}}(t_{\text{flight}})$$
$$= \begin{cases} \frac{\theta_{\text{pelvis\_ini}} - \theta_{\text{pelvis\_off}}}{T_{\text{landing}}} t_{\text{flight}} & \text{while reaching initial angle} \\ \theta_{\text{pelvis\_ini}} & \text{after reaching} \end{cases}$$

where $\theta_{\text{pelvis\_ini}}$ is the initial angle of the pelvis at landing; $\theta_{\text{pelvis\_off}}$, the angle of the pelvis at takeoff; and t_{flight}, the time of the flight phase.

Foot Placement Control

We used a foot placement controller to control running speed in the manner of Raibert (1986). This is achieved by moving the leg to an appropriate position with respect to the body during the flight phase in the SLIP model. The position of the foot with respect to the body when landing has a great influence on the running speed. In steady running without changing the running speed, the SLIP model should land its foot at the mid-point of the distance that the SLIP model moves while standing (**Figure 4A**). When the SLIP model accelerates, the SLIP model should land in front of the mid-point for utilizing forward-fall rotation (**Figure 4B**). In simulated research, Wensing et al., extended this method for 3D running (Wensing and Orin, 2013), and Zhao et al. used same control method for locomotion in rough terrains (Zhao and Sentis, 2012). As mentioned above, running can be modeled by a SLIP or SLIP^2. Some researchers developed the running control without foot placement controller (Lee and Goswami, 2012). However, we assume that the foot placement controller used for the SLIP model embodies one of the key principles of human running. In fact, humans change their foot position at the landing phase depending on the running speed (Cavagna et al., 1988). To implement this principle of human running, we used a foot placement controller. The foot position is given by:

$$x_{\text{f}} = \frac{\dot{x}T}{2} + K(\dot{x} - \dot{x}_{\text{ref}})$$

where x_{f} is the foot placement; x, the body mass placement; T, the stance time; $\dot{x}_{\text{ref}}$, the reference running speed, and K, the control gain.

Finally, the robot determined the angle of its hip pitch joint for landing according to the foot position and moved the hip pitch joint by landing. The angle of the hip pitch joint for landing is given by:

$$\theta_{\text{hip\_pitch}} = \arcsin\left(\frac{x_{\text{f}}}{l_{\text{leg}}}\right)$$

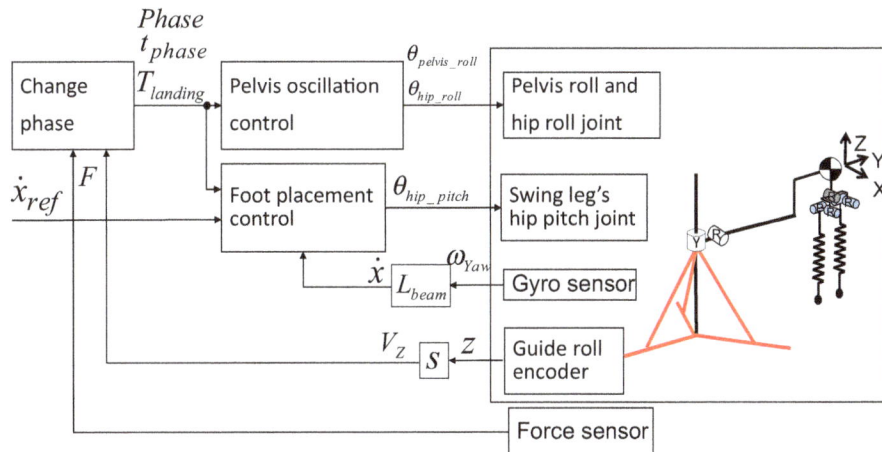

FIGURE 5 | Block diagram of running control. We implemented a pelvic oscillation controller for attaining jumping power and a foot placement controller for controlling the running speed.

where l_{leg} is the leg length. The control block diagram is shown in **Figure 5**.

Development of Running Robot with linear spring leg

Next, we sought to develop a running robot that could successfully execute running motion based on the SLIP[2] model. We determined the requirements for angular velocity and torque in the hip roll joint on the basis of human running data acquired by Ferber et al. (2003). We also fixed the requirements for the angular velocity of the pelvis roll joint on the basis of human running data acquired by Schache et al. (2002) and the requirements for the hip pitch joint based on our human running data. To the best of our knowledge, no work has previously been conducted on the torque of the pelvis roll joint. We calculated these requirements by substituting appropriate values in the equation of motion. The requirements for the angular velocity and torque of the pelvis and hip joints are summarized in **Table 1**.

We chose 150-W DC motors (Maxon Co., Ltd.), timing belts, and harmonic drives to actuate the pelvis and hip joints (**Figure 6**). The developed robot mimicked the human's parameters about the mass of the whole body, the mass of the leg, the width between the hip joints, the length of the leg, and the leg stiffness (**Table 2**). These parameters are significant in the SLIP[2] model. To adjust the mass, masses can be mounted on the upper part of the robot. We selected a compression spring, shaft, set collar, and linear bush for the leg spring. Owing to this mechanism, the compression spring is not detached from its upper and lower parts when the spring is at free length. In addition, the compression springs can be replaced with others by adjusting the distance between the set collar and the linear bush. We fixed the range of the spring's stiffness from 16 to 40 kN/m on the basis of the previous study (Dalleau et al., 1998). The developed robot is also equipped with an inertial measurement unit (IMU) named Waseda bioinstrumentation-4 (WB-4) on the body for measuring its posture. WB-4 has a 3-axis accelerometer sensor; 3-axis gyroscope; and 3-axis magnetometer (**Figure 7**) (Lin et al., 2011). The developed robot with a pelvis is shown in **Figure 8**.

TABLE 1 | Requirements.

	Pelvis roll joint	Hip roll joint	Hip pitch joint
Max. velocity (rad/s)	2.3	1.2	2.9
Max. peak torque (Nm)	44	113	45

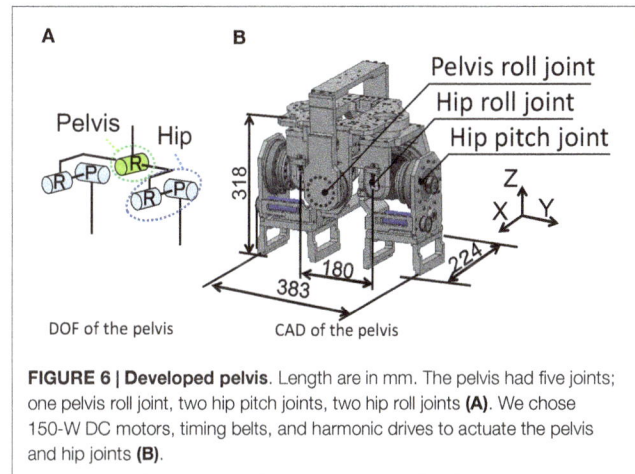

FIGURE 6 | Developed pelvis. Length are in mm. The pelvis had five joints; one pelvis roll joint, two hip pitch joints, two hip roll joints **(A)**. We chose 150-W DC motors, timing belts, and harmonic drives to actuate the pelvis and hip joints **(B)**.

TABLE 2 | Configuration of the developed robot and human.

	Robot	Human
Mass of whole body[a], m (kg)	55	64
Mass of leg	10	10
Pelvis width, l_{pw} (m)	0.18	0.18
Leg length[b], $l_k(t)$ (m)	0.7	0.7
Leg stiffness, k (kN/m)	16–40	16–40
Pelvic rotation amplitude, A (°)	5.0	5.3

[a]Without additional mass for adjustment.
[b]Leg spring is at free-length.

Results

Hopping Experiment

To evaluate the pelvic oscillation controller for attaining jumping power, we performed a hopping experiment by using the

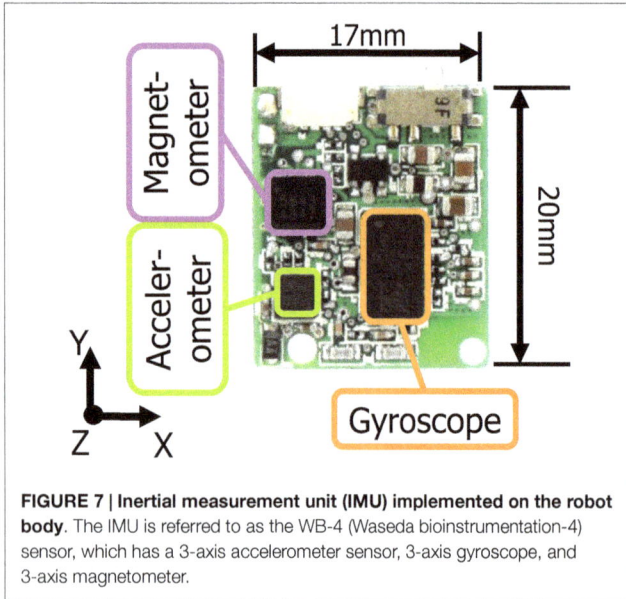

FIGURE 7 | Inertial measurement unit (IMU) implemented on the robot body. The IMU is referred to as the WB-4 (Waseda bioinstrumentation-4) sensor, which has a 3-axis accelerometer sensor, 3-axis gyroscope, and 3-axis magnetometer.

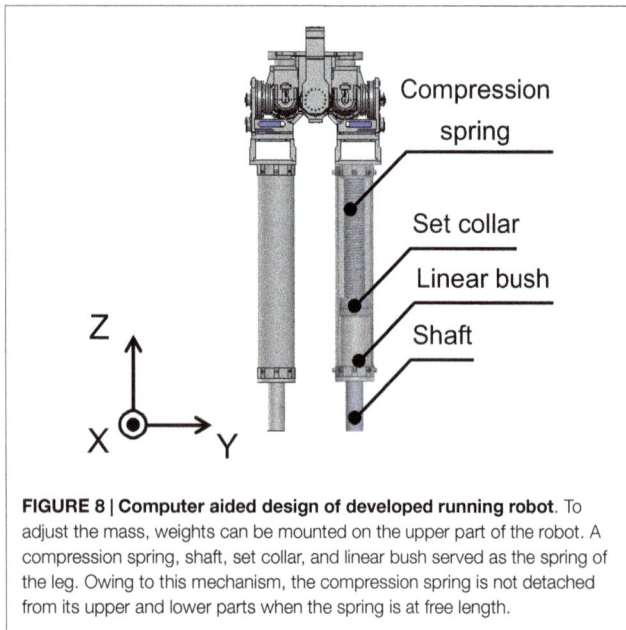

FIGURE 8 | Computer aided design of developed running robot. To adjust the mass, weights can be mounted on the upper part of the robot. A compression spring, shaft, set collar, and linear bush served as the spring of the leg. Owing to this mechanism, the compression spring is not detached from its upper and lower parts when the spring is at free length.

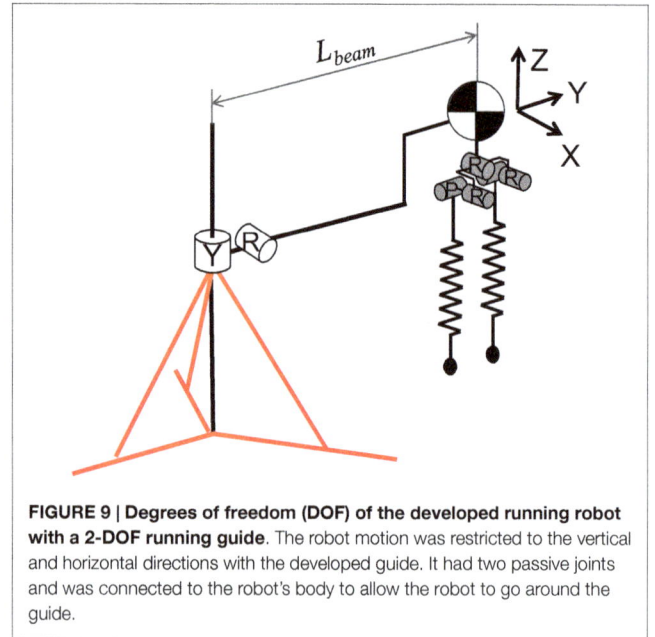

FIGURE 9 | Degrees of freedom (DOF) of the developed running robot with a 2-DOF running guide. The robot motion was restricted to the vertical and horizontal directions with the developed guide. It had two passive joints and was connected to the robot's body to allow the robot to go around the guide.

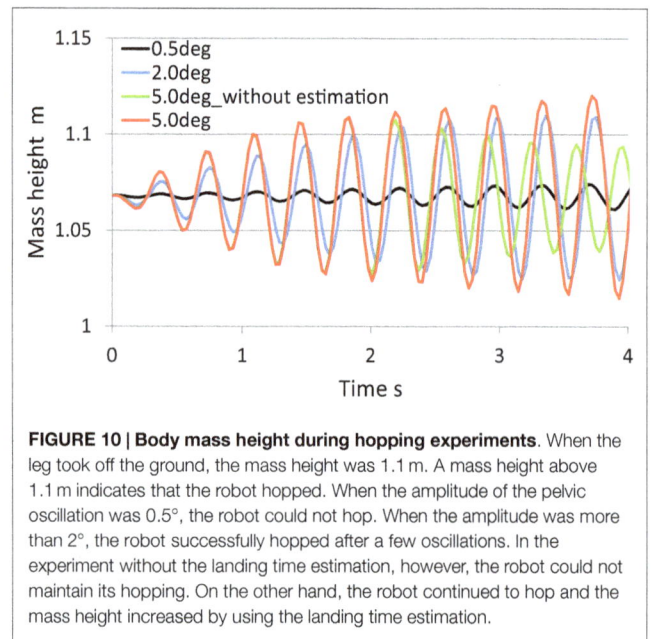

FIGURE 10 | Body mass height during hopping experiments. When the leg took off the ground, the mass height was 1.1 m. A mass height above 1.1 m indicates that the robot hopped. When the amplitude of the pelvic oscillation was 0.5°, the robot could not hop. When the amplitude was more than 2°, the robot successfully hopped after a few oscillations. In the experiment without the landing time estimation, however, the robot could not maintain its hopping. On the other hand, the robot continued to hop and the mass height increased by using the landing time estimation.

developed robot. In this experiment, the robot began the experiment in the standing position and started to move its pelvis according to the pelvic oscillation controller. The robot could detect whether it stood on the ground by force sensors implemented under the floor. On the basis of the ground reaction force data, the robot oscillated its pelvis in the stance phase and moved the pelvis to the landing angle by next landing in the flight phase. The hip roll joints were controlled to move in the opposite direction of the pelvis to maintain the leg perpendicular to the ground at all times. The hip pitch joint was controlled according to the foot placement controller with a reference running speed of 0 m/s to maintain vertical hopping. The robot motion was restricted to the vertical and horizontal directions with a developed guide (**Figure 9**), which had two passive joints, and was connected to the body of the robot. Because this allowed the robot to go around

the guide, we could calculate its running speed by the angular velocity of the body in the yaw direction, ω_{Yaw}, measured by a gyro sensor implemented in the robot's IMU. The running speed was given by:

$$\dot{x} = L_{\text{beam}} \omega_{\text{Yaw}}$$

where L_{beam} is the length of the beam connected to the robot's body and the pole of the guide. In this research, it was 1.5 m.

To measure the vertical displacement of the robot, a rotational encoder was implemented on the roll joint of the guide. When the leg spring is at free length, the body mass height is 1.1 m. We calculated the vertical velocity of the body mass from its displacement data in this experiment. The leg stiffness was based

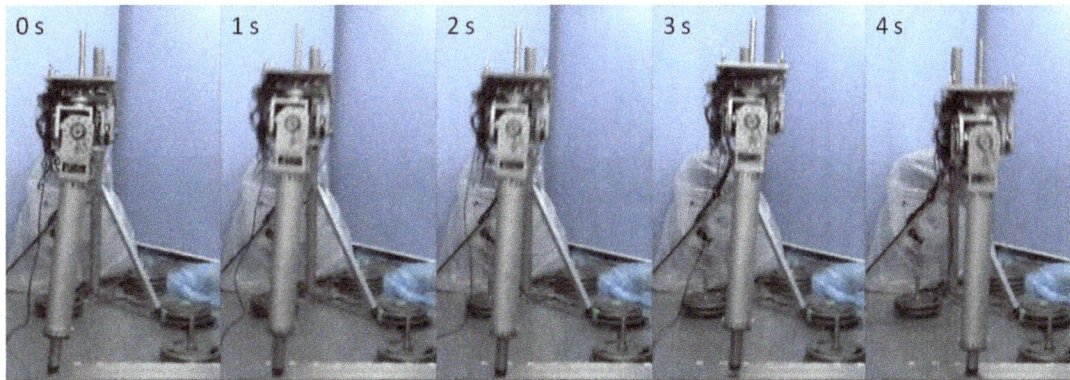

FIGURE 11 | Running experiment. The robot began its pelvic oscillation at 0 s, and started to hop and run after a few oscillations at 1 s. Then, the robot accelerated in forward direction by the foot placement control. In this experiment, the robot used only its pelvic motion for attaining jumping power.

on the human running data. And the phase difference was also determined from the human running data; $\frac{3}{8}\pi$. For evaluating the effect of the pelvic movement, we had this experiment under the condition that the amplitude of the pelvic oscillation was 0.5°, 2°, 5° with the landing time estimation, and 5° without the landing time estimation.

Figure 10 shows the mass height of the body in the hopping experiment. When the leg spring is at free length, the body mass height is 1.1 m. A body mass height that exceeded 1.1 m indicated that the robot hopped. When the amplitude of the pelvic oscillation was 0.5°, the robot could not hop. When the amplitude was more than 2°, the robot successfully hopped after a few oscillations. From these results, the robot could hop higher according to the amplitude of the pelvic oscillation. In the experiment without the landing time estimation, however, the robot could not maintain its hopping. After three hops, the mass height of the body decreased. On the other hand, by using the landing time estimation, the robot maintained its hopping and the mass height increased.

This experimental result suggests that pelvic movement in human running contributes to the increase in takeoff force. And that the amplitude of the pelvic oscillation and the difference in the pelvic rotational phase influence the effectiveness for the increase in takeoff forces. Nevertheless, it is common to instruct superior runners to emulate a running form in which the pelvis is stationary. This is because the pelvic movement in running can easily hurt the pelvis (Schache et al., 2002). To use the pelvic movement effectively and safely, it is assumed that runners should train their muscles around the trunk and pelvis. As in this example, it is possible that some motions posing a risk of injury improve human physical performance; if so, using a humanoid robot would allow us to verify the effects safely. It is also possible that humans use resonance by not only the pelvis but also other parts of the body (e.g., arms).

Running Experiment

We performed a running experiment by using the developed robot to evaluate the pelvic oscillation and foot placement controllers. The experimental conditions were identical to those for the hopping experiment, except for the reference running speed of

0.1 m/s. The pelvic oscillation amplitude was 5.0°, and the control gain of the foot placement control was 0.15.

Figure 11 shows photographs of the experiment, and **Figure 12** shows the running speed. We measured the running speed from gyro sensor data during each flight phase because the foot placement controller mainly affects the running speed during the flight phase. The robot started its pelvic oscillation and started to hop and run after a few oscillations. The robot succeeded in running at approximately 0.1 m/s.

Although the developed robot has the potential to run at a speed of up to roughly1 m/s, it cannot run as fast as a human. This is because the robot used only its pelvis while standing and its hip pitch joint during the flight phase. It did not kick the ground by actively moving its hip joint. This movement and its stabilization are required to run at a higher speed. In human running, the trunk and arms are said to be used for stabilization (Collins et al., 2009). Therefore, the upper body of the robot should be developed to allow it to actively kick the ground with its hip joint. As mentioned above, successful running has been demonstrated in some humanoid robots. However, these robots use only the hip joint, and not pelvic movement, for kicking. The results of this study suggest that pelvic movement could be effective for improving the running speed of these humanoid robots.

Conclusion and Future Work

Here, we described an analysis of human running motion focused on pelvic movement in the frontal plane, and developed the SLIP^2 model and a running control system. From the analysis, we concluded that pelvic movement in the frontal plane, which resembles a sine wave during the stance phase, can help to increase takeoff forces and absorb landing impacts. We then newly developed the SLIP^2 model, composed of the SLIP model, as well as a pelvis equipped with running control methods including pelvic oscillation and foot placement controllers. To evaluate the SLIP^2 model and the running control methods, we executed hopping and running experiments with the developed robot. The results indicated that pelvic movement can help to increase the jumping force and absorb the impact. The pelvic rotational phase difference had an influence on the increase in takeoff forces.

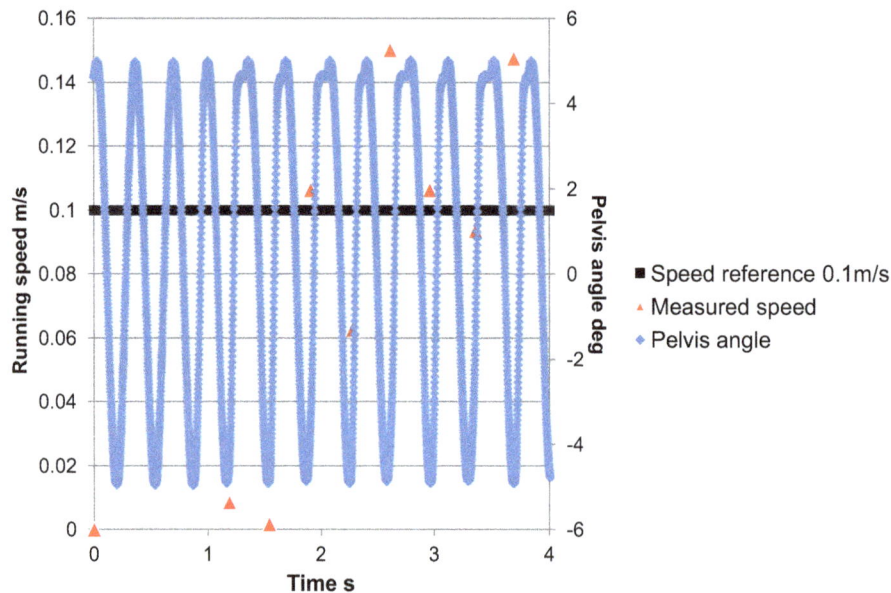

FIGURE 12 | Running speed transition. The robot successfully ran according to the reference running speed of approximately 0.1 m/s.

These results suggest that the running robot can run at a higher speed by using kicking and pelvic movement. Humans can run stably at high speeds because the moment generated by the ground reaction force and their leg movement are compensated for by the use of their upper body and arms. In future studies, we intend to combine the SLIP[2] model with an upper body to construct a new full-body model that mimics this characteristic of human running. Finally, we will develop a new stabilization control method that uses the upper body and arms for high speed running by means of kicking and pelvic movement.

Acknowledgments

This study was conducted with the support of the Research Institute for Science and Engineering, Waseda University; Institute of Advanced Active Aging Research, Waseda University, and as part of the humanoid project at the Humanoid Robotics Institute, Waseda University. It was also supported in part by the MEXT/JSPS KAKENHI Grant No. 25220005 and 25709019; Mizuho Foundation for the Promotion of Sciences; SolidWorks Japan K.K.; DYDEN Corporation and Cybernet Systems Co., Ltd.; we thank all of them for the financial and technical support provided.

References

Blickhan, R. (1989). The spring-mass model for running and hopping. *J. Biomech.* 22, 1217–1227. doi:10.1016/0021-9290(89)90224-8

Cavagna, G. A., Franzetti, P., Heglund, N. C., and Willems, P. (1988). The determinants of the step frequency in running, trotting and hopping in man and other vertebrates. *J. Physiol.* 29, 81–92.

Chapman, A. E., and Caldwell, G. E. (1983). Factors determining changes in lower limb energy during swing in treadmill running. *J. Biomech.* 16, 69–77. doi:10.1016/0021-9290(83)90047-7

Collins, S. H., Adamczyk, P. G., and Kuo, A. D. (2009). Dynamic arm swinging in human walking. *Proc. Biol. Sci.* 276, 3679–3688. doi:10.1098/rspb.2009.0664

Dalleau, G., Belli, A., Bourdin, M., and Lacour, J. R. (1998). The spring-mass model and the energy cost of treadmill running. *Eur. J. Appl. Physiol.* 77, 257–263. doi:10.1007/s004210050330

Farley, C. T., and González, O. (1996). Leg stiffness and stride frequency in human running. *J. Biomech.* 29, 181–186. doi:10.1016/0021-9290(95)00029-1

Ferber, R., Davis, I. M., and Williams, D. S. III (2003). Gender differences in lower extremity mechanics during running. *Clin. Biomech.* 18, 350–357. doi:10.1016/S0268-0033(03)00025-1

Grizzle, J. W., Hurst, J., Morris, B., Park, H. W., and Sreenath, K. (2009). "MABEL, a new robotic bipedal walker and runner," in *2009 American Control Conf.*, St. Louis, MO, 2030–2036.

Gunther, M., and Blickhan, R. (2002). Joint stiffness of the ankle and the knee in running. *J. Biomech.* 35, 1459–1474. doi:10.1016/S0021-9290(02)00183-5

Hashimoto, K., Hattori, K., Otani, T., Lim, H. O., and Takanishi, A. (2014). Foot placement modification for a biped humanoid robot with narrow feet. *ScientificWorldJournal* 2014, 259570. doi:10.1155/2014/259570

Hashimoto, K., Takezaki, Y., Lim, H. O., and Takanishi, A. (2013). Walking stabilization based on gait analysis for biped humanoid robot. *Adv. Robot.* 27, 541–551. doi:10.1080/01691864.2013.777015

Honda Motor Company Ltd. *ASIMO*, Honda Motor Company Ltd. Available at: http://www.honda.co.jp/ASIMO/

Hyon, S., Emura, T., and Mita, T. (2003). Dynamics-based control of a one-legged hopping robot. *J. Syst. Control Eng. Proc. Inst. Mech. Eng. I* 217, 83–98. doi:10.1177/095965180321700203

Kajita, S., and Espiau, B. (2008). "Chapter 16 legged robots," in *Springer Handbook of Robotics*, Vol. C, eds B. Siciliano and O. Khatib (Berlin: Springer), 361–389.

Lee, S. H., and Goswami, A. (2012). A momentum-based balance controller for humanoid robots on non-level and non-stationary ground. *Auton. Robots* 33, 399–414. doi:10.1007/s10514-012-9294-z

Lin, Z., Zecca, M., Sessa, S., Bartolomeo, L., Ishii, H., and Takanishi, A. (2011). "Development of the wireless ultra-miniaturized inertial measurement unit WB-4: preliminary performance evaluation," in *33rd Annual International Conference of the IEEE EMBS*, Boston, MA, 6927–6930.

McMahon, T., and Cheng, G. (1990). The mechanics of running: how does stiffness couple with speed? *J. Biomech.* 23, 65–78. doi:10.1016/0021-9290(90)90042-2

Niiyama, R., Nishikawa, S., and Kuniyoshi, Y. (2012). Biomechanical approach to open-loop bipedal running with a musculoskeletal athlete robot. *Adv. Robot.* 26, 383–398. doi:10.1163/156855311X614635

Ogura, Y., Shimomura, K., Kondo, H., Morishima, A., Okubo, T., Momoki, S., et al. (2006). "Human-like walking with knee stretched, heel-contact and toe-off motion by a humanoid robot," in *2006 IEEE/RSJ International Conference on Intelligent Robots and Systems*, Beijing, 3976–3981.

Pozzo, T., Berthoz, A., and Lefort, L. (1990). Head stabilization during various locomotor tasks in humans. *Exp. Brain Res.* 82, 97–106. doi:10.1007/BF00230842

Raibert, M. H. (1986). *Legged Robots that Balance*. Cambridge MA: MIT Press.

Rose, J., and Gamble, G. J. (2005). *Human Walking*. Cambridge MA: Lippincott Williams & Wilkins.

Schache, A. G., Branch, P., Rath, D., Wrigley, T., and Bennell, K. (2002). Three-dimensional angular kinematics of the lumbar spine and pelvis during running. *Hum. Mov. Sci.* 21, 273–293. doi:10.1016/S0167-9457(02)00080-5

Tajima, R., Honda, D., and Suga, K. (2009). "Fast running experiments involving a humanoid robot," in *2009 IEEE International Conference on Robotics and Automation*, Kobe, 1571–1576.

Takenaka, T., Matsumoto, T., Yoshiike, T., and Shirokura, S. (2011). Running gait generation for biped robot with horizontal force limit. *JRSJ* 29, 93–100. doi:10.7210/jrsj.29.849

Wensing, M. P., and Orin, E. D. (2013). "High-speed humanoid running through control with a 3D-SLIP model," in *2013 IEEE/RSJ International Conference on Intelligent Robots and Systems*, Tokyo, 5134–5140.

World Medical Association. (1964). *Declaration of Helsinki*. Helsinki, FL: World Medical Association.

Zhao, Y., and Sentis, L. (2012). "A three dimensional foot placement planner for locomotion in very rough terrains," in *2012 12th IEEE-RAS International Conference on. Humanoid Robots*, Osaka, 726–733.

Conflict of Interest Statement: The authors declare that the research was conducted in the absence of any commercial or financial relationships that could be construed as a potential conflict of interest.

3

Personalized social network activity feeds for increased interaction and content contribution

Shlomo Berkovsky* and Jill Freyne

Digital Productivity Flagship, CSIRO, Sydney, NSW, Australia

Online social networks were originally conceived as means of sharing information and activities with friends, and their success has been one of the primary contributors of the tremendous growth of the Web. Social network activity feeds were devised as a means to aggregate recent actions of friends into a convenient list. But the volume of actions and content generated by social network users is overwhelming, such that keeping users up-to-date with friend activities is an ongoing challenge for social network providers. Personalization has been proposed as a solution to combat social network information overload and help users to identify the nuggets of relevant information in the incoming flood of network activities. In this paper, we propose and thoroughly evaluate a personalized model for predicting the relevance of the activity feed items, which informs the ranking of the feeds and facilitates personalization. Results of a live study show that the proposed feed personalization approach successfully identifies and promotes relevant feed items and boosts the uptake of the feeds. In addition, it increases the contribution of user-generated content to the social network and spurs interaction between users.

Keywords: social network feed, feed personalization, online evaluation, user engagement, content contribution

Edited by:
David Balduzzi,
ETH Zurich, Switzerland

Reviewed by:
Seyed Mohammad Hadi
Daneshmand,
ETH Zurich, Switzerland
Utkarsh Upadhyay,
Max Planck Institute for Software
Systems, Germany

*Correspondence:
Shlomo Berkovsky,
Digital Productivity Flagship, CSIRO,
PO Box 76, Epping, NSW 1710,
Australia
shlomo.berkovsky@csiro.au

Introduction

The growth of the Web is relentless and set to continue and accelerate in the near future, as the Web continues to accommodate new forms of centralized and user-generated content (Susarla et al., 2012). Online social networks (in short, SNs) have recently experienced remarkable popularity and they are fast becoming the place where information is shared and found. Designed to allow people to create and share textual and multimedia content, SNs have become rich and diverse information sources, often competing with conventional websites and search engines in the dispersion of information. SNs have billions of users and the volume of content that can be found therein is astounding. Facebook alone reports more than a billion users, with an average user connected to hundreds friends, and using the system for more than 1 hour a day[1].

Most SNs allow their users to tune into streams of information and updates from other users, which act as virtual information filters for the incoming information. These streams, or *activity feeds*, typically contain a summary of the actions taken by other SN users, broadly defined as connections (friends, followers, contacts, articulated connections, and so on). This natural filter, where information items contributed by the trusted users are aggregated in the feed and presented in reverse chronological order, allows SN users to quickly discover updates and content of interest. The

---

[1] Facebook Statistics, available at: http://newsroom.fb.com/company-info/

popularity of SN, their ubiquity, and the ease of content generation and sharing, however, has swamped the simple aggregation mechanism of the feed, as the number of connections made and the volume of content contribute has increased. While simple and easy to understand, the standard information aggregation mechanism of the feed can hardly cope with the sheer volume and diversity of content contributed by SN users, and it crumbled under the pressure being placed (Berkovsky and Freyne, 2015). Users could, in principle, remove undesired users their feed, but the personal and social unease at removing online connections overweighed the benefits and precluded many users from actively curating their feeds.

Automatic re-organization of feeds, aimed at filtering out irrelevant or less interesting updates and highlighting updates of a particularly high importance, offers a solid alternative to manual filtering (Chen et al., 2010; Freyne et al., 2010). Research at the intersection of the research areas of data mining, machine learning, natural language processing, and social sciences turned their focus to the problem of the SN feed filtering. Several orthogonal solutions to the problem were proposed: what intrinsic and extrinsic factors make SN posts valuable (Hurlock and Wilson, 2011; Lage et al., 2013), how can the feeds be ranked in a domain-agnostic manner (Das Sarma et al., 2010; Duan et al., 2010; Huang et al., 2012), and what approaches from the Semantic Web realm can alleviate the ranking task (Bontcheva and Rout, 2012). However, much of this work faced a major obstacle, the perceived importance of the feed items was found to be user dependent, which brought to the fore a rather complex challenge of filtering the feed in a personalized manner.

In response to this emergent challenge, in this work, we investigate the application of established personalization techniques, widely recognized solutions in other information overload situations, to the task of identifying interesting content in SN activity feeds (Berkovsky and Freyne, 2015). We capitalize primarily on earlier works of Gilbert and Karahalios (2009) and Wu et al. (2010), and propose a model that leverages observable SN activities, such as users' interactions with content and other users, in order to elicit user preferences, predict relevance of feed items, and subsequently personalize activity feeds. Specifically, we judge the relevance of candidate feed items using two principal parameters: user-to-user relationship strengths and user-to-action interest score. The former incorporates 53 fine-grained factors reflecting the individual and mutual activity of users, which jointly quantify the degree of user-to-user closeness. The latter focuses on the actions performed by SN users and aims to derive individual action importance scores. The two parameters are combined in a linear manner, such that every candidate feed item is scored in a personalized manner tailored to the feed recipient. Finally, the feed gets re-ordered, such that high-scoring items appear on top of the feed, thus, highlighting relevant SN activities.

This paper extends our earlier work (Berkovsky et al., 2011; 2012). Specifically, we present in greater detail the developed model for feed scoring and personalization, and more thoroughly evaluate the proposed feed personalization mechanism. Initially, we outline the feed scoring mechanism, which capitalizes on prior works of Gilbert and Karahalios (2009) and Wu et al. (2010), but also contextualizes them to the target domain of the SN under

investigation. Then, we present an elaborate evaluation conducted as part of a live study involving users of the Online Total Wellbeing Diet (TWD) portal designed to support diet and lifestyle program participants. The portal incorporated a dedicated SN component and the activity feed was personalized according to our methodology. The analysis touches upon several aspects of personalization: (i) general uptake of the feed, (ii) temporal evolution of the feeds, (iii) ranking of the feed items, (iv) impact of the feed on user activities, and (v) relationships between the feed and online user friending. Out of these aspects, the temporal evolution of the feeds was not addressed at all in (Berkovsky et al., 2012), whereas the other four aspects are evaluated more thoroughly in this paper. The results show a clear evidence supporting our argument that feed personalization is a valuable tool supporting the success and popularity of SNs. With regards to the above-mentioned five aspects, the evaluation results show that personalization (i) highlighted important SN activities, (ii) improved in accuracy over time, (iii) diverted user activities toward highly ranked feed items, (iv) increased the contribution of user-generated content, and (v) assisted users in establishing and maintaining online friendship links. Hence, we conclude that the personalization of activity feeds is an important means for sustaining the engagement of SN users and increasing content contribution and online friending.

The rest of this paper is structured as follows. Section 2 surveys related work on personalization of SN feeds. Section 3 details our evaluation platform, the Online TWD portal, and the proposed feed personalization mechanism. Section 4 presents the experimental evaluation conducted and discusses the obtained results. Finally, Section 5 concludes the paper and outlines future research directions.

Related Work

To facilitate a re-organization of the SN feed, a robust mechanism for scoring the relevance of the feed items is required (De Choudhury et al., 2011). In this section, we survey prior approaches to scoring feed items: we first focus on the user-to-user relationships; we then examine work that incorporates text and content factors; and, finally, we turn to SN- and graph-related considerations.

User-to-User Relationships

Perhaps, the central factor when scoring feed items is the strength of relationship between the user who performed the activity and the feed recipient (Berkovsky and Freyne, 2015). Several works looked into the quantification of the strength of ties between SN users. The first work in this area was conducted by Gilbert and Karahalios (2009). They modeled the tie-strength using seven dimensions of features: intensity of communication, use of intimacy language, duration of online ties, resources and information shared, common groups and communities, gifts or congratulations exchanged, and demographic similarity. The overall tie-strength score was computed as a linear combination of 70 individual features instantiating these dimensions. They found that the intimacy language dimension accounted for more than 30% of weight in the tie-strength score, whereas the most important individual features reflected the duration of communication between the two users.

A similar work, aimed at predicting the closeness of enterprise SN users, was conduct by Wu et al. (2010). They derived 60 predictive features for their model and split these into five categories: user who performed the activity, user receiving the feed, direct interaction between the two, indirect interaction through common friends, and distance between the two in the enterprise. The tie-strength score was computed as a linear combination of the features. Here, the dominant group of features was the direct interaction between the users, which accounted for 40–50% of weight, depending on the type of closeness being predicted. Jacovi et al. (2011) studied implicit indicators of interest of one user in another in an enterprise SN. They proposed four indicators that may signal interest: following the user, tagging the user in a people-tagging service, viewing content generated by the user, commenting on the user's posts. Out of these, tagging was found to most strongly correlate with interest, followed by direct following.

Text and Content Factors

In addition to the tie-strength between the users, another fundamental predictor of interest is the content of (or, contributed by) the activity, which includes both the immediate text generated by the user and another auxiliary information, e.g., URLs, tags, user mentions, and so on. Paek et al. (2010) collected information on users' perceived importance of feed entries and developed an ensemble model for predicting the binary importance of Facebook feed items. The model incorporated 50 features that were partitioned into three high-level groups: Facebook metadata (number of comments/views/likes, inclusion of URLs, and temporal information), textual contribution corresponding to the item, and background information (location, activities, interests, and more). The evaluation clearly highlighted the importance of the textual content of the feed items for the scoring model.

A similar model for ranking tweets on Twitter was developed and evaluated by Uysal and Croft (2011). They derived a set of predictive features, which were partitioned into four categories: reputation and activity of the tweet poster, inclusion of hashtags and user mentions, textual model of the tweet content, and past interactions between the recipient and the poster of the tweet. These categories of features were used individually as well as in combination, but the highest accuracy was demonstrated by the combined model. The included hashtags and mentions were found to be the top-performing category of features, whereas the performance of the textual content model was surprisingly poor (presumably, due to the very short and noisy nature of tweets).

Shen et al. (2013) proposed a method for personalized reordering of tweets. User interests were determined by analyzing the tweets published and consumed by users with respect to five parameters: freshness of the tweet, authority of the poster, length and inclusion of hashtags, match to the recipient's interests, and interaction between the poster and recipient. A personalized ensemble model incorporating the features was trained and evaluated. The model outperformed a non-personalized model and simple temporal predictors, whereas the most important features were found to be the freshness of the tweet and the number of poster's followers.

SN and Graph Factors

Considering that the value of textual content in Twitter-like microblogs is typically lower than in general-purpose SNs, Chen et al. (2012) proposed the incorporation of information pertaining to the structure of the SN user graph in predictive models. The authors devised a personalized tweet ranking model, which encapsulated a range of features that were categorized into four groups: graph-based similarity of the user nodes, relevance of the tweet content to the recipient, hashtags, and URLs included in the tweet, and the poster's authority. These features were used to train a combined latent factor model, which was found to outperform several baseline models and the models exploiting the above groups of features individually.

The work of Feng and Wang (2013) also leveraged the graph-based representation of the Twitter graph to rank tweets. The nodes of the graph encapsulated the posters and recipient users as well as the tweets themselves, while the graph edges reflected the poster-recipient and recipient-tweet relationships. Additional features about the tweets (hashtags, URLs, age, popularity), users (similarity, mentions, reputation), and user-tweet relationships (content similarity, mentions, hashtags) were also considered. All the features were used to train a factorization model for predicting the probability of retweets. The accuracy of the model was high, while the importance of the poster-recipient and recipient-tweet relationship edges was found to dominate that of the poster and recipient nodes, highlighting the value encapsulated in the Twitter graph structure.

Yan et al. (2012) proposed a graph-theoretic model for tweet recommendations. The recommender leveraged a heterogeneous network model consisting of a graph of users and a graph of tweets. In both sub-graphs, the nodes represented the users and the tweets, respectively, while the edges reflected the degree of their similarity. The user-to-user similarity was established based on the commonality of followees, while the similarity of tweets was computed using their textual and semantic content. The nodes of the two sub-graphs were scored using the personalized PageRank algorithm, and then co-ranked, such that the tweet scores corresponded to the scores of their poster and retweeters, and vice versa. The model was evaluated using a very large corpus of retweets and the performance of the model was found to outperform several personalized ranking competitors.

Activity Feed Personalization

The TWD Online Portal

We now turn to the presentation of our feed personalization approach, which was deployed and evaluated within a live study of a diet and lifestyle intervention website, called the Online TWD portal. The goal of the portal was to support people embarking on a validated weight loss and maintenance program, the TWD diet (Noakes and Clifton, 2005). The portal contained dietary information, support tools, and a social component with typical SN functionalities. **Figure 1** shows a coarse-grain layout of the portal's front page. The static information presented included recipes (organized as per the content of the book, e.g., salads, chicken, beef, and deserts), exercises, menu plans, shopping lists, and links to additional health-related sites and material. The portal

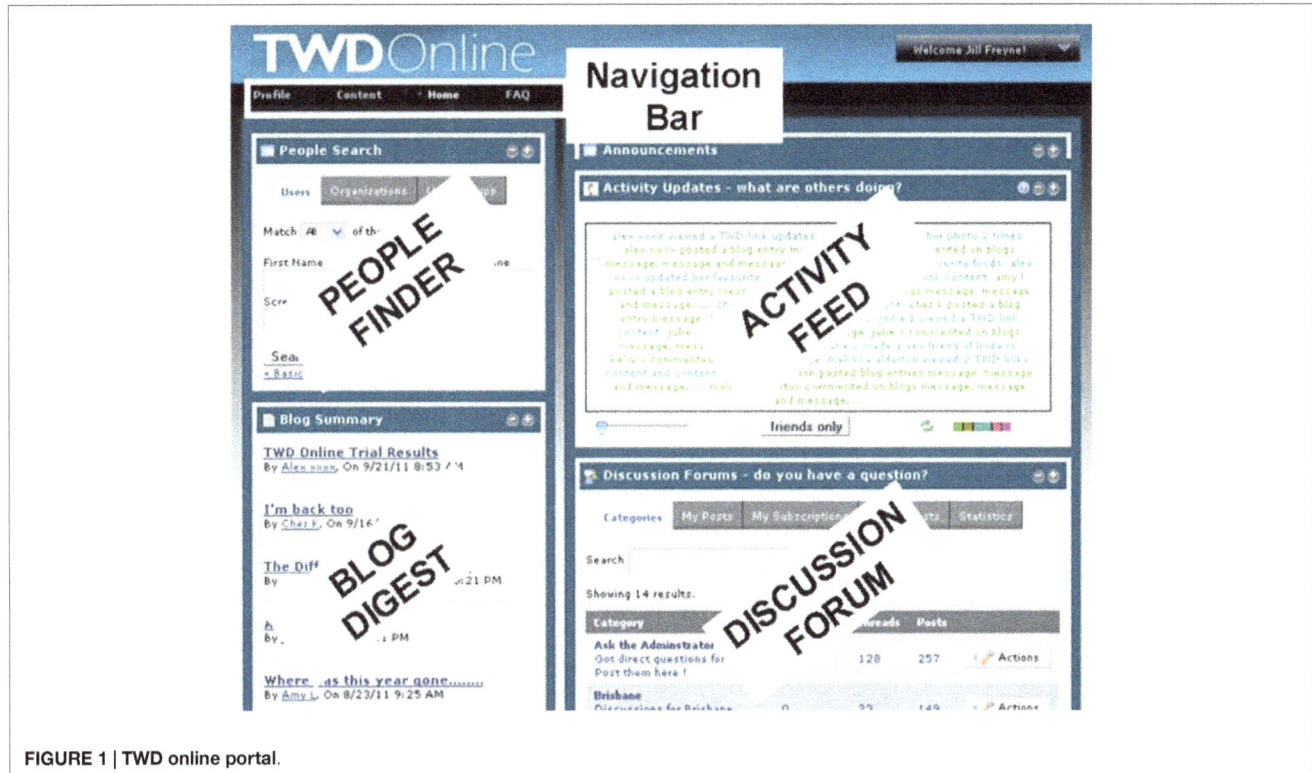

FIGURE 1 | TWD online portal.

also included a suite of interactive tools aimed at supporting the TWD participants and strengthening their engagement with the portal, such as a meal planer and weight tracker. These provided to users real-time feedback on their decisions and their progress on the diet program (Freyne et al., 2011; Brindal et al., 2012).

The goal of the SN component of the portal was to provide an online social support for program participants. Users registered for the portal were represented by a dedicated personal profile page, which contained some demographic information, an image gallery, a personal message board (wall) and a blog. Users were not issued with instructions on how the wall or the blog should be used as such they could freely contribute content as often as they wished on any topic they wished. Access to blogs was controlled by their respective owners. To facilitate community-based information sharing, the portal also contained a discussion forum open to all users. Here, users could ask questions, seek advice, provide support to others, and discuss ideas and thoughts with the community. The forum was moderated by domain experts, who responded to health-, exercise-, and nutrition-related questions raised by users.

Note that the SN of the Online TWD portal differed from other general-purpose SNs in a sense that it was not a familiarity-based SN. Indeed, most commercial SNs, such as Facebook or LinkedIn, reflect existing offline familiarity links; that is, online links and friendships mirror real-life links. On the contrary, Online TWD users were recruited online and had no offline familiarity with each other. Thus, the bootstrapping of the SN and the establishment of links between the users might have been slower than in general-purpose SNs. At the same time, this emphasized the importance of the TWD portal in supporting diet participants

and strengthening their engagement with the diet through social learning and comparison, i.e., by exposing them to the thoughts and actions of other dieters. By highlighting activities, such as meal planning, weighing in, browsing recipes and exercises, and reading/writing blog posts, we aimed to persuade users to also carry out these activities (Fogg, 2003).

To this end, our portal included a dedicated SN activity feed, which aggregated the activities of other users with the portal, such as content being viewed, use of the support tools, content generation, and interactions taking place between the users (friending, commenting, and so on). The feed was displayed on the portal's front page as a list of textual items corresponding to the observed SN activities. Feed items contained details of the user who performed the activity and the activity itself (or the content produced by the activity), as shown in **Figure 2**. Both the user and the activity were hyperlinked and could be clicked. By default, the feeds included 20 items, and the users had the opportunity to adjust the size of the feed. In practice, very few users utilized this feature such that we consider the size of the feed to be fixed across all the users for the entire duration of the study.

Feed Personalization Mechanism

In this subsection, we outline the feed scoring and personalization mechanism. The feed presents to a target user u_t a list of SN activities performed by other users. Each item i_x included in the feed references two components: the subject user u_x who performed the activity and the action a_x that was performed (or the content resulting from performing a_x), e.g., wall comments, forum posting, or content viewing. When the feed is visualized, both the name of u_x and the action a_x are hyperlinked, such that

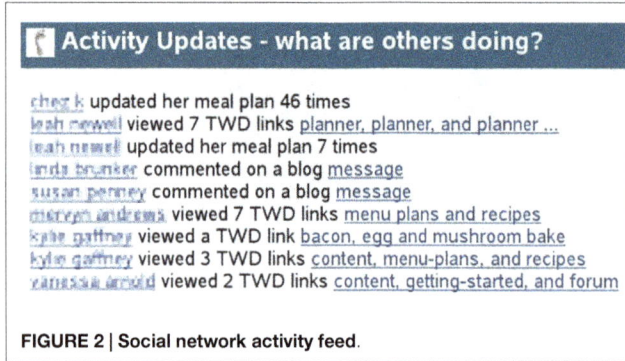

FIGURE 2 | Social network activity feed.

clicking on u_x, i.e., user click, redirects the target user u_t to the profile page of u_x, whereas clicking on a_x, i.e., action click, leads directly to the content viewed or contributed by the activity.

Our personalization mechanism assigns to each feed item a user-dependent relevance score $S(u_t, i_x)$ that represents the predicted level of interest of the target user u_t in item i_x. We model $S(u_t, i_x)$ as a linear combination of two parameters: a user-to-user score $S_u(u_t, u_x)$ and a user-to-action score $S_a(u_t, a_x)$. These parameters are weighted in a linear manner according to their relative importance, ω_u and ω_a, respectively, as shown in Eq. 1:

$$S(u_t, i_x) = \omega_u S_u(u_t, u_x) + \omega_a S_a(u_t, a_x) \qquad (1)$$

We presume that feed items corresponding to activities of users with which u_t has close online relationships, would attract more interest than feed items involving the actions of importance for u_t, regardless of the user who conducted the actions. Hence, we parameterize the weights by static values of $\omega_u = 0.8$ and $\omega_a = 0.2$, which prioritize the activities performed by relevant users over relevant actions. The rationale of this weighting scheme is in line with several previous studies (Gilbert and Karahalios, 2009; Wu et al., 2010; Jacovi et al., 2011; Uysal and Croft, 2011; Feng and Wang, 2013; Shen et al., 2013).

The user-to-user relevance score $S_u(u_t, u_x)$ reflects the degree of closeness between the target user u_t and the subject user u_x, as derived from their observable online interactions. To compute this relevance score, we deploy a modified variant of the tie-strength model developed by Gilbert and Karahalios (2009) and adapt it according to the closeness factors and weighting schema proposed by Wu et al. (2010). It should be noted that some factors outlined in Wu et al. (2010) are restricted to the enterprise environment considered in that work and are inapplicable to the setting of the Online TWD portal. Thus, we consider in this work four categories of factors:

- User factors (UF) – online behavior and activity of the target user u_t.
- Subject user factors (SUF) – online behavior and activity of the subject user u_x.
- Direct interaction factors (DIF) – direct online interaction between u_t and u_x.
- Mutual connection interaction factors (MCIF) – indirect interaction between u_t and u_x, i.e., interactions between u_t and $\{u_y\}$ and between u_x and $\{u_y\}$, where u_y is a mutual friend of u_t and u_x.

Thus, the user-to-user relevance score $S_u(u_t, u_x)$ is computed as a weighted linear combination of the scores of these four categories of factors, as shown in Eq. 2:

$$\begin{aligned} S_u(u_t, u_x) = {} & \omega_{uf} S_{uf}(u_t, u_x) + \omega_{suf} S_{suf}(u_t, u_x) \\ & + \omega_{dif} S_{dif}(u_t, u_x) + \omega_{mcif} S_{mcif}(u_t, a_x) \end{aligned} \qquad (2)$$

As the functionality and the components of the SN presented in Wu et al. (2010) resemble those offered by the TWD Online portal, we assign to the scores of these four categories relative weights that are proportional to the original weights derived in Wu et al. (2010): $\omega_{uf} = 0.178$, $\omega_{suf} = 0.079$, $\omega_{dif} = 0.610$, and $\omega_{mcif} = 0.133$.

The category scores $S_{uf}(u_t, u_x)$, $S_{suf}(u_t, u_x)$, $S_{dif}(u_t, u_x)$, and $S_{mcif}(u_t, u_x)$ are computed as a combination of the scores of the individual factors belonging to each category. For the UF and SUF categories, we derived 26 factors that reflect the individual observable behavior of u_t and u_x. The behavior factors include the number of days they logged-in, number of pages they viewed, number of forum/blog/wall posts/threads/comments they added/deleted/viewed/rated, number of chat sessions they participated in, number of meals they planned, number of times they updated/viewed user profiles, number of images they viewed/updated, and more (see **Table 1** for a complete list of UF and SUF factors). These factors are computed separately for the target user u_t and for the subject user u_x, and then weighted according to ω_{uf} and ω_{suf} for the $S(u_t, u_x)$ computation. Also, note that the weights of the factors are set *a priori* for the entire community of users and vary neither across the users nor over time. Adaptive setting of the factor weights was left beyond the scope of this work.

Likewise, we derived 27 factors for the DIF and MCIF categories that, respectively, reflect the direct interaction between u_t and u_x, and their interaction with the set of mutual friends $\{u_y\}$, i.e., other users friended by both u_t and u_x. The interaction factors include the direct friending, chat sessions, and blog subscription between the users, number of mutual/joint forum/blog/profile views/ratings/comments, number of appearances/selections in each other's social comparison questions, number of joint friends, number of days the users interacted on the portal, number of days since friending and last interaction, and more (see **Table 2** for a complete list of DIF and MCIF factors). It should be highlighted that for the MCIF factors, we compute individually the DIF factor scores for the user u_t (or u_x) and each of their mutual friends u_y, and then average these across the entire set of mutual users $\{u_y\}$. Eventually, the scores are weighted according to the parameters ω_{dif} and ω_{mcif}. Also, the weights of these factors were set *a priori* and were fixed for all the users and for the entire duration of the study.

Note that the factor scores are computed using the observed frequencies of user interactions with the TWD Online portal and other users, and then normalized to the [0,1] range. Specifically, the scores of the UF and SUF factors are normalized by dividing the observed frequency by the maximal frequency observed for the relevant action and any other user of the Online TWD portal. The scores of the DIF and MCIF factors, which involve multiple users, are normalized using Jaccard's similarity coefficient. Also,

TABLE 1 | Target and subject user factors and their weights.

UF		SUF	
Factor	**Weight (%)**	**Factor**	**Weight (%)**
Number of days u_t logged-in	0.061	Number of days u_x logged-in	0.027
Number of portal pages viewed by u_t	0.203	Number of portal pages viewed by u_x	0.090
Number of forum posts added by u_t	2.031	Number of forum posts added by u_x	0.900
Number of forum posts deleted by u_t	0.406	Number of forum posts deleted by u_x	0.180
Number of forum threads added by u_t	2.031	Number of forum threads added by u_x	0.900
Number of forum threads viewed by u_t	0.203	Number of forum threads viewed by u_x	0.090
Number of forum threads subscribed by u_t	0.609	Number of forum threads subscribed by u_x	0.270
Number of forum categories created by u_t	0.406	Number of forum categories created by u_x	0.180
Number of own blog posts added by u_t	2.031	Number of own blog posts added by u_x	0.900
Number of own blog posts deleted by u_t	0.203	Number of own blog posts deleted by u_x	0.090
Number of blog comments (overall) by u_t	1.015	Number of blog comments (overall) by u_x	0.450
Number of blog views (overall) by u_t	0.203	Number of blog views (overall) by u_x	0.090
Number of blog ratings (overall) by u_t	0.406	Number of blog ratings (overall) by u_x	0.180
Number of wall posts (overall) added by u_t	1.015	Number of wall posts (overall) added by u_x	0.450
Number of wall posts (overall) deleted by u_t	0.203	Number of wall posts (overall) deleted by u_x	0.090
Number of quizzes answered by u_t	1.422	Number of quizzes answered by u_x	0.630
Number of questions answered by u_t	0.406	Number of social questions answered by u_x	0.180
Number of chat sessions joined by u_t	2.031	Number of chat sessions joined by u_x	0.900
Number of meals planned by u_t	0.305	Number of meals planned by u_x	0.135
Number of own profile updates by u_t	0.406	Number of own profile updates by u_x	0.180
Number of own image uploads by u_t	1.015	Number of own image uploads by u_x	0.450
Number of own profile views by u_t	0.203	Number of own profile views by u_x	0.090
Number of profile views (overall) by u_t	0.203	Number of profile views (overall) by u_x	0.090
Number of own image views by u_t	0.203	Number of own image views by u_x	0.090
Number of image views (overall) by u_t	0.203	Number of image views (overall) by u_x	0.090
Number of images categories created by u_t	0.406	Number of images categories created by u_x	0.180
ω_{uf}	**17.829**	ω_{suf}	**7.902**

note that the three most dominant factors in the personalization model: online friendship between the two users, friend request sent by one user to the other, and chat session between the two users, which account for, respectively, 7.628%, 5.340%, and 6.102% of the overall user-to-user score. All the three belong to the direct interaction group and naturally reflect the strength of the tie between the two users and the heavy weight assigned to ω_{dif}.

We also calculate the user-to-action interest score $S_a(u_t,a_x)$, which is rarely taken into consideration in other feed personalization works. $S_a(u_t,a_x)$ reflects the importance of action a_x for the target user u_t and is informed by the frequency of u_t (as well as the other users) performing the action a_x and the frequencies of performing other actions (Bohnert et al., 2008). Specifically, we calculate the user-to-action relevance score $S_a(u_t,a_x)$ as shown in Eq. 3. There, $f(u_t,a_x)$ denotes the frequency of the user u_t performing the action a_x, $f(u_t,\cdot)$ denotes the average frequency of all actions performed by u_t, $f(\cdot,a_x)$ is the average frequency of all the portal users performing a_x, and $f(\cdot,\cdot)$ is the average frequency of all actions performed by all users.

$$S_a\left(u_t,a_x\right) = \frac{f\left(u_t,a_x\right)}{f\left(u_t,\cdot\right)} \bigg/ \frac{f\left(\cdot,a_x\right)}{f\left(\cdot,\cdot\right)} \qquad (3)$$

In a nutshell, this computation quantifies the relative importance of a_x for u_t and normalizes it by the relative importance of a_x for all the portal users. The user-to-action score $S_a(u_t,a_x)$ computed using and the user-to-user score $S_u(u_t,u_x)$ computed using Eq. 2 are finally aggregated into the overall feed item score $S(u_t,i_x)$, as shown by Eq. 1. This scoring is performed for any candidate item for inclusion in the activity feed and items having the highest predicted scores are included in the feed. The items are shown in a decreasing order of scores, such that the highest-scoring items appear at the top of the feed.

Evaluation

More than 5000 individuals across Australia participated in the live evaluation of the TWD Online portal. The duration of the study was 12 weeks and it was conducted synchronously, such that all the users commenced and completed the study at the same time. The study mainly focused on health-related outcomes, such as weight loss, interactions with the portal, and user engagement with the diet. In this paper, we focus solely on the observed interaction of the users with the SN feeds. Details of user engagement and more general health-related trial outcomes can be found in Brindal et al. (2012).

At the beginning of the study, users were randomly assigned into one of two experimental groups: control group and personalized group. Users in the personalized group received personalized activity feeds, in which the items were scored as per the model described in Section "Activity Feed Personalization" and sorted in a decreasing order of relevance. In the non-personalized control group, the users were presented with chronologically ordered feeds, where the top items were the most recently occurred SN activities. In the first week of the study, the personalization was not applied and both groups were equal. We considered this period to be the bootstrapping of the SN, establishment of

TABLE 2 | Direct and mutual connection interaction factors and their weights.

DIF		MCIF	
Factor	Weight (%)	Factor	Weight (%)
Number of u_x's forum posts rated by u_t (or vv)	1.526	Number of u_y's forum posts rated by u_t (or vv)	0.331
Number of forum posts rated by u_x and u_t	0.763	Number of forum posts rated by u_y and u_t	0.166
Number of forum threads contributed by u_x and u_t	0.763	Number of forum threads contributed by u_y and u_t	0.166
Is u_x subscribed to blog of u_t (or vv)	2.289	Is u_y subscribed to blog of u_t (or vv)	0.497
Number of u_x's comments in blog of u_t (or vv)	3.814	Number of u_y's comments in blog of u_t (or vv)	0.829
Number of blog posts commented by u_x and u_t	1.907	Number of blog posts commented by u_y and u_t	0.414
Number of u_x's views in blog of u_t (or vv)	0.763	Number of u_y's views in blog of u_t (or vv)	0.166
Number of blogs viewed by u_x and u_t	0.381	Number of blogs viewed by u_y and u_t	0.083
Number of u_x's ratings in blog of u_t (or vv)	1.526	Number of u_y's ratings in blog of u_t (or vv)	0.331
Number of blogs rated by u_x and u_t	0.763	Number of blogs rated by u_y and u_t	0.166
Number of questions with u_x for u_t (or vv)	0.763	Number of questions with u_y for u_t (or vv)	0.166
Number of questions with u_x and u_t	0.381	Number of questions with u_y and u_t	0.083
Number of times u_x chose u_t in questions (or vv)	1.526	Number of times u_y chose u_t in questions (or vv)	0.331
Number of times u_x and u_t chose same user	0.763	Number of times u_y and u_t chose same user	0.166
Did u_x have chat session with u_t	6.102	Was chat session between u_y and u_t	1.325
Number of chat sessions joined by u_x and u_t	3.051	Number of chat sessions joined by u_y and u_t	0.663
Number of u_x's views in profile of u_t (or vv)	0.763	Number of u_y's views in profile of u_t (or vv)	0.166
Number of profiles viewed by u_x and u_t	0.381	Number of profiles viewed by u_y and u_t	0.083
Number of u_x's views in images of u_t (or vv)	0.763	Number of u_y's views in images of u_t (or vv)	0.166
Number of images viewed by u_x and u_t	0.381	Number of images viewed by u_y and u_t	0.083
Did u_x friend u_t	7.628	Did u_y friend u_t	1.656
Did u_x get friend request by u_t (or vv)	5.340	Did u_y get friend request by u_t (or vv)	1.159
Number of users friended by u_x and u_t	3.814	Number of users friended by u_y and u_t	0.829
Number of users got friend requests by u_x and u_t	2.670	Number of users got friend requests by u_y and u_t	0.580
Number of days since u_x friended u_t	3.051	Number of days since u_y friended u_t	0.663
Number of days u_x interacted with u_t (or vv)	4.578	Number of days u_y interacted with u_t (or vv)	0.994
Number of days since u_x interacted with u_t (or vv)	4.578	Number of days since u_y interacted with u_t (or vv)	0.994
ω_{dif}	**61.028**	ω_{mcif}	**13.256**

the portal interaction habits and social ties between the users, and initial content posting. At the end of the first week, all the behavior and interaction factors were computed and fed into the personalization model. Following this, the feeds of the users in the personalized group were scored and re-ranked. The scores of all the factors in the personalization model were dynamically re-computed on a nightly basis.

Feed Characterization

The overall level of user interaction with the feeds was lower than anticipated, with 167 users generating 679 feed clicks over the course of the study. **Table 3** summarizes the number of unique users who interacted with the feed, i.e., clicked on feed items, number of sessions that included feed clicks, the number of logged clicks and their breakdown into the components selected (user clicks on the user who performed the activity and action clicks on the action that was carried out or the outcome of the action), and the user- and session-based click-through rates, CTR_u and CTR_s (Shani and Gunawardana, 2011). CTR_u and CTR_s were computed as the ratio between the number of clicks and the number of users and sessions, respectively. This metrics are compared for the groups of users that were exposed to the personalized and non-personalized feed.

We note that users interacted with the personalized feeds slightly more than with the non-personalized: the overall number of clicks was 10.9% higher and the number of unique users was slightly higher, whereas the number of sessions involving feed clicks was comparable. Both the user- and the session-based click-through rates were higher for the personalized group (with more pronounced difference observed for CTR_s), although the differences between the two groups were not statistically significant[2]. Out of the personalized clicks, the distribution between user clicks and actions clicks was 65.0% vs. 35.0%, whereas for the non-personalized clicks it stood at 53.1% vs. 46.9%. That is, users in the personalized group were more interested in the subject users who performed the activities than users in the non-personalized group. This should be attributed to the weighting mechanism of Eq. 1, assigning 80% of the overall weight to $S_u(u_b, u_x)$ and 20% to $S_a(u_b, a_x)$. Breaking all the observed feed clicks into user clicks and action clicks, we note statistical difference between the groups with respect to user clicks.

Table 4 focuses on the actions corresponding to the feed items that were clicked by users and lists the six most popular actions in each group. As shown, the most popular actions were the same across the groups and their distribution was also very similar. The three most popular actions were content viewing, commenting on blog posts, and updating meal plans, which summed up to 70.6% and 71.1% of actions in the personalized and non-personalized groups, respectively. The next actions in both groups were recipe viewing, posting content in blogs, and commenting on forum posts. Altogether, these six actions account for a vast majority of actions of the clicked feed items – close to 93.8% of actions, both in the personalized and the non-personalized group.

---

[2] All statistical significance results hereafter refer to a two tailed t-test assuming equal distribution. In the comparative tables, we mark significance ($p < 0.05$) with * and strong significance ($p < 0.01$) with **.

TABLE 3 | Feed uptake statistics.

	Unique users	Number of sessions	Feed clicks			Click-through rate	
			User*	Action	Total	User	Session
Personalized	85	181	232	125	357	4.200	1.972
Non-personalized	82	183	171	151	322	3.927	1.760

TABLE 4 | Clicked activities.

	Content view	Comment in blog	Meal plan update	Recipe view	New blog post	Comment in forum	Overall (%)
Personalized	27.451	24.930	18.207	10.644	6.723	5.882	93.838
Non-personalized	29.193	28.571	13.354	13.044	6.832	2.795	93.789

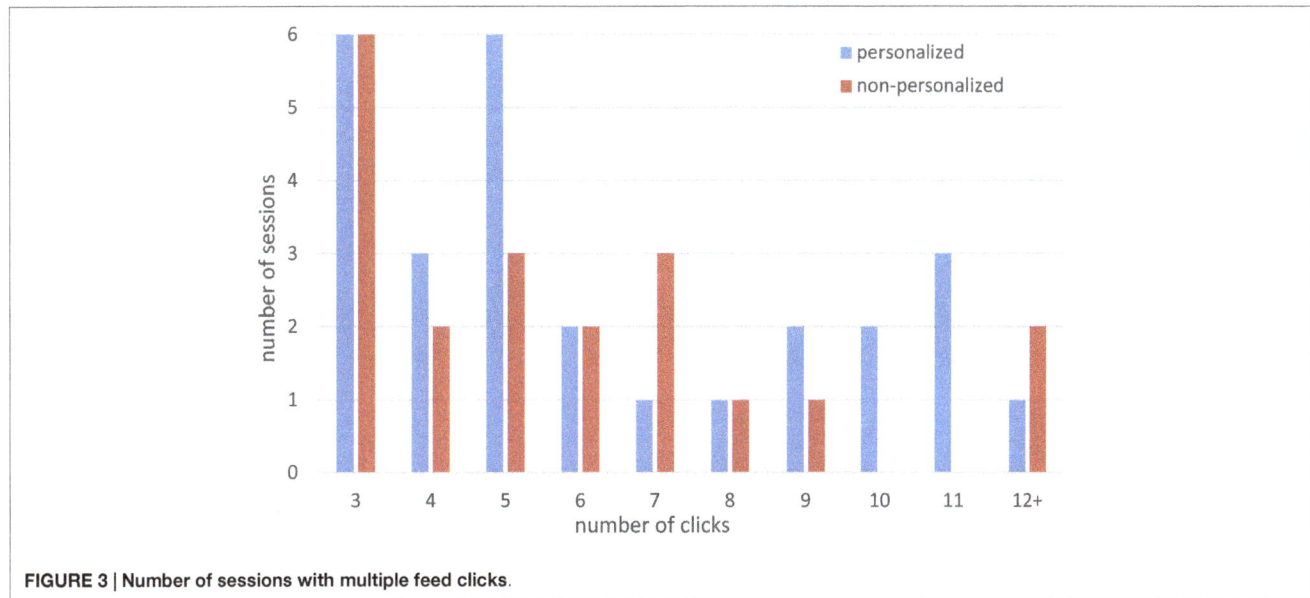

FIGURE 3 | Number of sessions with multiple feed clicks.

We also analyzed the sessions that included multiple feed clicks. For this analysis, we concentrated on sessions with three feed clicks or more, as an indicator of a strong user engagement with the feed. Overall, we observed 28 such sessions in the personalized group and 18 sessions in the non-personalized group. **Figure 3** shows for each group the number of sessions, in which multiple feed clicks were observed. The number of clicks ranges from 3 to 11, with sessions with 12 clicks or more being aggregated in the last pair of bars. The number of sessions with multiple clicks in the personalized group was higher than in the non-personalized group for 4, 9, 10, and 11 clicks, the numbers were equal for 3, 6, and 8 clicks, and more sessions were observed in the non-personalized group for 7 and 12 or more clicks. This shows that feed personalization led to a stronger user engagement with the feed, especially in sessions with high number of clicks (eight and more), where the superiority of the personalized feeds was more pronounced.

Temporal Analysis

As discussed, the feeds were personalized from week 2 till the end of the study, whereas the data from the first week was used for the computation of the personalization model. Hence, it is of interest to analyze the evolution of the personalized feeds over time and compare it to the baseline behavior of the non-personalized feeds, which should not change over time.

We plot in **Figure 4** the average weekly user-to-user score $S_u(u_b, u_x)$, user-to-action score $S_a(u_b, a_x)$, and the overall score $S(u_b, i_x)$ of the clicked feed items. These are shown for 8 weeks only: from week 2 to 9, since the volume of clicks for weeks 10, 11, and 12 was not sufficient for the analyses. A range of values of $S_u(u_b, u_x)$, $S_a(u_b, a_x)$, and $S(u_b, i_x)$ exists, such that the user and overall score values are shown on the left axis, while the action scores appear on the right axis. The user-to-action scores $S_u(u_b, u_x)$ of the clicked items steadily increase meaning that as more data becomes available to the personalization mechanism, the $S_u(u_b, u_x)$ computation becomes more reliable and activities performed by more relevant users are clicked. The observed user-to-action scores $S_a(u_b, a_x)$ are stable over time. Since the overall item score $S(u_b, i_x)$ is dominated by the user-to-user score (recall the 80–20% weight distribution in Eq. 1), the behavior of the overall item score resembles that of the user-to-action score and steadily increases over time.

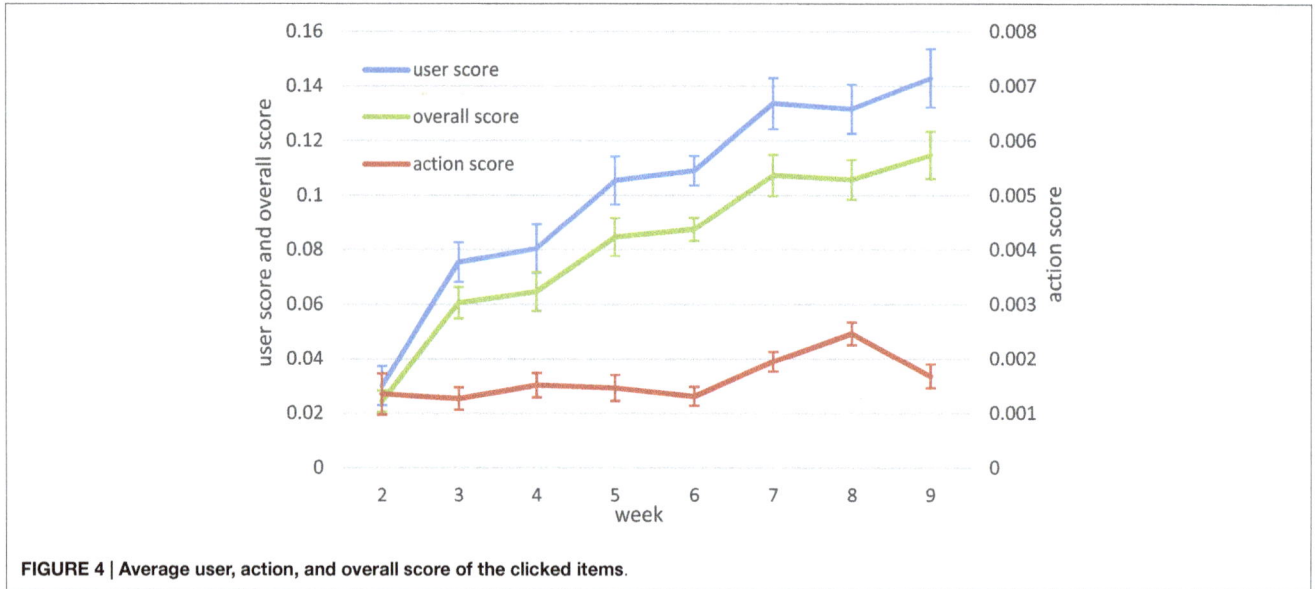

FIGURE 4 | Average user, action, and overall score of the clicked items.

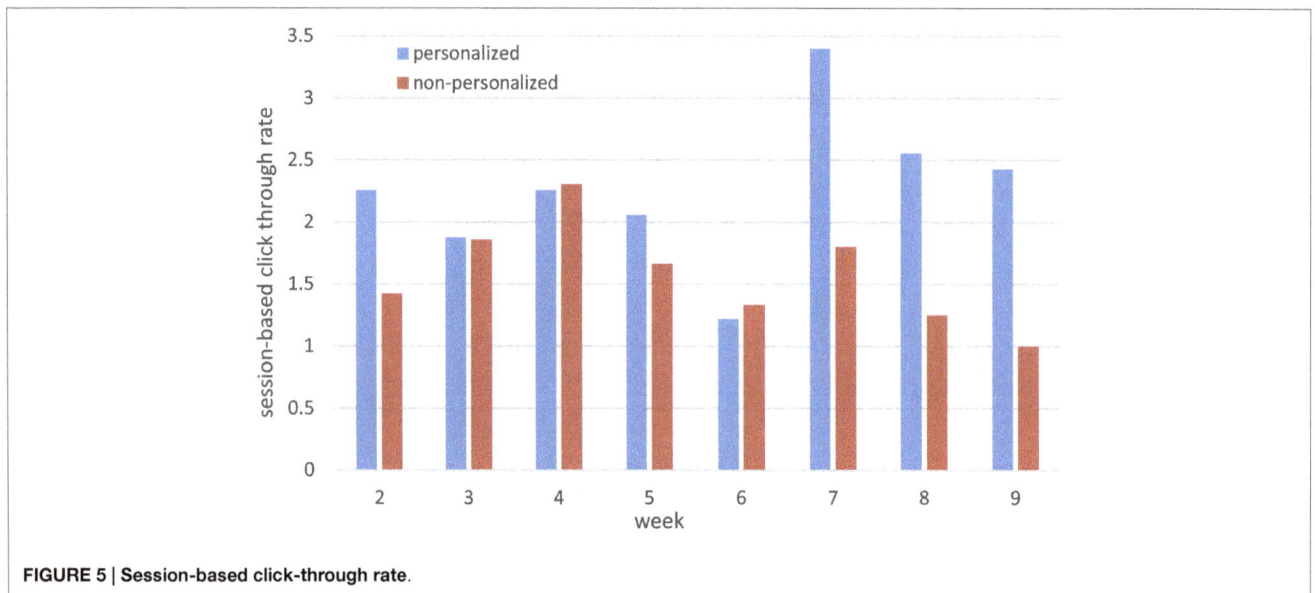

FIGURE 5 | Session-based click-through rate.

Another important dimension in the temporal analysis refers to the changes in the observed session-based click-through rate, CTR_s. We focus here solely on CTR_s since the difference between the groups with respect to CTR_s were more pronounced than with CTR_u (see **Table 3**). **Figure 5** shows the average CTR_s of the personalized and non-personalized feeds for weeks 2–9 of the study. Personalized feeds generally demonstrate higher CTR_s than non-personalized feeds, with the differences between the two groups was more pronounced at weeks 7–9. This finding is in agreement with the observed overall uptake of the feeds toward the end of the study, when personalized feeds substantially outperformed the non-personalized feeds. Hence, users who were exposed to the personalized feeds discovered more content to engage with in the feeds, on a session basis, than the users exposed to non-personalized feeds.

Feed Ranking

We now turn to the analysis of the rank of the clicked items within the feed, i.e., the position of the selected feed items. The average rank of the clicked items in the personalized feeds was 4.35 in comparison to 5.66 in the non-personalized feeds (smaller numbers indicate higher ranks closer to the top of the feed) and the difference between the two was statistically significant, $p < 0.05$. Similarly to the previous evolution analysis, we computed the average rank of the clicked items observed for weeks 2–9 of the study. As shown in **Figure 6**, the personalized feeds consistently placed relevant items higher than the non-personalized feeds, with the difference between the groups being statistically significance, $p < 0.01$.

We assessed the impact of feed personalization on the distribution of the clicked items. **Figures 7** and **8** demonstrate the distribution of feed clicks observed for items in positions 1–15 in

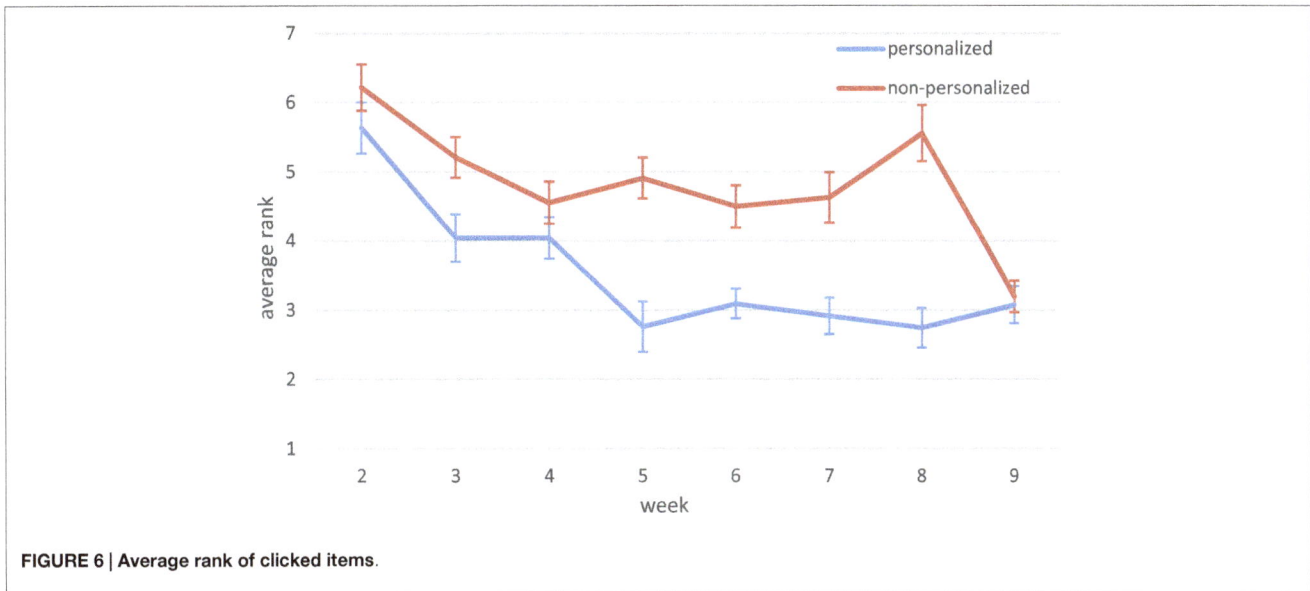

FIGURE 6 | Average rank of clicked items.

the feed for both groups: **Figure 7** shows the probability density function (PDF) and **Figure 8** – cumulative distribution function (CDF). Focusing on the PDF, we observe a strong dominance of the top two positions in the personalized feeds – they accounted for 29.2% and 15.2% of clicks each. Then we observe a decline in the clicks, such that items in positions 3–8 achieved about 7% of clicks each and items in further positions – only 2–3% each. However, the distribution is more balanced in the non-personalized feeds: top four positions account for 12–14% of clicks each, positions 5–9 – for 6–8% each, and then it decreases to 2–3% for further positions. Thus, users in the non-personalized group exhibited a comparable level of interest in a large group of items. On the contrary, in the personalized feeds we observed a stronger preference for items in the top positions.

This finding is supported also by the CDF plot shown in **Figure 8**. For example, the three highest positions in the personalized feeds accounted in summary for 51.4% of clicks, compared to 39.1% of clicks in the non-personalized feeds. The differences are less evident, although still noticeable, when progressing further down the feed, e.g., for the top 6 positions we observed 72.8% vs. 67.8% of clicks and for top 10 only 93.0% vs. 92.0%. The difference between the distributions observed in the two groups was statistically significant, $p < 0.05$. Interpreting the observed user clicks as implicit indicators of relevance, we conclude that the personalized feeds presented promoted relevant items to the top of the feed more effectively than the non-personalized feeds. The observed click distributions clearly show that users found items in the top positions in the personalized feeds more attractive than those in the non-personalized feeds.

Impact on User Activities

One of the goals of the activity feed, especially in the context of the portal, was to highlight activities of other portal users, in order to trigger social learning and encourage users to perform activities that others already perform. To understand the extent to which this was achieved, we examined the actions carried out by users in sessions, when they interacted with the feed. Focusing on

the interactions with content, an important indicator of activity for SNs, we examined the number of blog and forum activities, wall posts, and user profile views, and wall posts observed in each session that also included feed interactions (see **Table 5**). Furthermore, we split the blog and forum activities into contribution activities (posting of posts, responding to posts of others) and consumption activities (viewing of posts), and split the profile views into views of the own user profile and views of profiles of other users (see **Table 6**).

As **Table 5** indicates, the application of personalization to the feeds increased, although not significantly, from 45.1 to 48.2 activities, the overall length of user sessions. This was reflected by an increase in blog, forum, wall, and profile viewing activities. Of these, the increase in wall activities and profile viewing was statistically significant, $p < 0.01$ and $p < 0.05$, respectively. It should be noted that these two activities were not independent, as the wall was located on a user's profile page, such that any wall post was necessarily preceded by a profile view. That said, <10% of profile views resulted in wall posts, showing that the majority of profile views were not a consequence of an intended wall posts. Furthermore, **Table 6** clearly shows that significant increases in the contribution to blogs and forums, and in the views of other users' profiles were observed in the personalized group. Thus, personalization was found to trigger an increased social interaction and to encourage contribution of user-generated content to blogs, walls, and forum.

Feeds and Friending

The proposed feed item scoring mechanism assigns more weight to activities of relevant users in general, and to activities of online friends in particular. To this end, we examined the interplay between online friending and feed clicks. Over the course of the study, more than 4500 online friendships were established, out of more than 9000 friend requests that had been initiated. This friendship rate may be lower than those observed in general-purpose SNs. However, it should be highlighted that our portal differs from the general-purpose SNs; indeed, the users are part of

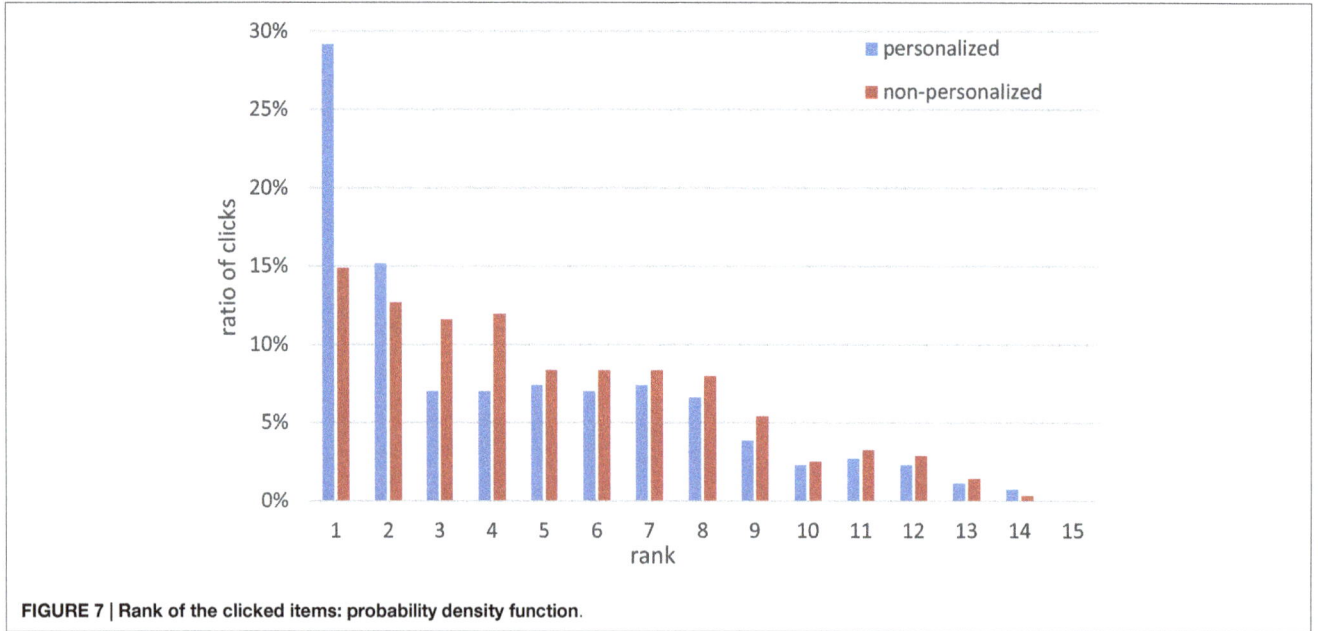

FIGURE 7 | Rank of the clicked items: probability density function.

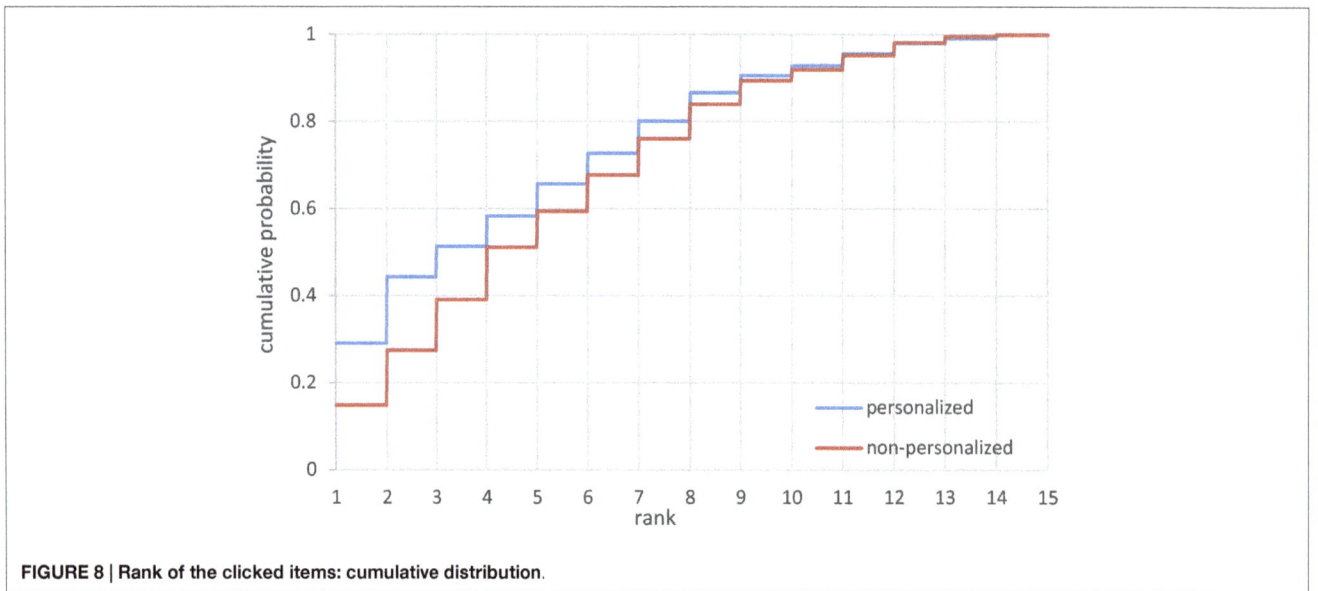

FIGURE 8 | Rank of the clicked items: cumulative distribution.

TABLE 5 | Sessions with feed clicks: statistics.

	Blog activities	Forum activities	Wall posts**	Profile views*	Session length
Personalized	10.352	5.584	0.712	12.720	48.192
Non-personalized	9.767	5.038	0.308	8.962	45.120

a health-driven virtual community and their links are not based on existing offline familiarity.

Figure 9 shows the portion of feed items corresponding to activities performed by online friends of the target user. In the personalized group, this was consistently higher than in the non-personalized group, and the difference between the groups was

statistically significant, $p < 0.01$. Notably, the portion of friend activities in the feed steadily increased over time for both the personalized and non-personalized group, such that in weeks 8 and 9 more than 50% of activities included in the personalized feeds were performed by online friends, whereas in the non-personalized feeds it hovered around the 25% mark. We hypothesize that this finding should be attributed to the higher density of the SN graph toward the end of the study: less users remained active and more friendship links were established at the later stages, such that the portion of feed activities performed by online friends naturally increased in both groups. On top of this, recall the high weights assigned to the direct friendship link established between the users and the friendship request sent by one user to the other, and the dominance of friends' activities in the personalized feeds becomes clear.

TABLE 6 | Sessions with feed clicks: contribution vs. consumption.

	Blog		Forum		Profile views	
	Contribution*	Consumption	Contribution**	Consumption	Own	Other**
Personalized	5.544	4.808	1.000	4.584	5.368	7.352
Non-personalized	3.925	5.843	0.277	4.761	4.799	4.164

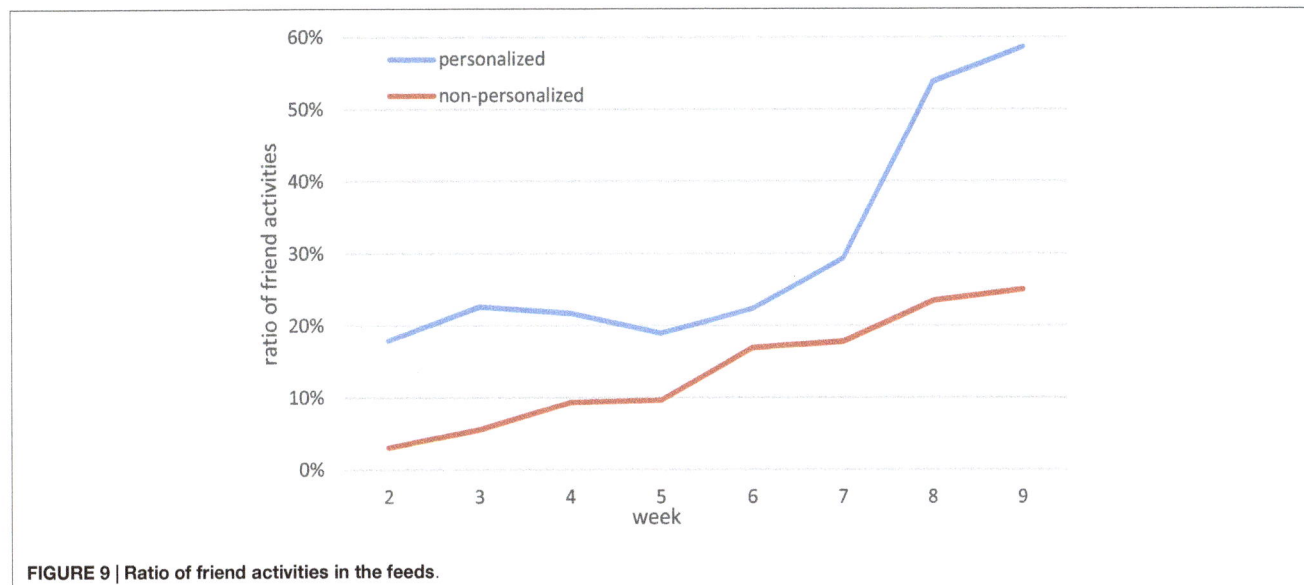

FIGURE 9 | Ratio of friend activities in the feeds.

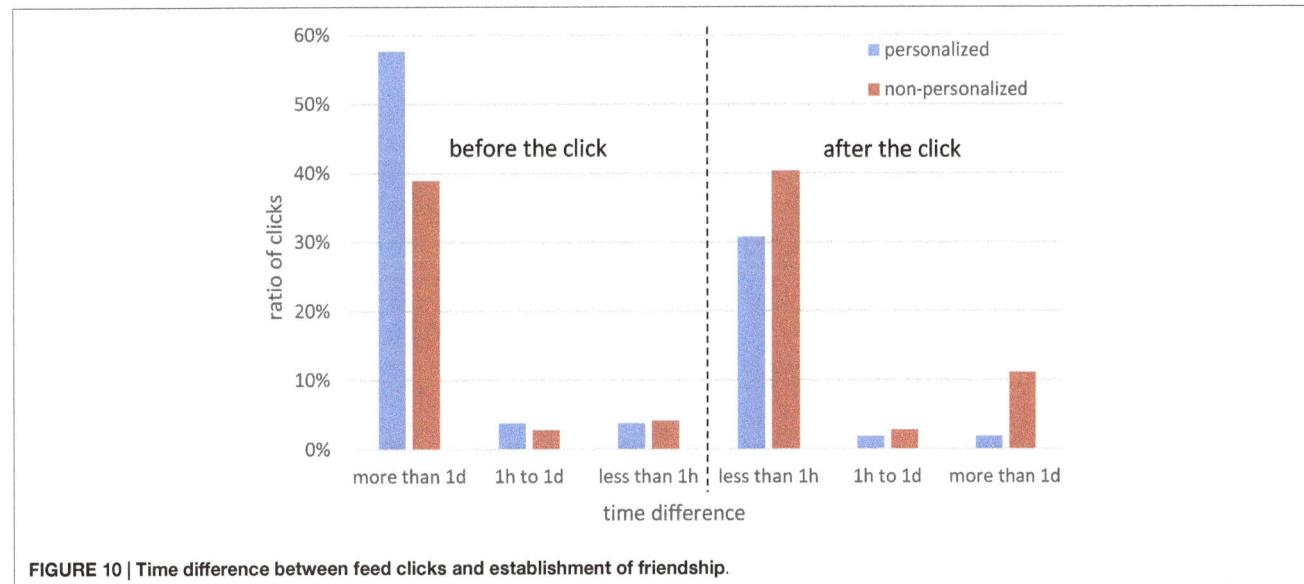

FIGURE 10 | Time difference between feed clicks and establishment of friendship.

Overall, in 32.1% of clicks in the personalized feeds, the target and subject user established an online friendship over the course of the study, whereas in the non-personalized feeds, this was observed in 27.4% of clicks. We also attribute this to the weight assigned to the friendship between the users, which brought the actions of friends to the top of the feed. **Figure 10** further details this analysis and shows the distribution of time between the clicks on activities of friends and the establishment of the friendship, as observed for both groups. We observe two dominant windows: in the personalized feeds 57.7% of friendship links were established more than 1 day before the click and 30.8% within 1 hour of the click, whereas in the non-personalized feeds, the corresponding numbers were 38.9 and 40.3%. This means that the majority of clicks in the personalized feeds were on activities performed by existing online friends and about 30% of clicks were likely to trigger new friendships. On the contrary, in the non-personalized feeds both the activities of existing friends and of those users who were friended shortly after the click accounted for about 40% of clicks. That is, activities of online friends were of relevance for the target user, but the lack of personalization required users to look these up in the feed. This exploration was found to boost the post-click friending rate, which was higher than in the personalized group.

Conclusion and Discussion

This work was motivated by the information overload problem in SNs, which is exacerbated by the simplistic nature of activity feeds. We suggested that the problem can be addressed through the application of personalization to feeds, which has the potential to identify activities of particular interest for the feed recipient. To this end, we developed a model for personalized predictions of the relevance of feed items and applied this model to generate personalized feeds for the users of the Online TWD portal. The model incorporated a suite of observable features, primarily corresponding to the online activity of users and the level of their direct and indirect interaction. In this paper, we report on a thorough evaluation of the uptake and evolution of the personalized feed and its impact on the contribution of user-generated content and user interactions with the portal and other users.

The results show that the overall uptake of the personalized feeds was higher than of the non-personalized feeds. The most pronounced difference was observed for the clicks on users who performed the activities. Focusing on the evolution of the personalized feeds, we noted that over time they attracted clicks on activities performed by closer users, as more and more information facilitating accurate predictions was available to the model. We also observed session-based click-through rates of the personalized feeds outperforming those of the non-personalized feeds, especially toward the end of the study, when the SN became denser. Turning to the rank of the clicked items, we observed that the clicked items in the personalized feeds were ranked higher, i.e., closer to the top of the feed, than in the non-personalized feeds. That is, personalization successfully promoted relevant activities to re-organize the feed, assisting the users in accessing activities and people of interest.

We also investigated the impact of personalization on the observed SN activities, content contribution, and online friending. We found that the personalization notably increased the contribution of user-generated content; in particular, of blog and wall posts. Also, it increased the volume of profile viewing activities, specifically, of viewing profiles of other users. Not only the viewing was boosted but also an increase in online friending was triggered by the feeds. The personalized feeds were found to highlight the activities performed by online friends, while not limiting user awareness of the activities of other SN users. Hence, the application of personalization to activity feeds had a prolific impact on user engagement and content contribution, playing an important role in the sustainability of the social features of the Online TWD portal.

Considering the features used by the personalization model, we highlight the importance of strength of the user-to-user relationships. Although some insight can be obtained from the established SN links and friendship, a more fine-grained tie-strength quantification can be derived from their observable SN interactions, e.g., viewing contributed content, mentioning each other, sending direct messages, or even interacting with the same set of users. Being able to extrapolate the strength and context of relationships between users is a valuable means for not only filtering untrusted users from the feeds but also highlighting specific activities or content contributions, in order to satisfy the differences in user interests.

It should be noted that neither the textual content of the posts nor the network structure were exploited by our personalization model (Berkovsky and Freyne, 2015). The textual content of the posts was discarded primarily due to the fact that the majority of user activities in the Online TWD portal refer to content contribution rather than to content consumption, which is in line with trends observed on larger general-purpose SNs (Muller et al., 2009). The absence of network structure features is explained by the small scale and relative sparsity of the friendship graph, as the SN of the Online TWD portal differed from the usual familiarity-based SN. However, we note the great potential of features reflecting the network structure, which can be projected onto the reputation or authority of users, or their salience on the SN.

Another issue that deserves attention is the applicability of the feed personalization mechanism to other (and particularly general-purpose) SNs. Although this has not been done yet, we would like to point out two works that applied similar techniques. In Freyne et al. (2010) a similar personalization model was applied to IBM's enterprise SN. Similarly to our work, this feed personalization incorporated user-to-user and user-to-action scores as the key relevance predictors. The evaluation focused on the temporal dimension and concluded that short-term models were suitable for the user-to-user relevance, whereas long-term were more appropriate for the user-to-action relevance. Another similar application of feed personalization was done in (Agarwal et al., 2015), where LinkedIn activity feeds were personalized. In addition to the above two relevance predictors, their model also included the fine-grained user-to-user-and-action score, reflecting the interest of the feed recipient in a specific action of a specific user. The evaluation found that each of these factors, deployed individually and exposed to a large number of users, caused a moderate improvement in the observed click-through rate. The greatest improvement was obtained when deploying the user-to-user relevance predictor. Nevertheless, broader applicability of feed personalization to other SNs and application domains still remains unclear.

In the context of SN activity feeds, an important consideration pertaining to their uptake is their contribution to the filter bubble (Pariser, 2011), i.e., situation where due to the SN information filtering some users may become isolated from cultural or ideological circles different to their opinions. This is unacceptable from an ethical perspective and SNs should strike a balance between personalizing feeds and promoting information exploration by their users. The feeds may also pose a privacy threat, as they may inadvertently expose potentially sensitive information about user activities to the user's social circles, which can potentially be accessed by untrusted parties or malicious users, and then be used inappropriately. Hence, user privacy considerations and privacy-preserving mechanism should also be taken into account when filtering SN feeds (Xu et al., 2010).

All in all, we feel that the research into SN feed personalization is still in its infancy. Several solid algorithmic techniques have been developed and evaluated thus far. However, much of the feed personalization work focuses on the accuracy of identifying relevant posts, little work considers the user needs associated with the personalized feeds. SN designers should keep in mind that SNs are, in essence, user-facing systems (Burke et al., 2010). Hence, more attention should be devoted to user

aspects of feed personalization, including what do users find valuable; what should the presentation mode be; how do users prefer to interact with the feed (Wang et al., 2014). We believe that there is ample space for research into interfaces, visualizations, and control mechanisms that can make user interactions with personalized feeds more intuitive and productive, and we conjecture that these topics will receive an increased attention in the future.

References

Agarwal, D., Chen, B. C., He, Q., Hua, Z., Lebanon, G., Ma, Y., et al. (2015). "Personalizing LinkedIn feed," in *Proceedings of the 21th ACM SIGKDD International Conference on Knowledge Discovery and Data Mining* (ACM), 1651–1660.

Berkovsky, S., and Freyne, J. (2015). "Personalized network activity feeds: finding needles in the haystacks," in *Mining, Modeling, and Recommending 'Things' in Social Media* (Springer International Publishing), 21–34.

Berkovsky, S., Freyne, J., Kimani, S., and Smith, G. (2011). "Selecting items of relevance in social network feeds," in Proceedings of UMAP (Girona).

Berkovsky, S., Freyne, J., and Smith, G. (2012). "Personalized network updates: increasing social interactions and contributions in social networks," in *Proceedings of UMAP* (Montreal).

Bohnert, F., Zukerman, I., Berkovsky, S., Baldwin, T., and Sonenberg, E. (2008). "Using collaborative models to adaptively predict visitor locations in museums," in *Proceedings of AH* (Hannover).

Bontcheva, K., and Rout, D. P. (2012). Making sense of social media streams through semantics: a survey. *Semant Web* 1, 1–31. doi:10.3233/FSW-130110

Brindal, E., Freyne, J., Saunders, I., Berkovsky, S., Smith, G., and Noakes, M. (2012). Features predicting weight loss in overweight or obese participants in a web-based intervention: randomized trial. *J. Med. Internet Res.* 14. doi:10.2196/jmir.2156

Burke, M., Marlow, C., and Lento, T. (2010). "Social network activity and social well-being," in *Proceedings of the SIGCHI Conference on Human Factors in Computing Systems* (ACM), 1909–1912.

Chen, J., Nairn, R., Nelson, L., Bernstein, M. S., and Chi, E. H. (2010). "Short and tweet: experiments on recommending content from information streams," in *Proceedings of CHI* (Atlanta, GA).

Chen, K., Chen, T., Zheng, G., Jin, O., Yao, E., and Yu, Y. (2012). "Collaborative personalized tweet recommendation," in *Proceedings of the 35th International ACM SIGIR Conference on Research and Development in Information Retrieval* (ACM), 661–670.

Das Sarma, A., Das Sarma, A., Gollapudi, S., and Panigrahy, R. (2010). "Ranking mechanisms in twitter-like forums," in *Proceedings of the Third ACM International Conference on Web Search and Data Mining* (ACM), 21–30.

De Choudhury, M., Counts, S., and Czerwinski, M. (2011). "Identifying relevant social media content: leveraging information diversity and user cognition," in *Proceedings of the 22nd ACM Conference on Hypertext and Hypermedia* (ACM), 161–170.

Duan, Y., Jiang, L., Qin, T., Zhou, M., and Shum, H. Y. (2010). "An empirical study on learning to rank of tweets," in *Proceedings of the 23rd International Conference on Computational Linguistics* (Association for Computational Linguistics), 295–303.

Feng, W., and Wang, J. (2013). "Retweet or not?: personalized tweet re-ranking," in *Proceedings of the Sixth ACM International Conference on Web Search and Data Mining* (ACM), 577–586.

Fogg, B. J. (2003). *Persuasive Technology: Using Computers to Change What We Think and Do.* Morgan Kaufmann Publishing.

Freyne, J., Berkovsky, S., Baghaei, N., Kimani, S., and Smith, G. (2011). "Personalized techniques for lifestyle change," in *Artificial Intelligence in Medicine* (Berlin, Heidelberg: Springer), 139–148.

Freyne, J., Berkovsky, S., Daly, E. M., and Geyer, W. (2010). "Social networking feeds: recommending items of interest," in *Proceedings of the Fourth ACM Conference on Recommender Systems* (ACM), 277–280.

Gilbert, E., and Karahalios, K. (2009). "Predicting tie strength with social media," in *Proceedings of the SIGCHI Conference on Human Factors in Computing Systems* (ACM), 211–220.

Huang, H., Zubiaga, A., Ji, H., Deng, H., Wang, D., Le, H. K., et al. (2012). "Tweet ranking based on heterogeneous networks," in *Proceedings of COLING* (Mumbai).

Hurlock, J., and Wilson, M. L. (2011). "Searching twitter: separating the tweet from the chaff," in *ICWSM*, 161–168.

Jacovi, M., Guy, I., Ronen, I., Perer, A., Uziel, E., and Maslenko, M. (2011). "Digital traces of interest: deriving interest relationships from social media interactions," in *ECSCW 2011: Proceedings of the 12th European Conference on Computer Supported Cooperative Work, 24–28 September 2011, Aarhus Denmark* (London: Springer), 21–40.

Lage, R., Denoyer, L., Gallinari, P., and Dolog, P. (2013). "Choosing which message to publish on social networks: a contextual bandit approach," in *2013 IEEE/ACM International Conference on Advances in Social Networks Analysis and Mining* (ASONAM), (IEEE), 620–627.

Muller, M. J., Freyne, J., Dugan, C., Millen, D. R., and Thom-Santelli, J. (2009). "Return on contribution (ROC): a metric for enterprise social software," in *ECSCW 2009* (London: Springer), 143–150.

Noakes, M., and Clifton, P. (2005). *The CSIRO Total Wellbeing Diet.* UK: Penguin Publishing.

Paek, T., Gamon, M., Counts, S., Chickering, D. M., and Dhesi, A. (2010). "Predicting the importance of newsfeed posts and social network friends," in *Proceedings of AAAI* (Atlanta, GA).

Pariser, E. (2011). *The Filter Bubble: What the Internet Is Hiding from You.* Penguin.

Shani, G., and Gunawardana, A. (2011). "Evaluating recommendation systems," in *Recommender Systems Handbook* (Springer), 257–297.

Shen, K., Wu, J., Zhang, Y., Han, Y., Yang, X., Song, L., et al. (2013). Reorder user's tweets. *ACM Trans Intell Syst Technol* 4, 6. doi:10.1145/2414425.2414431

Susarla, A., Oh, J., and Tan, Y. (2012). Social networks and the diffusion of user-generated content: evidence from youtube. *Inf Syst Res* 23, 23–41. doi:10.1287/isre.1100.0339

Uysal, I., and Croft, W. B. (2011). "User oriented tweet ranking: a filtering approach to microblogs," in *Proceedings of the 20th ACM International Conference on Information and Knowledge Management* (ACM), 2261–2264.

Wang, B., Sun, Y., Tang, C., and Liu, Y. (2014). "A visualization toolkit for online social network propagation and influence analysis with content features," in *Proceedings of ICOT* (Xian).

Wu, A., DiMicco, J. M., and Millen, D. R. (2010). "Detecting professional versus personal closeness using an enterprise social network site," in *Proceedings of the SIGCHI Conference on Human Factors in Computing Systems* (ACM), 1955–1964.

Xu, H., Parks, R., Chu, C. H., and Zhang, X. L. (2010). "Information disclosure and online social networks: from the case of Facebook news feed controversy to a theoretical understanding," in *AMCIS*, 503.

Yan, R., Lapata, M., and Li, X. (2012). "Tweet recommendation with graph co-ranking," in *Proceedings of the 50th Annual Meeting of the Association for Computational Linguistics: Long Papers-Volume 1* (Association for Computational Linguistics), 516–525.

Conflict of Interest Statement: The authors declare that the research was conducted in the absence of any commercial or financial relationships that could be construed as a potential conflict of interest.

4

Sensitivity analysis of compressive sensing solutions

Liyi Dai*

Computing Sciences Division, U.S. Army Research Office, Research Triangle Park, NC, USA

The compressive sensing framework finds a wide range of applications in signal processing, data analysis, and fusion. Within this framework, various methods have been proposed to find a sparse solution x from a linear measurement model $y = Ax$. In practice, the linear model is often an approximation. One basic issue is the robustness of the solution in the presence of uncertainties. In this paper, we are interested in compressive sensing solutions under a general form of measurement $y = (A + B)x + v$ in which B and v describe modeling and measurement inaccuracies, respectively. We analyze the sensitivity of solutions to infinitesimal modeling error B or measurement inaccuracy v. Exact solutions are obtained. Specifically, the existence of sensitivity is established and the equation governing the sensitivity is obtained. Worst-case sensitivity bounds are derived. The bounds indicate that sensitivity is linear to measurement inaccuracy due to the linearity of the measurement model, and roughly proportional to the solution for modeling error. An approach to sensitivity reduction is subsequently proposed.

Keywords: compressive sensing, sparse solutions, sensitivity analysis, robustness, gradient method

Edited by:
Soumik Sarkar,
Iowa State University, USA

Reviewed by:
Soumalya Sarkar,
Pennsylvania State University, USA
Han Guo,
Iowa State University, USA

Correspondence:
Liyi Dai,
Computing Sciences Division, U.S.
Army Research Office, 4300 S. Miami
Blvd, Research Triangle Park,
NC 27703, USA
liyi.dai.civ@mail.mil

1. Introduction

Consider the following minimization problem from perfect measurements

$$\min ||x||_{l_0}, \text{ subject to } y = Ax, \tag{1}$$

where $y \in R^m$ is a vector of measurements, $x \in R^n$ is the vector to be solved, $A \in R^{m \times n}$ is a matrix, and l_0 denotes the l_0 norm, i.e., the number of non-zero entries. In the compressive sensing framework, the number of measurements available is far smaller than the dimension of the solution x, i.e., $m \ll n$. Because the l_0 norm is not convex, equation (1) is a combinatorial optimization problem, solving, which directly is computationally intensive and often prohibitive for problems of practical interest. Therefore, equation (1) is replaced with the following l_1 minimization problem

$$\min ||x||_{l_1}, \text{ subject to } y = Ax, \tag{2}$$

where the l_1 norm is defined as $||x||_{l_1} = \sum_i |x_i|$. Note that the matrix A in equations (1) or (2) is assumed to be known exactly and that y is free from measurement inaccuracy. In practice, the problem formulation equation (2) is often an approximation because there may exist modeling errors in A and measurement inaccuracies in y. Therefore, a realistic measurement equation would be

$$y = [A + B(\theta)]x + v(\theta), \tag{3}$$

where $\theta = [\theta_j] \in R^p$ is a vector of unknown parameters, $B(\theta) \in R^{m \times n}$ is a matrix describing modeling errors, and $v(\theta) \in R^m$ is measurement noise. One appealing feature of the measurement form equation (3) is that it can incorporate prior knowledge of the inaccuracies. For example, we may have

$$B(\theta) = \sum_j \theta_j B_j, v(\theta) = \sum_j \theta_j v_j, \qquad (4)$$

where B_j and v_j are known matrices or vectors of appropriate dimensions from prior knowledge. When no prior knowledge is available, B_j and v_j can be chosen as the entry indicator matrix or vector, e.g., v_j has 1 for the j-th entry of $v(\theta)$ and 0 everywhere else. The form equation (3) can be used to describe a wide range of inaccuracies. In equation (4), θ_j is the unknown magnitude of inaccuracy. Another interpretation of the measurement form equation (3) is that it represents structured uncertainties and can be used to describe the characteristics of inaccuracies for different measurements. For example, in multi-spectrum imaging, the noise characteristics are different for different spectrum. Images taken from different views or at different times may exhibit different noise characteristics. In the setting of a sensor network, a particular sensor may produce bad data due to malfunctioning, which likely leads to different characteristics of inaccuracy compared with other functioning sensors. Without loss of generality, it is assumed that the nominal value of θ is 0, which corresponds to the case with perfect modeling and measurements. Under measurement equation (3), the problem equation (2) can be recast as

$$\min ||x||_{l_1}, \text{ subject to } y = [A + B(\theta)]x + v(\theta). \qquad (5)$$

A natural question is how sensitive is the solution of equation (5) to small perturbations at θ, without loss of generality, particularly around $\theta = 0$? Because equation (2) is often only an approximation for problems of practical interest, such sensitivity characterizes the robustness of the solution to modeling or measurement inaccuracies. Among the vast literature on compressive sensing, there has been significant interest in analyzing the robustness of compressive sensing solutions in the presence of measurement inaccuracies. One widely adopted approach, such as those in Candes et al. (2006) and Candes (2008), is to reformulate the problem equation (2) as

$$\min ||x||_{l_1}, \text{ subject to } ||y - Ax||_{l_2} \leq \epsilon, \qquad (6)$$

in which ϵ is a tolerable bound of solution inaccuracy, and the l_2 norm is defined as $||x||_{l_2} = (\sum_i x_i^2)^{1/2}$. Existing literature on equation (6) is extensive. Interested readers are referred to Donoho (2006), Candes and Wakin (2008), and Eldar and Kutyniok (2012) for a comprehensive treatment and literature review. Of particular relevance to this paper, Donoho et al. (2011) (Donoho and Reeew, 2012) analyzed sensitivity of signal recovery solutions to the relaxation of sparsity under the Least Absolute Shrinkage and Selection Operator (LASSO) problem formulation and obtained asymptotic performance bounds in terms of underlying parameters in the methods of finding the solutions. Herman

and Strohmer (2010) derived l_2 bounds between the solutions of equations (2) and (5) in terms of perturbation bounds of unknown $B(\theta)$ and $v(\theta)$. Only the magnitudes (i.e., $||.||_{l_2}$ bounds) of $B(\theta)$ and $v(\theta)$ are assumed known. Chi et al. (2011) analyzed solution bounds to modeling errors of unknown $B(\theta)$ and $v(\theta)$. The goal was to investigate the effects of basis mismatch since the matrix A is also known as the basis matrix in compressive sensing. Upper bounds of solution deviation were obtained. The treatments in those papers considered both strictly sparse signals and compressible signals, i.e., the ordered entries of the signal vector decay exponentially fast. The bounds are of the following form Upper bounds of solutions are obtained (Candes et al., 2006; Candes, 2008)

$$||x(\theta) - x||_{l_2} \leq C_1 ||x - x_k||_{l_1} + C_2 \epsilon,$$

where C_1 and C_2 are constants dependent of A, k is the sparsity of x, and x_k is x with all but the k-largest entries set to zero. Recent publications (Arias-Castro and Eldar, 2011; Davenport et al., 2012; Tang et al., 2013) analyzed the statistical bounds when measurement inaccuracies are modeled as Gaussian noises. A recent publication (Moghadam et al., 2014) after this paper was written-derived sensitivity bounds while this paper seeks exact solutions to sensitivity.

The objective of this paper is to analyze the sensitivity of compressive sensing solutions to perturbations (inaccuracies) in matrix A and measurement y, i.e., the sensitivity of solutions to equation (5) to the unknown parameter θ at $\theta = 0$. Such sensitivity characterizes the effects of infinitesimal perturbations in modeling and measurements. Deterministic problem setting is adopted, and the true solution is assumed to be k-sparse. The sensitivity is local at $\theta = 0$. Exact expressions of the sensitivity are derived. The results of this paper provide complementary insights regarding the effects of modeling and measurement inaccuracies on compressive sensing solutions.

The rest of the paper is arranged as the follows: in Section 2, we examine the continuity of the solution $x(\theta)$ at $\theta = 0$. In Section 3, we establish the existence, finiteness, and equation for the sensitivity of $x(\theta)$ at $\theta = 0$. Bounds for worst-case infinitesimal perturbations are derived in Section 4. An approach to sensitivity reduction is proposed in Section 5. A numerical example is provided in Section 6 to illustrate the sensitivity reduction algorithms. Finally, concluding remarks are provided in Section 7.

In this paper, we use lower case letters to denote column vectors, upper case letters to denote matrices. For parameterized vectors, the following notational conventions, if exist, will be adopted.

$$\nabla v(\theta) = \begin{bmatrix} \frac{\partial v(\theta)}{\partial \theta_1} & \frac{\partial v(\theta)}{\partial \theta_2} & \cdots & \frac{\partial v(\theta)}{\partial \theta_p} \end{bmatrix} \in R^{m \times p},$$

$$\nabla B(\theta) = \begin{bmatrix} \frac{\partial B(\theta)}{\partial \theta_1} & \frac{\partial B(\theta)}{\partial \theta_2} & \cdots & \frac{\partial B(\theta)}{\partial \theta_p} \end{bmatrix} \in R^{m \times np}.$$

Consequently, for a vector $x \in R^n$ independent of θ,

$$\nabla [B(\theta)x] = \begin{bmatrix} \frac{\partial B(\theta)}{\partial \theta_1}x & \frac{\partial B(\theta)}{\partial \theta_2}x & \cdots & \frac{\partial B(\theta)}{\partial \theta_p}x \end{bmatrix} \in R^{m \times p}.$$

For notational consistence, the l_2 norm of a matrix $A = [a_{i,j}]$ is defined as the Frobenius norm, $||A||_{l_2} = (\sum_{i,j} a_{i,j}^2)^{1/2}$.

2. Solution Continuity

Let $x(\theta)$ denote the solution to equation (5). Without confusion, we reserve $x = x(0)$ to denote a solution to the unperturbed problem equation (2). In this section, we examine the continuity of $x(\theta)$ in the neighborhood of $\theta = 0$. A vector $x \in R^n$ is said to be k-sparse, $k \ll n$, if it has at most k non-zero entries (Candes et al., 2006). In the compressive sensing framework, we are interested in k-sparse solutions.

Without further assumption, the existence of a (sparse) solution to the optimization problem equation (5) is not sufficient in guaranteeing the continuity of $x(\theta)$ near $\theta = 0$. To illustrate this issue, consider the following example.

Example 2.1. Consider the problem equation (5) with

$$y = 1, A = \begin{bmatrix} 1 & 1+\theta \end{bmatrix}$$

near $\theta = 0$. Then the minimum $||x||_{l_1}$ is achieved by

$$x(\theta) = \begin{cases} \begin{bmatrix} 1 \\ 0 \end{bmatrix}, & \text{if } \theta < 0 \\[2ex] \alpha \begin{bmatrix} 1 \\ 0 \end{bmatrix} + (1-\alpha) \begin{bmatrix} 0 \\ 1 \end{bmatrix}, & \text{for any } 0 \le \alpha \le 1 \quad \text{if } \theta = 0 \\[2ex] \begin{bmatrix} 0 \\ 1/(1+\theta) \end{bmatrix}, & \text{if } \theta > 0. \end{cases}$$

It's clear that the solution $x(\theta)$ is not continuous at $\theta = 0$.

In Example 2.1, the solution is not unique at $\theta = 0$. It turns out that the uniqueness of the solution is critical in guaranteeing the continuity of the solution near $\theta = 0$.

One popular approach to finding the solution to equation (5) is to solve the following LASSO problem.

$$\min_{x(\theta)} \left\{ \frac{1}{2} ||y - [(A + B(\theta))x(\theta) + v(\theta)]||_{l_2}^2 + \tau ||x(\theta)||_{l_1} \right\} \quad (7)$$

which $\tau > 0$ is a weighting parameter.

Theorem 2.1. Assume that both $B(\theta)$ and $v(\theta)$ are continuous at $\theta = 0$, $B(0) = 0$, $v(0) = 0$, AA^T is positive definite, and that the solution $x(\theta)$ to equation (7) is unique at $\theta = 0$. Then, $x(\theta)$ is continuous at $\theta = 0$, i.e.,

$$\lim_{\theta \to 0} x(\theta) = x, \quad (8)$$

where x is the solution of equation (7) at $\theta = 0$.

Proof. Consider

$$\hat{x}(\theta) = [A + B(\theta)][(A + B(\theta))(A + B(\theta))^T]^{-1}(y - v(\theta)).$$

Then $\hat{x}(\theta)$ is the Moore–Penrose pseudoinverse solution to $y = [A + B(\theta)]\hat{x}(\theta) + v(\theta)$. Under the assumptions that AA^T is positive definite and that both $B(\theta)$ and $v(\theta)$ are continuous at $\theta = 0$ with $B(0) = 0$, $v(0) = 0$, we know that $\hat{x}(\theta)$ is continuous and uniformly bounded in a small neighborhood of $\theta = 0$, i.e., there

exist $c > 0, \delta > 0$ such that $||\hat{x}(\theta)||_{l_1} \le c$ for all $||\theta||_{l_2} \le \delta$. Because $x(\theta)$ is the optimal solution to equation (7) while $\hat{x}(\theta)$ may not, we have

$$\tau ||x(\theta)||_{l_1} \le \frac{1}{2} ||y - [(A + B(\theta))x(\theta) + v(\theta)]||_{l_2}^2 + \tau ||x(\theta)||_{l_1}$$

$$\le \frac{1}{2} ||y - [(A + B(\theta))\hat{x}(\theta) + v(\theta)]||_{l_2}^2 + \tau ||\hat{x}(\theta)||_{l_1} = \tau ||\hat{x}(\theta)||_{l_1}.$$

or

$$||x(\theta)||_{l_1} \le ||\hat{x}(\theta)||_{l_1} \le c$$

for all $||\theta||_{l_2} \le \delta$. Therefore, $x(\theta)$ is uniformly bounded for all $||\theta||_{l_2} \le \delta$.

We next prove equation (8) by contradiction. Assume that equation (8) is not true. Because $x(\theta)$ is uniformly bound for all $||\theta||_{l_2} \le \delta$, there exists a sequence θ_i, $i = 1, 2, \ldots$, such that

$$\lim_{i \to \infty} \theta_i = 0 \text{ and } \lim_{i \to \infty} x(\theta_i) = \tilde{x} \ne x. \quad (9)$$

Note that $x(\theta)$ is the unique solution to equation (7). It must be that

$$\frac{1}{2} ||y - [(A + B(\theta_i))x(\theta_i) + v(\theta_i)]||_{l_2}^2 + \tau ||x(\theta_i)||_{l_1}$$

$$< \frac{1}{2} ||y - [(A + B(\theta_i))x + v(\theta_i)]||_{l_2}^2 + \tau ||x||_{l_1}.$$

By setting $i \to \infty$, also noting the continuity of $B(\theta)$ and $v(\theta)$ at $\theta = 0$, we obtain

$$\frac{1}{2} ||y - A\tilde{x}||_{l_2}^2 + \tau ||\tilde{x}||_{l_1} \le \frac{1}{2} ||y - Ax||_{l_2}^2 + \tau ||x||_{l_1}.$$

Therefore, we must have $\tilde{x} = x$ because the solution to equation (7) at $\theta = 0$ is unique, which contradicts equation (9). The contradiction establishes equation (8). ∎

Note that the number of rows of A is far smaller than the number of columns, $m = n$. The positive definiteness of AA^T is equivalent to that all measurements y are not redundant, which is technical and mild. Theorem 2.1 states that the solution to equation (7) is continuous if it is unique at $\theta = 0$. Similarly, we can establish the continuity of the solution to equation (5). The proof is similar to that for Theorem 2.1 and omitted to avoid repetitiveness.

Theorem 2.2. Assume that both $B(\theta)$ and $v(\theta)$ are continuous at $\theta = 0$, $B(0) = 0$, $v(0) = 0$, AA^T is positive definite, and that the solution $x(\theta)$ to equation (5) is unique at $\theta = 0$. Then $x(\theta)$ is continuous at $\theta = 0$, i.e.,

$$\lim_{\theta \to 0} x(\theta) = x. \quad (10)$$

We introduce the following concept, which is a one-sided relaxation of the *Restricted Isometry Property (RIP)* (Candes et al., 2006).

Definition 2.1. A matrix $A \in R^{m \times n}$ is said to be k-sparse positive definite if there exists a constant $c > 0$ such that

$$||Ax||_{l_2}^2 \ge c ||x||_{l_2}^2 \quad (11)$$

for any k-sparse vector $x \in R^n$.

The $2k$-sparse positive definiteness is a sufficient condition for guaranteeing the uniqueness of the optimal solution to equation (5) (Candes, 2008).

3. Solution Sensitivity

In this section, we are interested in the gradient information

$$\nabla x(\theta) = \begin{bmatrix} \frac{\partial x(\theta)}{\partial \theta_1} & \frac{\partial x(\theta)}{\partial \theta_2} & ... & \frac{\partial x(\theta)}{\partial \theta_p} \end{bmatrix} \qquad (12)$$

which, if exists, is an $n \times p$ matrix in $R^{n \times p}$.

The next theorem establishes the existence of the gradient equation (12).

Theorem 3.1. Consider problem equation (5). Assume that both $B(\theta)$ and $v(\theta)$ are differentiable at $\theta = 0$, $B(0) = 0$, $v(0) = 0$, and that there exists a $\delta > 0$ such that A is $3k$-sparse positive definite for all $||\theta||_{l_2} \leq \delta$. Then at $\theta = 0$, the gradient $\nabla x(\theta)$ exists, is finite, and satisfies

$$AZ + EX + W = 0, \qquad (13)$$

where

$$Z = \nabla x(\theta)|_{\theta=0} \in R^{n \times p}, E = \nabla B(\theta)|_{\theta=0} \in R^{m \times np},$$
$$W = \nabla v(\theta)|_{\theta=0} \in R^{m \times p},$$
$$X = \begin{bmatrix} x & 0 & 0 & ... & 0 \\ 0 & x & 0 & ... & 0 \\ ... & ... & ... & ... & ... \\ 0 & 0 & 0 & ... & x \end{bmatrix} \in R^{np \times p},$$

and x is the solution to equation (2).

Proof. For the sake of simple notations, we only prove the theorem for scalar θ. The proof can be extended component wisely to the vector case.

We first establish the existence of $\nabla x(\theta)$. Let $h(\theta) = \theta^{-1}[x(\theta) - x]$ where $x(\theta)$ and x are solutions to the optimization problems equations (5) and (2), respectively.

Under the continuity of $B(\theta)$, $A + B(\theta)$ is $3k$-sparse positive definite with a constant $\tilde{c}$ independent of θ in a small neighborhood of $\theta = 0$ if A is. Therefore,

$$\tilde{c}||x - x(\theta)||_{l_2}^2 \leq ||[A + B(\theta)][x - x(\theta)]||_{l_2}^2 \leq ||B(\theta)||_{l_2}^2||x||_{l_1}^2 + ||v(\theta)||_{l_2}^2. \qquad (14)$$

or

$$||h(\theta)||^2 \leq \tilde{c}^{-1}[||\theta^{-1}B(\theta)||_{l_2}^2||x||_{l_1}^2 + ||\theta^{-1}v(\theta)||_{l_2}^2].$$

By assumption, $B(\theta)$ and $v(\theta)$ are differentiable at $\theta = 0$. The previous inequality shows that $h(\theta)$ is uniformly bounded in a small neighborhood of $\theta = 0$. Consequently, there exists a sequence $\theta^{(i)} \to 0$ such that

$$\hat{h} \triangleq \lim_{i \to \infty} h(\theta^{(i)}) \qquad (15)$$

exists and is finite. Recall that

$$[A + B(\theta)]h(\theta) = -\theta^{-1}B(\theta)x - \theta^{-1}v(\theta)$$

and

$$[A + B(\theta^{(i)})]h(\theta^{(i)}) = -(\theta^{(i)})^{-1}B(\theta^{(i)})x - (\theta^{(i)})^{-1}v(\theta^{(i)}).$$

Their difference gives

$$A[h(\theta) - h(\theta^{(i)})] = -B(\theta)h(\theta) + B(\theta^{(i)})h(\theta^{(i)}) - \theta^{-1}B(\theta)x$$
$$+ (\theta^{(i)})^{-1}B(\theta^{(i)})x - \theta^{-1}v(\theta) + (\theta^{(i)})^{-1}v(\theta^{(i)})$$

or

$$||A[h(\theta) - h(\theta^{(i)})]||_{l_2}^2 \leq ||B(\theta)h(\theta)||_{l_2}^2 + ||B(\theta^{(i)})h(\theta^{(i)})||_{l_2}^2$$
$$+ ||[\theta^{-1}B(\theta) - (\theta^{(i)})^{-1}B(\theta^{(i)})]x||_{l_2}^2$$
$$+ ||\theta^{-1}v(\theta) - (\theta^{(i)})^{-1}v(\theta^{(i)})||_{l_2}^2.$$

Since A is $3k$-sparse positive definite, there exists a constant $c > 0$ such that

$$c||h(\theta) - h(\theta^{(i)})||_{l_2}^2 \leq ||A[h(\theta) - h(\theta^{(i)})]||_{l_2}^2.$$

Consequently,

$$c||h(\theta) - h(\theta^{(i)})||_{l_2}^2 \leq ||B(\theta)h(\theta)||_{l_2}^2 + ||B(\theta^{(i)})h(\theta^{(i)})||_{l_2}^2$$
$$+ ||\theta^{-1}B(\theta) - (\theta^{(i)})^{-1}B(\theta^{(i)})||_{l_2}^2||x||_{l_1}^2$$
$$+ ||\theta^{-1}v(\theta) - (\theta^{(i)})^{-1}v(\theta^{(i)})||_{l_2}^2. \qquad (16)$$

The right hand side of equation (16) goes to zero as $\theta \to 0$, $i \to \infty$. Therefore,

$$\lim_{\theta \to 0, i \to \infty} ||h(\theta) - h(\theta^{(i)})||_{l_2}^2 = 0.$$

Or equivalently,

$$\lim_{\theta \to 0} h(\theta) = \lim_{i \to \infty} h(\theta^{(i)}) = \hat{h}, \qquad (17)$$

which establishes that $\nabla x(\theta)$ exists at $\theta = 0$ and is finite.

Finally, the fact that $x(\theta)$ and x are the solutions to problems equations (5) and (2), respectively, gives

$$[A + B(\theta)]h(\theta) + \theta^{-1}B(\theta)x + \theta^{-1}v(\theta) = 0.$$

Denote $Z \triangleq \hat{h} = \lim_{\theta \to 0} h(\theta)$ which exists and is finite. By setting $\theta \to 0$ in the above equation, we obtain

$$AZ + EX + W = 0$$

which is equation (13). This completes the proof of the theorem. ∎

Since the solution $x(\theta)$ is k-sparse for all θ in a small neighborhood of $\theta = 0$, by focusing on the non-zero entries of x and $x(\theta)$, the proof of Theorem 3.1 also shows that each column of Z is k-sparse. The support of Z is the same as that of x.

Equation (13) indicates that only those perturbations in E that corresponds to non-zero entries of x contribute to the sensitivity.

A note is needed regarding the requirement of the $3k$-sparse positive definiteness in Theorem 3.1. It has been known in the literature [e.g., Candes (2008) in the form of the Restricted Isometry Property] that $2k$-sparse positive definiteness is needed for guaranteeing the uniqueness of a k-sparse solution. The $3k$-sparse positive definiteness is needed for ensuring the differentiability of the solution, a higher order property. This is a sufficient condition, and its relaxation is a subject of current effort.

4. Sensitivity Analysis

Equation (13) gives an explicit expression of the sensitivity $\bigtriangledown x(\theta)$ at $\theta = 0$. Theorem 3.1 shows that the k-sparse solution Z is unique for given x, E, and W. In this section, we consider the worst-case perturbations, i.e., the value(s) of E and W that maximizes $||Z||_{l_2}$. It's obvious that the sensitivity increases when either $||E||_{l_2}$ or $||W||_{l_2}$ increases. We analyze the sensitivity based on normalized perturbations. Specifically, we consider the following

$$\max_{E:||E||_{l_2}=1;W=0}||Z||_{l_2}, \quad \max_{E=0;W:||W||_{l_2}=1}||Z||_{l_2}, \text{ and}$$

$$\max_{E,W:||[E\ W]||_{l_2}=1}||Z||_{l_2}.$$

Lemma 4.1. The sensitivity Z satisfies the following relationship

$$\max_{E:||E||_{l_2}=1;W=0}||AZ||_{l_2} = ||x||_{l_2}, \tag{18}$$

$$\max_{E=0;W:||W||_{l_2}=1}||AZ||_{l_2} = 1, \tag{19}$$

$$\max_{E,W:||[E\ W]||_{l_2}=1}||AZ||_{l_2} = (||x||_{l_2}^2 + 1)^{1/2}. \tag{20}$$

Proof. According to equation (13), Z satisfies

$$AZ = -EX - W.$$

Next we consider each of the three cases.
Case I: $W = 0$. In this case,

$$||AZ||_{l_2} = ||EX||_{l_2}.$$

Recall that

$$EX = [E_1 x \quad E_2 x \quad ... \quad E_p x].$$

By applying the Cauchy-Schwarz inequality $||Ax||_{l_2} \leq ||A||_{l_2} ||x||_{l_2}$, we obtain

$$||EX||_{l_2}^2 = \sum_{j=1}^{p}||E_j x||_{l_2}^2 \leq \sum_{j=1}^{p}||E_j||_{l_2}^2 ||x||_{l_2}^2$$

$$= \left(\sum_{j=1}^{p}||E_j||_{l_2}^2\right)||x||_{l_2}^2 = ||E||_{l_2}^2 ||x||_{l_2}^2.$$

Therefore,

$$\sup_{E:||E||_{l_2}=1;W=0}||AZ||_{l_2} = \sup_{E:||E||_{l_2}=1;W=0}||EX||_{l_2}$$

$$\leq \max_{E:||E||_{l_2}=1}||E||_{l_2}||x||_{l_2} = ||x||_{l_2}. \tag{21}$$

Next, we prove that the upper bound is achievable for a specifically chosen E. Let

$$\hat{E} = [\hat{E}_1 \quad \hat{E}_2 \quad ... \quad \hat{E}_p], \hat{E}_j = \alpha e_m x^T, \alpha = (mp)^{-1/2}||x||_{l_2}^{-1},$$

$$e_m = [1 \quad 1 \quad ... \quad 1]^T \in R^m.$$

Then

$$||\hat{E}_j||_{l_2}^2 = \text{trace}(\hat{E}_j\hat{E}_j^T) = \alpha^2 ||x||_{l_2}^2 \text{trace}(e_m e_m^T) = m\alpha^2 ||x||_{l_2}^2$$

and

$$||\hat{E}||_{l_2}^2 = \sum_{j=1}^{p}||\hat{E}_j||_{l_2}^2 = \sum_{j=1}^{p}m\alpha^2 ||x||_{l_2}^2 = \alpha^2(mp||x||_{l_2}^2) = 1.$$

For this $E = \hat{E}$, we have

$$||\hat{E}X||_{l_2}^2 = \sum_{j=1}^{p}||\hat{E}_j x||_{l_2}^2 = \sum_{j=1}^{p}||\alpha e_m x^T x||_{l_2}^2$$

$$= \alpha^2 ||x||_{l_2}^4 mp = ||x||_{l_2}^2,$$

or equivalently

$$||AZ||_{l_2} = |\hat{E}X||_{l_2} = ||x||_{l_2}. \tag{22}$$

The combination of equations (21) and (22) gives (18).
Case II: $E = 0$. In case,

$$||AZ||_{l_2} = ||W||_{l_2}.$$

from which equation (19) follows.
Case III: In general,

$$||AZ||_{l_2} = ||EX + W||_{l_2} = ||\tilde{E}\tilde{X}||_{l_2}$$

where

$$\tilde{E} = [\tilde{E}_1\tilde{x} \quad \tilde{E}_2\tilde{x} \quad ... \quad \tilde{E}_p\tilde{x}], \tilde{x} = \begin{bmatrix} x \\ 1 \end{bmatrix},$$

$$\tilde{E}_j = [E_j \quad W_j], j = 1, 2, ..., p$$

and W_j is the j-th column of W. By following the proof of Case I, we know that

$$\max_{E,W:||[E\ W]||_{l_2}=1}||AZ||_{l_2} = \max_{\tilde{E}:||\tilde{E}||_{l_2}=1}||\tilde{E}\tilde{X}||_{l_2}$$

$$= ||\tilde{x}||_{l_2} = (||x||_{l_2}^2 + 1)^{1/2}$$

which establishes equation (20). ∎

Equations (18)–(20) show that the worst-case $||AZ||_{l_2}$ is a constant for measurement noise but proportional to the solution vector for modeling error E.

Theorem 4.1. Under the assumptions in Theorem 3.1, the sensitivity Z satisfies the following bounds

$$\max_{E:||E||_{l_2}=1;W=0}||Z||_{l_2} \leq \sigma_{\min}^{-1}(A)||x||_{l_2}, \tag{23}$$

$$\max_{E=0;:||W||_{l_2}=1}||Z||_{l_2} \leq \sigma_{\min}^{-1}(A), \tag{24}$$

$$\max_{E:||E||_{l_2}=1;W:||W||_{l_2}=1}||Z||_{l_2} \leq \sigma_{\min}^{-1}(A)(||x||_{l_2}^2 + 1)^{1/2}. \tag{25}$$

in which $\sigma_{\min}(A) > 0$ is the minimal singular value of A.

Proof. Without loss of generality, assume θ is a scalar. Let S be the collection of indices of possibly non-zero entries of Z, $Z_S \in \mathbb{R}^{|S|}$ be the vector of non-zero entries of Z corresponding to the indices in S, and A_S be the matrix consisting of corresponding columns of A. Then, A_S is of full column rank if it is k-sparse positive definite, $\sigma_{\min}(A_S) \geq \sigma_{\min}(A)$, and

$$AZ = A_S Z_S.$$

Consider the matrix AZ,

$$\begin{aligned}
\|AZ\|_{l_2}^2 &= \text{trace}(Z^T A^T A Z) = \text{trace}(Z_S^T A_S^T A_S Z_S) \\
&\geq \text{trace}(Z_S^T (\sigma_{\min}(A_S))^2 Z_S) \\
&\geq (\sigma_{\min}(A))^2 \|Z_S\|_{l_2}^2 = (\sigma_{\min}(A))^2 \|Z\|_{l_2}^2,
\end{aligned}$$

which gives

$$\|Z\|_{l_2} \leq \sigma_{\min}^{-1}(A) \|AZ\|_{l_2}. \tag{26}$$

The rest follows from combining the previous inequality with equations (18)–(20) in Lemma 4.1. ∎

If $\sigma_{\min}(A)$ is achieved over k columns of A that correspond to the k-sparse signal to be recovered, then the equality in equation (26) holds and the bounds in Theorem 4.1 could be tight. The bounds are not tight in general because the k columns of A corresponding to the indices in S are unknown *a priori*.

5. Sensitivity Reduction

One natural way of reducing the sensitivity of compressive sensing solutions is to carefully select the basis matrix A so that Z is as small as possible. There has been extensive research on how to select A (Donoho et al., 2006). The general idea is to improve the incoherence of the matrix A, for example, by making A as close as possible to the identity matrix in terms of the RIP property. The problem could be difficult because selecting the correct columns of A from available choices is a combinatorial optimization problem. In this section, we consider an alternative. We assume that the matrix A is given. The objective is to find a solution x that is least sensitive to perturbations B and v. One example where this problem arises is in pattern recognition where each column of A represents a vector in a feature space. The issue is how to classify the pattern based on features of the training data so that the solution is robust to potential perturbations (i.e., noise in data).

We consider to improve the robustness of a signal recovery solution by reducing its sensitivity to model error and measurement inaccuracy. Toward that goal, we reformulate the l_1 optimization problem equation (5) by including a term to penalize high sensitivity. This approach offers an alternative to the conventional approach of basis matrix selection and could be performed after the matrix A is selected.

Solution robustness can be improved by reducing the magnitude of sensitivity. The l_∞ norm is a natural choice because it characterizes the largest entry of a vector or a matrix. Therefore, we consider the following optimization problem.

$$\min \|x\|_{l_1} + \lambda \|Z\|_{l_\infty}, \text{ subject to } y = Ax, AZ + EX + W = 0, \tag{27}$$

where $\lambda > 0$ is a penalizing weight and the l_∞ norm is defined as $\|Z\|_{l_\infty} = \max_{i,j} |z_{ij}|$. Note that $Z \in \mathbb{R}^{n \times p}$ is a matrix. We convert Z into a vector by defining

$$\tilde{z} = \begin{bmatrix} z_1 \\ z_2 \\ \vdots \\ z_p \end{bmatrix} \in \mathbb{R}^{np},$$

where for each j, $1 \leq j \leq p$, z_j is the j-th column of Z. Then problem equation (27) can be re-written as

$$\min \|x\|_{l_1} + \lambda \|\tilde{z}\|_{l_\infty}, \text{ subject to } Cx + D\tilde{z} = f, \tag{28}$$

in which

$$C = \begin{bmatrix} A \\ E_1 \\ E_2 \\ \vdots \\ E_p \end{bmatrix} \in \mathbb{R}^{(m+1)p \times n}, D = \begin{bmatrix} 0 & 0 & \dots & 0 \\ A & 0 & \dots & 0 \\ 0 & A & \dots & 0 \\ \dots & \dots & \dots & \dots \\ 0 & 0 & \dots & A \end{bmatrix} \in \mathbb{R}^{(m+1)p \times np},$$

$$f = \begin{bmatrix} y \\ -w_1 \\ -w_2 \\ \vdots \\ -w_p \end{bmatrix} \in \mathbb{R}^{(m+1)p},$$

and w_j is the j-th column of W, $1 \leq j \leq p$.

Several methods are available to solve the optimization problem equation (28), one of which is the Alternating Direction Method of Multipliers (ADMM) (Boyd et al., 2011). In fact, equation (28) is in the standard form of the ADMM problem formulation. There is a trade-off in selecting the method or algorithm to solve the optimization problem equation (28) in terms of accuracy, efficiency, and reliability. We adopt the gradient projection algorithms proposed by Figueiredo et al. (2007) for solving LASSO problems. Toward that end, we consider two modifications: The first modification is to replace the l_∞ norm with the l_ν norm defined as $\|Z\|_{l_\nu} = (\sum_{i,j} |z_j(i)|^\nu)^{1/\nu}$ where $z_j(i)$ is the ith entry of z_j-the jth column of Z. It is known that $\lim_{\nu \to \infty} \|Z\|_{l_\nu} = \|Z\|_{l_\infty}$. So a large ν is chosen. To ensure the smoothness of the objective function, ν is chosen as an even number. The second modification is to reformulate equation (28) as a modified LASSO problem so that the efficient gradient projection algorithms are applicable with minimum modifications. Finally, it is noted that for each j, $z_j(i) = 0$ for all i not in the support of x. Consequently, we consider the following optimization problem for sensitivity reduction.

$$\min \tau \|x\|_{l_1} + \lambda \|\tilde{z}\|_{l_\nu} + \frac{1}{2} \|Cx + D\tilde{z} - f\|_{l_2}^2 \tag{29}$$

$$\text{subject to } S(x) = S(z_j), j = 1, 2, \dots, p$$

where $\nu > 1$ is a large even number, $\tau > 0$ and $\lambda > 0$ are user-selected weights, $S(x)$ is the set of indices corresponding to possibly non-zero entries of a sparse vector x, i.e., the support of x.

We next provide the algorithmic details of implementing the gradient projection method. First, the problem equation (29) is equivalent to the following constrained optimization problem (Figueiredo et al., 2007).

$$\min F(u,r,\tilde{z}) \triangleq \frac{\tau}{2}e^T(u+r) + \lambda||\tilde{z}||_{l_\nu}$$
$$+ \frac{1}{2}||C(r-u)+D\tilde{z}-f||^2_{l_2} \qquad (30)$$
$$\text{subject to } u \geq 0, r \geq 0, S(r-u) = S(z_j), j = 1,2,...,p$$

in which e is the vector of all ones, $u = [u_j] \geq 0$ (or $r = [r_j] \geq 0$) denotes $u_j \geq 0$ (or $r_j \geq 0$, respectively) for all $j = 1, 2,..., n$. The sparse solution x is given by

$$x = \frac{1}{2}(r-u),$$

and thus $S(r-u) = S(x)$ in equation (30). The gradient of the objective function $F(u,r,\tilde{z})$ in equation (30) is

$$\nabla F(u,r,\tilde{z}) = \begin{bmatrix} \nabla_u F(u,r,\tilde{z}) \\ \nabla_r F(u,r,\tilde{z}) \\ \nabla_{\tilde{z}} F(u,r,\tilde{z}) \end{bmatrix}, \qquad (31)$$

where

$$\nabla_u F(u,r,\tilde{z}) = \frac{\tau}{2}e - \frac{1}{2}C^T[C(r-u)+D\tilde{z}-f],$$
$$\nabla_r F(u,r,\tilde{z}) = \frac{\tau}{2}e + \frac{1}{2}C^T[C(r-u)+D\tilde{z}-f],$$
$$\nabla_{\tilde{z}} F(u,r,\tilde{z}) = \lambda||\tilde{z}||^{1-\nu}_{l_\nu}\tilde{z}^{\nu-1} + D^T[\frac{1}{2}C(r-u)+D\tilde{z}-f],$$

in which $\tilde{z}^{\nu-1} = [\tilde{z}^{\nu-1}_j]$ for $\tilde{z} = [\tilde{z}_j]$.

The gradient projection algorithms proposed in Figueiredo et al. (2007) find the optimal solutions iteratively. Let $u^{(i)}, r^{(i)}, \tilde{z}^{(i)}$ denote the values of $u, r, \tilde{z}$ at the ith iteration. Then, at the $(i+1)$th iteration, $u^{(i+1)}, r^{(i+1)}, \tilde{z}^{(i+1)}$ are updated according to

$$\begin{bmatrix} u^{(i+1)} \\ r^{(i+1)} \\ \tilde{z}^{(i+1)} \end{bmatrix} = \begin{bmatrix} u^{(i)} + \eta^{(i)}(u^{(i+1:i)} - u^{(i)}) \\ r^{(i)} + \eta^{(i)}(r^{(i+1:i)} - r^{(i)}) \\ \tilde{z}^{(i)} + \eta^{(i)}(\tilde{z}^{(i+1:i)} - \tilde{z}^{(i)}) \end{bmatrix} \qquad (32)$$

where $\eta^{(i)} \in [0,1]$ is a weighting factor, and the transition values $u^{(i+1:i)}, r^{(i+1:i)}, \tilde{z}^{(i+1:i)}$ are gradient projection updates

$$u^{(i+1:i)} = \Pi_{C^+}(u^{(i)} - \alpha^{(i)}\nabla_u F(u^{(i)}, r^{(i)}, \tilde{z}^{(i)})). \qquad (33)$$
$$r^{(i+1:i)} = \Pi_{C^+}(r^{(i)} - \alpha^{(i)}\nabla_r F(u^{(i)}, r^{(i)}, \tilde{z}^{(i)})). \qquad (34)$$
$$\tilde{z}^{(i+1:i)} = \Pi_{S(r^{(i+1)}-u^{(i+1)})}(\tilde{z}^{(i)} - \alpha^{(i)}\nabla_{\tilde{z}} F(u^{(i)}, r^{(i)}, \tilde{z}^{(i)})). \qquad (35)$$

In equations (33)–(35), $\alpha^{(i)} > 0$ is the step-size of the gradient algorithm, $C^+ = \{u = [u_j] \in R^n | u_j \geq 0, j = 1,2,...n\}$ is the non-negative subspace, $S(x)$ is the support of x as defined in equation (29), and $\Pi_C(u)$ is the projection operator that projects u onto the subspace C. Because $x = 1/2(r-u)$, the subspace

$S[r^{(i+1)} - u^{(i+1)}]$ in equation (35) is the subspace of x at the $(i+1)$th iteration. To increase the efficiency of the algorithm, the step-size $\alpha^{(i)}$ is selected according to the Armijo rule, i.e., $\alpha^{(i)} = \alpha_0\beta^{i_0}$ in which α_0 is the initial step-size, $\beta \in (0,1)$, and i_0 is the first number in the sequence of $\alpha_0, \alpha\beta, \alpha\beta^2, \alpha\beta^3, \ldots$ that achieves

$$\min_i F(\Pi_{C^+}(u^{(i)} - \alpha_0\beta^i\nabla_u F), \Pi_{C^+}(r^{(i)} - \alpha_0\beta^i\nabla_r F),$$
$$\Pi_{S(r^{(i+1)}-u^{(i+1)})}(\tilde{z}^{(i)} - \alpha_0\beta^i\nabla_{\tilde{z}} F)),$$

in which the values of ∇F are evaluated at $(u^{(i)}, r^{(i)}, \tilde{z}^{(i)})$. Note that the projection of $\tilde{z}$ onto the support of x in equation (35) is unique to solving the optimization problem equation (30).

The value of x at the ith iteration is given by $\frac{1}{2}(r^{(i)} - u^{(i)})$.

The dimension of z is np which could be high for large p since n typically is a large number. However, the matrix multiplications in updating the gradient equation (31) could be done offline. Updating equations (32)–(35) is straightforward.

The algorithm described by equations (32)–(35) has been observed with fast convergence for the numerical example in Section 6. It is nevertheless a basic version of the gradient projection algorithms. Our purpose is to illustrate the incorporation sensitivity information in improving the robustness of compressive sensing solutions through sensitivity reduction. Readers interested in the gradient projection approach are referred to Figueiredo et al. (2007) for a comprehensive treatment and improvements.

6. A Numerical Example

The goal of this section is to illustrate the performance of the sensitivity reduction algorithm described in Section 5 through a numerical example. In this example, the matrix A and measurement y are taken from the numerical example described in the software package accompanying (Candes and Romberg, 2005). In this example, $n = 1024$, $m = 512$. The matrix A and measurement y are, respectively, a random matrix and vector drawn from uniform distributions. Sparsity factor $k = 102$, about 10% of the entries of x. The sparse solution x is shown in **Figure 1**.

We consider the sensitivity analysis and reduction for the following perturbation

$$B = \theta a_{i_0}, v = \theta \begin{bmatrix} 1 \\ 0 \\ \vdots \\ 0 \end{bmatrix} \in R^m, \qquad (36)$$

where $i_0 = 959$, a_{i_0} is the i_0-th column of A. Note that θ is a scalar, thus $p = 1$. Perturbation equation (36) leads to a spike in sensitivity as shown in **Figure 2**. For pattern classification, each column of A represents a feature vector of a training data point. The choice of B implies that we are interested in the sensitivity of a bad training data point and seek the reduction of its effects to pattern classification. Other user-selected parameters are $\nu = 50$, $\tau = 0.85$, $\eta^{(i)} = 0.4$ for all i, $\alpha_0 = 0.9$, and $\beta = 0.8$. For comparison, we show the sparse solution x and the corresponding sensitivity for two cases.

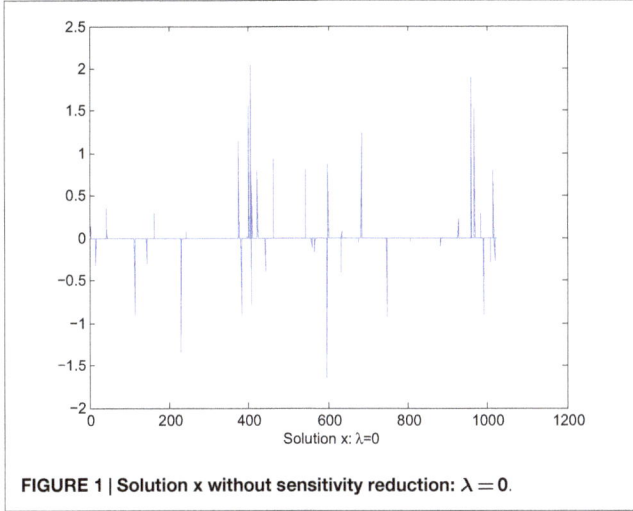

FIGURE 1 | Solution x without sensitivity reduction: $\lambda = 0$.

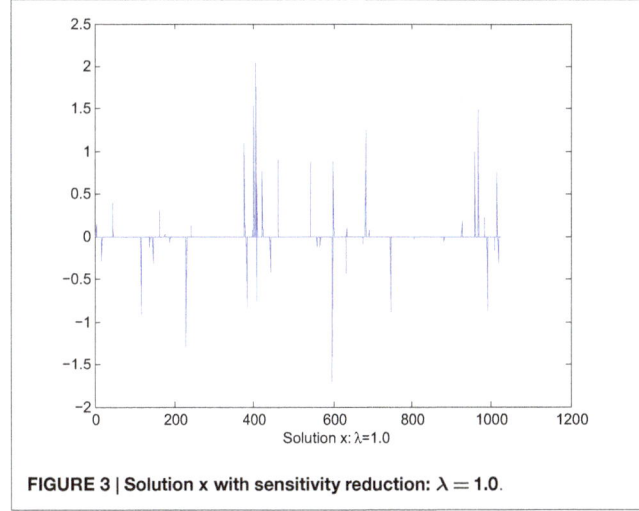

FIGURE 2 | Sensitivity Z without sensitivity reduction: $\lambda = 0$.

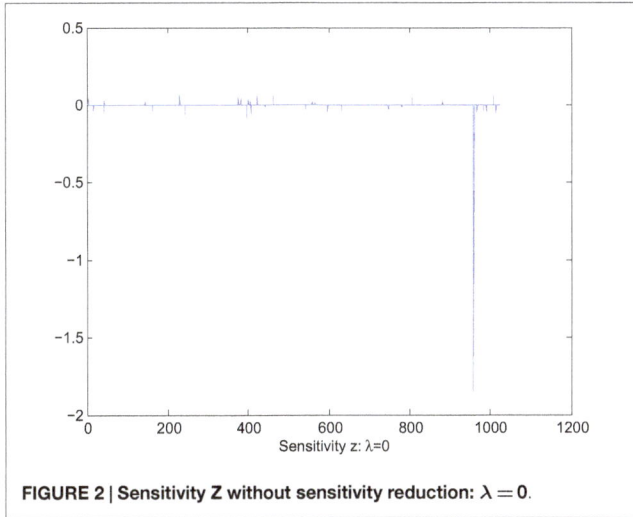

FIGURE 3 | Solution x with sensitivity reduction: $\lambda = 1.0$.

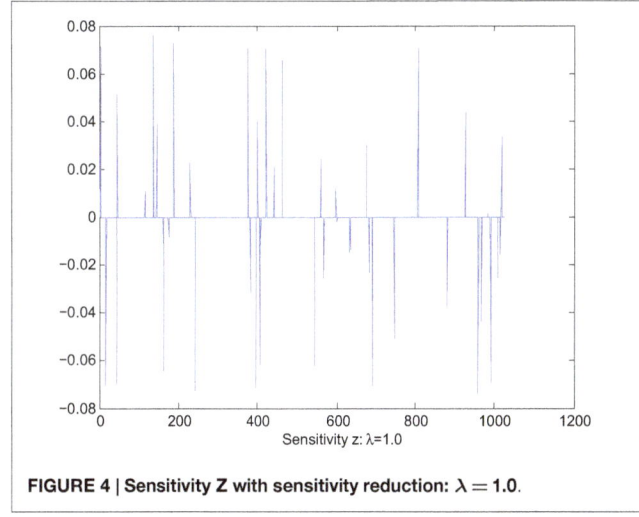

FIGURE 4 | Sensitivity Z with sensitivity reduction: $\lambda = 1.0$.

Case 1: $\lambda = 0$. This case corresponds to the sparse solution to the following LASSO optimization:

$$\min \tau ||x||_{l_1} + \frac{1}{2}||y - Ax||^2_{l_2}.$$

Figure 1 shows the values of the solution x, which clearly indicates the sparseness of the solution. **Figure 2** shows the sensitivity, Z displayed in the vector form $\tilde{z}$ ($Z = \tilde{z}$ for $p = 1$), of x with respect to the parameter θ. The largest value of sensitivity is $||\tilde{z}||_{l_\infty} = 1.8443$, which corresponds to $x(i_0) = x(959)$. The largest value of sensitivity is orders of magnitude larger than the rest of sensitivity values, which indicates the solution x is sensitive to the value of the i_0th column of A.

Case 2: $\lambda = 1.0$. This case corresponds to the sparse solution with sensitivity reduction through the gradient projection algorithm described in Section 5. The values of x and $\tilde{z}$ are shown in **Figures 3** and **4**. **Figure 3** indicates that $x(i_0)$ is now set to 0 to remove the effect of the i_0th column of A. For pattern classification, this would mean that the feature vector corresponding to the i_0 column of A is sensitive to noise and thus is removed for this

particular case. Such feature vector may be resulted from biased data or incorrect choice, depending on the particular problem of concern or scenario. The largest value of sensitivity is $||Z||_{l_\infty} = 0.0761$, which represents orders of magnitude reduction from $||Z||_{l_\infty} = 1.8443$.

7. Conclusion

Robustness of compressive sensing solutions has attracted extensive interest. Existing efforts have been focused on obtaining analytical bounds between the solutions of the perturbed and unperturbed linear measurement models. The perturbation is unknown but finite with a known upper bound. In this paper, we consider solution sensitivity to infinitesimal perturbations, the "other" side of perturbation has opposed to finite perturbations. The problem formulation enables the derivation of exact solutions. The results show that solution sensitivity is linear to measurement noise and proportional to the solution. We have also demonstrated how the sensitivity information can be incorporated in problem formulation to improve solution robustness. The new problem formulation provides a trade-off (i.e., by adjusting the parameter λ) between accuracy and robustness of compressive sensing solutions. One future research direction would be using the sensitivity information to adaptive optimization of parameterized compressive sensing problems. In computer vision, it is well recognized that the performance, e.g., the probability of detection, of object recognition is sensitive to feature selection and training data. The sensitivity information may be used to reduce such sensitivity, for example, in compressive sensing-based approaches to object recognition.

References

Arias-Castro, E., and Eldar, Y. C. (2011). Noise folding in compressed sensing. *IEEE Signal Process. Lett.* 18, 478–481. doi:10.1109/LSP.2011.2159837

Boyd, S., Parikh, N., Chu, E., Peleato, B., and Eckstein, J. (2011). Distributed optimization and statistical learning via the alternating direction method of multipliers. *Found. Trends Mach. Learn.* 3, 1–122. doi:10.1561/2200000016

Candes, E. J. (2008). The restricted isometry property and its implications for compressed sensing. *C.R. Acad. Sci. Ser. I* 346, 589–592. doi:10.1016/j.crma.2008.03.014

Candes, E. J., and Romberg, J. (2005). l_1-MAGIC: Recovery of Sparse Signals via Convex Programming. Available at: http://users.ece.gatech.edu/~justin/l1magic/

Candes, E. J., Romberg, J., and Tao, T. (2006). Stable signal recovery from incomplete and inaccurate measurements. *Commun. Pure Appl. Math.* 59, 1207–1223. doi:10.1002/cpa.20124

Candes, E. J., and Wakin, M. B. (2008). An introduction to compressive sampling. *IEEE Signal Process. Mag.* 25, 21–30. doi:10.1109/MSP.2007.914731

Chi, Y., Scharf, L. L., Pezeshki, A., and Calderbank, A. R. (2011). Sensitivity to basis mismatch in compressed sensing. *IEEE Trans. Signal Process.* 59, 2182–2195. doi:10.1109/TSP.2011.2112650

Davenport, M. A., Laska, J. N., Treichler, J. R., and Baraniuk, R. G. (2012). The pros and cons of compressive sensing for wideband signal acquisition: noise folding versus dynamic range. *IEEE Trans. Signal Process.* 60, 4628–4642. doi:10.1109/TSP.2012.2201149

Donoho, D., and Reevew, G. (2012). "The sensitivity of compressed sensing performance to relaxation of sparsity," in *Proceedings of the 2012 IEEE International Symposium on Information Theory* (Cambridge, MA: IEEE Conference Publications), 2211–2215. doi:10.1109/ISIT.2012.6283846

Donoho, D. L. (2006). Compressed sensing. *IEEE Trans. Inf. Theory* 52, 1289–1306. doi:10.1109/TIT.2006.871582

Acknowledgments

This work is part of the author's staff research at the North Carolina State University. The author would like to thank Prof. Dror Baron, Jin Tan, and Nikhil Krishnan for their helpful comments and constructive suggestions. Their comments lead to the correction of an error in Lemma 3.1 in an earlier version. This work is supported in part by the U.S. Army Research Office under agreement W911NF-04-D-0003.

Donoho, D. L., Elad, M., and Temlyakov, V. (2006). Stable recovery of sparse overcomplete representations in the presence of noise. *IEEE Trans. Inf. Theory* 52, 6–18. doi:10.1109/TIT.2005.860430

Donoho, D. L., Maleki, A., and Montanari, A. (2011). The noise-sensitivity phase transition in compressed sensing. *IEEE Trans. Inf. Theory* 57, 6920–6941. doi:10.1109/TIT.2011.2165823

Eldar, Y. C., and Kutyniok, G. (2012). *Compressed Sensing: Theory and Applications.* New York: Cambridge University Press.

Figueiredo, M. A. T., Nowak, R. D., and Wright, S. J. (2007). Gradient projection for sparse reconstruction: application to compressed sensing and other inverse problems. *IEEE J. Sel. Top. Signal Process.* 1, 586–597. doi:10.1109/JSTSP.2007.910281

Herman, M. A., and Strohmer, T. (2010). General deviants: an analysis of perturbations in compressed sensing. *IEEE J. Sel. Top. Signal Process.* 4, 342–349. doi:10.1109/JSTSP.2009.2039170

Moghadam, A. A., Aghagolzadeh, M., and Radha, H. (2014). "Sensitivity analysis in RIPless compressed sensing," in *2014 52nd Annual Allerton Conference on Communication, Control, and Computing (Allerton)* (Monticello, IL: IEEE Conference Publications), 881–888. doi:10.1109/ALLERTON.2014.7028547

Tang, Y., Chen, L., and Gu, Y. (2013). On the performance bound of sparse estimation with sensing matrix perturbation. *IEEE Trans. Signal Process.* 61, 4372–4386. doi:10.1109/TSP.2013.2271481

Conflict of Interest Statement: The author declares that the research was conducted in the absence of any commercial or financial relationships that could be construed as a potential conflict of interest.

5

Improving Evolvability of Morphologies and Controllers of Developmental Soft-Bodied Robots with Novelty Search

Michał Joachimczak*, Reiji Suzuki and Takaya Arita

Graduate School of Information Science, Nagoya University, Nagoya, Japan

Novelty search is an evolutionary search algorithm based on the superficially contradictory idea that abandoning goal-focused fitness function altogether can lead to the discovery of higher fitness solutions. In the course of our work, we have created a biologically inspired artificial development system with the purpose of automatically designing complex morphologies and controllers of multicellular, soft-bodied robots. Our goal is to harness the creative potential of *in silico* evolution, so that it can provide us with novel and efficient designs that are free of any preconceived notions a human designer would have. In order to do so, we strive to allow for the evolution of arbitrary morphologies. Using a fitness-driven search algorithm, the system has been shown to be capable of evolving complex multicellular solutions consisting of hundreds of cells that can walk, run, and swim; yet, the large space of possible designs makes the search expensive and prone to getting stuck in local minima. In this work, we investigate how a developmental approach to the evolution of robotic designs benefits from abandoning objective fitness function. We discover that novelty search produced significantly better performing solutions. We then discuss the key factors of the success in terms of the phenotypic representation for the novelty search, the deceptive landscape for co-designing morphology/brain, and the complex development-based phenotypic encoding.

Keywords: novelty search, artificial development, soft-robotics, body–brain co-evolution, evolutionary algorithm, artificial life

Edited by:
Zdzisław Kowalczuk,
Gdansk University of Technology,
Poland

Reviewed by:
Dimitrije Markovic,
Dresden University of Technology,
Germany
Luís Correia,
University of Lisbon, Portugal

*Correspondence:
Michał Joachimczak
mjoach@gmail.com

1. INTRODUCTION

The potential to automatically design whole robots with their morphology and control system specialized for a particular task has been one of the most exciting promises of evolutionary robotics, ever since Karl Sims presented his seminal work (Sims, 1994). Over the years, a variety of approaches have been proposed to achieve this goal, often becoming staples of the field. Just like Sim's work, most of them assumed some kind of indirect phenotypic encoding, based on a high level abstraction of development, such as the grammatical approaches [used by Komosinski and Ulatowski (1999), Lipson and Pollack (2000), and Pilat et al. (2012)] or abstractions, such as the CPPN (Stanley, 2007) used by Cheney et al. (2013) and Auerbach and Bongard (2012). Less commonly, and that includes our approach, morphologies and controllers were evolved using a more direct abstraction of biological development, where bodies progressively build themselves through subsequent cellular divisions, deaths, and realignments (Bongard and Pfeifer, 2003; Kowaliw et al., 2004; Meng et al., 2011; Schramm et al., 2011; Joachimczak et al., 2013).

While the two decades of evolutionary robotics have advanced the field to the level where evolutionary approaches excel at designing controllers for robot gaits [see, e.g., Boddhu and Gallagher (2010) and Lee et al. (2013)] or whole complex behaviors [see, e.g., Lessin et al. (2014)], co-evolving bodies and brains *de novo*, even despite famous physical implementations (Lipson and Pollack, 2000) remains largely at the proof-of-concept stage. This should be hardly surprising, given the exploded search space that covers possible pairs of morphologies and controllers and its potentially highly deceptive structure coming from continuous interactions between the two. Naturally, limiting the search space to the area of interest (such as by assuming that bodies consist of sticks or boxes) is essential to solve any problem through evolutionary optimization and means that a knowledge about the expected types of solutions is incorporated into the search algorithm. There is a trade-off, however, evolution, just like a human brain, can be creative and the more we limit the search space, the less likely an evolutionary search algorithm is to come up with a novel and unexpected solution. Thus, by carefully restricting the amount of knowledge about the problem domain that is transferred to the search algorithm, we give the evolutionary process a chance to offer new insights and inspire us with original ways to solve the problem [see, e.g., Hornby et al. (2010) for a classic example]. In particular, in our work, we strive to avoid restrictions put on morphologies and gaits that can evolve, and instead of assembling bodies from primitives, we allow them to grow cell-by-cell and form arbitrary shapes made from hundreds of cells.

One of the research areas where we think it would be beneficial to allow for a possibly unconstrained evolution of morphologies and gaits is soft-robotics. Soft-bodied robotics is a very recent and quickly developing branch of robotics that abandons the idea of robots made of rigid parts. Instead, robots are assumed to be made of elastic material and can deform itself in order to produce gaits or to dynamically adapt to an environment (e.g., crawl through a small opening). This makes them much more similar to animals and, in particular, invertebrates, such as cephalopods. While the field is still in its early stage, there were already multiple successful demonstrations of physical implementations of such entities. Some most prominent examples include robots relying on fluid or air-filled cavities (Steltz et al., 2009; Shepherd et al., 2011), materials that differentially respond to external pressures (Hiller and Lipson, 2012) or actual, 3-D printed biological tissue (Chan et al., 2012).

Following the inspiration from nature and, in particular, the increasing understanding of the role of development in evolution [see, e.g., Carroll et al. (2004)] and how their interactions produce what Darwin called "endless forms most beautiful," we have proposed a biologically inspired, artificial development system that allows to evolve morphologies and controllers for 2-D, soft-bodied animats (Joachimczak et al., 2014). In our approach, animats grow from a single cell through subsequent divisions with each cell controlled by a copy of the same gene regulatory network (GRN) encoded in individual's genotype. Other than being the

way nature produces animal forms, developmental systems are well known for having higher evolvability and scalability than direct encodings [see a direct comparison in, e.g., Komosinski and Rotaru-Varga (2002) and Cheney et al. (2013)]. They also display useful properties, such as robustness, to damage during development or ability to self-repair (Andersen et al., 2009; Joachimczak and Wróbel, 2012b). We use morphologies that emerged through development as a template for a physical model of an animat, which is evaluated in a virtual environment for its performance on a given task. Movement is achieved by contracting and expanding regions of the body, with each body region (originally a cell) making independent decisions about its behavior (though potentially in communication with neighboring areas). In that way, gaits emerge as a product of a distributed control mechanism, being a continuation of the distributed self-assembly process that creates the multicellular morphology.

Using fitness-driven evolutionary search, we were successful in producing complex morphologies that consist of hundreds of cells, walk, run, and swim (Joachimczak et al., 2014) or even reshape their bodies when changing environments through the process of metamorphosis (Joachimczak et al., 2015). Importantly, we were able to show how such a fine-grained approach development leads to the emergence of higher level structural features, such as simple appendages that function as legs, fins, or tails. As a method of fitness-driven search, we have employed the NEAT algorithm [Stanley and Miikkulainen (2002), see Section 2.7.1], one of the most successful method of evolving neural networks (Mouret and Doncieux, 2011). Despite promising results, simulating multicellular growth (with hundreds of cells interacting) is very computationally expensive. Furthermore, the large search space and a complex, highly indirect relation between genotype and phenotype makes evolutionary search prone to getting stuck in local minima.

In this work, we show and analyze how a fine-grained developmental system evolving morphologies and behaviors of robots can be improved by the use of novelty search algorithm (Lehman and Stanley, 2011a; Stanley and Lehman, 2015). This seems superficially counterintuitive as the algorithm entirely abandons the use of an objective fitness function. However, it is found to improve quality of solutions as well as diversity of candidate morphologies that are evaluated during the evolutionary process. The former is needed for the approach to be useful, whereas the latter is essential given our overall aim of harnessing the creative potential of the evolutionary process.

Novelty search is an evolutionary search algorithm based on the radical idea that abandoning objective, goal-focused fitness function altogether can actually lead to a discovery of higher fitness solutions. To do so, novelty search replaces the concept of objective fitness function with the concept of novelty, a scalar value corresponding to how much a given phenotype differs from phenotypes in the current population as well as from phenotypes that have been found to be novel in previous generations, stored in a dedicated archive. This causes the evolutionary search to pursue phenotypes that are different from the already discovered ones instead of phenotypes that have higher fitnesses. While methods that increase genetic diversity have long been demonstrated to be useful in evolutionary algorithms [see, e.g., Mahfoud (1995) and

Sareni and Krahenbuhl (1998)], novelty search differs from them by focusing entirely on increasing the phenotypic, not genotypic diversity.

Novelty search had been suggested to improve evolvability in problems that are deceptive in nature, i.e., where greedily focusing the search on improving fitness will likely lead population into local minima in the fitness landscape that are difficult to escape from. It is suggested that the pressure to produce novel phenotypes will instead lead to the discovery of progressively more and more complex solutions and, among them, the evolutionary stepping stones that open access to new regions of higher fitness in the solution space (Lehman and Stanley, 2011a). Importantly, it is argued that most problems of interest for evolutionary algorithms are deceptive in nature, as non-deceptive problems are simply easy to solve. To what extent different problem domains benefit from the use of novelty search is continuously being explored, with a particular focus on the evolution of robotic controllers. It has been shown to improve evolvability in Lehman and Stanley (2011b), Krčah (2012), Gomes et al. (2013), Lehman et al. (2013), and Urbano and Georgiou (2013). It has, however, been shown to decrease evolvability in problems with very large solution spaces (Cuccu and Gomez, 2011), to which co-evolution of morphologies and control likely belongs to.

In the next section, we provide a concise description of the developmental model that we used. However, we refer the reader to the original paper (Joachimczak et al., 2014) for more details and an overview of what kinds of structures it can evolve. In this work, we focus on investigating how and why novelty search contributes to evolvability in this problem domain.

2. DEVELOPMENTAL APPROACH

Following the biological inspiration, we have attempted to design a possibly simple model of the developmental process, in which arbitrary morphologies could self-assemble through multicellular growth. The underlying assumptions of the approach are

- a genome encodes a control network that commands each cell's behavior,

- self-assembly starts from a single cell, proceeds through subsequent cellular divisions and deaths, and
- all cells share the same control network and respond to local signals.

The fitness evaluation is a two-step process (**Figure 1**). First, an animat undergoes the developmental stage during which its morphology and controller forms. Next, the resultant morphology is used as a template for a soft-bodied model that is simulated in a physical environment for a fixed number of time steps, where it undergoes only elastic changes. We discuss each of these steps in more detail next.

2.1. Growth Stage

Cells are controlled by a simple abstraction of gene regulatory network (GRN) in the form of a neural network where nodes are meant to represent genes and their state rather than neurons. To update state of each node, we used a sigmoidal transfer function (tanh) with output values within $[-1, 1]$ applied to the weighted sum of incoming nodes outputs, as implemented by the MultiNEAT library (Chervenski and Ryan, 2014). The network determines each cell behavior during growth as well as during the locomotion stage, where cells act as "muscles." Ultimately, the behavior of each cell depends on its internal state and the external signals received by the network, such as positional information.

Development takes place in a continuous 2-D space, where cells are represented as disks and undergo elastic collisions simulated with springs that connect them (**Figure 2B**). While there is nothing that would in principle prevent implementing the presented approach in 3-D [in fact, we have implemented an earlier version of the system in 3-D, see, e.g., Joachimczak and Wróbel (2011, 2012c)], a 2-D model is much less computationally demanding. As the body length generally scales with the cube root of the number of cells in 3-D and a square root in 2-D, a fine-grained 2-D development allows to produce structures of higher apparent complexity with lower numbers of cells.

A cell physical state is defined by its position, its velocity, and orientation vector, which determines the direction of division. Springs connect only the nearest neighbors and are determined

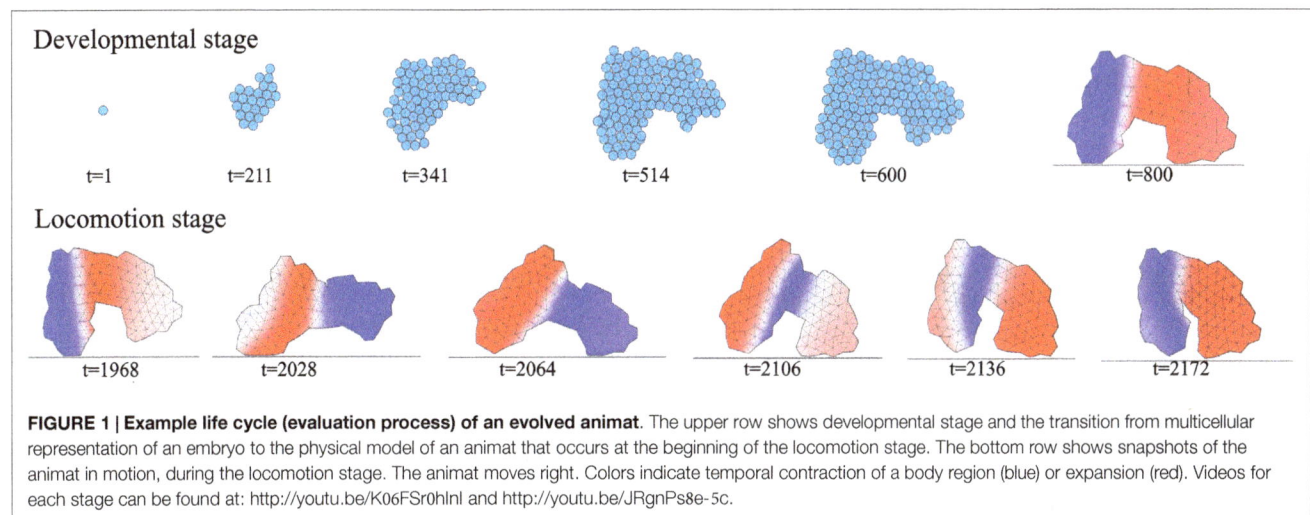

FIGURE 1 | Example life cycle (evaluation process) of an evolved animat. The upper row shows developmental stage and the transition from multicellular representation of an embryo to the physical model of an animat that occurs at the beginning of the locomotion stage. The bottom row shows snapshots of the animat in motion, during the locomotion stage. The animat moves right. Colors indicate temporal contraction of a body region (blue) or expansion (red). Videos for each stage can be found at: http://youtu.be/K06FSr0hlnl and http://youtu.be/JRgnPs8e-5c.

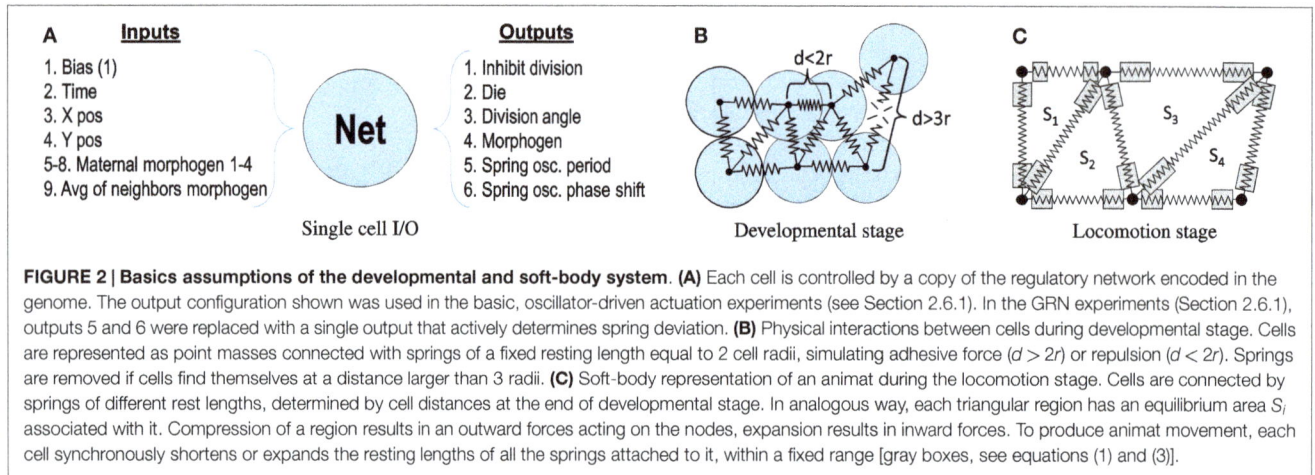

FIGURE 2 | Basics assumptions of the developmental and soft-body system. (A) Each cell is controlled by a copy of the regulatory network encoded in the genome. The output configuration shown was used in the basic, oscillator-driven actuation experiments (see Section 2.6.1). In the GRN experiments (Section 2.6.1), outputs 5 and 6 were replaced with a single output that actively determines spring deviation. **(B)** Physical interactions between cells during developmental stage. Cells are represented as point masses connected with springs of a fixed resting length equal to 2 cell radii, simulating adhesive force ($d > 2r$) or repulsion ($d < 2r$). Springs are removed if cells find themselves at a distance larger than 3 radii. **(C)** Soft-body representation of an animat during the locomotion stage. Cells are connected by springs of different rest lengths, determined by cell distances at the end of developmental stage. In analogous way, each triangular region has an equilibrium area S_i associated with it. Compression of a region results in an outward forces acting on the nodes, expansion results in inward forces. To produce animat movement, each cell synchronously shortens or expands the resting lengths of all the springs attached to it, within a fixed range [gray boxes, see equations (1) and (3)].

dynamically, as the embryo grows. We use Delaunay triangulation to determine the connectivity between cells and then remove links longer than 150% of cell diameter (to allow for non-convex shapes). The resting length of a spring is equal to the sum of attached cells radii. As this simple approach can, however, produce a disjoint structure (e.g., after cells in a central part of an elongated embryo die out), a spring between a pair of nodes is removed only when some other path between these two nodes can be found. Generally, the physics of development was made to resemble soft interactions between cells suspended in a fluid environment.

To reduce computational time, the neighborhood relation was recalculated every 10 steps of physics simulation. The use of Delaunay triangulation as the way to determine neighborhood in the growing embryo is likely not essential for producing current system dynamics and other neighborhood definitions (such as distance based) can be expected to produce similar results. Delaunay was chosen for the conceptual simplicity of spring-based handling of cell adhesion and repulsion, its quasi-linear complexity of determining neighborhood for all cells, and the fact that it allowed us to use the same definition of neighborhood we later used for the locomotion stage. Additionally, the developmental physics was updated at a higher update rate than the state of control networks. More precisely, we allowed for 600 developmental steps and updated networks every 30 steps.

2.2. Network Inputs

Morphogen gradients have long been known to play a fundamental role in development and, in particular, in establishing the basic body plan of animals (Carroll et al., 2004). Thus, to facilitate differentiation of cellular fates, we provided cells with positional information, both direct and indirect (**Figure 2A**). The direct one was in the form of X and Y coordinate directly fed as input to a network. The indirect information would come in the form of four virtual maternal morphogens at prespecified positions in the environment. Additionally, as a simple mechanism that substitutes for morphogens produced by cells, the control network had one "morphogen" output and an associated morphogen input. For any given cell, the activation of the latter was set to the average activation of corresponding morphogen outputs of its neighbors. Finally, all cells were provided with two global signals: a bias input

(set to 1 in every cell) and a time signal (with a value increasing linearly from 0 to 1 throughout development).

2.3. Cell Division and Death

All cells were bound to divide with each subsequent update of the control network unless the output interpreted as the inhibitor of division had activation above 0. Furthermore, division was allowed to occur if and only if space in the direction of the division was not entirely occupied already by other cells [see Joachimczak et al. (2014) for the rationale of this approach].

As long as some space was available, the newly created cell was placed next to the original cell in the direction determined by the network output representing the division angle. The angle was determined at the moment of division and was relative to the mean orientation angle of the cell nearest neighbors. Unless the state was different from zero, all cells would simply divide in the same direction.

Apoptosis (cellular death) occurred whenever activation of the associated network output in a cell was found to be above zero. Such cell was immediately removed from the embryo.

2.4. Termination of Development

To prevent a trivial scenario in which cells divide in an uncontrolled manner until the hard limit of embryo size is reached, we required development never to reach the limit of 256 cells. Individuals who would not fulfill this criterion (even if temporarily) would have their fitnesses set to 0. Moreover, we added a limit on the total number of cells that could be created during the development of an embryo to be no larger than 1024. Such individuals were penalized by having their fitness multiplied by 0.1. We did so in order to limit the occurrence of rather unrealistic solutions in which cells would be continuously created and destroyed (see Section 2.7.1 for a complete list of penalties applied).

2.5. Locomotion Stage

The morphology of an embryo in the final developmental step was used as a template for a soft-bodied animat that was then evaluated for its capability to produce gaits (see an example of such transformation in **Figure 1**). The animat was represented in the physics engine as a 2-D spring-mass system, with point masses located at cell centers and springs forming a triangular mesh.

The resting length of springs was assigned based on the distances between cell centers at the end of the developmental stage (**Figure 2C**). Additionally, each triangular region had its equilibrium pressure (determined based on its surface area at the end of development) providing animat with a hydrostatic skeleton and preventing excessive compression or stretching of body regions.

Springs were governed by the Hooke's law with damping. For simplicity, all springs shared the same Hooke's constant value k but could have different resting lengths.

To avoid self-penetration of animat bodies, masses representing cells would undergo elastic collisions with springs. Movement during locomotion stage was simulated using a custom soft-body engine that we have built on top of the rigid-body physics part of the Bullet Physics Library 2.81 (2013).[1]

The terrestrial environment was constructed by placing animats on top of a flat surface and introducing gravity and friction between animat's nodes and the surface. To prevent sudden changes in resting length for cells with a non-zero phase shift at the start of a simulation, the amplitude was progressively increased during the initial 200 steps of the locomotion stage. Furthermore, to prevent evolution from exploiting any initial motion that would come from relaxation of the animat body at the beginning of the locomotion stage (when gravity, not present during development, was enabled), before actuation would start, we first waited for the structure to stabilize. This was implemented by making sure that the speeds of the nodes are sustained under a preset threshold for 800 time steps.

For an aquatic environment, gravity was disabled and fluid drag was introduced. We used the fluid drag model based on the work by Sfakiotakis and Tsakiris (2006), which assumes that fluid is stationary and that the force acting on a single edge on the outline of the body is a sum of tangential and normal drag components for the motion of this edge against the fluid [see also Joachimczak and Wróbel (2012a) for details].

Actuation meant modifying resting length of springs attached to a given cell. This resulted in the body region contracting or expanding. The total change of resting length of a spring would depend on both cells it connects. The maximum possible range of change was from -40 to $+40\%$ in the aquatic environment and from -30 to $+30\%$ in terrestrial (the maximum was reached assuming both cells acted in accord). However, due to the hydrostatic pressures and other forces, the change of resting length of springs would not result in the change of spring length of the same magnitude.

How the resting length of springs was modified to produce gaits depend on which of the two investigated control approaches were used, discussed next.

2.6. Two Approaches to Actuation

In the course of our earlier work, we have developed two different approaches to gait control for the presented model, one being simple and inspired by SodaPlay approach (Burton, 2007) and a more complex one, employing networks to actively control the state of actuators.

2.6.1. Oscillator-Driven Actuation

The simple one assumed that all "muscular" activity during locomotion stage is determined by sinusoidal patterns of contraction and expansion of body regions surrounding cells, with each cell having its own oscillation period and phase assigned. Both period and phase shift would evolve and be determined by the control network; we realized this by setting the parameters for each cell oscillations based on the state of two corresponding outputs in the given cell network at the final step of development. The state of control networks would then no longer be updated during the locomotion stage. More precisely, during the locomotion stage the default resting length L_0 of a given spring was modified according to the oscillation parameters of the two cells it connects:

$$L_t = \left(1 + A\sin\left(\frac{2\pi t}{T_1} + \phi_1\right) + A\sin\left(\frac{2\pi t}{T_2} + \phi_2\right)\right) \cdot L_0 \quad (1)$$

where t was simulation time, A was the predefined amplitude, T_1, T_2 were periods of oscillation (scaled to span a predefined range), and ϕ_1, ϕ_2 were phase shifts (scaled to $-\pi$ to π) determined by the activation of corresponding network outputs at the end of development.

2.6.2. GRN-Driven Actuation

The second approach allowed for a more fine-grained control over actuation, such as using different oscillation patterns for each cell or making only a subset of cells take part in gait generation. Instead of two outputs determining the period and phase shift of muscular contractions at the end of development, we used one output which determined whether a cell contracts or expands its springs at a given point in time (and therefore the area that surrounds it), represented as an output value between -1 and 1 (0 meaning neutral). Importantly, this meant that the control networks in the cells would continue to be updated during the locomotion stage. The state of the corresponding output was, however, used indirectly. First, to limit the maximum possible frequency of changes in the spring resting lengths, the state of control networks was updated every 50 time steps of the physics simulation (which lasted for 4000 time steps in total). Second, in order to avoid strong forces that are generated if the resting lengths change in a stepwise manner after a network update, the resting length of springs was changed progressively. More precisely, in between network updates, the state of rest-length changing actuation signal $a_{i,t}$ in a cell i at the time t was a linear interpolation of the network's spring output s_i at the time of a previous network update $(t_u - 50)$ and the new desired state determined by the most recent output at a time t_u:

$$a_{i,t} = s_{i,t_u-50} + \frac{(t-t_u)\left(s_{i,t_u} - s_{i,t_u-50}\right)}{50}, \quad \text{where} \ \ t_u \le t < t_u + 50$$
$$(2)$$

Just like in the previous approach [equation (1)], the amount of change applied to resting length of a spring was limited by a globally set amplitude A, and a spring's resting length would be set according to the actuation signals $a_{1,t}$, $a_{2,t}$ coming from both cells connected by the spring:

$$L_t = \left(1 + A\frac{(a_{1,t} + a_{2,t})}{2}\right) \cdot L_0 \quad (3)$$

---

[1]http://www.bulletphysics.org

Previously, we have observed that the oscillator-driven method has lower computational cost and tends to evolve gaits faster. We think it is because the search space is considerably smaller: some type of a cyclic gait was guaranteed to emerge and the evolutionary search only had to optimize the periods and phase shifts of each body region. The second method allows for more fine-grained control, but it was up to evolution to discover that repetitive actuation patterns are the way to produce gaits. It is thus suitable for more complex tasks that require either some degree of irregularity or reactiveness (such as maze navigation, obstacles avoidance, and chemotaxis), but comes at the cost of a more complex and more deceptive fitness landscape. An example of such deceptiveness is structures that collapse after a single pulse of contractions: they score high on the distance measure initially but are local minima that hinder the discovery of cyclic motion patterns. This minimum does not exist in the oscillator-driven variant, as each cell will actuate periodically by default.

Finally, seeing the two approaches as a simpler and a more complex variant of control network evolution, we introduced one more change between them. Namely, in the simple oscillator-driven scenario, we only allowed for feed-forward control networks. This meant that during development, the networks were stateless and all cellular actions would depend entirely on the external signals provided for each cell. With each update of the network, the state of the inputs was set and the signals were propagated the number of steps equal to the precalculated depth of the network. In the GRN-driven variant, we allowed for recurrent connections and, therefore, the networks were no longer stateless during development and locomotion. Here, however, as the network graph would potentially contain cycles, with every network update, we propagated the signals in the network only by a distance of 1.

2.7. Evolution

Networks controlling development and determining gaits were encoded as a list of nodes with their respective types (input, output, and normal gene) and a list of connections. As our goal was to compare directly evolvability of a fitness-driven search with that of novelty search, we have performed repeated evolutionary runs using each of the two following algorithms.

2.7.1. NEAT Evolutionary Algorithm

As a method of fitness-driven search, we have employed the NEAT algorithm [neuroevolution of augmenting topologies (Stanley and Miikkulainen, 2002)], which is often considered the most successful method of evolving artificial neural networks (Mouret and Doncieux, 2011). It starts from simple topologies and grows them over evolutionary time through complexification. To do so efficiently, it relies on two mechanisms. First, it keeps tracks of new genes (new connections and new neurons) and uses history markers to perform a meaningful crossover between genomes, aware of which genes correspond to which in the two parents. Second, it uses speciation, with crossover occurring only within species (as it would likely be destructive otherwise). Speciation is based on the similarity between genomes, calculated by taking into account the number of disjoint and excess neurons and the difference between corresponding connection weights that both networks have. Finally, as means of promoting genetic diversity,

it uses fitness sharing (the larger the species, the lower fitness score its members receive). While the NEAT approach turned out to be extremely fruitful and led to the development of multiple related [e.g., CPPN (Stanley, 2007), HyperNEAT (Stanley et al., 2009)] and hybrid approaches [e.g., Mouret and Clune (2012)] shown to improve over it in various domains, we used here the original version of the NEAT method. Given that it is simple, well understood and available out-of-the-box in various evolutionary optimization libraries, we think it establishes a sensible baseline for the performance of a fitness-driven search.

As a fitness function, we used the displacement of an animat's center of mass between the time the body was found to be at rest after the simulation was started (to avoid profiting from relaxation under gravity, see also Section 2.5) and the final step of simulation:

$$f = \left(\sqrt{\left(x_f - x_s\right)^2 + \left(y_f - y_s\right)^2} + 0.1 b_a \right) \cdot p \qquad (4)$$

where (x_f, y_f) represents the position of the animat's center of mass at the end of locomotion stage, (x_s, y_s) represent its position when it came to rest after initial relaxation under gravity. The actuation bonus term b_a was used only in the GRN-driven actuation experiments and was equal to:

$$b_a = \sum_{u=2}^{u_{cnt}} \sum_{i=1}^{c_{cnt}} \frac{|s_{i,u} - s_{i,u-1}|}{2 \cdot c_{cnt} \cdot (u_{cnt} - 1)} \qquad (5)$$

where u_{cnt} was the total number of network updates during the locomotion stage, c_{cnt} was the number of cells in the embryo, and $s_{i,u}$ was the state of the network output that modifies the resting lengths of springs attached to cell i at the time of update u. It represents a small bonus to the fitness for having some change occur at the actuators with subsequent network updates, and it was used to promote discovery of individuals that actuate. It had a maximum possible value of 1 in the case when each cell switched from maximum contraction to maximum expansion (or vice versa) with each subsequent update of network's output during the locomotion stage.

Finally, p represents applied fitness penalties, where:

$$p = \begin{cases} 0 & \text{if the number of cells less than 8, or} \\ & \text{if the number of cells hit maximum allowed body} \\ & \quad \text{size during development, or} \\ & \text{if animat never stabilized during initialization of the} \\ & \quad \text{locomotion stage} \\ 0.1 & \text{if the number of cells created during growth is equal} \\ & \quad \text{or greater than the maximum allowed} \end{cases}$$

$$(6)$$

We used the MultiNEAT implementation of the NEAT algorithm (Chervenski and Ryan, 2014). The configuration file specifying the parameters of the library is linked within the reference.

2.7.2. Novelty Search

Lehman and Stanley (2011a) define novelty of a phenotype m as proportional to how sparsely the phenotype space surrounding m has been explored so far:

$$\rho = \frac{1}{k} \sum_{i=1}^{k} d(m, \mu_i) \qquad (7)$$

where μ_i is the i-th nearest individual to m out of k according to the distance metric d.

Introducing the novelty search algorithm into an existing, NEAT-based system required only replacing the fitness function computation with calculation of the novelty value, disabling the fitness sharing, introduction of the novelty archive that stores past novel individuals and the algorithm for dynamic updating of the novelty threshold value (the novelty value at which an individual is added to the archive). We chose to increase the threshold by 10% if more than eight individuals were added to the archive one after another and to decrease it by 10% if no individuals were added within 50 generations. We used the final coordinates of an animat's center of mass P at the end of locomotion stage as its phenotypic representation, where $P \in R^2$. We then used Euclidean metric as a measure of distance d between two phenotypes, with $k = 15$ [see equation (7)]. Therefore, the novelty of an individual was highest if it finished its movement in a location in which other individuals did not finish yet. Note that while the aquatic environment allowed to easily vary the phenotypic representation vector along both X and Y coordinate, in the terrestrial gait problem the variation could almost solely come from the X coordinate, as the horizontal movement of the body would be many times larger than the vertical, and we did not normalize each dimension in the phenotype characterization vector. Thus, novelty search algorithm was forced to introduce variation almost entirely by modifying the X coordinate.

Since novelty search required only the change of how the fitness function is calculated (by replacing it with a measure of novelty), we could keep the remaining properties and settings of both algorithms identical (including speciation for the meaningful crossover). Naturally, the fitness adjustments normally performed by NEAT (fitness sharing based on genetic similarity) were not applied in the case of novelty search as it would imply changing its core behavior. We used a population size of 300 individuals, and evolutionary runs lasted for 2000 generations. The initial population was created as a fully connected network with inputs directly connected to outputs and random weights.

3. RESULTS

We compared the performance of the novelty search algorithm to the fitness-driven search (represented by the NEAT algorithm), using four different scenarios: we combined two separate tasks:

evolution of aquatic and terrestrial gaits with two approaches to actuation. We used the same method of evaluating performance both in aquatic and terrestrial environments, that is, for NEAT-search experiments, fitness represented the displacement of the center of mass during locomotion stage and, in novelty search experiments, phenotype was characterized by the final location of an animat.

To allow for meaningful comparisons between the two approaches, for each of the four experimental settings, we repeated evolution using a given search algorithm 20 times, using different random seed values. To provide the reader with an overview of what kinds of designs emerge in our system, we first present examples of evolved solutions. We then compare the performance of each algorithm. Finally, we analyze how each of the algorithms explores the phenotypic space in order to understand the reason behind their different behavior and performance in this problem domain.

3.1. Evolved Morphologies

With only two exceptions, all evolutionary runs resulted in individuals capable of producing repetitive gaits in their target environments. The two exceptions were a result of fitness-driven search attempting to evolve GRN-controlled gaits and occurred both in the swimming and walking tasks. The terrestrial individual was vertically elongated and would fall after expanding its muscle to its side at the beginning of the simulation. A swimming individual would rapidly contract its body once and rely on the momentum produced by this single contraction. Thus, none of these individuals evolved a cyclic motion pattern. As this was only possible in the case of GRN-controlled actuation, this type of local minima would not be an issue in the experiments relying on oscillator-driven actuation.

In the case of oscillator-driven actuation and aquatic gait evolution, morphologies produced by each type of evolutionary search were found to be visually very similar: best evolutionary runs resulted in individuals who had elongated, snake-like morphologies, and moved with undulatory gaits. Wider, more fish-like morphologies also emerged, yet the runs that produced them reached lower fitnesses. Comparing morphologies obtained using fitness driven (**Figure 3A**) with individuals obtained using novelty search (**Figure 3B**), there was no obvious visual difference, though the top five best-performing individuals of the two experiments were obtained with novelty search (three of which are shown in

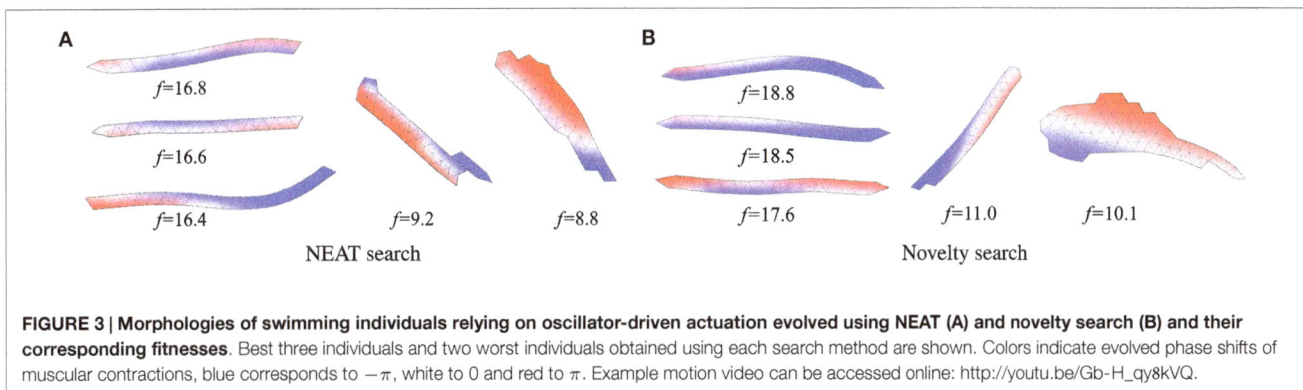

A $f=16.8$ $f=16.6$ $f=16.4$ $f=9.2$ $f=8.8$

NEAT search

B $f=18.8$ $f=18.5$ $f=17.6$ $f=11.0$ $f=10.1$

Novelty search

FIGURE 3 | Morphologies of swimming individuals relying on oscillator-driven actuation evolved using NEAT (A) and novelty search (B) and their corresponding fitnesses. Best three individuals and two worst individuals obtained using each search method are shown. Colors indicate evolved phase shifts of muscular contractions, blue corresponds to $-\pi$, white to 0 and red to π. Example motion video can be accessed online: http://youtu.be/Gb-H_qy8kVQ.

Figure 3B). Similarly, in the case of ground-based locomotion (**Figure 4**), all evolutionary experiments converged to similar morphologies, in that case supporting themselves on two (occasionally more) appendages. Here, again, no immediately obvious visual difference was observed between novelty- and fitness-driven searches, although one more time, the top three individuals came from the novelty search experiment.

Similarly, in the experiments with GRN-controlled terrestrial gaits, the results obtained using each search method were visually similar (cf. **Figures 5A,B**), with individuals running on two "legs" and highly convergent morphologies. The shapes were, however, visually quite different from those obtained with the oscillation-driven actuation approach. Likewise, in the case of individuals evolved for swimming, both search algorithms produced similar morphologies (**Figures 5C,D**), though this time a wider and more fish-like type of morphology dominated. Two exceptions displaying snake-like morphologies, however, emerged in the fitness-driven search experiments. They were also the two most successful swimmers obtained in both GRN-driven actuation experiments. While the investigation of how different types of genetic control

over development and behavior lead to different shapes is beyond the scope of this paper, we note that a likely significant factor were different time constants determining how quickly muscular contractions could occur in each of the two types of actuation mechanisms. In any case, differences in evolutionary trajectories that emerge depending on the type of genetic control employed were the very reason we were interested in testing if novelty search behaves consistently in different setups.

3.2. Performance: Oscillator-Driven Actuation

Figure 6A compares the performance of the two investigated search algorithms on the problem of evolving aquatic animats using the simpler approach to control. We found that novelty search clearly outperformed the NEAT algorithm (median of expected achieved distance was different, Wilcoxon two-tailed rank-sum test, $p = 0.01$). On average, novelty search not only produced better individuals given 2000 generations but would also do so regardless on the number of generations.

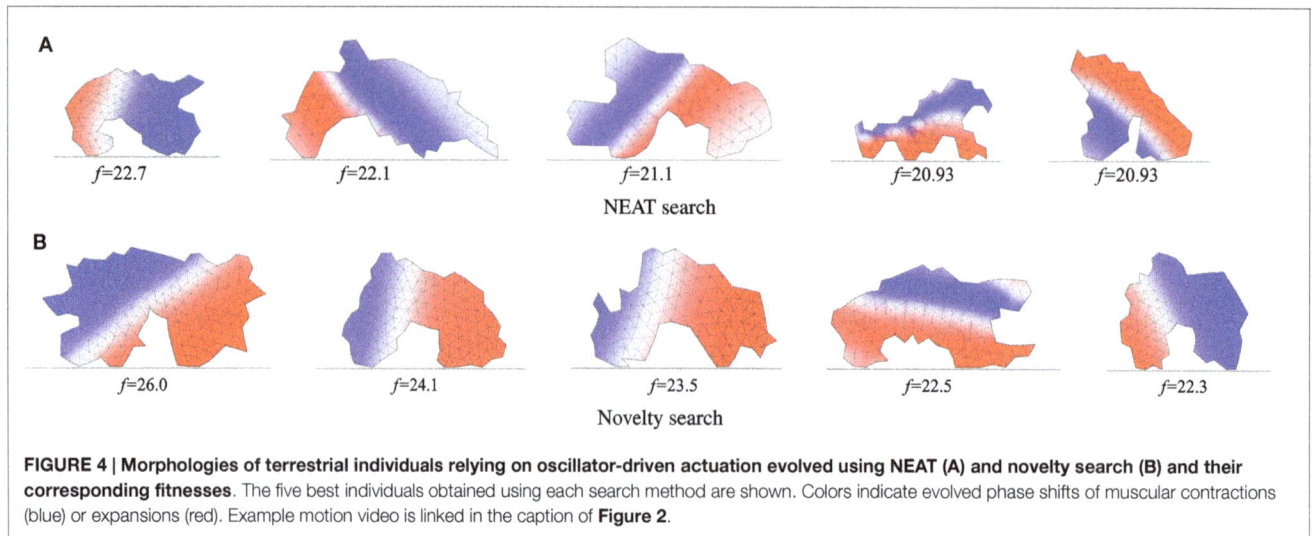

FIGURE 4 | Morphologies of terrestrial individuals relying on oscillator-driven actuation evolved using NEAT (A) and novelty search (B) and their corresponding fitnesses. The five best individuals obtained using each search method are shown. Colors indicate evolved phase shifts of muscular contractions (blue) or expansions (red). Example motion video is linked in the caption of **Figure 2**.

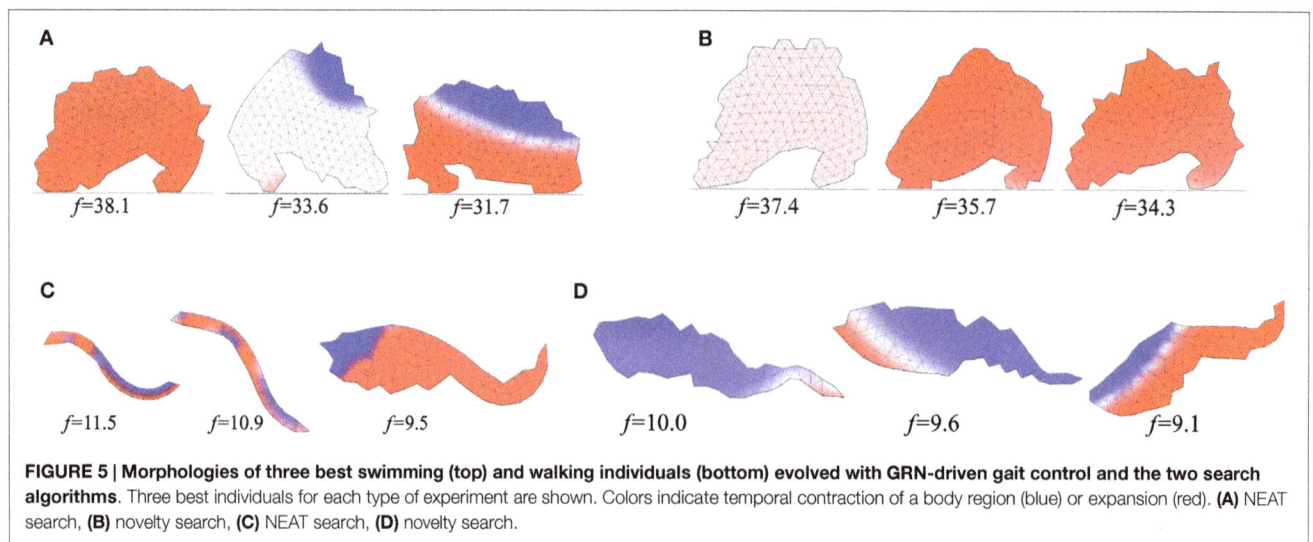

FIGURE 5 | Morphologies of three best swimming (top) and walking individuals (bottom) evolved with GRN-driven gait control and the two search algorithms. Three best individuals for each type of experiment are shown. Colors indicate temporal contraction of a body region (blue) or expansion (red). **(A)** NEAT search, **(B)** novelty search, **(C)** NEAT search, **(D)** novelty search.

To investigate whether the advantage of novelty search extended to another problem, we next compared evolvability of terrestrial gaits. As novelty search could now only act on the first dimension of the phenotypic representation vector (i.e., the X coordinate of the final position), we expected novelty search advantage to diminish as the measure of novelty becomes largely synonymous with the objective fitness. Indeed, after 2000 generations (**Figure 6B**), novelty search seemed to hold only a small advantage over NEAT and, in fact, the median of expected achieved distance in generation 2000 was no longer significantly different (Wilcoxon two-tailed rank-sum test, $p = 0.30$). On average, novelty search did, however, produce higher fitness individuals regardless of the number of generations and was much faster at finding good swimmers early on (e.g., the expected median fitness of best individuals at generation 400 is significantly better, $p = 0.02$).

Overall, we found these results surprising since the NEAT method is by itself a state of the art evolutionary algorithm specifically tuned to evolve network topologies and employs various techniques that prevent premature convergence. Also, while developing the model, we tuned the parameters of the system while optimizing its performance under the NEAT algorithm [see, e.g., fitness function tweaks, equations (5) and (6)], which was likely to introduce a bias in favor of this search method. Nonetheless, a direct and trivial replacement of NEAT with novelty search resulted with a clear improvement, even though novelty search did not even attempt to explicitly optimize for distances.

3.3. Performance: GRN-Driven Actuation

To see if the above results hold for a more complex, but more powerful version of the model, we performed an analogous pair of experiments, this time with development controlled by a recurrent network and motion patterns generated by continuous activity of the gene regulatory network. Here, a cyclic pattern of locomotion was no longer a default and evolution had to discover that repetitive patterns of actuation are the way to produce sustained gaits. Other than increasing the search space considerably, we suspected this scenario to be less suited for novelty search, as new final

positions could now be easily generated by producing movement patterns that sustain themselves only through part of the total evaluation time. To our surprise, the average performance of novelty search was one more time higher in both aquatic (**Figure 7A**) and terrestrial environment (**Figure 7B**). In the case of the former, the advantage of novelty search was significant during the first few hundreds of generations (difference in medians at generation 800, $p = 0.008$), but decreased over evolutionary time (the medians were no longer significantly different at generation 2000).

The final experiment in which GRN-controlled terrestrial animats were evolved produced similar results: novelty search converged faster and produced higher average fitnesses. The medians were clearly different at generation 800 ($p = 0.01$), but again, not significantly so at generation 2000 ($p = 0.10$).

Overall, seeing how novelty search found good individuals much faster than NEAT search and sustained average advantage throughout the full length of evolutionary runs, we consider these results consistent with superiority of novelty search method in this problem domain.

3.4. Exploration of the Phenotypic Space

To understand how novelty search improves over NEAT in our problem domain, we visualized how the phenotypic space is being explored by each of the algorithms in the case of evolving swimming animats using oscillator-driven actuation. Although a direct visualization of complex phenotypic space would be impossible, we chose to visualize each phenotype by using the same representation that was employed to compute phenotypic distances in novelty search, i.e., as points in R^2 corresponding to the final locations of individuals. This allowed us to plot the final positions of all individuals who existed during a given evolutionary run on a surface. To reduce visual clutter, we have plotted only 1% of (randomly sampled) individuals. We selected two runs from each type of experiment for visualization: a run that led to the best individual obtained in a given type of experiment and a median quality individual (the best individual having fitness below the median). Inspection of the remaining runs confirmed that the patterns observed in the selected examples are representative for

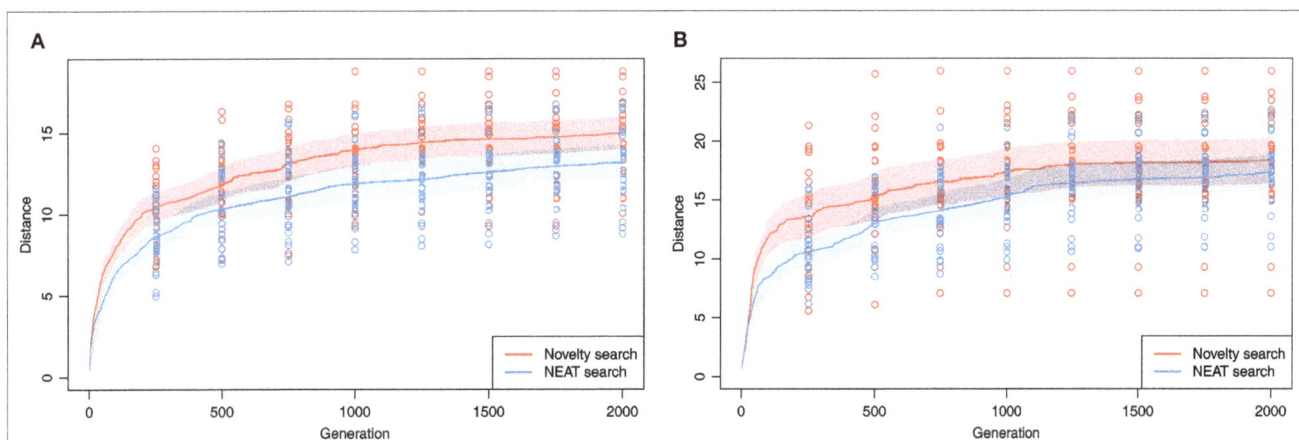

FIGURE 6 | Oscillator-driven actuation: performance of novelty search compared to the NEAT algorithm on the problem of co-evolving morphology and controller of soft-bodied, swimming (A) and walking (B) animats. Solid lines show mean best fitness in a given generation from 20 evolutionary runs of each type of experiment. Dashed lines show 95%, bootstrapped confidence intervals for the means. **(A)** Aquatic environment, **(B)** terrestrial environment.

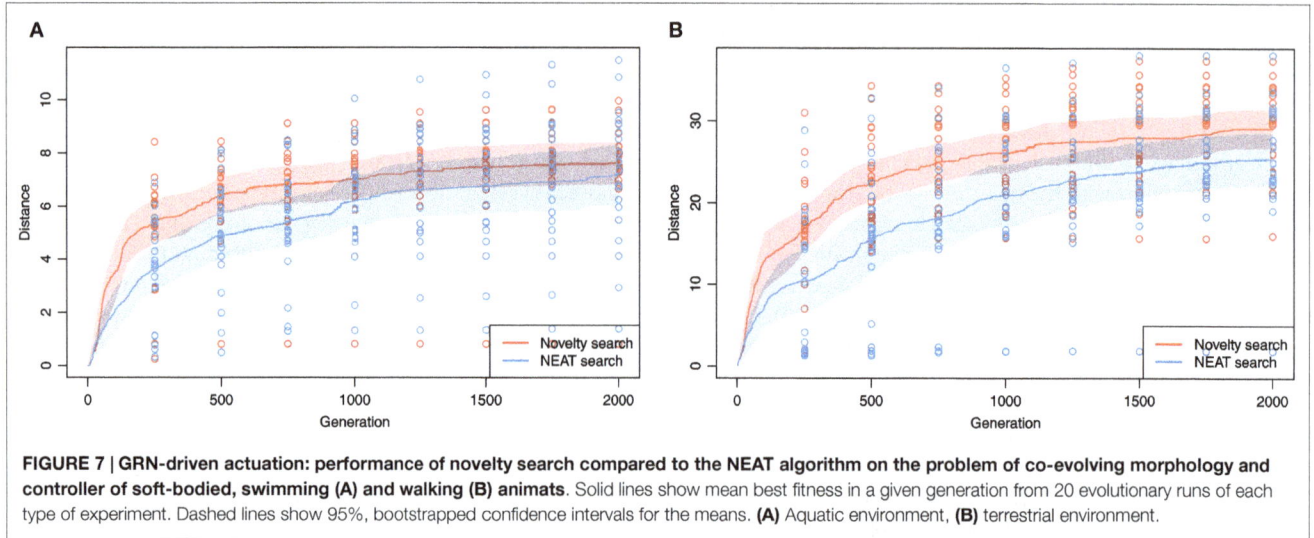

FIGURE 7 | GRN-driven actuation: performance of novelty search compared to the NEAT algorithm on the problem of co-evolving morphology and controller of soft-bodied, swimming (A) and walking (B) animats. Solid lines show mean best fitness in a given generation from 20 evolutionary runs of each type of experiment. Dashed lines show 95%, bootstrapped confidence intervals for the means. **(A)** Aquatic environment, **(B)** terrestrial environment.

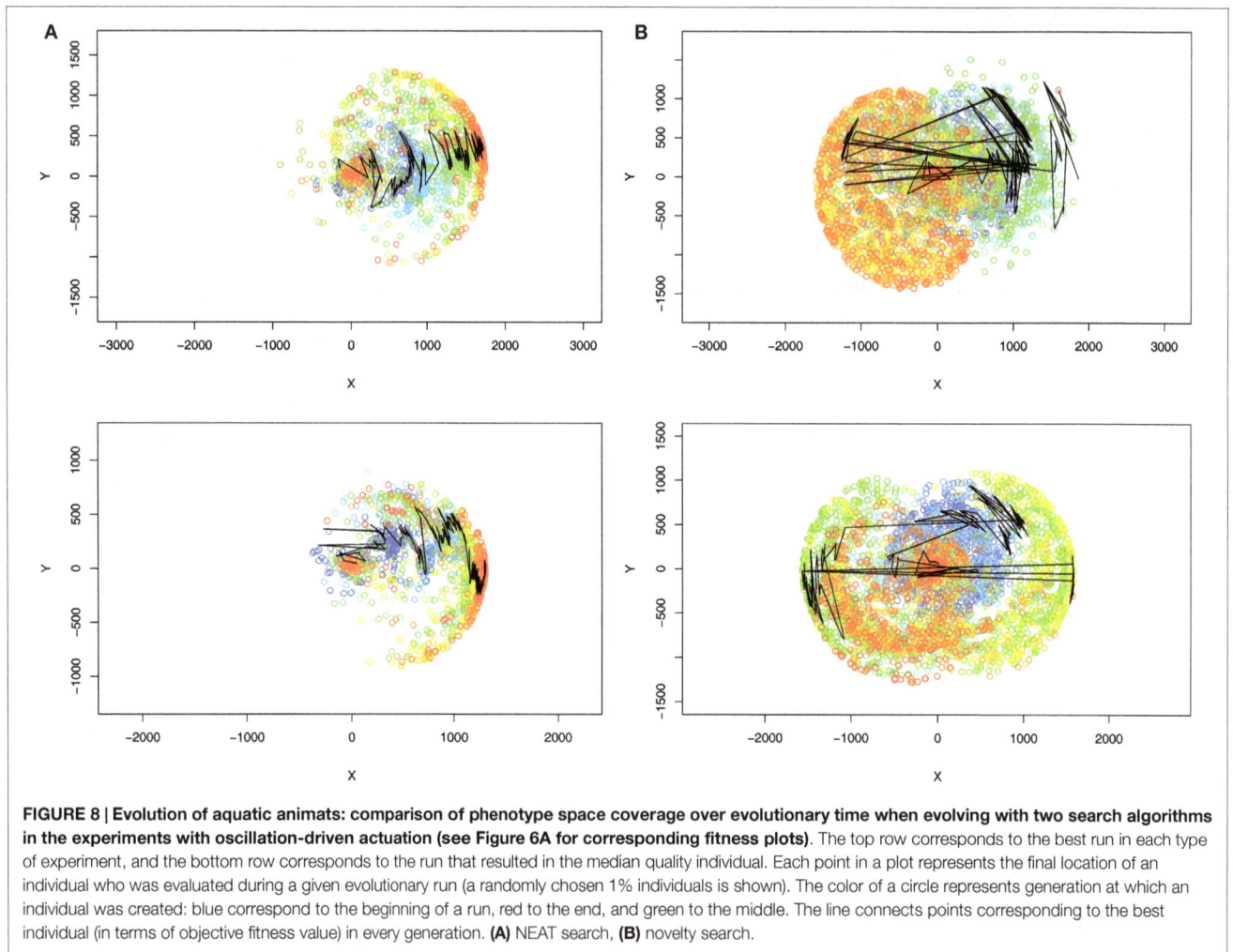

FIGURE 8 | Evolution of aquatic animats: comparison of phenotype space coverage over evolutionary time when evolving with two search algorithms in the experiments with oscillation-driven actuation (see Figure 6A for corresponding fitness plots). The top row corresponds to the best run in each type of experiment, and the bottom row corresponds to the run that resulted in the median quality individual. Each point in a plot represents the final location of an individual who was evaluated during a given evolutionary run (a randomly chosen 1% individuals is shown). The color of a circle represents generation at which an individual was created: blue correspond to the beginning of a run, red to the end, and green to the middle. The line connects points corresponding to the best individual (in terms of objective fitness value) in every generation. **(A)** NEAT search, **(B)** novelty search.

the experiments as a whole. The comparative results for each search algorithm are shown in **Figure 8**. It can be immediately seen that the way each algorithm has explored the space of possible phenotypes is qualitatively different. While the novelty search progressively discovers solutions that swim in every possible direction, the NEAT algorithm shows a clear focus toward exploring only a subregion of the phenotypic space, with individuals moving in a similar direction. This exploitation of a single direction is most likely explained by population becoming dominated by descendants of a particular good design that happened to be successful early on. This then leads to subregions being overexplored through the creation of large numbers of individuals who differ only slightly in their final position. On the other hand, novelty search algorithm clearly explores the phenotypic space in a much more even manner, without particular motion directionality and mostly avoiding overexploration of phenotypic subregions. This interpretation is further reinforced by the overlaid visualization of the trajectory of the best (in terms of objective fitness) individual in every generation (solid line in **Figures 8A,B**). While in the fitness-driven runs, subsequent best individuals are close neighbors in the phenotypic space, novelty search is capable of exploring different scenarios at the same time, visible as long "jumps" through the phenotypic space between generations. We note, however, that these jumps are unlikely to imply that mutations caused large phenotypic changes, rather, parallel exploration of diverse designs allows new best individuals to emerge in any of the regions of the phenotypic space that are being explored.

Next, we attempted to visualize how each search algorithm behaved on the problem of evolving terrestrial animats. Due to

the 1-dimensional characterization of phenotypes, we were able to more easily visualize the progression of the populations over evolutionary times (**Figure 9**). Here, fitness-driven search focused almost entirely on either left or right direction of motion, determined at the beginning of an evolutionary run. Also, most of the individuals being created seem to be minor modifications to the winning design, visible as a high-density region (dark blue) overlapping with the red line representing best solution found so far. On the other hand, novelty search seems to explore scenarios in which individuals move either left or right evenly, and it was not uncommon to observe how a winning left-running solution is replaced by right-running one (or vice versa), often multiple times during a single evolutionary run.

The observation above is further supported by inspection of the contents of the novelty search archives in novelty-driven runs (consisting of individuals who were found to be novel at the time they were created, **Figure 10**). It can be seen how, in the case of swimming individuals, novelty search discovers new concentric "layers" of solution space, rather than focusing on exploitation of a particular subregion of the search space (which, in that case, would manifest as discovering subsequent solutions that move roughly in the same direction). In the case of walking individuals, during the first few hundred generations, subsequent

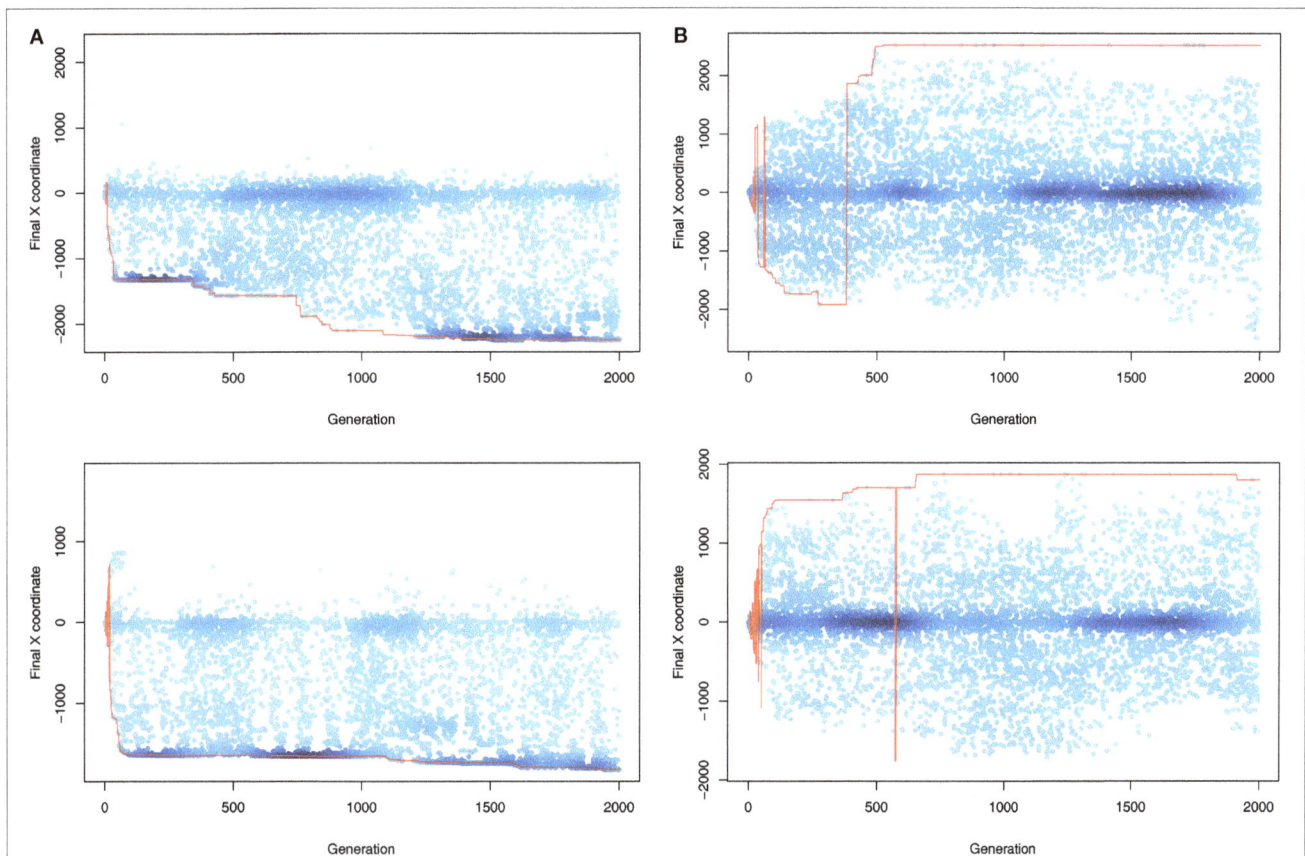

FIGURE 9 | Evolution of terrestrial animats: comparison of phenotype space coverage over evolutionary time when evolving with two search algorithms in experiments with oscillation-driven actuation (see Figure 6B for corresponding fitness plots). The top row corresponds to the best run in each type of experiment, and the bottom row corresponds to the run that resulted in the median quality individual. Each point in a plot represents the final position of an individual who was evaluated during a given evolutionary run (a randomly chosen 1% individuals is shown). Color intensity corresponds to points density. The line connects points corresponding to the best individual (in terms of objective fitness value) in every generation. (A) NEAT search, (B) novelty search.

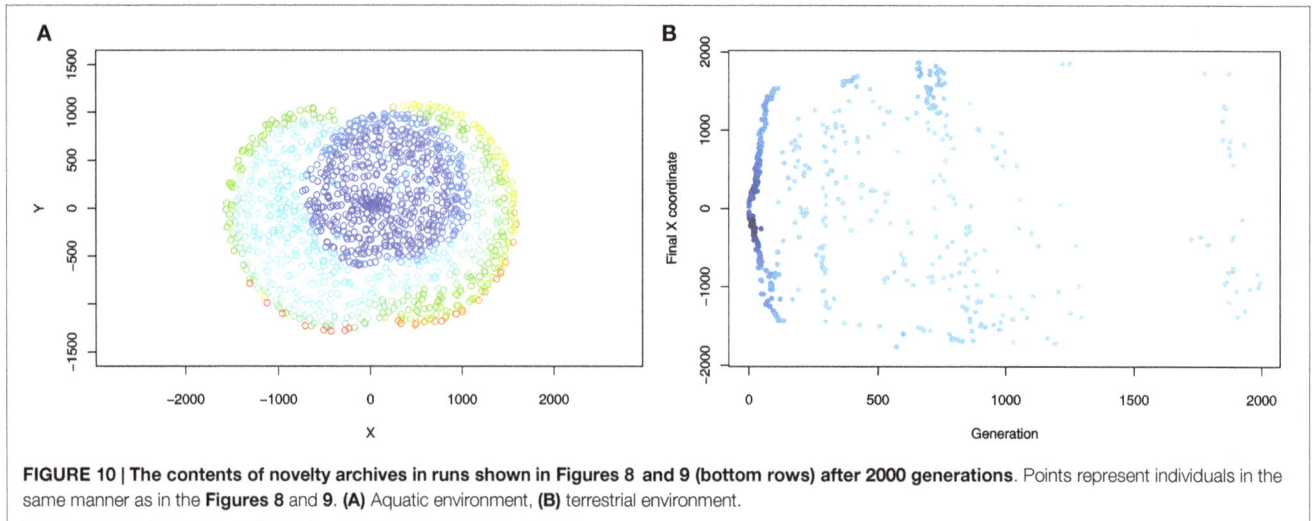

FIGURE 10 | The contents of novelty archives in runs shown in Figures 8 and 9 (bottom rows) after 2000 generations. Points represent individuals in the same manner as in the **Figures 8** and **9**. **(A)** Aquatic environment, **(B)** terrestrial environment.

novel individuals tend to have higher objective fitnesses and are found both among left-running and right-running individuals, without any clear preference. After "low-hanging fruit" solutions are found, novelty archive is slowly populated with individuals ending at almost any possible location in the region between current best left-running and right-running solution. Ultimately, this also leads to discovering improvements to the best individuals as well.

Finally, we have inspected evolutionary runs that relied on the more complex, GRN-driven actuation and found out that they displayed qualitatively similar properties (not shown).

3.5. Phenotypic Diversity in Populations

Phenotypic diversity is a necessary condition of an evolutionary process and hence sustaining it is of primary importance to many evolutionary algorithms. The NEAT algorithm attempts to sustain diversity at the genetic level by using speciation and fitness sharing, in which genotypes grouped into species receive lower fitnesses as the species size increases, consequently promoting novel genotypes. Grouping into species is based on the genetic distance between individuals. The genetic distance, however, does not necessarily reflect phenotypic distance (in the extreme case, neutral mutations lead to two different genomes with identical phenotypes). To understand what kind of population structure emerges during evolution, we investigated fitness distributions in the final generations and observed radical differences between the two search algorithms. Populations evolved using NEAT would progressively lose fitness diversity over evolutionary time, as the populations were slowly more and more dominated by an elite individual and its variations that differ only slightly in fitness. This pattern was universally observed in all types of experiments, as confirmed by averaging normalized fitness distributions in the final generation from independent evolutionary runs (**Figure 11**). The difference behavior of these algorithms can be understood if we consider that novelty search, by its very definition, attempts to diversify phenotypes. At the same time, developmental systems are known to evolve toward both environmental and mutational robustness, a process known as genetic canalization [see, e.g., Federici and Ziemke (2006), Basanta et al. (2008), Andersen et al.

(2009), and Joachimczak and Wróbel (2012b)], which likely drives the phenotypic diversity down.

3.6. Morphological Diversity

While the results discussed in the Section 3.4 clearly show that novelty search explores the space of possible solutions in a more even and thorough manner than the fitness-driven search, our assessment of phenotypic diversity was based solely on the final position of an animat being evaluated. Although the quality of solutions obtained with novelty search was also higher on average, we were interested whether its success can be attributed to actually having explored a greater range of potential morphologies. After all, a larger diversity of final positions does not automatically imply a larger diversity of morphologies. In particular, larger diversity of final positions could simply be a result of mutations that only influence behavior control (e.g., resulting in a change of swimmer's angle), providing evolution with a cheap and endless source of phenotypic variation. This would, however, be strongly against our goal of creating a system that discovers innovative morphologies and controllers without any prior knowledge of desired solutions, as being able to explore a large range of morphologies is essential to its creativeness.

Measuring the diversity of arbitrary morphologies requires a way to quantify dissimilarities between them. This is generally a non-trivial problem that can be approached in multiple ways, taking things like scale or rotation invariance into consideration [e.g., through the use of shape histograms (Ankerst et al., 1999)]. Since our goal here was mainly to make sure that the higher phenotypic diversity of the novelty search scenario is not just a result of evolution tweaking the controllers, we decided to rely on a much simpler approach. We calculated diversity of individuals' sizes (represented by their body cell count) during an evolutionary run. While a very crude approach: a potentially extremely large numbers of morphologies can be made of the same number of cells, body sizes are sometimes used in biological [e.g., Imroze and Prasad (2011)] and artificial embryogeny [e.g., Matos et al. (2009)] experiments as a way to estimate phenotypic diversity. Here, we expected that if morphologies tend to repeat more often in some experiments, we should also observe more

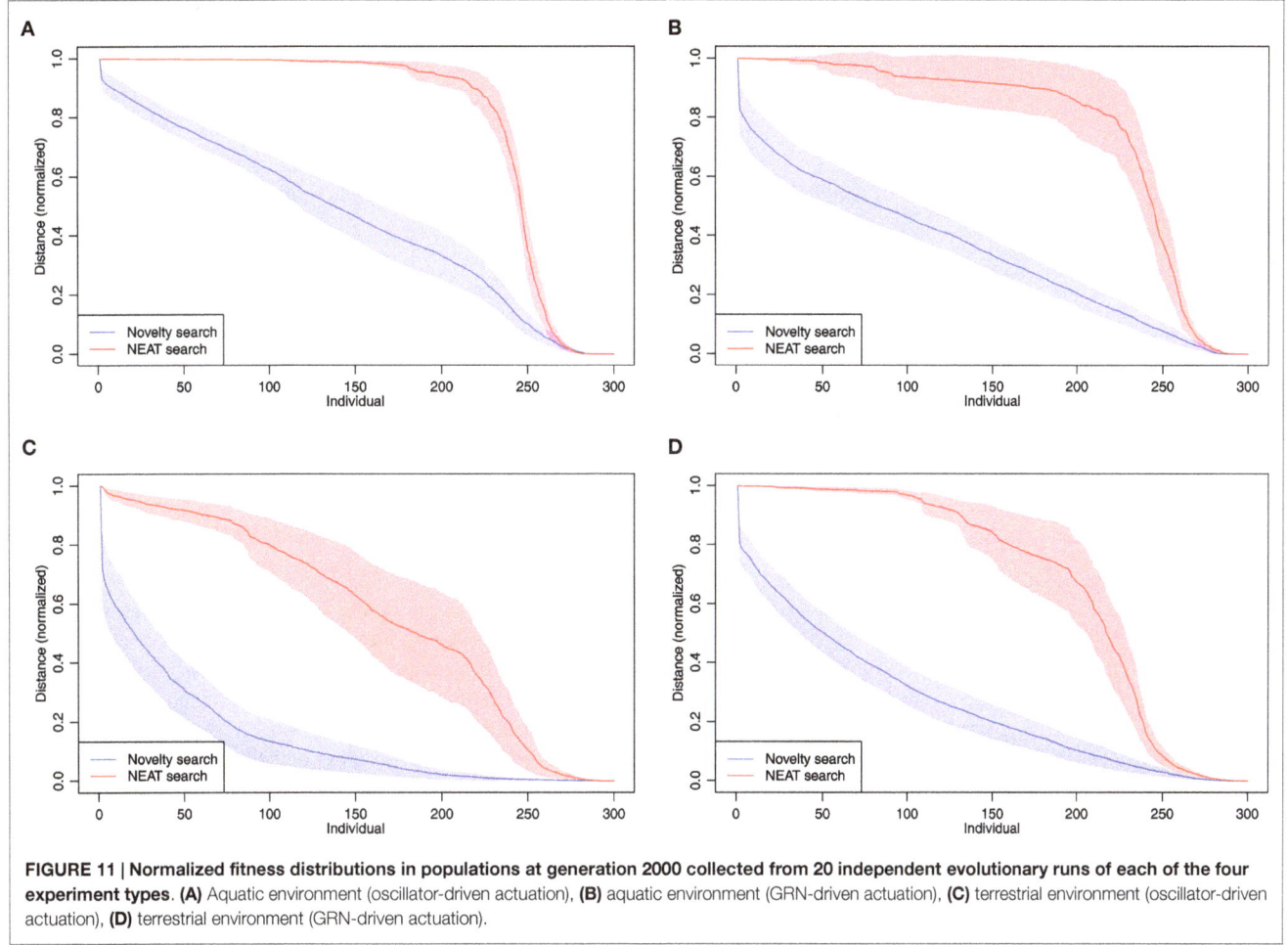

FIGURE 11 | Normalized fitness distributions in populations at generation 2000 collected from 20 independent evolutionary runs of each of the four experiment types. (A) Aquatic environment (oscillator-driven actuation), **(B)** aquatic environment (GRN-driven actuation), **(C)** terrestrial environment (oscillator-driven actuation), **(D)** terrestrial environment (GRN-driven actuation).

frequent repetitions of identical cell counts in the populations. Thus, as a simple proxy of morphological diversity, we decided to apply one of the measures commonly employed to calculate ecological diversities, namely the Shannon entropy, to a set of body sizes that occur throughout an evolutionary run. For each run, we calculated Shannon entropy of body sizes over the whole evolutionary history:

$$H(\mathbb{S}) = -\sum_{s \in \mathbb{S}} p(s) \ln p(s) \qquad (8)$$

where $\mathbb{S}$ was the set of sizes (numbers of cells in the body) of all animats that existed during a given evolutionary run and were considered viable (i.e., their fitness was not zero). We calculated morphological diversity for all evolutionary runs and all types of experiments that we performed and present summary results in **Figure 12**. Regardless of whether we used the oscillator-driven actuation or the more complex, GRN scenario, novelty search produced much higher morphological diversity, with the difference being even more pronounced in the latter case. This result strongly suggests that novelty search indeed manages to produce significantly higher morphological diversity and increased variation of animat final positions was associated with an actually higher diversity of morphologies explored.

4. DISCUSSION AND FUTURE WORK

Throughout our work, we aim to harness the creative power of the evolutionary process in order to either inspire or to fully replace a human designer. While we were able to reach very promising results using the classical, fitness-driven evolutionary search algorithm, the large and complex search space stemming from the weakly constrained range of possible morphologies, and complex interactions between morphologies and controllers made evolutionary runs expensive and prone to being stuck in local minima. After introducing a very low cost change to the search method, a one that abandons optimization for the fitness altogether, we were surprised to see it immediately improve over the fitness-driven search, even though the novelty search method was not explicitly optimizing for gait performance.

A more extensive analysis revealed that the result was robust in respect to two types of tasks and two versions of control mechanism that we used; in all investigated scenarios, novelty search produced on average higher fitness individuals. The advantage of novelty search was especially visible within the first few hundreds of generations and given limited time for an evolutionary run, it was clearly superior at finding gaits. Performing longer evolutionary experiments (2000 generations) allowed the fitness-driven search to get closer to the results obtained with novelty,

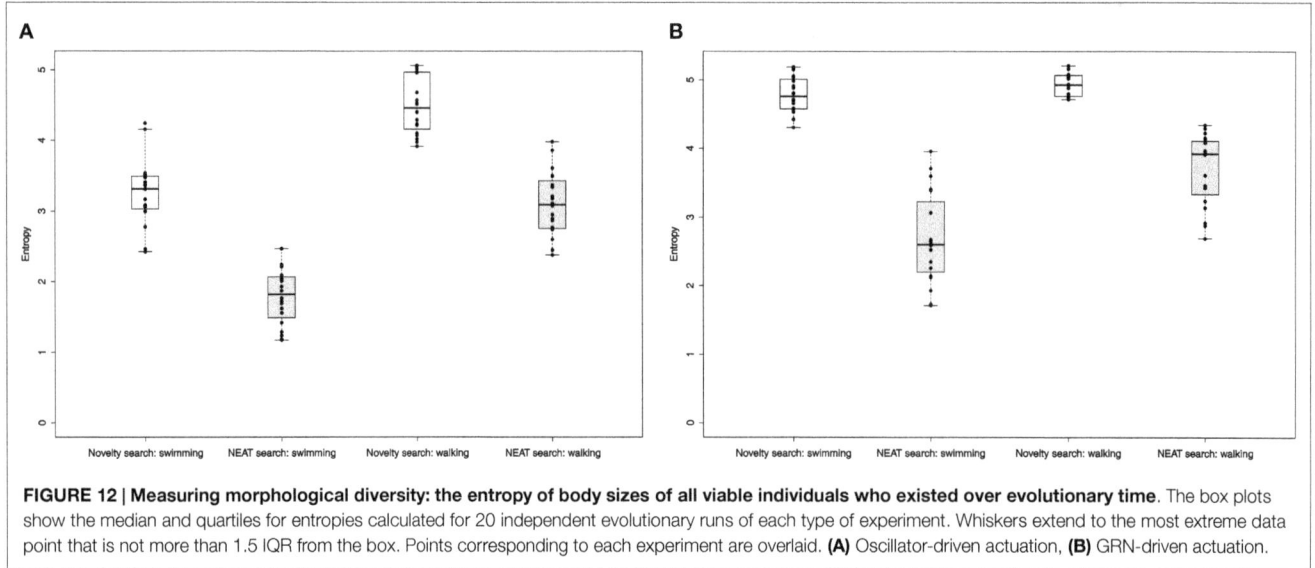

FIGURE 12 | Measuring morphological diversity: the entropy of body sizes of all viable individuals who existed over evolutionary time. The box plots show the median and quartiles for entropies calculated for 20 independent evolutionary runs of each type of experiment. Whiskers extend to the most extreme data point that is not more than 1.5 IQR from the box. Points corresponding to each experiment are overlaid. **(A)** Oscillator-driven actuation, **(B)** GRN-driven actuation.

though it remained, on average, inferior. The reduced difference in performance of each algorithm given longer search can be explained by the fact that each algorithm ultimately attempts to converge on similar designs; indeed, the best solutions obtained using each search method had similar morphologies.

Importantly, our analysis revealed how both algorithms explore the space of possible solutions very differently: novelty search was far better at maintaining population diversity and explored the space of possible solutions in a much more even manner. Fitness-driven search, on the other hand, despite the use of mechanisms that promote genotypic diversity would heavily focus on subregions of the phenotypic space. Furthermore, our analysis suggests that novelty search, despite simply searching for novel end positions of locomoting animats, actually evaluated a more diverse set of morphological designs.

As novelty search is believed to show its strengths in deceptive fitness landscapes, this also suggests that the problem we are working on, i.e., evolution of morphologies and controllers based on artificial development exhibits a highly deceptive fitness landscape. Indeed, one potential source of such a deceptiveness could be constant interactions between the controller and morphology being simultaneously optimized. More precisely, when a controller becomes fine-tuned to a given morphology over evolutionary time, many potentially useful modifications of this morphology will likely turn out to be detrimental, as they will lead to a mismatch between the new morphology and a controller. Thus, only after fine-tuning of the latter, the performance of a modified morphology can be ascertained: a new hill of increased fitness landscape can only be accessed by crossing a lower fitness valley, a telltale of a deceptive fitness landscape.

Another reason we suspect novelty search turned out to be highly successful on this problem may be the difficulty of sustaining phenotypic diversity when using development-based encodings. Developmental systems are known to evolve toward being robust to perturbations, both environmental and mutational [see, e.g., Federici and Ziemke (2006), Basanta et al. (2008), Andersen et al. (2009), and Joachimczak and Wróbel (2012b)], and this

robustness is one of the very reasons behind the interest in them. However, the propensity to sustain phenotype despite genetic perturbations should also mean that such systems are partially robust to methods that attempt to boost genetic diversity (e.g., through fitness sharing, as in this work), weakening their effect. Since novelty search promotes only the phenotypic novelty, it is immune to this effect, rewarding only increase in phenotypic diversity. This suggests that either novelty search or methods of promoting phenotypic diversity should be, in general, a more meaningful way to increase evolvability of developmental systems and we hope to investigate this more in depth in our future work. Furthermore, recent evidence from the domain of evolving robot behaviors (without morphological evolution) suggests that promoting phenotypic diversity also improves search performance (Gomez, 2009; Mouret and Doncieux, 2011; Trujillo et al., 2011; Lehman et al., 2013). As the presented system deals both with the evolution of development and evolution of behavior, it suggests that phenotypic diversity maintenance may have contributed at both levels.

Lacking a predefined objective, pure novelty search is unlikely to succeed at higher complexity tasks and the reason it performs so well at gait evolution is likely related to the coupling between novelty measure and the distance measure: exploring further areas of the phenotypic space means finding solutions of higher objective fitness. It has, however, already been shown that the performance of novelty search can be improved on certain problems if a novelty measure is combined with an objective fitness function through the use of multi-criteria optimization (Mouret and Doncieux, 2011; Lehman et al., 2013). Hence, as a straightforward extension of the current work, we plan to incorporate an objective fitness function into the search algorithm, through the use of a Pareto-based multi-objective approach. We hope that this will allow us to evolve robots capable of performing more complex and multi-step behaviors, such as reshaping to crawl through an opening or grabbing objects with an elastic appendage. Furthermore, a fitness component would enable evolution of behaviors that optimize energy used to perform a task or the amount of active and passive material in a design [e.g., as in Cheney et al. (2013)]. Finally, having observed the extreme importance of sustaining phenotypic diversity in an artificial embryogeny model, we are interested in finding better ways of generating the initial population. After all, the morphological diversity of multicellular life did not emerge out of random gene regulatory networks but was preceded by an extremely long period of complexification of unicellular forms. Thus, one way we hope we can further improve evolvability is to first discover promising regions of genotypic space by pre-evolving initial population toward general morphological diversity, and only later attempting to co-evolve the form and behavior. This would require developing a much less simplistic measure of morphological diversity than the proxy used in Section 3.6 but would also enable us to perform direct comparisons of morphological diversities produced by different search algorithms.

With the low-effort modifications needed to adapt an objective fitness-driven genetic algorithm into novelty search and limited number of parameters that need to be set up (novelty search being able to adjust most of its parameters dynamically), we believe novelty search is likely to be of high utility for encodings based on artificial development in general. The results presented in this work also show the overall importance of focusing on maintaining phenotypic diversity in systems with highly indirect genotype–phenotype mappings, novelty search being only one of the easiest methods to achieve it.

AUTHOR CONTRIBUTIONS

Concept and design of experiments: MJ, RS, and TA. Implementation, running experiments, and analysis: MJ. Writing paper: MJ, RS, and TA.

ACKNOWLEDGMENTS

This work was supported by the Japan Society for the Promotion of Science (JSPS) through the JSPS Fellowship for Foreign Researchers and JSPS KAKENHI Grant Number 26-04349. High-performance computing resources were provided by the Interdisciplinary Center for Molecular and Mathematical Modeling (ICM, University of Warsaw) and the Tri-city Academic Computer Center (TASK). CGAL library (CGAL, 2015) was used.

REFERENCES

Andersen, T., Newman, R., and Otter, T. (2009). Shape homeostasis in virtual embryos. *Artif. Life* 15, 161–183. doi:10.1162/artl.2009.15.2.15201

Ankerst, M., Kastenmüller, G., Kriegel, H.-P., and Seidl, T. (1999). "3D shape histograms for similarity search and classification in spatial databases," in *Advances in Spatial Databases, Volume 1651 of LNCS*, eds R. H.Güting, D. Papadias, and F. Lochovsky (Berlin: Springer), 207–226. doi:10.1007/3-540-48482-5_14

Auerbach, J. E., and Bongard, J. C. (2012). "On the relationship between environmental and mechanical complexity in evolved robots," in *Artificial Life 13*, eds C. Adami, D. M. Bryson, C. Ofria, and R. T. Pennock (Cambridge, MA: MIT Press), 309–316. doi:10.7551/978-0-262-31050-5-ch041

Basanta, D., Miodownik, M., and Baum, B. (2008). The evolution of robust development and homeostasis in artificial organisms. *PLoS Comput. Biol.* 4:e1000030. doi:10.1371/journal.pcbi.1000030

Boddhu, S. K., and Gallagher, J. C. (2010). Evolving neuromorphic flight control for a flapping-wing mechanical insect. *Int. J. Intell. Comput. Cybern.* 3, 94–116. doi:10.1108/17563781011028569

Bongard, J. C., and Pfeifer, R. (2003). "Evolving complete agents using artificial ontogeny," in *Morpho-Functional Machines: The New Species*, eds F. Hara and R. Pfeifer (Tokyo: Springer-Verlag), 237–258. doi:10.1007/978-4-431-67869-4_12

Burton, F. (2007). *Soda Constructor*. Available at: http://www.sodaplay.com

Carroll, S., Grenier, J., and Weatherbee, S. (2004). *From DNA to Diversity: Molecular Genetics and the Evolution of Animal Design*, 2nd Edn. Malden, MA: Wiley-Blackwell.

CGAL. (2015). *Computational Geometry Algorithms Library*. Available at: http://www.cgal.org

Chan, V., Park, K., Collens, M. B., Kong, H., Saif, T. A., and Bashir, R. (2012). Development of miniaturized walking biological machines. *Sci. Rep.* 2, 857. doi:10.1038/srep00857

Cheney, N., MacCurdy, R., Clune, J., and Lipson, H. (2013). "Unshackling evolution: evolving soft robots with multiple materials and a powerful generative encoding," in *Proc. of the 15th Annual Conference on Genetic and Evolutionary Computation, GECCO '13* (New York, NY: ACM), 167–174. doi:10.1145/2463372.2463404

Chervenski, P., and Ryan, S. (2014). *MultiNEAT*. Available at: http://multineat.com configuration file: http://pastebin.com/ZPNTa6kQ

Cuccu, G., and Gomez, F. (2011). "When novelty is not enough," in *Applications of Evolutionary Computation, Volume 6624 of LNCS*, eds C. Di Chio, S. Cagnoni, C. Cotta, M. Ebner, A. Ekárt, A. I. Esparcia-Alcázar et al. (Berlin: Springer), 234–243. doi:10.1007/978-3-642-20525-5_24

Federici, D., and Ziemke, T. (2006). "Why are evolved developing organisms also fault-tolerant?," in *From Animals to Animats 9: Proc. of the 9th International Conference on Simulation of Adaptive Behaviour (SAB 2006), Volume 4095 of LNCS* (Berlin: Springer), 449–460.

Gomes, J., Urbano, P., and Christensen, A. (2013). Evolution of swarm robotics systems with novelty search. *Swarm Intell.* 7, 115–144. doi:10.1007/s11721-013-0081-z

Gomez, F. J. (2009). "Sustaining diversity using behavioral information distance," in *Proc. of the 11th Annual Conference on Genetic and Evolutionary Computation, GECCO '09* (New York, NY: ACM), 113–120.

Hiller, J., and Lipson, H. (2012). Automatic design and manufacture of soft robots. *IEEE Trans. Robot.* 28, 457–466. doi:10.1109/TRO.2011.2172702

Hornby, G., Lohn, J. D., and Linden, D. S. (2010). Computer-automated evolution of an X-band antenna for NASA's space technology 5 mission. *Evol. Comput.* 19, 1–23. doi:10.1162/EVCO_a_00005

Imroze, K., and Prasad, N. G. (2011). Mating with large males decreases the immune defence of females in drosophila melanogaster. *J. Genet.* 90, 427–434. doi:10.1007/s12041-011-0105-7

Joachimczak, M., Kowaliw, T., Doursat, R., and Wróbel, B. (2013). Evolutionary design of soft-bodied animats with decentralized control. *Artif. Life Robot.* 18, 152–160. doi:10.1007/s10015-013-0121-1

Joachimczak, M., Suzuki, R., and Arita, T. (2014). "Fine grained artificial development for body-controller coevolution of soft-bodied animats," in *Artificial Life 14: Proc. of the 14th International Conference on the Synthesis and Simulation of Living Systems* (Cambridge, MA: The MIT Press), 239–246. doi:10.7551/978-0-262-32621-6-ch040

Joachimczak, M., Suzuki, R., and Arita, T. (2015). "From tadpole to frog: artificial metamorphosis as a method of evolving self-reconfiguring robots," in *Proc. of the 13th European Conference on the Synthesis and Simulation of Living Systems (ECAL 2015)*, eds P. Andrews, L. Caves, R. Doursat, S. Hickinbotham, F. Polack, S. Stepney, et al. (Cambridge, MA: The MIT Press), 51–58. doi:10.7551/978-0-262-33027-5-ch012

Joachimczak, M., and Wróbel, B. (2011). "Evolution of the morphology and patterning of artificial embryos: scaling the tricolour problem to the third dimension," in *Advances in Artificial Life. Darwin Meets Von Neumann: Proc. of the 10th European Conference on Artificial Life (ECAL 2009), Volume 5777 of LNCS* (Berlin: Springer), 35–43.

Joachimczak, M., and Wróbel, B. (2012a). "Co-evolution of morphology and control of soft-bodied multicellular animats," in *Proc. of the 14th International Conference on Genetic and Evolutionary Computation, GECCO '12* (New York, NY: ACM), 561–568. doi:10.1145/2330163.2330243

Joachimczak, M., and Wróbel, B. (2012b). Evolution of robustness to damage in artificial 3-dimensional development. *BioSystems* 109, 498–505. doi:10.1016/j.biosystems.2012.05.014

Joachimczak, M., and Wróbel, B. (2012c). "Open ended evolution of 3d multicellular development controlled by gene regulatory networks," in *Artificial Life XIII: Proc. of the 13th International Conference on the Simulation and Synthesis of Living Systems* (Cambridge, MA: MIT Press), 67–74. doi:10.7551/978-0-262-31050-5-ch010

Komosinski, M., and Rotaru-Varga, A. (2002). Comparison of different genotype encodings for simulated three-dimensional agents. *Artif. Life* 7, 395–418. doi:10.1162/106454601317297022

Komosinski, M., and Ulatowski, S. (1999). "Framsticks: towards a simulation of a nature-like world, creatures and evolution," in *Proc. of Fifth European Conference on Artificial Life (ECAL 1999), Volume 1674 of LNAI* (Berlin: Springer-Verlag), 261–265. doi:10.1007/3-540-48304-7_33

Kowaliw, T., Grogono, P., and Kharma, N. (2004). "Bluenome: a novel developmental model of artificial morphogenesis," in *Genetic and Evolutionary Computation GECCO '04* eds T. Kanade, J. Kittler, J. M. Kleinberg, F. Mattern, J. C. Mitchell, M. Naor, et al. (Berlin: Springer), 93–104. doi:10.1007/978-3-540-24854-5_9

Krčah, P. (2012). "Solving deceptive tasks in robot body-brain co-evolution by searching for behavioral novelty," in *Advances in Robotics and Virtual Reality, Volume 26 of Intelligent Systems Reference Library* eds T. Gulrez and A. E. Hassanien (Berlin: Springer), 167–186. doi:10.1007/978-3-642-23363-0_7

Lee, S., Yosinski, J., Glette, K., Lipson, H., and Clune, J. (2013). "Evolving gaits for physical robots with the hyperneat generative encoding: the benefits of simulation," in *Applications of Evolutionary Computation, Volume 7835 of LNCS* ed. A. Esparcia-Alcázar (Berlin: Springer), 540–549. doi:10.1007/978-3-642-37192-9_54

Lehman, J., and Stanley, K. O. (2011a). Abandoning objectives: evolution through the search for novelty alone. *Evol. Comput.* 19, 189–223. doi:10.1162/EVCO_a_00025

Lehman, J., and Stanley, K. O. (2011b). "Evolving a diversity of virtual creatures through novelty search and local competition," in *Proc. of the 13th annual conference on Genetic and evolutionary computation, GECCO '11* (New York, NY: ACM), 211–218. doi:10.1145/2001576.2001606

Lehman, J., Stanley, K. O., and Miikkulainen, R. (2013). "Effective diversity maintenance in deceptive domains," in *Proc. of the 15th Annual Conference on Genetic and Evolutionary Computation, GECCO '13* (New York, NY: ACM), 215–222. doi:10.1145/2463372.2463393

Lessin, D., Fussell, D., and Miikkulainen, R. (2014). "Adopting morphology to multiple tasks in evolved virtual creatures," in *Artificial Life 14: Proc. of the 14th International Conference on the Synthesis and Simulation of Living Systems* (Cambridge, MA: The MIT Press), 247–254. doi:10.7551/978-0-262-32621-6-ch041

Lipson, H., and Pollack, J. B. (2000). Automatic design and manufacture of robotic lifeforms. *Nature* 406, 974–978. doi:10.1038/35023115

Mahfoud, S. W. (1995). *Niching Methods for Genetic Algorithms*. Ph.D. thesis, University of Illinois at Urbana-Champaign, Champaign, IL.

Matos, A., Suzuki, R., and Arita, T. (2009). Heterochrony and artificial embryogeny: a method for analyzing artificial embryogenies based on developmental dynamics. *Artif. Life* 15, 131–160. doi:10.1162/artl.2009.15.2.15200

Meng, Y., Zhang, Y., and Jin, Y. (2011). Autonomous self-reconfiguration of modular robots by evolving a hierarchical mechanochemical model. *IEEE Comput. Intell. Mag.* 6, 43–54. doi:10.1109/MCI.2010.939579

Mouret, J.-B., and Clune, J. (2012). "An algorithm to create phenotype-fitness maps," in *Proceedings of the Thirteenth International Conference on the Simulation and Synthesis of Living Systems (ALIFE 13)* eds C. Adami, D. M. Bryson, C. Ofria, and R. T. Pennock, (Cambridge, MA: MIT Press), 593–594.

Mouret, J. B., and Doncieux, S. (2011). Encouraging behavioral diversity in evolutionary robotics: an empirical study. *Evol. Comput.* 20, 91–133. doi:10.1162/EVCO_a_00048

Pilat, M. L., Ito, T., Suzuki, R., and Arita, T. (2012). "Evolution of virtual creature foraging in a physical environment," in *Artificial Life XIII: Proc. of the 13th International Conference on the Simulation and Synthesis of Living Systems* (Cambridge, MA: MIT Press), 423–430. doi:10.7551/978-0-262-31050-5-ch056

Sareni, B., and Krahenbuhl, L. (1998). Fitness sharing and niching methods revisited. *IEEE Trans. Evol. Comput.* 2, 97–106. doi:10.1109/4235.735432

Schramm, L., Jin, Y., and Sendhoff, B. (2011). "Emerged coupling of motor control and morphological development in evolution of multi-cellular animats," in *Advances in Artificial Life. Darwin Meets Von Neumann: Proc. of the 10th European Conference on Artificial Life (ECAL 2009), Volume 5777 of LNCS* (Berlin: Springer), 27–34.

Sfakiotakis, M., and Tsakiris, D. (2006). SIMUUN: a simulation environment for undulatory locomotion. *Int. J. Model. Simul.* 26, 350–358. doi:10.2316/Journal.205.2006.4.205-4430

Shepherd, R. F., Ilievski, F., Choi, W., Morin, S. A., Stokes, A. A., Mazzeo, A. D., et al. (2011). Multigait soft robot. *Proc. Natl. Acad. Sci. U.S.A.* 108, 20400–20403. doi:10.1073/pnas.1116564108

Sims, K. (1994). "Evolving virtual creatures," in *Proc. of the 21st Annual Conference on Computer Graphics and Interactive Techniques, SIGGRAPH '94* (New York, NY: ACM Press), 15–22. doi:10.1145/192161.192167

Stanley, K. (2007). Compositional pattern producing networks: a novel abstraction of development. *Genet. Program. Evol. Mach.* 8, 131–162. doi:10.1007/s10710-007-9028-8

Stanley, K. O., D'Ambrosio, D. B., and Gauci, J. (2009). A hypercube-based encoding for evolving large-scale neural networks. *Artif. Life* 15, 185–212. doi:10.1162/artl.2009.15.2.15202

Stanley, K. O., and Lehman, J. (2015). *Why Greatness Cannot be Planned: The Myth of the Objective.* Springer Cham, Heidelberg, New York, Dordrecht, London: Springer International Publishing. doi:10.1007/978-3-319-15524-1

Stanley, K. O., and Miikkulainen, R. (2002). Evolving neural networks through augmenting topologies. *Evol. Comput.* 10, 99–127. doi:10.1162/106365602320169811

Steltz, E., Mozeika, A., Rodenberg, N., Brown, E., and Jaeger, H. M. (2009). "JSEL: jamming skin enabled locomotion," in *IEEE/RSJ International Conference on Intelligent Robots and Systems (IROS 2009)* (Piscataway, NJ: IEEE), 5672–5677. doi:10.1109/iros.2009.5354790

Trujillo, L., Olague, G., Lutton, E., Fernández de Vega, F., Dozal, L., and Clemente, E. (2011). Speciation in behavioral space for evolutionary robotics. *J. Intell. Robot. Syst.* 64, 323–351. doi:10.1007/s10846-011-9542-z

Urbano, P., and Georgiou, L. (2013). "Improving grammatical evolution in Santa Fe trail using novelty search," in *Advances in Artificial Life, ECAL 2013,* eds P. Liò, O. Miglino, G. Nicosia, S. Nolfi, and M. Pavone (Cambridge, CA: MIT Press), 917–924.

Conflict of Interest Statement: The authors declare that the research was conducted in the absence of any commercial or financial relationships that could be construed as a potential conflict of interest.

6

Object Detection: Current and Future Directions

Rodrigo Verschae[1]*[†] and Javier Ruiz-del-Solar[1,2]

[1] *Advanced Mining Technology Center, Universidad de Chile, Santiago, Chile,* [2] *Department of Electrical Engineering, Universidad de Chile, Santiago, Chile*

Object detection is a key ability required by most computer and robot vision systems. The latest research on this area has been making great progress in many directions. In the current manuscript, we give an overview of past research on object detection, outline the current main research directions, and discuss open problems and possible future directions.

Keywords: object detection, perspective, mini review, current directions, open problems

Edited by:
Venkatesh Babu Radhakrishnan,
Indian Institute of Science Bangalore,
India

Reviewed by:
Juxi Leitner,
Queensland University of Technology,
Australia
George Azzopardi,
University of Groningen, Netherlands
Soma Biswas,
Indian Institute of Science Bangalore,
India

*Correspondence:
Rodrigo Verschae
rodrigo@verschae.org

†Present address:
Rodrigo Verschae,
Graduate School of Informatics,
Kyoto University, Kyoto, Japan

1. INTRODUCTION

During the last years, there has been a rapid and successful expansion on computer vision research. Parts of this success have come from adopting and adapting machine learning methods, while others from the development of new representations and models for specific computer vision problems or from the development of efficient solutions. One area that has attained great progress is object detection. The present works gives a *perspective on object detection research*.

Given a set of object classes, *object detection* consists in *determining the location and scale of all object instances, if any, that are present in an image*. Thus, the objective of an object detector is to find all object instances of one or more given object classes regardless of scale, location, pose, view with respect to the camera, partial occlusions, and illumination conditions.

In many computer vision systems, object detection is the first task being performed as it allows to obtain further information regarding the detected object and about the scene. Once an object instance has been detected (e.g., a face), it is be possible to obtain further information, including: (i) to recognize the specific instance (e.g., to identify the subject's face), (ii) to track the object over an image sequence (e.g., to track the face in a video), and (iii) to extract further information about the object (e.g., to determine the subject's gender), while it is also possible to (a) infer the presence or location of other objects in the scene (e.g., a hand may be near a face and at a similar scale) and (b) to better estimate further information about the scene (e.g., the type of scene, indoor versus outdoor, etc.), among other contextual information.

Object detection has been used in many applications, with the most popular ones being: (i) human-computer interaction (HCI), (ii) robotics (e.g., service robots), (iii) consumer electronics (e.g., smart-phones), (iv) security (e.g., recognition, tracking), (v) retrieval (e.g., search engines, photo management), and (vi) transportation (e.g., autonomous and assisted driving). Each of these applications has different requirements, including: processing time (off-line, on-line, or real-time), robustness to occlusions, invariance to rotations (e.g., in-plane rotations), and detection under pose changes. While many applications consider the detection of a single object class (e.g., faces) and from a single view (e.g., frontal faces), others require the detection of multiple object classes (humans, vehicles, etc.), or of a single class from multiple views (e.g., side and frontal view of vehicles). In general, most systems can detect only a single object class from a restricted set of views and poses.

Several surveys on detection and recognition have been published during the last years [see Hjelmås and Low (2001), Yang et al. (2002), Sun et al. (2006), Li and Allinson (2008), Enzweiler and Gavrila (2009), Dollar et al. (2012), Andreopoulos and Tsotsos (2013), Li et al. (2015), and Zafeiriou et al. (2015)], and there are four main problems related to object detection. The first one is *object localization*, which consists of determining the location and scale of a single object instance known to be present in the image; the second one is *object presence classification*, which corresponds to determining whether at least one object of a given class is present in an image (without giving any information about the location, scale, or the number of objects), while the third problem is *object recognition*, which consist in determining if a specific object instance is present in the image. The fourth related problem is *view and pose estimation*, which consist of determining the view of the object and the pose of the object.

The problem of *object presence classification* can be solved using object detection techniques, but in general, other methods are used, as determining the location and scale of the objects is not required, and determining only the presence can be done more efficiently. In some cases, *object recognition* can be solved using methods that do not require detecting the object in advance [e.g., using methods based on Local Interest Points such as Tuytelaars and Mikolajczyk (2008) and Ramanan and Niranjan (2012)]. Nevertheless, solving the object detection problem would solve (or help simplifying) these related problems. An additional, recently addressed problem corresponds to *determining the "objectness"* of an image patch, i.e., measuring the likeliness for an image window to contain an object of any class [e.g., Alexe et al. (2010), Endres and Hoiem (2010), and Huval et al. (2013)].

In the following, we give a summary of past research on object detection, present an overview of current research directions, and discuss open problems and possible future directions, all this with a focus on the classifiers and architectures of the detector, rather than on the used features.

2. A BRIEF REVIEW OF OBJECT DETECTION RESEARCH

Early works on object detection were based on template matching techniques and simple part-based models [e.g., Fischler and Elschlager (1973)]. Later, methods based on statistical classifiers (e.g., Neural Networks, SVM, Adaboost, Bayes, etc.) were introduced [e.g., Osuna et al. (1997), Rowley et al. (1998), Sung and Poggio (1998), Schneiderman and Kanade (2000), Yang et al. (2000a,b), Fleuret and Geman (2001), Romdhani et al. (2001), and Viola and Jones (2001)]. This initial successful family of object detectors, all of them based on statistical classifiers, set the ground for most of the following research in terms of training and evaluation procedures and classification techniques.

Because face detection is a critical ability for any system that interacts with humans, it is the most common application of object detection. However, many additional detection problems have been studied [e.g., Papageorgiou and Poggio (2000), Agarwal et al. (2004), Alexe et al. (2010), Everingham et al. (2010), and Andreopoulos and Tsotsos (2013)]. Most cases correspond to objects that people often interact with, such as other humans [e.g., pedestrians (Papageorgiou and Poggio, 2000; Viola and Jones, 2002; Dalal and Triggs, 2005; Bourdev et al., 2010; Paisitkriangkrai et al., 2015)] and body parts [(Kölsch and Turk, 2004; Ong and Bowden, 2004; Wu and Nevatia, 2005; Verschae et al., 2008; Bourdev and Malik, 2009) e.g., faces, hands, and eyes], as well as vehicles [(Papageorgiou and Poggio, 2000; Felzenszwalb et al., 2010b), e.g., cars and airplanes], and animals [e.g., Fleuret and Geman (2008)].

Most object detection systems consider the same basic scheme, commonly known as *sliding window*: in order to detect the objects appearing in the image at different scales and locations, an exhaustive search is applied. This search makes use of a classifier, the core part of the detector, which indicates if a given image patch, corresponds to the object or not. Given that the classifier basically works at a given scale and patch size, several versions of the input image are generated at different scales, and the classifier is used to classify all possible patches of the given size, for each of the downscaled versions of the image.

Basically, three alternatives exist to the sliding window scheme. The first one is based on the use of bag-of-words (Weinland et al., 2011; Tsai, 2012), method sometimes used for verifying the presence of the object, and that in some cases can be efficiently applied by iteratively refining the image region that contains the object [e.g., Lampert et al. (2009)]. The second one samples patches and iteratively searches for regions of the image where it is likely that the object is present [e.g., Prati et al. (2012)]. These two schemes reduce the number of image patches where to perform the classification, seeking to avoid an exhaustive search over all image patches. The third scheme finds key-points and then matches them to perform the detection [e.g., Azzopardi and Petkov (2013)]. These schemes cannot always guarantee that all object's instances will be detected.

3. OBJECT DETECTION APPROACHES

Object detection methods can be grouped in five categories, each with merits and demerits: while some are more robust, others can be used in real-time systems, and others can be handle more classes, etc. **Table 1** gives a qualitative comparison.

3.1. Coarse-to-Fine and Boosted Classifiers

The most popular work in this category is the boosted cascade classifier of Viola and Jones (2004). It works by efficiently rejecting, in a cascade of test/filters, image patches that do not correspond to the object. Cascade methods are commonly used with boosted classifiers due to two main reasons: (i) boosting generates an additive classifier, thus it is easy to control the complexity of each stage of the cascade and (ii) during training, boosting can be also used for feature selection, allowing the use of large (parametrized) families of features. A coarse-to-fine cascade classifier is usually the first kind of classifier to consider when efficiency is a key requirement. Recent methods based on boosted classifiers include Li and Zhang (2004), Gangaputra and Geman (2006), Huang et al. (2007), Wu and Nevatia (2007), Verschae et al. (2008), and Verschae and Ruiz-del-Solar (2012).

3.2. Dictionary Based

The best example in this category is the Bag of Word method [e.g., Serre et al. (2005) and Mutch and Lowe (2008)]. This approach is basically designed to detect a single object per image, but after removing a detected object, the remaining objects can be detected [e.g., Lampert et al. (2009)]. Two problems with this approach are that it cannot robustly handle well the case of two instances of the object appearing near each other, and that the localization of the object may not be accurate.

3.3. Deformable Part-Based Model

This approach considers object and part models and their relative positions. In general, it is more robust that other approaches, but it is rather time consuming and cannot detect objects appearing at small scales. It can be traced back to the deformable models (Fischler and Elschlager, 1973), but successful methods are recent (Felzenszwalb et al., 2010b). Relevant works include Felzenszwalb et al. (2010a) and Yan et al. (2014), where efficient evaluation of deformable part-based model is implemented using a coarse-to-fine cascade model for faster evaluation, Divvala et al. (2012), where the relevance of the part-models is analyzed, among others [e.g., Azizpour and Laptev (2012), Zhu and Ramanan (2012), and Girshick et al. (2014)].

3.4. Deep Learning

One of the first successful methods in this family is based on convolutional neural networks (Delakis and Garcia, 2004). The key difference between this and the above approaches is that in this approach the feature representation is learned instead of being designed by the user, but with the drawback that a large number of training samples is required for training the classifier. Recent methods include Dean et al. (2013), Huval et al. (2013), Ouyang and Wang (2013), Sermanet et al. (2013), Szegedy et al. (2013), Zeng et al. (2013), Erhan et al. (2014), Zhou et al. (2014), and Ouyang et al. (2015).

3.5. Trainable Image Processing Architectures

In such architectures, the parameters of predefined operators and the combination of the operators are learned, sometimes considering an abstract notion of fitness. These are general-purpose architectures, and thus they can be used to build several modules of a larger system (e.g., object recognition, key point detectors and object detection modules of a robot vision system). Examples include trainable COSFIRE filters (Azzopardi and Petkov, 2013, 2014), and Cartesian Genetic Programming (CGP) (Harding et al., 2013; Leitner et al., 2013).

4. CURRENT RESEARCH PROBLEMS

Table 2 presents a summary of solved, current, and open problems. In the present section we discuss current research directions.

4.1. Multi-Class

Many applications require detecting more than one object class. If a large number of classes is being detected, the processing speed becomes an important issue, as well as the kind of classes that the system can handle without accuracy loss. Works that have addressed the multi-class detection problem include Torralba et al. (2007), Razavi et al. (2011), Benbouzid et al. (2012),

---

TABLE 1 | Qualitative comparison of object detection approaches.

Method	Coarse-to-fine and boosted classifiers	Dictionary based	Deformable part-based models	Deep learning	Trainable image processing architectures
Accuracy	++	+=	++	++	+=
Generality	==	++	+=	++	+=
Speed	++	+=	==	+=	+=
Advantages	Real-time, it can work at small resolutions	Representation can be shared across classes	It can handle deformations and occlusions	Representation can be transfered to other classes	General-purpose architecture that can be used is several modules of a system
Drawbacks/ requirements	Features are predefined	It may not detect all object instances	It can not detect small objects	Large training sets specialized hardware (GPU) for efficiency	The obtained system may be Too specialized for a particular setting
Typical applications	Robotics, security	Retrieval, search	Transportation pedestrian detection	Retrieval, search	HCI, health, robotics

Accuracy: ++, High; +=, Good; ==, Low.
Speed: ++, real-time (15 fps or more); +=, online (10–5 fps); ==, offline (5 fps or more).
Generality: ++ (+=), applicable to many (some) object classes; ==, depend on features designed for specific classes.

TABLE 2 | Summary of current directions and open problems.

Solved problems	Single-class	Single-view	Small deformations	Multi-scale
Current directions	Multi-class (scalability and efficiency)	Multi-view/pose Multi-resolution	Occlusions, deformable Interlaced object and background	Contextual information Temporal features
Open	Incremental learning	Object-part relation	Pixel-level detection Background objects	Multi-modal

Song et al. (2012), Verschae and Ruiz-del-Solar (2012), and Erhan et al. (2014). Efficiency has been addressed, e.g., by using the same representation for several object classes, as well as by developing multi-class classifiers designed specifically to detect multiple classes. Dean et al. (2013) presents one of the few existing works for very large-scale multi-class object detection, where 100,000 object classes were considered.

4.2. Multi-View, Multi-Pose, Multi-Resolution

Most methods used in practice have been designed to detect a single object class under a single view, thus these methods cannot handle multiple views, or large pose variations; with the exception of deformable part-based models which can deal with some pose variations. Some works have tried to detect objects by learning subclasses (Wu and Nevatia, 2007) or by considering views/poses as different classes (Verschae and Ruiz-del-Solar, 2012); in both cases improving the efficiency and robustness. Also, multi-pose models [e.g., Erol et al. (2007)] and multi-resolution models [e.g., Park et al. (2010)] have been developed.

4.3. Efficiency and Computational Power

Efficiency is an issue to be taken into account in any object detection system. As mentioned, a coarse-to-fine classifier is usually the first kind of classifier to consider when efficiency is a key requirement [e.g., Viola et al. (2005)], while reducing the number of image patches where to perform the classification [e.g., Lampert et al. (2009)] and efficiently detecting multiple classes [e.g., Verschae and Ruiz-del-Solar (2012)] have also been used. Efficiency does not imply real-time performance, and works such as Felzenszwalb et al. (2010b) are robust and efficient, but not fast enough for real-time problems. However, using specialized hardware (e.g., GPU) some methods can run in real-time (e.g., deep learning).

4.4. Occlusions, Deformable Objects, and Interlaced Object and Background

Dealing with partial occlusions is also an important problem, and no compelling solution exits, although relevant research has been done [e.g., Wu and Nevatia (2005)]. Similarly, detecting objects that are not "closed," i.e., where objects and background pixels are interlaced with background is still a difficult problem. Two examples are hand detection [e.g., Kölsch and Turk (2004)] and pedestrian detection [see Dollar et al. (2012)]. Deformable part-based model [e.g., Felzenszwalb et al. (2010b)] have been to some extend successful under this kind of problem, but further improvement is still required.

4.5. Contextual Information and Temporal Features

Integrating contextual information (e.g., about the type of scene, or the presence of other objects) can increase speed and robustness, but "when and how" to do this (before, during or after the detection), it is still an open problem. Some proposed solutions include the use of (i) spatio-temporal context [e.g., Palma-Amestoy et al. (2010)], (ii) spatial structure among visual words [e.g., Wu et al. (2009)], and (iii) semantic information

aiming to map semantically related features to visual words [e.g., Wu et al. (2010)], among many others [e.g., Torralba and Sinha (2001), Divvala et al. (2009), Sun et al. (2012), Mottaghi et al. (2014), and Cadena et al. (2015)]. While most methods consider the detection of objects in a single frame, temporal features can be beneficial [e.g., Viola et al. (2005) and Dalal et al. (2006)].

5. OPEN PROBLEMS AND FUTURE DIRECTIONS

In the following, we outline problems that we believe have not been addressed, or addressed only partially, and may be interesting relevant research directions.

5.1. Open-World Learning and Active Vision

An important problem is to incrementally learn, to detect new classes, or to incrementally learn to distinguish among subclasses after the "main" class has been learned. If this can be done in an unsupervised way, we will be able to build new classifiers based on existing ones, without much additional effort, greatly reducing the effort required to learn new object classes. Note that humans are continuously inventing new objects, fashion changes, etc., and therefore detection systems will need to be continuously updated, adding new classes, or updating existing ones. Some recent works have addressed these issues, mostly based on deep learning and transfer learning methods [e.g., Bengio (2012), Mesnil et al. (2012), and Kotzias et al. (2014)]. This open-world learning is of particular importance in robot applications, case where active vision mechanisms can aid in the detection and learning [e.g., Paletta and Pinz (2000) and Correa et al. (2012)].

5.2. Object-Part Relation

During the detection process, should we detect the object first or the parts first? This is a basic dilemma, and no clear solution exists. Probably, the search for the object and for the parts must be done concurrently where both processes give feedback to each other. How to do this is still an open problem and is likely related to how to use of context information. Moreover, in cases the object part can be also decomposed in subparts, an interaction among several hierarchies emerge, and in general it is not clear what should be done first.

5.3. Multi-Modal Detection

The use of new sensing modalities, in particular depth and thermal cameras, has seen some development in the last years [e.g., Fehr and Burkhardt (2008) and Correa et al. (2012)]. However, the methods used for processing visual images are also used for thermal images, and to a lesser degree for depth images. While using thermal images makes easier to discriminate the foreground from the background, it can only be applied to objects that irradiate infrared light (e.g., mammals, heating, etc.). Using depth images is easy to segment the objects, but general methods for detecting specific classes has not been proposed, and probably higher resolution depth images are required. It seems that depth and thermal cameras alone are not enough for object detection, at least with their current resolution, but further advances can be expected as the sensing technology improves.

5.4. Pixel-Level Detection (Segmentation) and Background Objects

In many applications, we may be interested in detecting objects that are usually considered as background. The detection of such "background objects," such as rivers, walls, mountains, has not been addressed by most of the here mentioned approaches. In general, this kind of problem has been addressed by first segmenting the image and later labeling each segment of the image [e.g., Peng et al. (2013)]. Of course, for successfully detecting all objects in a scene, and to completely understand the scene, we will need to have a pixel level detection of the objects, and further more, a 3D model of such scene. Therefore, at some point object detection and image segmentation methods may need to be integrated. We are still far from attaining such automatic understanding of the world, and to achieve this, active vision mechanisms might be required [e.g., Aloimonos et al. (1988) and Cadena et al. (2015)].

6. CONCLUSION

Object detection is a key ability for most computer and robot vision system. Although great progress has been observed in the last years, and some existing techniques are now part of many consumer electronics (e.g., face detection for auto-focus in smartphones) or have been integrated in assistant driving technologies, we are still far from achieving human-level performance, in particular in terms of open-world learning. It should be noted that object detection has not been used much in many areas where it could be of great help. As mobile robots, and in general autonomous machines, are starting to be more widely deployed (e.g., quad-copters, drones and soon service robots), the need of object detection systems is gaining more importance. Finally, we need to consider that we will need object detection systems for nano-robots or for robots that will explore areas that have not been seen by humans, such as depth parts of the sea or other planets, and the detection systems will have to learn to new object classes as they are encountered. In such cases, a real-time open-world learning ability will be critical.

ACKNOWLEDGMENTS

This research was partially funded by the FONDECYT Projects 3120218 and 1130153 (CONICYT, Chile).

REFERENCES

Agarwal, S., Awan, A., and Roth, D. (2004). Learning to detect objects in images via a sparse, part-based representation. *IEEE Trans. Pattern Anal. Mach. Intell.* 26, 1475–1490. doi:10.1109/TPAMI.2004.108

Alexe, B., Deselaers, T., and Ferrari, V. (2010). "What is an object?," in *Computer Vision and Pattern Recognition (CVPR), 2010 IEEE Conference on* (San Francisco, CA: IEEE), 73–80. doi:10.1109/CVPR.2010.5540226

Aloimonos, J., Weiss, I., and Bandyopadhyay, A. (1988). Active vision. *Int. J. Comput. Vis.* 1, 333–356. doi:10.1007/BF00133571

Andreopoulos, A., and Tsotsos, J. K. (2013). 50 years of object recognition: directions forward. *Comput. Vis. Image Underst.* 117, 827–891. doi:10.1016/j.cviu.2013.04.005

Azizpour, H., and Laptev, I. (2012). "Object detection using strongly-supervised deformable part models," in *Computer Vision-ECCV 2012* (Florence: Springer), 836–849.

Azzopardi, G., and Petkov, N. (2013). Trainable cosfire filters for keypoint detection and pattern recognition. *IEEE Trans. Pattern Anal. Mach. Intell.* 35, 490–503. doi:10.1109/TPAMI.2012.106

Azzopardi, G., and Petkov, N. (2014). Ventral-stream-like shape representation: from pixel intensity values to trainable object-selective cosfire models. *Front. Comput. Neurosci.* 8:80. doi:10.3389/fncom.2014.00080

Benbouzid, D., Busa-Fekete, R., and Kegl, B. (2012). "Fast classification using sparse decision dags," in *Proceedings of the 29th International Conference on Machine Learning (ICML-12), ICML '12*, eds J. Langford and J. Pineau (New York, NY: Omnipress), 951–958.

Bengio, Y. (2012). "Deep learning of representations for unsupervised and transfer learning," in *ICML Unsupervised and Transfer Learning, Volume 27 of JMLR Proceedings*, eds I. Guyon, G. Dror, V. Lemaire, G. W. Taylor, and D. L. Silver (Bellevue: JMLR.Org), 17–36.

Bourdev, L. D., Maji, S., Brox, T., and Malik, J. (2010). "Detecting people using mutually consistent poselet activations," in *Computer Vision – ECCV 2010 – 11th European Conference on Computer Vision, Heraklion, Crete, Greece, September 5-11, 2010, Proceedings, Part VI, Volume 6316 of Lecture Notes in Computer Science*, eds K. Daniilidis, P. Maragos, and N. Paragios (Heraklion: Springer), 168–181.

Bourdev, L. D., and Malik, J. (2009). "Poselets: body part detectors trained using 3d human pose annotations," in *IEEE 12th International Conference on Computer Vision, ICCV 2009, Kyoto, Japan, September 27 – October 4, 2009* (Kyoto: IEEE), 1365–1372.

Cadena, C., Dick, A., and Reid, I. (2015). "A fast, modular scene understanding system using context-aware object detection," in *Robotics and Automation (ICRA), 2015 IEEE International Conference on* (Seattle, WA).

Correa, M., Hermosilla, G., Verschae, R., and Ruiz-del-Solar, J. (2012). Human detection and identification by robots using thermal and visual information in domestic environments. *J. Intell. Robot Syst.* 66, 223–243. doi:10.1007/s10846-011-9612-2

Dalal, N., and Triggs, B. (2005). "Histograms of oriented gradients for human detection," in *Computer Vision and Pattern Recognition, 2005. CVPR 2005. IEEE Computer Society Conference on*, Vol. 1 (San Diego, CA: IEEE), 886–893. doi:10.1109/CVPR.2005.177

Dalal, N., Triggs, B., and Schmid, C. (2006). "Human detection using oriented histograms of flow and appearance," in *Computer Vision ECCV 2006, Volume 3952 of Lecture Notes in Computer Science*, eds A. Leonardis, H. Bischof, and A. Pinz (Berlin: Springer), 428–441.

Dean, T., Ruzon, M., Segal, M., Shlens, J., Vijayanarasimhan, S., Yagnik, J., et al. (2013). "Fast, accurate detection of 100,000 object classes on a single machine," in *Computer Vision and Pattern Recognition (CVPR), 2013 IEEE Conference on* (Washington, DC: IEEE), 1814–1821.

Delakis, M., and Garcia, C. (2004). Convolutional face finder: a neural architecture for fast and robust face detection. *IEEE Trans. Pattern Anal. Mach. Intell.* 26, 1408–1423. doi:10.1109/TPAMI.2004.97

Divvala, S., Hoiem, D., Hays, J., Efros, A., and Hebert, M. (2009). "An empirical study of context in object detection," in *Computer Vision and Pattern Recognition, 2009. CVPR 2009. IEEE Conference on* (Miami, FL: IEEE), 1271–1278. doi:10.1109/CVPR.2009.5206532

Divvala, S. K., Efros, A. A., and Hebert, M. (2012). "How important are deformable parts in the deformable parts model?," in *Computer Vision-ECCV 2012. Workshops and Demonstrations* (Florence: Springer), 31–40.

Dollar, P., Wojek, C., Schiele, B., and Perona, P. (2012). Pedestrian detection: an evaluation of the state of the art. *IEEE Trans. Pattern Anal. Mach. Intell.* 34, 743–761. doi:10.1109/TPAMI.2011.155

Endres, I., and Hoiem, D. (2010). "Category independent object proposals," in *Proceedings of the 11th European Conference on Computer Vision: Part V, ECCV'10* (Berlin: Springer-Verlag), 575–588.

Enzweiler, M., and Gavrila, D. (2009). Monocular pedestrian detection: survey and experiments. *IEEE Trans. Pattern Anal. Mach. Intell.* 31, 2179–2195. doi:10.1109/TPAMI.2008.260

Erhan, D., Szegedy, C., Toshev, A., and Anguelov, D. (2014). "Scalable object detection using deep neural networks," in *Computer Vision and Pattern Recognition*

(CVPR), 2014 IEEE Conference on (Columbus, OH: IEEE), 2155–2162. doi:10.1109/CVPR.2014.276

Erol, A., Bebis, G., Nicolescu, M., Boyle, R. D., and Twombly, X. (2007). Vision-based hand pose estimation: a review. Comput. Vis. Image Underst. 108, 52–73; Special Issue on Vision for Human-Computer Interaction. doi:10.1016/j.cviu.2006.10.012

Everingham, M., Van Gool, L., Williams, C. K. I., Winn, J., and Zisserman, A. (2010). The pascal visual object classes (voc) challenge. Int. J. Comput. Vis. 88, 303–338. doi:10.1007/s11263-009-0275-4

Fehr, J., and Burkhardt, H. (2008). "3d rotation invariant local binary patterns," in Pattern Recognition, 2008. ICPR 2008. 19th International Conference on (Tampa, FL: IEEE), 1–4. doi:10.1109/ICPR.2008.4761098

Felzenszwalb, P. F., Girshick, R. B., and McAllester, D. (2010a). "Cascade object detection with deformable part models," in Computer Vision and Pattern Recognition (CVPR), 2010 IEEE Conference on (San Francisco, CA: IEEE), 2241–2248.

Felzenszwalb, P., Girshick, R., McAllester, D., and Ramanan, D. (2010b). Object detection with discriminatively trained part-based models. IEEE Trans. Pattern Anal. Mach. Intell. 32, 1627–1645. doi:10.1109/TPAMI.2009.167

Fischler, M. A., and Elschlager, R. (1973). The representation and matching of pictorial structures. IEEE Trans. Comput. C-22, 67–92. doi:10.1109/T-C.1973.223602

Fleuret, F., and Geman, D. (2001). Coarse-to-fine face detection. Int. J. Comput. Vis. 41, 85–107. doi:10.1023/A:1011113216584

Fleuret, F., and Geman, D. (2008). Stationary features and cat detection. Journal of Machine Learning Research (JMLR) 9, 2549–2578.

Gangaputra, S., and Geman, D. (2006). "A design principle for coarse-to-fine classification," in Proc. of the IEEE Conference of Computer Vision and Pattern Recognition, Vol. 2 (New York, NY: IEEE), 1877–1884. doi:10.1109/CVPR.2006.21

Girshick, R., Donahue, J., Darrell, T., and Malik, J. (2014). "Rich feature hierarchies for accurate object detection and semantic segmentation," in Computer Vision and Pattern Recognition (CVPR), 2014 IEEE Conference on (Columbus, OH: IEEE), 580–587.

Harding, S., Leitner, J., and Schmidhuber, J. (2013). "Cartesian genetic programming for image processing," in Genetic Programming Theory and Practice X, Genetic and Evolutionary Computation, eds R. Riolo, E. Vladislavleva, M. D. Ritchie, and J. H. Moore (New York, NY: Springer), 31–44.

Hjelmås, E., and Low, B. K. (2001). Face detection: a survey. Comput. Vis. Image Underst. 83, 236–274. doi:10.1006/cviu.2001.0921

Huang, C., Ai, H., Li, Y., and Lao, S. (2007). High-performance rotation invariant multiview face detection. IEEE Trans. Pattern Anal. Mach. Intell. 29, 671–686. doi:10.1109/TPAMI.2007.1011

Huval, B., Coates, A., and Ng, A. (2013). Deep Learning for Class-Generic Object Detection. arXiv preprint arXiv:1312.6885.

Kölsch, M., and Turk, M. (2004). "Robust hand detection," in Proceedings of the Sixth International Conference on Automatic Face and Gesture Recognition (Seoul: IEEE), 614–619.

Kotzias, D., Denil, M., Blunsom, P., and de Freitas, N. (2014). Deep Multi-Instance Transfer Learning. CoRR, abs/1411.3128.

Lampert, C. H., Blaschko, M., and Hofmann, T. (2009). Efficient subwindow search: a branch and bound framework for object localization. IEEE Trans. Pattern Anal. Mach. Intell. 31, 2129–2142. doi:10.1109/TPAMI.2009.144

Leitner, J., Harding, S., Chandrashekhariah, P., Frank, M., Frster, A., Triesch, J., et al. (2013). Learning visual object detection and localisation using icvision. Biol. Inspired Cogn. Archit. 5, 29–41; Extended versions of selected papers from the Third Annual Meeting of the {BICA} Society (BICA 2012). doi:10.1016/j.bica.2013.05.009

Li, J., and Allinson, N. M. (2008). A comprehensive review of current local features for computer vision. Neurocomputing 71, 1771–1787; Neurocomputing for Vision Research Advances in Blind Signal Processing. doi:10.1016/j.neucom.2007.11.032

Li, S. Z., and Zhang, Z. (2004). Floatboost learning and statistical face detection. IEEE Trans. Pattern Anal. Mach. Intell. 26, 1112–1123. doi:10.1109/TPAMI.2004.68

Li, Y., Wang, S., Tian, Q., and Ding, X. (2015). Feature representation for statistical-learning-based object detection: a review. Pattern Recognit. 48, 3542–3559. doi:10.1016/j.patcog.2015.04.018

Mesnil, G., Dauphin, Y., Glorot, X., Rifai, S., Bengio, Y., Goodfellow, I. J., et al. (2012). "Unsupervised and transfer learning challenge: a deep learning approach," in JMLR W& CP: Proceedings of the Unsupervised and Transfer Learning Challenge and Workshop, Vol. 27, eds I. Guyon, G. Dror, V. Lemaire, G. Taylor, and D. Silver (Bellevue: JMLR.org) 97–110.

Mottaghi, R., Chen, X., Liu, X., Cho, N.-G., Lee, S.-W., Fidler, S., et al. (2014). "The role of context for object detection and semantic segmentation in the wild," in Computer Vision and Pattern Recognition (CVPR), 2014 IEEE Conference on (Columbus, OH: IEEE), 891–898. doi:10.1109/CVPR.2014.119

Mutch, J., and Lowe, D. G. (2008). Object class recognition and localization using sparse features with limited receptive fields. Int. J. Comput. Vis. 80, 45–57. doi:10.1007/s11263-007-0118-0

Ong, E.-J., and Bowden, R. (2004). "A boosted classifier tree for hand shape detection," in Proceedings of the Sixth International Conference on Automatic Face and Gesture Recognition (Seoul: IEEE), 889–894. doi:10.1109/AFGR.2004.1301646

Osuna, E., Freund, R., and Girosi, F. (1997). "Training support vector machines: an application to face detection," in Proc. of the IEEE Conference of Computer Vision and Pattern Recognition (San Juan: IEEE), 130–136. doi:10.1109/CVPR.1997.609310

Ouyang, W., and Wang, X. (2013). "Joint deep learning for pedestrian detection," in Computer Vision (ICCV), 2013 IEEE International Conference on (Sydney, VIC: IEEE), 2056–2063. doi:10.1109/ICCV.2013.257

Ouyang, W., Wang, X., Zeng, X., Qiu, S., Luo, P., Tian, Y., et al. (2015). "Deepidnet: deformable deep convolutional neural networks for object detection," in Proceedings of the IEEE Conference on Computer Vision and Pattern Recognition (Boston, MA: IEEE), 2403–2412.

Paisitkriangkrai, S., Shen, C., and van den Hengel, A. (2015). Pedestrian detection with spatially pooled features and structured ensemble learning. IEEE Trans. Pattern Anal. Mach. Intell. PP, 1. doi:10.1109/TPAMI.2015.2474388

Paletta, L., and Pinz, A. (2000). Active object recognition by view integration and reinforcement learning. Rob. Auton. Syst. 31, 71–86. doi:10.1016/S0921-8890(99)00079-2

Palma-Amestoy, R., Ruiz-del Solar, J., Yanez, J. M., and Guerrero, P. (2010). Spatiotemporal context integration in robot vision. Int. J. Human. Robot. 07, 357–377. doi:10.1142/S0219843610002192

Papageorgiou, C., and Poggio, T. (2000). A trainable system for object detection. Int. J. Comput. Vis. 38, 15–33. doi:10.1023/A:1008162616689

Park, D., Ramanan, D., and Fowlkes, C. (2010). "Multiresolution models for object detection," in Computer Vision ECCV 2010, Volume 6314 of Lecture Notes in Computer Science, eds K. Daniilidis, P. Maragos, and N. Paragios (Berlin: Springer), 241–254.

Peng, B., Zhang, L., and Zhang, D. (2013). A survey of graph theoretical approaches to image segmentation. Pattern Recognit. 46, 1020–1038. doi:10.1016/j.patcog.2012.09.015

Prati, A., Gualdi, G., and Cucchiara, R. (2012). Multistage particle windows for fast and accurate object detection. IEEE Trans. Pattern Anal. Mach. Intell. 34, 1589–1604. doi:10.1109/TPAMI.2011.247

Ramanan, A., and Niranjan, M. (2012). A review of codebook models in patch-based visual object recognition. J. Signal Process. Syst. 68, 333–352. doi:10.1007/s11265-011-0622-x

Razavi, N., Gall, J., and Van Gool, L. (2011). "Scalable multi-class object detection," in Computer Vision and Pattern Recognition (CVPR), 2011 IEEE Conference on (Providence, RI: IEEE), 1505–1512. doi:10.1109/CVPR.2011.5995441

Romdhani, S., Torr, P., Scholkopf, B., and Blake, A. (2001). "Computationally efficient face detection," in Computer Vision, 2001. ICCV 2001. Proceedings. Eighth IEEE International Conference on, Vol. 2 (Vancouver, BC: IEEE), 695–700. doi:10.1109/ICCV.2001.937694

Rowley, H. A., Baluja, S., and Kanade, T. (1998). Neural network-based detection. IEEE Trans. Pattern Anal. Mach. Intell. 20, 23–28. doi:10.1109/34.655647

Schneiderman, H., and Kanade, T. (2000). "A statistical model for 3D object detection applied to faces and cars," in Proc. of the IEEE Conf. on Computer Vision and Pattern Recognition (Hilton Head, SC: IEEE), 746–751.

Sermanet, P., Eigen, D., Zhang, X., Mathieu, M., Fergus, R., and LeCun, Y. (2013). Overfeat: Integrated Recognition, Localization and Detection Using Convolutional Networks. arXiv preprint arXiv:1312.6229.

Serre, T., Wolf, L., and Poggio, T. (2005). "Object recognition with features inspired by visual cortex," in CVPR (2) (San Diego, CA: IEEE Computer Society), 994–1000.

Song, H. O., Zickler, S., Althoff, T., Girshick, R., Fritz, M., Geyer, C., et al. (2012). "Sparselet models for efficient multiclass object detection," in Computer Vision-ECCV 2012 (Florence: Springer), 802–815.

Sun, M., Bao, S., and Savarese, S. (2012). Object detection using geometrical context feedback. *Int. J. Comput. Vis.* 100, 154–169. doi:10.1007/s11263-012-0547-2

Sun, Z., Bebis, G., and Miller, R. (2006). On-road vehicle detection: a review. *IEEE Trans. Pattern Anal. Mach. Intell.* 28, 694–711. doi:10.1109/TPAMI.2006.104

Sung, K.-K., and Poggio, T. (1998). Example-based learning for viewed-based human face detection. *IEEE Trans. Pattern Anal. Mach. Intell.* 20, 39–51. doi:10.1109/34.655648

Szegedy, C., Toshev, A., and Erhan, D. (2013). "Deep neural networks for object detection," in *Advances in Neural Information Processing Systems 26*, eds C. Burges, L. Bottou, M. Welling, Z. Ghahramani, and K. Weinberger (Harrahs and Harveys: Curran Associates, Inc), 2553–2561.

Torralba, A., Murphy, K. P., and Freeman, W. T. (2007). Sharing visual features for multiclass and multiview object detection. *IEEE Trans. Pattern Anal. Mach. Intell.* 29, 854–869. doi:10.1109/TPAMI.2007.1055

Torralba, A., and Sinha, P. (2001). "Statistical context priming for object detection," in *Computer Vision, 2001. ICCV 2001. Proceedings. Eighth IEEE International Conference on*, Vol. 1 (Vancouver, BC: IEEE), 763–770. doi:10.1109/ICCV.2001.937604

Tsai, C.-F. (2012). Bag-of-words representation in image annotation: a review. *ISRN Artif. Intell.* 2012, 19. doi:10.5402/2012/376804

Tuytelaars, T., and Mikolajczyk, K. (2008). Local invariant feature detectors: a survey. *Found. Trends Comput. Graph. Vis.* 3, 177–280. doi:10.1561/0600000017

Verschae, R., and Ruiz-del-Solar, J. (2012). "Tcas: a multiclass object detector for robot and computer vision applications," in *Advances in Visual Computing, Volume 7431 of Lecture Notes in Computer Science*, eds G. Bebis, R. Boyle, B. Parvin, D. Koracin, C. Fowlkes, S. Wang, et al. (Berlin: Springer), 632–641.

Verschae, R., Ruiz-del-Solar, J., and Correa, M. (2008). A unified learning framework for object detection and classification using nested cascades of boosted classifiers. *Mach. Vis. Appl.* 19, 85–103. doi:10.1007/s00138-007-0084-0

Viola, P., and Jones, M. (2001). "Rapid object detection using a boosted cascade of simple features," in *Proc. of the IEEE Conf. on Computer Vision and Pattern Recognition* (Kauai: IEEE), 511–518. doi:10.1109/CVPR.2001.990517

Viola, P., and Jones, M. (2002). "Fast and robust classification using asymmetric adaboost and a detector cascade," in *Advances in Neural Information Processing System 14* (Vancouver: MIT Press), 1311–1318.

Viola, P., Jones, M., and Snow, D. (2005). Detecting pedestrians using patterns of motion and appearance. *Int. J. Comput. Vis.* 63, 153–161. doi:10.1007/s11263-005-6644-8

Viola, P., and Jones, M. J. (2004). Robust real-time face detection. *Int. J. Comput. Vis.* 57, 137–154. doi:10.1023/B:VISI.0000013087.49260.fb

Weinland, D., Ronfard, R., and Boyer, E. (2011). A survey of vision-based methods for action representation, segmentation and recognition. *Comput. Vis. Image Underst.* 115, 224–241. doi:10.1016/j.cviu.2010.10.002

Wu, B., and Nevatia, R. (2005). "Detection of multiple, partially occluded humans in a single image by bayesian combination of edgelet part detectors," in *ICCV '05: Proceedings of the 10th IEEE Int. Conf. on Computer Vision (ICCV'05) Vol 1* (Washington, DC: IEEE Computer Society), 90–97.

Wu, B., and Nevatia, R. (2007). "Cluster boosted tree classifier for multi-view, multi-pose object detection," in *ICCV* (Rio de Janeiro: IEEE), 1–8.

Wu, L., Hoi, S., and Yu, N. (2010). Semantics-preserving bag-of-words models and applications. *IEEE Trans. Image Process.* 19, 1908–1920. doi:10.1109/TIP.2010.2045169

Wu, L., Hu, Y., Li, M., Yu, N., and Hua, X.-S. (2009). Scale-invariant visual language modeling for object categorization. *IEEE Trans. Multimedia* 11, 286–294. doi:10.1109/TMM.2008.2009692

Yan, J., Lei, Z., Wen, L., and Li, S. Z. (2014). "The fastest deformable part model for object detection," in *Computer Vision and Pattern Recognition (CVPR), 2014 IEEE Conference on* (Columbus, OH: IEEE), 2497–2504.

Yang, M.-H., Ahuja, N., and Kriegman, D. (2000a). "Mixtures of linear subspaces for face detection," in *Proc. Fourth IEEE Int. Conf. on Automatic Face and Gesture Recognition* (Grenoble: IEEE), 70–76.

Yang, M.-H., Roth, D., and Ahuja, N. (2000b). "A SNoW-based face detector," in *Advances in Neural Information Processing Systems 12* (Denver: MIT press), 855–861.

Yang, M.-H., Kriegman, D., and Ahuja, N. (2002). Detecting faces in images: a survey. *IEEE Trans. Pattern Anal. Mach. Intell.* 24, 34–58. doi:10.1109/34.982883

Zafeiriou, S., Zhang, C., and Zhang, Z. (2015). A survey on face detection in the wild: past, present and future. *Comput. Vis. Image Underst.* 138, 1–24. doi:10.1016/j.cviu.2015.03.015

Zeng, X., Ouyang, W., and Wang, X. (2013). "Multi-stage contextual deep learning for pedestrian detection," in *Computer Vision (ICCV), 2013 IEEE International Conference on* (Washington, DC: IEEE), 121–128.

Zhou, B., Khosla, A., Lapedriza, À., Oliva, A., and Torralba, A. (2014). *Object Detectors Emerge in Deep Scene Cnns.* CoRR, abs/1412.6856.

Zhu, X., and Ramanan, D. (2012). "Face detection, pose estimation, and landmark localization in the wild," in *Computer Vision and Pattern Recognition (CVPR), 2012 IEEE Conference on* (Providence: IEEE), 2879–2886.

Conflict of Interest Statement: The authors declare that the research was conducted in the absence of any commercial or financial relationships that could be construed as a potential conflict of interest.

7

Bounded Rationality, Abstraction, and Hierarchical Decision-Making: An Information-Theoretic Optimality Principle

Tim Genewein[1,2,3]*, Felix Leibfried[1,2,3], Jordi Grau-Moya[1,2,3] and Daniel Alexander Braun[1,2]

[1] *Max Planck Institute for Intelligent Systems, Tübingen, Germany,* [2] *Max Planck Institute for Biological Cybernetics, Tübingen, Germany,* [3] *Graduate Training Centre of Neuroscience, Tübingen, Germany*

Edited by:
Joschka Boedecker,
University of Freiburg, Germany

Reviewed by:
Dimitrije Markovic,
Dresden University of Technology,
Germany
Sam Neymotin,
State University of New York
Downstate Medical Center, USA

*Correspondence:
Tim Genewein
tim.genewein@tuebingen.mpg.de

Abstraction and hierarchical information processing are hallmarks of human and animal intelligence underlying the unrivaled flexibility of behavior in biological systems. Achieving such flexibility in artificial systems is challenging, even with more and more computational power. Here, we investigate the hypothesis that abstraction and hierarchical information processing might in fact be the consequence of limitations in information-processing power. In particular, we study an information-theoretic framework of bounded rational decision-making that trades off utility maximization against information-processing costs. We apply the basic principle of this framework to perception-action systems with multiple information-processing nodes and derive bounded-optimal solutions. We show how the formation of abstractions and decision-making hierarchies depends on information-processing costs. We illustrate the theoretical ideas with example simulations and conclude by formalizing a mathematically unifying optimization principle that could potentially be extended to more complex systems.

Keywords: information theory, bounded rationality, computational rationality, rate-distortion, decision-making, hierarchical architecture, perception-action system, lossy compression

1. INTRODUCTION

A key characteristic of intelligent systems, both biological and artificial, is the ability to flexibly adapt behavior in order to interact with the environment in a way that is beneficial to the system. In biological systems, the ability to adapt affects the fitness of an organism and becomes key to survival not only of individual organisms but species as a whole. Both in the theoretical study of biological systems and in the design of artificial intelligent systems, the central goal is to understand adaptive behavior formally. A formal framework for tackling the problem of general adaptive systems is decision-theory, where behavior is conceptualized as a series of optimal decisions or actions that a system performs in order to respond to changes to the input of the system. An important idea, originating from the foundations of decision-theory, is the maximum expected utility (MEU) principle (Ramsey, 1931; Von Neumann and Morgenstern, 1944; Savage, 1954). Following MEU, an intelligent system is formalized as a decision-maker that chooses actions in order to maximize the desirability of the expected outcome of the action, where the desirability of an outcome is quantified by a utility function.

A fundamental problem of MEU is that the computation of an optimal action can easily exceed the computational capacity of a system. It is for example in general prohibitive trying to compute an optimal chess move due to the large number of possibilities. One way to deal with such problems is to study optimal decision-making with information-processing constraints. Following the pioneering work of Simon (1955, 1972) on bounded rationality, decision-making with limited information-processing resources has been studied extensively in psychology (Gigerenzer and Todd, 1999; Camerer, 2003; Gigerenzer and Brighton, 2009), economics (McKelvey and Palfrey, 1995; Rubinstein, 1998; Kahneman, 2003; Parkes and Wellman, 2015), political science (Jones, 2003), industrial organization (Spiegler, 2011), cognitive science (Howes et al., 2009; Janssen et al., 2011), computer science, and artificial intelligence research (Horvitz, 1988; Lipman, 1995; Russell, 1995; Russell and Subramanian, 1995; Russell and Norvig, 2002; Lewis et al., 2014). Conceptually, the approaches differ widely ranging from heuristics (Tversky and Kahneman, 1974; Gigerenzer and Todd, 1999; Gigerenzer and Brighton, 2009; Burns et al., 2013) to approximate statistical inference schemes (Levy et al., 2009; Vul et al., 2009, 2014; Sanborn et al., 2010; Tenenbaum et al., 2011; Fox and Roberts, 2012; Lieder et al., 2012).

In this study, we use an information-theoretic model of bounded rational decision-making (Braun et al., 2011; Ortega and Braun, 2012, 2013; Braun and Ortega, 2014; Ortega and Braun, 2014; Ortega et al., 2014) that has precursors in the economic literature (McKelvey and Palfrey, 1995; Mattsson and Weibull, 2002; Sims, 2003, 2005, 2006, 2010; Wolpert, 2006) and that is closely related to recent advances in the information theory of perception-action systems (Todorov, 2007, 2009; Still, 2009; Friston, 2010; Peters et al., 2010; Tishby and Polani, 2011; Daniel et al., 2012, 2013; Kappen et al., 2012; Rawlik et al., 2012; Rubin et al., 2012; Neymotin et al., 2013; Tkačik and Bialek, 2014; Palmer et al., 2015). The basis of this approach is formalized by a free energy principle that trades off expected utility, and the cost of computation that is required to adapt the system accordingly in order to achieve high utility. Here, we consider an extension of this framework to systems with multiple information-processing nodes and in particular discuss the formation of information-processing hierarchies, where different levels in the hierarchy represent different levels of abstraction. The basic intuition is that information-processing nodes with little computational resources can adapt only a little for different inputs and are therefore forced to treat different inputs in the same or a similar way, that is the system has to abstract (Genewein and Braun, 2013). Importantly, abstractions arising in decision-making hierarchies are a core feature of intelligence (Kemp et al., 2007; Braun et al., 2010a,b; Gershman and Niv, 2010; Tenenbaum et al., 2011) and constitute the basis for flexible behavior.

The paper is structured as follows. In Section 2, we recapitulate the information-theoretic framework for decision-making and show its fundamental connection to a well-known trade-off in information theory (the rate-distortion problem for lossy compression). In Section 3, we show how the extension of the basic trade-off principle leads to a theoretically grounded design principle that describes how perception is shaped by action. In Section 4, we apply the basic trade-off between expected utility and computational cost to a two-level hierarchy and show how this leads to emergent, bounded-optimal hierarchical decision-making systems. In Section 5, we present a mathematically unifying formulation that provides a starting point for generalizing the principles presented in this paper to more complex architectures.

2. BOUNDED RATIONAL DECISION-MAKING

2.1. A Free Energy Principle for Bounded Rationality

In a decision-making task with context, an actor or agent is presented with a world-state w and is then faced with finding an optimal action a_w^* out of a set of actions $\mathcal{A}$ in order to maximize the utility $U(w, a)$:

$$a_w^* = \arg\max_a U(w, a). \quad (1)$$

If the cardinality of the action-set is large, the search for the single best action can become computationally very costly. For an agent with limited computational resources that has to react within a certain time-limit, the search problem can potentially become infeasible. In contrast, biological agents, such as animals and humans, are constantly confronted with picking an action out of a very large set of possible actions. For instance, when planning a movement trajectory for grasping a certain object with a biological arm with many degrees of freedom, the number of possible trajectories is infinite. Yet, humans are able to quickly find a trajectory that is not necessarily optimal but good enough. The paradigm of picking a good enough solution that is actually computable has been termed *bounded rational* acting (Simon, 1955, 1972; Horvitz, 1988; Horvitz et al., 1989; Horvitz and Zilberstein, 2001). Note that bounded rational policies are in general stochastic and thus expressed as a probability distribution over actions given a world-state $p(a|w)$.

We follow the work of Ortega and Braun (2013), where the authors present a mathematical framework for bounded rational decision-making that takes into account computational limitations. Formally, an agent's initial behavior (or search strategy through action-space) is described by a prior distribution $p_0(a)$. The agent transforms its behavior to a posterior $p(a|w)$ in order to maximize expected utility $\Sigma_a p(a|w)U(w, a)$ under this posterior policy. The computational cost of this transformation is measured by the KL-divergence between prior and posterior and is upper-bounded in case of a bounded rational actor. Decision-making with limited computational resources can then be formalized with the following constrained optimization problem:

$$p^*(a|w) = \arg\max_{p(a|w)} \sum_a p(a|w)U(w, a)$$

$$\text{s.t. } D_{\mathrm{KL}}(p(a|w)||p_0(a)) \leq K. \quad (2)$$

This principle models bounded rational actors that initially follow a prior policy $p_0(a)$ and then use information about the world-state w to adapt their behavior to $p(a|w)$ in a way that optimally trades off the expected gain in utility against the transformation costs for adapting from $p_0(a)$ to $p(a|w)$. The constrained optimization problem in equation (2) can be rewritten as an unconstrained variational problem using the method of Lagrange multipliers:

$$p^*(a|w) = \underset{p(a|w)}{\arg\max} \underbrace{\sum_a p(a|w) U(w,a)}_{\mathbf{E}_{p(a|w)}[U(w,a)]} - \frac{1}{\beta} \underbrace{\sum_a p(a|w) \log \frac{p(a|w)}{p_0(a)}}_{D_{KL}(p(a|w)||p_0(a))},$$

(3)

where β is known as the *inverse temperature*. The inverse temperature acts as a conversion-factor, translating the amount of information imposed by the transformation (usually measured in nats or bits) into a cost with the same units as the expected utility (utils). The distribution $p^*(a|w)$ that maximizes the variational principle is given by

$$p^*(a|w) = \frac{1}{Z(w)} p_0(a) e^{\beta U(w,a)},$$

(4)

with *partition sum* $Z(w) = \Sigma_a \, p_0(a) \, e^{\beta U(w,a)}$. Evaluating equation (3) with the maximizing distribution $p^*(a|w)$ yields the *free energy difference*

$$\Delta F(w) = \max_{p(a|w)} \mathbf{E}_{p(a|w)}[U(w,a)] - \frac{1}{\beta} D_{KL}(p(a|w)||p_0(a))$$

$$= \frac{1}{\beta} \log Z(w),$$

(5)

which is well known in thermodynamics and quantifies the energy of a system that can be converted to work. $\Delta F(w)$ is composed of the expected utility under the posterior policy $p^*(a|w)$ minus information processing cost that is required for computing the posterior policy measured as the Kullback-Leibler (KL) divergence between the posterior $p^*(a|w)$ and the prior $p_0(a)$.

The inverse temperature β governs the influence of the transformation cost and thus the boundedness of the actor which determines the maximally allowed deviation of the final behavior $p^*(a|w)$ from the initial behavior $p_0(a)$. A perfectly rational actor that maximizes its utility can be recovered as the limit case $\beta \to \infty$ where transformation cost is ignored. This case is identical to equation (1) and simply reflects maximum utility action selection, which is the foundation of most modern decision-making frameworks. Note that the optimal policy $p^*(a|w)$ in this case collapses to a delta over the best action $p^*(a|w) = \delta_{aa_w^*}$. In contrast, $\beta \to 0$ corresponds to an actor that has infinite transformation cost or no computational resources and thus sticks with its prior policy $p_0(a)$. An illustrative example is given in **Figure 1**.

Interestingly, the free energy principle for bounded rational acting can also be used for inference problems. In particular if the utility is chosen as a log-likelihood function $U(w,a) = \log q(w|a)$ and the inverse temperature β is set to one, Bayes' rule is recovered as the optimal bounded rational solution [by plugging into equation (4)]:

$$p^*(a|w) = \frac{p_0(a)q(w|a)}{\sum_a p_0(a)q(w|a)}.$$

Importantly, the inverse temperature β can also be interpreted in terms of computational or sample complexity (Braun and Ortega, 2014; Ortega and Braun, 2014; Ortega et al., 2014).

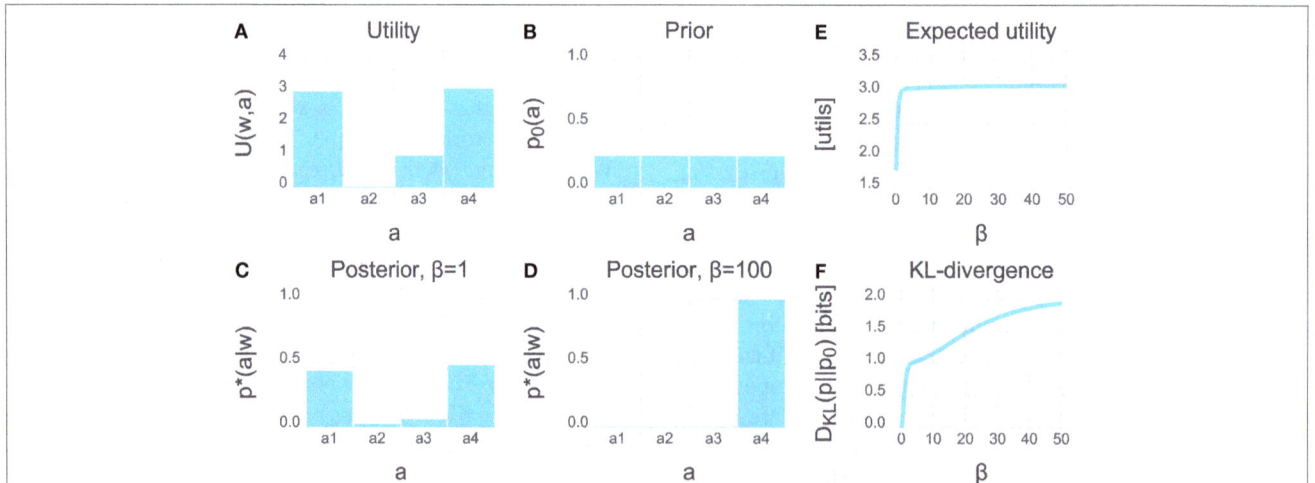

FIGURE 1 | Bounded rationality and the free energy principle. Imagine an actor that has to grasp a particular cup w from a table. There are four options $a1$ to $a4$ to perform the movement and the utility $U(w, a)$ shown in **(A)** measures the performance of each option. There are two actions $a1$ and $a4$ that lead to a successful grasp without spilling, and $a4$ is minimally better. Action $a3$ leads to a successful grasp but spills half the cup, and $a2$ represents an unsuccessful grasp. **(B)** Prior distribution over actions $p_0(a)$: no preference for a particular action. **(C)** Posterior $p^*(a|w)$ [equation (4)] for an actor with limited computational capacity. Due to the computational limits, the posterior cannot deviate from the prior arbitrarily far, otherwise the KL-divergence constraint would be violated. The computational resources are mostly spent on increasing the chance of picking one of the two successful options and decreasing the chance of picking $a2$ or $a3$. The agent is almost indifferent between the two successful options $a1$ and $a4$. **(D)** Posterior for an actor with large computational resources. Even though $a4$ is only slightly better than $a1$, the agent is almost unbounded and can deviate a lot from the prior. This solution is already close to the deterministic maximum expected utility solution and incurs a large KL-divergence from prior to posterior. **(E)** Expected utility $\mathbf{E}_{p*(a|w)}[U(w, a)]$ as a function of the inverse temperature β. Initially, allowing for more computational resources leads to a rapid increase in expected utility. However, this trend quickly flattens out into a regime where small increases in expected utility imply large increases in β. **(F)** KL-divergence D_{KL} $(p^*(a|w)||p_0(a))$ as a function of the inverse temperature β. In order to achieve an expected utility of $\approx 95\%$ of the maximum utility roughly 1 bit suffices [leading to a posterior similar to **(C)**]. Further increasing the performance by 5% requires twice the computational capacity of 2 bits [leading to a posterior similar to **(D)**]. A bounded rational agent that performs reasonably well could thus be designed at half the cost (in terms of computational capacity) compared to a fully rational maximum expected utility agent. An interactive version of this plot where β can be freely changed is provided in the Supplementary Jupyter Notebook "1-FreeEnergyForBoundedRationalDecisionMaking."

The basic idea is that in order to make a decision, the bounded rational decision-maker needs to generate a sample from the posterior $p^*(a|w)$. Assuming that the decision-maker can draw samples from the prior $p_0(a)$, samples from the posterior $p^*(a|w)$ can be generated by rejecting any samples from $p_0(a)$ until one sample is accepted as a sample of $p^*(a|w)$ according to the acceptance rule $u \leq \exp(\beta(U(w,a) - T(w)))$, where u is drawn from the uniform distribution over the unit interval $[0;1]$ and $T(w)$ is the aspiration level or acceptance target value with $T(w) \geq \max_a U(w, a)$. This is known as rejection sampling (Neal, 2003; Bishop, 2006). The efficiency of the rejection sampling process depends on how many samples are needed on average from $p_0(a)$ to obtain one sample from $p^*(a|w)$. This average number of samples $\overline{\#Samples}(w)$ is given by the mean of a geometric distribution

$$\overline{\#Samples}(w) = \frac{1}{\sum_a p_0(a) \exp(\beta(U(w,a) - T(w)))}$$
$$= \frac{\exp(\beta T(w))}{Z(w)}, \tag{6}$$

where the partition sum $Z(w)$ is defined as in equation (4). The average number of samples increases exponentially with increasing resource parameter β when $T(w) > \max_a U(w, a)$. It is also noteworthy that the exponential of the Kullback-Leibler divergence provides a lower bound for the required number of samples that is $\overline{\#Samples}(w) \geq \exp(D_{KL}(p^*(a|w)\|p_0(a)))$ (see Section 6 in the Supplementary Methods for a derivation). Accordingly, a decision-maker with high β can manage high sampling complexity, whereas a decision-maker with low β can only process a few samples.

2.2. From Free Energy to Rate-Distortion: The Optimal Prior

In the free energy principle in equation (3), the prior $p_0(a)$ is assumed to be given. A very interesting question is which prior distribution $p_0(a)$ maximizes the free energy difference $\Delta F(w)$ for all world-states w *on average* (assuming that $p(w)$ is given). To formalize this question, we extend the variational principle in equation (3) by taking the expectation over w and the arg max over $p_0(a)$

$$\arg\max_{p_0(a)} \sum_w p(w)$$
$$\times \left[\arg\max_{p(a|w)} \mathbf{E}_{p(a|w)}[U(w,a)] - \frac{1}{\beta} D_{KL}\left(p(a|w)\|p_0(a)\right) \right].$$

The inner arg max-operator over $p(a|w)$ and the expectation over w can be swapped because the variation is not over $p(w)$. With the KL-term expanded this leads to

$$\arg\max_{p_0(a),p(a|w)} \sum_{w,a} p(w,a)U(w,a)$$
$$- \frac{1}{\beta} \sum_w p(w) \sum_a p(a|w) \log \frac{p(a|w)}{p_0(a)}.$$

The solution to the arg max over $p_0(a)$ is given by $p_0^*(a) = \sum_w p(w)p(a|w) = p(a)$. [see Section 2.1.1 in Tishby et al. (1999)

or Csiszár and Tusnády (1984)]. Plugging in the marginal $p(a)$ as the optimal prior $p_0^*(a)$ yields the following variational principle for bounded rational decision-making

$$\arg\max_{p(a|w)} \underbrace{\sum_{w,a} p(w,a)U(w,a)}_{\mathbf{E}_{p(a|w)}[U(w,a)]} - \frac{1}{\beta} \underbrace{\sum_w p(w)D_{KL}(p(a|w)\|p(a))}_{I(W;A)}$$
$$= \arg\max_{p(a|w)} J_{RD}(p(a|w)), \tag{7}$$

where $I(W; A)$ is the *mutual information* between actions A and world-states W. The mutual information $I(W; A)$ is a measure of the reduction in uncertainty about the action a after having observed w or vice versa since the mutual information is symmetric

$$I(W;A) = H(W) - H(W|A) = H(A) - H(A|W) = I(A; W),$$

where $H(L) = -\Sigma_l p(l)\log p(l)$ is the Shannon entropy of random variable L.

The exact same variational problem can also be obtained as the Langragian for maximizing expected utility with an upper bound on the mutual information

$$p^*(a|w) = \arg\max_{p(a|w)} \sum_{w,a} p(w,a)U(w,a) \quad \text{s.t. } I(W;A) \leq R \tag{8}$$

or in the dual point of view, as minimizing the mutual information between actions and world-states with a lower bound on the expected utility. Thus, the problem in equation (7) is equivalent to the problem formulation in rate-distortion theory (Shannon, 1948; Cover and Thomas, 1991; Tishby et al., 1999; Yeung, 2008), the information-theoretic framework for lossy compression. It deals with the problem that a stream of information must be transmitted over a channel that does not have sufficient capacity to transmit all incoming information – therefore some of the incoming information must be discarded. In rate-distortion theory, the distortion $d(w, a)$ quantifies the recovery error of the output symbol a with respect to the input symbol w. Distortion corresponds to a negative utility which thus leads to an arg min instead of an arg max and a positive sign for the mutual information term in the optimization problem. In this case, a maximum expected utility decision-maker would minimize the expected distortion which is typically achieved by a one-to-one mapping between w and a, which implies that the compression is not lossy. From this, it becomes obvious why MEU decision-making might be problematic: if the MEU decision-maker requires a rate of information processing that is above channel capacity, it simply cannot be realized with the given system.

The solution that extremizes the variational problem of equation (7) is given by the self-consistent equations [see Tishby et al. (1999)]

$$p^*(a|w) = \frac{1}{Z(w)}p(a)e^{\beta U(w,a)}, \tag{9}$$

$$p(a) = \sum_w p(w)p^*(a|w), \tag{10}$$

with *partition sum* $Z(w) = \Sigma_a p(a)e^{\beta U(w,a)}$.

In the limit case $\beta \to \infty$ where transformation costs are ignored, $p^*(a|w) = \delta_{aa_w^*}$ is the perfectly rational policy for each value of w *independent* of any of the other policies and $p(a)$ becomes a mixture of these solutions. Importantly, due to the low price of information processing $\frac{1}{\beta}$, high values of the mutual information term in equation (7) will not lead to a penalization, which means that actions a can be very informative about the world-state w. The behavior of an actor with infinite computational resources will thus in general be very world-state-specific.

In the case where $\beta \to 0$ the mutual information between actions and world-states is minimized to $I(W; A) = 0$, leading to $p^*(a|w) = p(a) \forall w$, the maximal abstraction where all w elicit the same response. Within this limitation, the actor will, however, emit actions that maximize the expected utility $\sum_{w,a} p(w)p(a) U(w, a)$ using the same policy for all world-states.

For values of the rationality parameter β in between these limit cases, that is $0 < \beta < \infty$, the bounded rational actor trades off *world-state-specific* actions that lead to a higher expected utility for particular world-states (at the cost of an increased information processing rate), against more robust or *abstract* actions that yield a "good" expected utility for many world-states (which allows for a decreased information processing rate).

Note that the solution for the conditional distribution $p^*(a|w)$ in the rate-distortion problem [equation (9)] is the same as the solution in the free energy case of the previous section [equation (4)], except that the prior $p_0(a)$ is now defined as the marginal distribution $p_0(a) = p(a)$ [see equation (10)]. This particular prior distribution minimizes the average relative entropy between $p(a|w)$ and $p(a)$ which is the mutual information between actions and world-states $I(W; A)$.

An alternative interpretation is that the decision-maker is a channel that transmits information from w to a according to $p(a|w)$. The channel has a limited capacity, which could arise from the agent not having a "brain" that is powerful enough, but a limited channel capacity could also arise from noise that is induced into the channel, i.e., an agent with noisy sensors or actuators. For a large capacity, the transmission is not severely influenced and the best action for a particular world-state can be chosen. For smaller capacities, however, some information must be discarded and robust (or abstract) actions that are "good" under a number of world-states must be chosen. This is possible by lowering β until the required rate $I(W; A)$ does no longer exceed the channel capacity. The notion that a decision-maker can be considered as an information processing channel is not new and goes back to the cybernetics movement (Ashby, 1956; Wiener, 1961). Other recent applications of rate-distortion theory to decision-making problems can be found for example in Sims (2003, 2006) and Tishby and Polani (2011).

2.3. Computing the Self-Consistent Solution

The self-consistent solutions that maximize the variational principle in equation (7) can be computed by starting with an initial distribution $p_{init}(a)$ and then iterating equations (9) and (10) in an alternating fashion. This procedure is well known in the rate-distortion framework as a Blahut-Arimoto-type algorithm (Arimoto, 1972; Blahut, 1972; Yeung, 2008). The iteration is guaranteed to converge to a unique maximum [see Section 2.1.1 in Tishby et al. (1999) and Csiszár and Tusnády (1984) and Cover and Thomas (1991)]. Note that $p_{init}(a)$ has to have the same support as $p(a)$. Implemented in a straightforward manner, the Blahut-Arimoto iterations can become computationally costly since the iterations involve evaluating the utility function for every action-world-state-pair (w, a) and computing the normalization constant $Z(w)$. In case of continuous-valued random variables, closed-form analytic solutions exist only for special cases. Extending the sampling approach presented at the end of Section 2.1 could be one potential alleviation. A proof-of-concept implementation of the extended sampling scheme is provided in the Supplementary Jupyter Notebook "S1-SampleBasedBlahutArimoto."

2.4. Emergence of Abstractions

The rate-distortion objective for decision-making [equation (7)] penalizes high information processing demand measured in terms of the mutual information between actions and world-states $I(W; A)$. A large mutual information arises when actions are very informative about the world-state which is the case when a particular action is mostly chosen under a particular world-state and is rarely chosen otherwise. Policies $p(a|w)$ with many world-state-specific actions are thus more demanding in terms of informational cost and might not be affordable by an agent with limited computational capacity. In order to keep informational costs low while at the same time optimizing expected utility, actions that yield a "good" expected utility for *many* different world-states must be favored. This leads to abstractions in the sense that the agent does not discriminate between different world-states out of a subset of all world-states, but rather responds with the same policy for the entire subset. Importantly, these abstractions are driven by the agent-environment structure encoded through the utility function $U(w, a)$. Limits in computational resources thus lead to abstractions where different world-states are treated as if they were the same.

To illustrate the influence of different degrees of computational limits and the resulting emergence of abstractions we constructed the following example. The goal is to design a recommender system that observes an item bought w and then recommends another item a. In this example the system can either recommend another concrete item or the best-selling item of a certain category or the best-selling item of a super-category which subsumes several categories (see **Table 1**). An illustration of the example is shown in **Figure 2A**. The possible items bought are shown on the x-axis and possible recommendations are shown on the y-axis. The super-categories and categories as well as the corresponding bought items can be seen in **Table 1** where each bought item also indicates the corresponding concrete item that scores highest when recommended.

The utility of each (w, a)-pair is color-coded in blue in **Figure 2A**. For each possible world-state there is one concrete item that can be recommended that will (deterministically) yield the highest possible utility of 3 utils. Further, each bought item belongs to a category and recommending the best-selling item of the corresponding category leads to a utility of 2.2 utils. Finally, recommending the best-selling item of the corresponding super-category yields a utility of 1.6 utils. For each world-state there is

TABLE 1 | Recommender system example.

Super-category	Category	Bought item	Best recommended item
Electric devices and electronics	Computers	Laptop Monitor Game pad	Laptop sleeve Monitor cable Video game
	Small appliances	Coffee machine Vacuum cleaner Electric toothbrush	Coffee capsules Vacuum cleaner bags Brush heads
Food and cooking	Fruit	Grapes Strawberries Limes	Cheese Cream Cane sugar
	Baking	Pancake mix Baking soda Baker's yeast Muffin cups	Maple syrup Vinegar Flour Flour and chocolate chips

The system observes an item bought w and can then recommend another item to buy a. For each bought item w, there is one other concrete item a that yields the maximum utility when recommended (indicated in the last column of the table). Additionally, each bought item belongs to a category and a less specific super-category. Recommending the best-selling item of the corresponding category or super-category yields sub-optimal but non-zero utility values. A depiction of the utility function U(w, a) is shown in Figure 2A.

one specific action that leads to the highest possible utility but zero utility for all other world-states. At the same time there exist more abstract actions that are sub-optimal but still "good" for a set of world-states. See the legend of **Figure 2** for more details on the example.

Figure 2B shows the result $p^*(a|w)$ obtained through Blahut-Arimoto iterations [equations (9) and (10)] for $\beta = 1.3$. For each world-state (on the x-axis) the probability over all actions (y-axis) corresponds to one column in the plot and is color-coded in red. For this particular value of β the agent cannot afford to pick the specific actions for most of the world-states (except for the last three world-states) in order to stay within the limit on the maximum allowed rate. Rather, the agent recommends the best-selling items of the corresponding category which allows for a lower rate by having identical policies (i.e., columns in the plot) for sets of world-states. The optimal policies thus lead to abstractions, where several different world-states elicit identical responses of the agent. Importantly, the abstractions are not induced because some stimuli are more similar than others under some utility-free measure and they are also not the result of a *post hoc* aggregation or clustering scheme. Rather, the abstractions are shaped by the utility function and appear as a consequence of bounded rational decision-making in the given task.

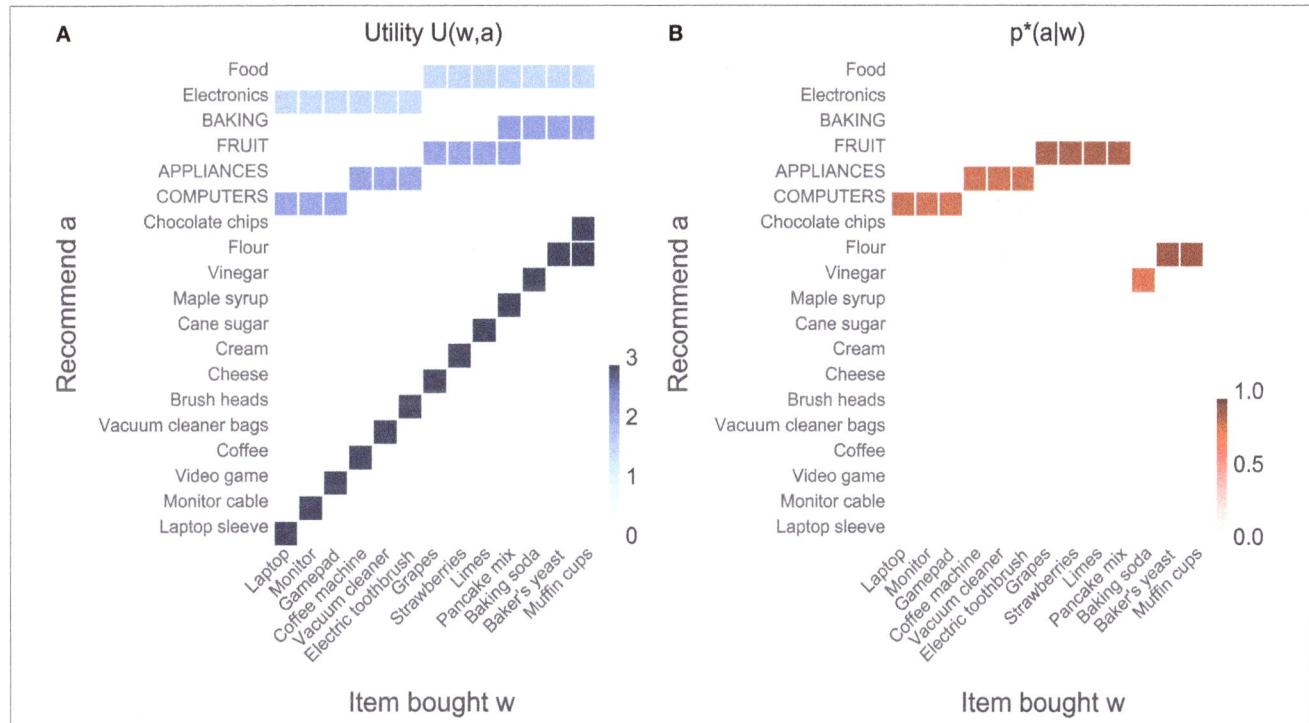

FIGURE 2 | Task setup and solution $p^*(a|w)$ for $\beta = 1.3$. (A) Utility function $U(w, a)$ for the recommender system task. The recommender system observes an item w bought by a customer and recommends another item a to buy to the customer. For each item bought there is another concrete item that has a high chance of being bought by the customer. Therefore, recommending the correct concrete item leads to the maximum utility of 3. However, each item also belongs to a category (indicated by capital letters) and recommending the best-selling item of the corresponding category leads to a utility of 2.2. Finally, each item also belongs to a super-category (either "food" or "electronics") and recommending the best-selling item of the corresponding super-category leads to a utility of 1.6. There is one item (muffin cups) where two concrete items can be recommended and both yield maximum utility. Additionally there is one item (pancake mix) where the recommendation of the best-selling item of both categories "fruit" and "baking" yields the same utility. **(B)** Solution $p^*(a|w)$ [equation (9)] for $\beta = 1.3$. Due to the low β, the computational resources of the recommender system are quite limited and it cannot recommend the highest scoring items (except in the last three columns). Instead, it saves computational effort by applying the same policy to multiple items.

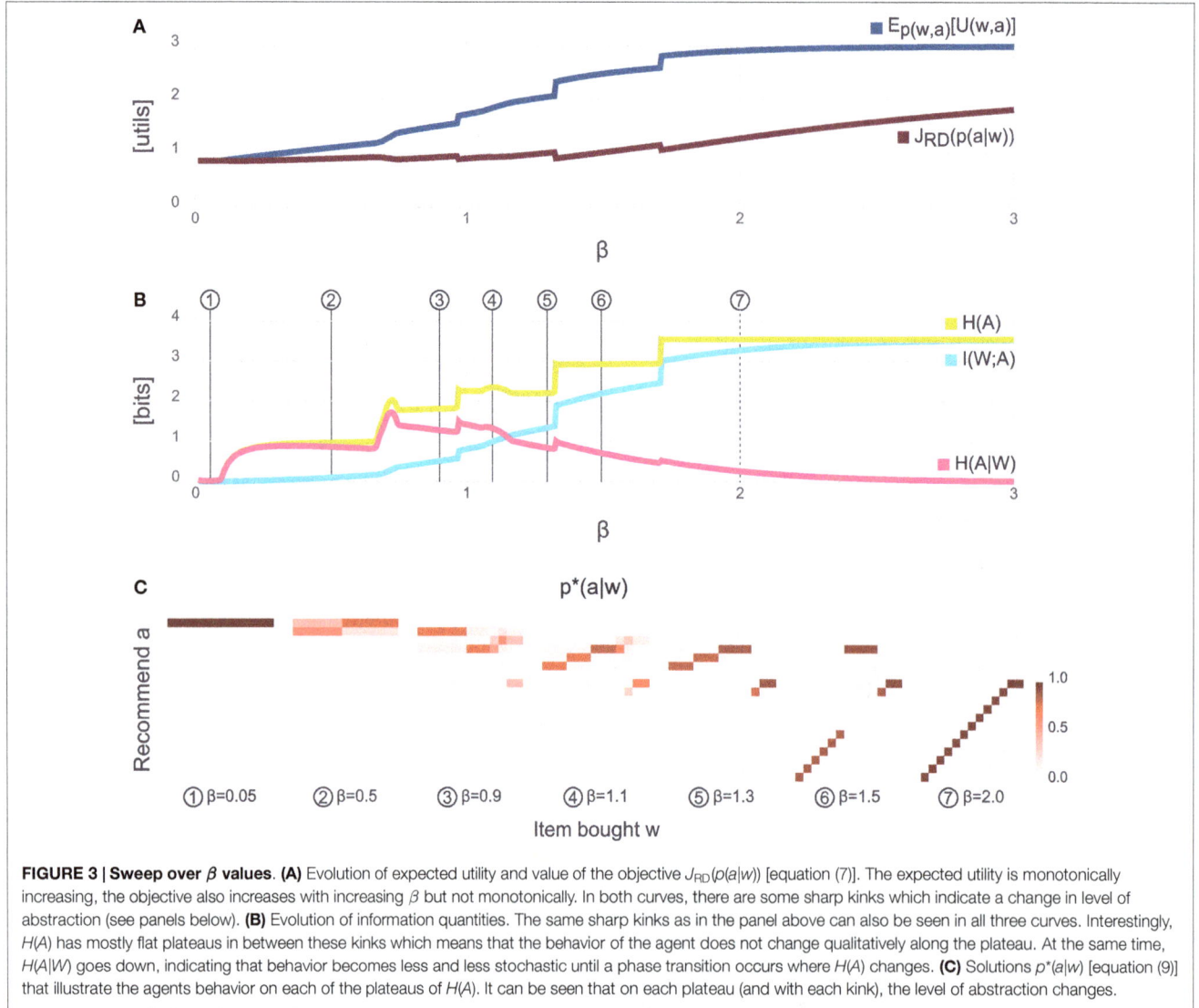

FIGURE 3 | Sweep over β values. (A) Evolution of expected utility and value of the objective $J_{RD}(p(a|w))$ [equation (7)]. The expected utility is monotonically increasing, the objective also increases with increasing β but not monotonically. In both curves, there are some sharp kinks which indicate a change in level of abstraction (see panels below). **(B)** Evolution of information quantities. The same sharp kinks as in the panel above can also be seen in all three curves. Interestingly, $H(A)$ has mostly flat plateaus in between these kinks which means that the behavior of the agent does not change qualitatively along the plateau. At the same time, $H(A|W)$ goes down, indicating that behavior becomes less and less stochastic until a phase transition occurs where $H(A)$ changes. **(C)** Solutions $p^*(a|w)$ [equation (9)] that illustrate the agents behavior on each of the plateaus of $H(A)$. It can be seen that on each plateau (and with each kink), the level of abstraction changes.

Figure 3A shows the expected utility $\mathbf{E}_{p(w,a)}[U(w,a)]$ and the rate-distortion objective $J_{RD}(p(a|w))$ as a function of the inverse temperature β. The plot shows that by increasing β the expected utility increases monotonically, whereas the objective $J_{RD}(p(a|w))$ also shows a trend to increase but not monotonically. Interestingly, there are a few sharp transitions at the same points in both curves. The same steep transitions are also found in **Figure 3B**, which shows the mutual information and its decomposition into the entropic terms $I(W, A) = H(A) - H(A|W)$ as a function of β. The line corresponding to the entropy over actions $H(A)$ shows flat plateaus in between these phase transitions. **Figure 3C** illustrates solutions $p^*(a|w)$ for β values corresponding to points on each of the plateaus (labels for bought and recommended items have been omitted for visual compactness but are identical to the plot in **Figure 2B**). Surprisingly, most of the solutions correspond to different levels of abstraction – from fully abstract for $\beta \to 0$, then going through several levels of abstraction and getting more and more specific up to the case $\beta \to \infty$ where the conditional entropy $H(A|W)$ goes to zero implying that the

conditionals $p^*(a|w)$ become deterministic and identical to the maximum expected utility solutions. Within a plateau of $H(A)$, the entropy over actions does not change but the conditional entropy $H(A|W)$ tends to decrease with increasing β. This means that qualitatively the behavior along a plateau does not change in the sense that across all world-states the same subset of actions is used. However, the stochasticity within this subset of actions decreases with increasing β (until at some point a phase-transition occurs). Changing the temperature leads to a natural emergence of different levels of abstraction – levels that emerge from the agent-environment interaction structure described by the utility function. Each level of abstraction corresponds to one plateau in $H(A)$.

In general, abstractions are formed by reducing the information content of an entity until it only contains relevant information. For a discrete random variable $w \in \mathcal{W}$, this translates into forming a partitioning over the space $\mathcal{W}$ where "similar" elements are grouped into the same subset of $\mathcal{W}$ and become indistinguishable within the subset. In physics, changing the granularity of a partitioning to a coarser level is known as *coarse-graining* which reduces the resolution of the space W in a non-uniform manner. Here, the partitioning emerges in $p^*(a|w)$ as a *soft-partitioning* (see Still and Crutchfield, 2007), where "similar" world-states w get mapped to an action a (or a subset of actions) and essentially become indistinguishable. Readers are encouraged to interactively explore the example in the Supplementary Jupyter Notebook "2-RateDistortionForDecisionMaking."

In analogy to rate-distortion theory where the rate-distortion function serves as an information-theoretic characterization of a system, one can define the *rate-utility function*

$$U(R) = \max_{p(a|w):I(W;A) \leq R} \mathbf{E}_{p(w,a)}[U(w,a)]. \tag{11}$$

where the expected utility is a function of the information processing rate $I(W; A)$. If the decision-maker is conceptualized as a communication channel between world-states and actions, the rate $I(W; A)$ defines the minimally required capacity of that channel. The rate-utility function thus specifies the minimum required capacity for computing actions given a certain expected utility target, or analogously the maximally achievable expected utility given a certain information processing capacity. The rate-utility curve is obtained by varying the inverse temperature β (corresponding to different values of R) and plotting the expected utility as a function of the rate. The resulting plot is shown in **Figure 4**, where the solid line denotes the rate-utility curve and the shaded region corresponds to systems that are theoretically infeasible and cannot be achieved regardless of the implementation. Systems in the white region are sub-optimal, meaning that they could either achieve the same performance with a lower rate or given their limits on computational capacity they could theoretically achieve higher performance. This curve is interesting for both designing systems as well as characterizing the degree of sub-optimality of given systems.

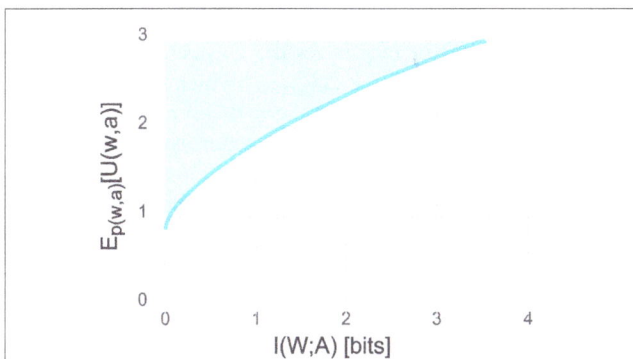

FIGURE 4 | Rate-utility curve. Analogously to the rate-distortion curve in rate-distortion theory, the rate-utility curve shows the minimally required information processing rate to achieve a certain level of expected utility or dually, the maximally achievable expected utility, given a certain rate. Systems that optimally trade-off expected utility against cost of computation lie exactly on the curve. Systems in the shaded region are theoretically impossible and cannot be realized. Systems that lie in the white region of the figure are sub-optimal in the sense that they could achieve a higher expected utility given their computational resources or they could achieve the same expected utility with lower resources.

3. SERIAL INFORMATION-PROCESSING HIERARCHIES

In this section, we apply the rate-distortion principle for decision-making to a serial perception-action system. We design two stages: a perceptual stage $p(x|w)$ that maps world-states w to observations x and an action stage $p(a|x)$ that maps observations x to actions a. Note that the world-state w does not necessarily have to be considered as a latent variable but could in general also be an observation from a previous processing stage. The action stage implements a bounded rational decision-maker (similar to the one presented in the previous section) that optimally trades off expected utility against cost of computation [see equation (7)]. Classically, the perceptual stage might be designed to represent w as faithfully as possible, given the computational limitations of the perceptual stage. Here, we show that trading off expected utility against the cost of information processing on *both* the perceptual and the action stage leads to bounded-optimal perception that does not necessarily represent w as faithfully as possible but rather extracts the most relevant information about w such that the action stage can work most efficiently. As a result, bounded-optimal perception will be tightly coupled to the action stage and will be shaped by the utility function as well as the computational capacity of the action channel.

3.1. Optimal Perception is Shaped by Action

To model a perceptual channel we extend the model from Section 2.2 as follows: The agent is no longer capable of fully observing the state of the world W but using its sensors it is capable to form a percept X as $p(x|w)$ which then allows for adaptation of behavior according to $p(a|x)$. The three random variables for world-state, percept, and action form a serial chain of channels, one channel from world-states to percepts expressed by $p(x|w)$ and another channel from percepts to actions expressed by $p(a|x)$ which implies the following conditional independence

$$p(w, x, a) = p(w)p(x|w)p(a|x),$$

that is also expressed by the graphical model $W \to X \to A$. We assume that $p(w)$ is given and the utility function depends on the world-state and the action $U(w, a)$. Note that mathematically, the results are identical for $U(w, x, a)$, but in this paper we consider the utility independent of the internal percept x.

Classically, inference and decision-making are separated – for instance, by first performing Bayesian inference over the state of the world w using the observation x and then choosing an action a according to the maximum expected utility principle. The MEU action-selection principle can be replaced by a bounded rational model for decision-making that takes into account the computational cost of transforming a (optimal) prior behavior $p_0(a)$ to a posterior behavior $p(a|x)$ as shown in Section 2.

$$\text{Bayesian inference:} \qquad p(w|x) = \frac{p(w)p(x|w)}{\sum_w p(w)p(x|w)} \tag{12}$$

$$\text{Bounded rational decision:} \quad p^*(a|x) = \frac{1}{Z(x)}p(a)e^{\beta_2 U(x,a)} \tag{13}$$

where $U(x, a) = \Sigma_w p(w|x) U(w, a)$ is the expectation of the utility under the Bayesian posterior over w given x. Note that the bounded rational decision-maker in equation (13) is identical to the rate-distortion decision-maker introduced in Section 2 that minimizes the trade-off given by equation (7) by implementing equation (9). It includes the MEU solution as a special case for $\beta_2 \to \infty$. Here, the inverse temperature is denoted by β_2 (instead of β as in the previous section) for notational reasons that ensure consistency with later results of this section.

In equation (12), the choice of the likelihood model $p(x|w)$ remains unspecified and the question is where does it come from? In general, it is chosen by the designer of a system and the choice is often driven by bandwidth or memory constraints. In purely descriptive scenarios, the likelihood model is determined by the sensory setup of a given system and $p(x|w)$ is obtained by fitting it to data of the real system. In the following, we present a particular choice of $p(x|w)$ that is fundamentally grounded on the principle that any transformation of behavior or beliefs is costly (which is identical to the assumption of limited-rate information processing channels) and this cost should be traded off against gains in expected utility. Remarkably, equations (12) and (13) drop out naturally from the principle.

Given the graphical model: $W \to X \to A$, we consider an information processing channel between W and X and another one between X and A and introduce different rate-limits on these channels, i.e., the information processing price on the perceptual level $\frac{1}{\beta_1}$ can be different from the price of information processing on the action level $\frac{1}{\beta_2}$. Formally, we set up the following variational problem:

$$
\begin{aligned}
\underset{p(x|w),p(a|x)}{\arg\max} \ & \mathbf{E}_{p(w,x,a)}[U(w,a)] - \frac{1}{\beta_1} I(W;X) - \frac{1}{\beta_2} I(X;A) \\
= \ & \underset{p(x|w),p(a|x)}{\arg\max} \ J_{\mathrm{ser}}(p(x|w), p(a|x)).
\end{aligned} \tag{14}
$$

Similar to the rate-distortion case, the solution is given by the following set of four self-consistent equations:

$$
p^*(x|w) = \frac{1}{Z(w)} p(x) exp\left(\beta_1 \Delta F_{\mathrm{ser}(w,x)}\right) \tag{15}
$$

$$
p(x) = \sum_w p(w) p^*(x|w) \tag{16}
$$

$$
p^*(a|x) = \frac{1}{Z(x)} p(a) exp\left(\beta_2 \sum_w p(w|x) U(w,a)\right) \tag{17}
$$

$$
p(a) = \sum_{w,x} p(w) p^*(x|w) p^*(a|x), \tag{18}
$$

where $Z(w)$ and $Z(x)$ denote the corresponding normalization constants or partition sums. The conditional probability $p(w|x)$ is given by Bayes' rule $p(w|x) = \frac{p(w)p^*(x|w)}{p(x)}$ and $\Delta F_{\mathrm{ser}}(w, x)$ is the free energy difference of the action stage:

$$
\Delta F_{\mathrm{ser}(w,x)} := \mathbf{E}_{p^*(a|x)}[U(w,a)] - \frac{1}{\beta_2} D_{\mathrm{KL}}(p^*(a|x)||p(a)), \tag{19}
$$

see also equation (5). More details on the derivation of the solution equations can be found in the Supplementary Methods Section 2.

The bounded-optimal perceptual model is given by equation (15). It follows the typical structure of a bounded rational solution consisting of a prior times the exponential of the utility multiplied by the inverse temperature. Compare equation (9) to see that the downstream free-energy trade-off $\Delta F_{\mathrm{ser}(w,x)}$ now plays the role of the utility function for the perceptual model. The distribution $p^*(x|w)$ thus optimizes the downstream free-energy difference in a bounded rational fashion, that is taking into account the computational resources of the perceptual channel. Therefore, the optimal percept becomes tightly coupled to the agent-environment interaction structure as described by the utility function or in other words: the optimal percept is shaped by the embodiment of the agent and, importantly, is not simply a maximally faithful representation of W through X given the limited rate of the perceptual channel. A second interesting observation is that the action stage given by equation (17) turns out to be a bounded rational decision-maker using the Bayesian posterior $p(w|x)$ for inferring the true world-state w given the observation x. This is identical to equation (13) (using the optimal prior $p(a) = \Sigma_{w,x} p(w) p^*(x|w) p^*(a|x)$), even though the latter was explicitly modeled by first performing Bayesian inference over the world-state w given the percept x [equation (12)] and then performing bounded rational decision-making [equation (13)], whereas the same principle drops out naturally in equation (17) as a result of optimizing equation (14).

3.2. Illustrative Example

In this section, we design a hand-crafted perceptual model $p_\lambda(x|w)$ with precision-parameter λ, that drives a subsequent bounded rational decision-maker that maps an observation x to a distribution over actions $p(a|x)$ in order to maximize expected utility while not exceeding a constraint on the rate of the action channel. The latter is implemented by following equation (13) and setting β_2 according to the limit on the rate $I(X; A)$. We compare the bounded rational actor with hand-crafted perception against a bounded-optimal actor that maximizes equation (14) by implementing the four corresponding self-consistent equations (15)–(18). Importantly, the perceptual model $p^*(x|w)$ of the bounded-optimal actor maximizes the downstream free-energy trade-off of the action stage $\Delta F_{\mathrm{ser}(w,x)}$ which leads to a tight coupling between perception and action that is not present in the hand-crafted model of perception. The action stage is identical in both models and given by equation (17).

We designed the following example where the actor is an animal in a predator-prey scenario. The actor has sensors to detect the size of other animals it encounters. In this simplified scenario, animals can only belong to one of three size-groups and their size correlates with their hearing-abilities:

- Small animals (insects): either 2, 3, or 4 size-units cannot hear very well.
- Medium-sized animals (rodents): either 6, 7, or 8 size-units can hear quite well.
- Large animals (cats of prey): either 10, 11, or 12 size-units can hear quite well.

The actor has a sensor for detecting the size of an animal, however, depending on the capacity of the perceptual channel this sensor will either be more or less noisy. To survive, the actor can

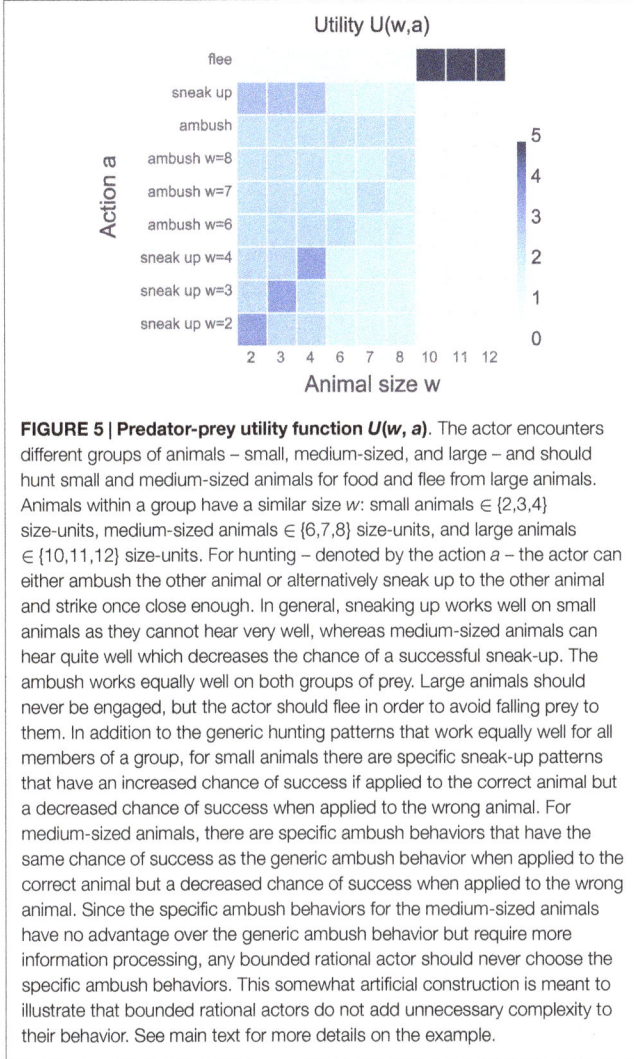

FIGURE 5 | Predator-prey utility function *U(w, a)*. The actor encounters different groups of animals – small, medium-sized, and large – and should hunt small and medium-sized animals for food and flee from large animals. Animals within a group have a similar size *w*: small animals ∈ {2,3,4} size-units, medium-sized animals ∈ {6,7,8} size-units, and large animals ∈ {10,11,12} size-units. For hunting – denoted by the action *a* – the actor can either ambush the other animal or alternatively sneak up to the other animal and strike once close enough. In general, sneaking up works well on small animals as they cannot hear very well, whereas medium-sized animals can hear quite well which decreases the chance of a successful sneak-up. The ambush works equally well on both groups of prey. Large animals should never be engaged, but the actor should flee in order to avoid falling prey to them. In addition to the generic hunting patterns that work equally well for all members of a group, for small animals there are specific sneak-up patterns that have an increased chance of success if applied to the correct animal but a decreased chance of success when applied to the wrong animal. For medium-sized animals, there are specific ambush behaviors that have the same chance of success as the generic ambush behavior when applied to the correct animal but a decreased chance of success when applied to the wrong animal. Since the specific ambush behaviors for the medium-sized animals have no advantage over the generic ambush behavior but require more information processing, any bounded rational actor should never choose the specific ambush behaviors. This somewhat artificial construction is meant to illustrate that bounded rational actors do not add unnecessary complexity to their behavior. See main text for more details on the example.

hunt animals from both the small and the medium-sized group for food. On the other hand, it can fall prey to animals of the large group. The actor has three basic actions:

- Ambush: steadily wait for the other animal to get close and then strike.
- Sneak-up: slowly move closer to the animal and then strike.
- Flee: quickly move away from the other animal.

The advantage of the ambush is that it is silent, however, the risk is that the animal might not move toward the position of the ambush – it works equally well on animals from the small and medium-sized group. The sneak-up is not silent but does not rely on the other animal coincidentally getting closer – it works better than the ambush for small-sized animals but the opposite is true for medium-sized animals. If the actor encounters a large animal the only sensible action is to flee in order to avoid falling prey to the large animal. Besides these generic actions, the actor also has a repertoire of more specific hunting patterns – see **Figure 5** which shows the full details of the utility function for the predator-prey scenario. The exact numeric values are found in the Supplementary Jupyter Notebook "3-SerialHierarchy."

The hand-crafted model of perception is specified by $p_\lambda(x|w)$, where the observed size x corresponds to the actual size of the animal w corrupted by noise. The precision-parameter λ governs the noise-level and thus the quality of the perceptual channel which can be measured with $I(W; X)$. In particular, the observation o is a discretized noisy version of w with precision λ:

$$x|w, \lambda \sim \text{round}(\mathcal{N}_{\text{trunc}}(w, 1/\lambda)), \qquad (20)$$

where the set of world-states is given by all possible animal sizes $w \in \mathcal{W} = \{2,3,4,6,7,8,10,11,12\}$ and the set of possible observations is given by $x \in \mathcal{X} = \{1,2,3,\ldots,11,12,13\}$. To avoid a boundary-bias due to the limited interval $\mathcal{X}$ we reject and re-sample all values of x that would fall outside of $\mathcal{X}$. For $\lambda \to \infty$, the perceptual channel is very precise, and there is no uncertainty about the true value of w after observing x. However, such a channel incurs a large computational effort as the mutual information $I(W; X)$ is maximal in this case. If the perceptual channel has a smaller capacity than required to uniquely map each w to an x, the rate must be reduced by lowering the precision λ. Medium precision will mostly lead to within-group confusion whereas low precision will also lead to across-group confusion and corresponds to perceptual channels with a very low rate $I(W; X)$.

The results in **Figure 6** show solutions when having large computational resources on both the perception and action channel. As the figure clearly shows, the hand-crafted model $p_\lambda(x|w)$ looks quite different from the bounded-optimal solution $p^*(x|w)$, even though the rate on the perceptual channel is identical in both cases (given by the mutual information $I(W; X) \approx 2$ bits). The difference is that the bounded-optimal percept spends the two bits mainly on discriminating between specific animals of the small group and on discriminating between medium-sized and large animals. It does not discriminate between specific sizes within the latter two groups. This makes sense, as there is no gain in utility by applying any specific actions to specific animals in the medium- or large-sized group. **Figure 6** also shows the overall-behavior from the point of view of an external observer $p(a|w)$, which is computed as follows

$$p_\lambda(a|w) = \sum_x p_\lambda(x|w)p_\lambda(a|x) \quad \text{and}$$

$$p^*(a|w) = \sum_x p^*(x|w)p^*(a|x) \qquad (21)$$

The overall-behavior in the bounded-optimal case is more deterministic, leading to a higher expected utility in the bounded-optimal case. The distributions $p_\lambda(a|x)$ and $p^*(a|w)$ are not shown in the figure but can easily be inspected in the Supplementary Jupyter Notebook "3-SerialHierarchy." If the price of information processing on the perceptual channel in the hand-crafted model is the same as in the bounded-optimal model (given by β_1), then the overall objective $J_{\text{ser}}(p(x|w), p(a|x))$ is larger for the bounded-optimal case compared to the hand-crafted case, implying that the bounded optimal actor achieves a better trade-off between expected utility and computational cost. The crucial insight of this example is that the optimal percept depends on the utility function, where in this particular case it does for instance make no sense to waste computational resources on discriminating

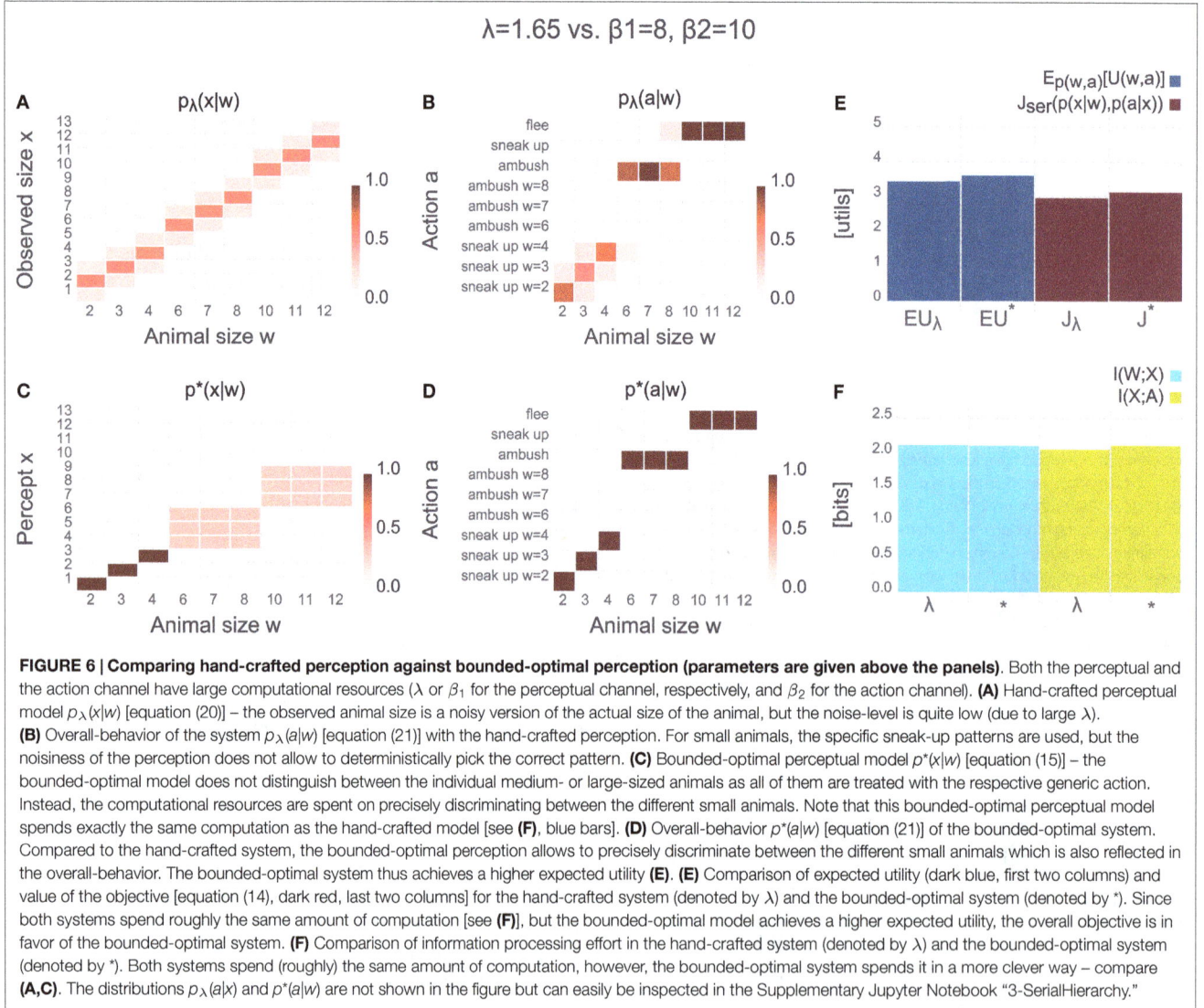

FIGURE 6 | Comparing hand-crafted perception against bounded-optimal perception (parameters are given above the panels). Both the perceptual and the action channel have large computational resources (λ or β_1 for the perceptual channel, respectively, and β_2 for the action channel). **(A)** Hand-crafted perceptual model $p_\lambda(x|w)$ [equation (20)] – the observed animal size is a noisy version of the actual size of the animal, but the noise-level is quite low (due to large λ). **(B)** Overall-behavior of the system $p_\lambda(a|w)$ [equation (21)] with the hand-crafted perception. For small animals, the specific sneak-up patterns are used, but the noisiness of the perception does not allow to deterministically pick the correct pattern. **(C)** Bounded-optimal perceptual model $p^\star(x|w)$ [equation (15)] – the bounded-optimal model does not distinguish between the individual medium- or large-sized animals as all of them are treated with the respective generic action. Instead, the computational resources are spent on precisely discriminating between the different small animals. Note that this bounded-optimal perceptual model spends exactly the same computation as the hand-crafted model [see **(F)**, blue bars]. **(D)** Overall-behavior $p^\star(a|w)$ [equation (21)] of the bounded-optimal system. Compared to the hand-crafted system, the bounded-optimal perception allows to precisely discriminate between the different small animals which is also reflected in the overall-behavior. The bounded-optimal system thus achieves a higher expected utility **(E)**. **(E)** Comparison of expected utility (dark blue, first two columns) and value of the objective [equation (14), dark red, last two columns] for the hand-crafted system (denoted by λ) and the bounded-optimal system (denoted by $\star$). Since both systems spend roughly the same amount of computation [see **(F)**], but the bounded-optimal model achieves a higher expected utility, the overall objective is in favor of the bounded-optimal system. **(F)** Comparison of information processing effort in the hand-crafted system (denoted by λ) and the bounded-optimal system (denoted by $\star$). Both systems spend (roughly) the same amount of computation, however, the bounded-optimal system spends it in a more clever way – compare **(A,C)**. The distributions $p_\lambda(a|x)$ and $p^\star(a|w)$ are not shown in the figure but can easily be inspected in the Supplementary Jupyter Notebook "3-SerialHierarchy."

between the specific animals of the large group because the optimal response (flee with certainty) is identical to all of them. In the Supplementary Jupyter Notebook "3-SerialHierarchy" the utility function can easily be switched while keeping all other parameters identical in order to observe how the bounded-optimal percept changes accordingly. Note that the bounded-optimal behavior $p^\star(a|w)$ shown in **Figure 6D** yields the highest possible expected utility in this task setup – there is no behavior that would lead to a higher expected utility (though there are other solutions that lead to the same expected utility).

The bounded-optimal percept depends not only on the utility function but also on the behavioral richness of the actor which is governed by the rate on the action channel $I(X; A)$. In **Figure 7** we show the results of the same setup as in **Figure 6** with the only change being the significantly increased price for information processing in the action stage (as specified by $\beta_2 = 1$ bit per util whereas it used to be $\beta_2 = 10$ bits per util in the previous figure). The hand-crafted perceptual model is unaffected by this change of the action stage, but the bounded-optimal model of perception has changed compared to the previous figure and now reflects

the limited behavioral richness. As shown in $p^\star(a|w)$ in **Figure 7**, the actor is no longer capable of applying different actions to animals of the small group and animals of the medium-sized group. Accordingly, the bounded-optimal percept does not waste computational resources for discriminating between small and medium-sized animals since the downstream policy is identical for both groups of animals. In terms of expected utility, both the hand-crafted model as well as the bounded-optimal decision-maker score equally at ≈ 3 utils. However, the bounded-optimal model does so by using lower computational resources and thus scoring better on the overall trade-off $J_{\text{ser}}(p(x|w), p(a|x))$.

In **Figure 8** we again use large resources on the action channel $\beta_2 = 10$ (as in the first example in **Figure 6**), but now the resources on the perceptual channel are limited by setting $\beta_1 = 1$ (compared to $\beta_1 = 8$ in the first case). Accordingly, the precision of the hand-crafted perceptual model is tuned to $\lambda = 0.4$ (compared to $\lambda = 1.65$ in the first case) such that it has the same rate $I(W; X)$ as the bounded-optimal model. By comparing the two panels for $p_\lambda(x|w)$ and $p^\star(x|w)$, it can clearly be seen that the bounded-optimal perceptual model now spends its scarce

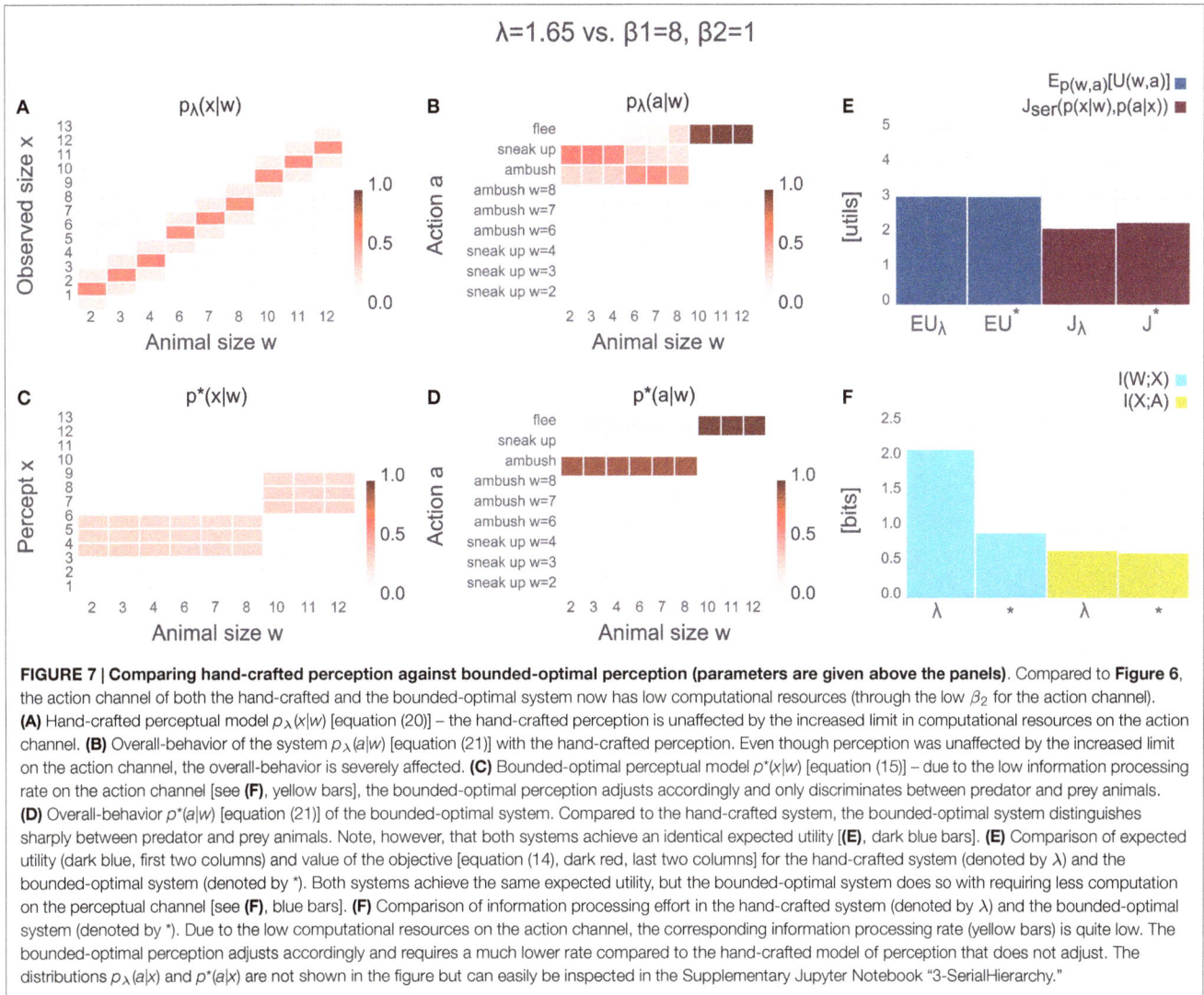

FIGURE 7 | Comparing hand-crafted perception against bounded-optimal perception (parameters are given above the panels). Compared to **Figure 6**, the action channel of both the hand-crafted and the bounded-optimal system now has low computational resources (through the low β_2 for the action channel). **(A)** Hand-crafted perceptual model $p_\lambda(x|w)$ [equation (20)] – the hand-crafted perception is unaffected by the increased limit in computational resources on the action channel. **(B)** Overall-behavior of the system $p_\lambda(a|w)$ [equation (21)] with the hand-crafted perception. Even though perception was unaffected by the increased limit on the action channel, the overall-behavior is severely affected. **(C)** Bounded-optimal perceptual model $p^*(x|w)$ [equation (15)] – due to the low information processing rate on the action channel [see **(F)**, yellow bars], the bounded-optimal perception adjusts accordingly and only discriminates between predator and prey animals. **(D)** Overall-behavior $p^*(a|w)$ [equation (21)] of the bounded-optimal system. Compared to the hand-crafted system, the bounded-optimal system distinguishes sharply between predator and prey animals. Note, however, that both systems achieve an identical expected utility [**(E)**, dark blue bars]. **(E)** Comparison of expected utility (dark blue, first two columns) and value of the objective [equation (14), dark red, last two columns] for the hand-crafted system (denoted by λ) and the bounded-optimal system (denoted by *). Both systems achieve the same expected utility, but the bounded-optimal system does so with requiring less computation on the perceptual channel [see **(F)**, blue bars]. **(F)** Comparison of information processing effort in the hand-crafted system (denoted by λ) and the bounded-optimal system (denoted by *). Due to the low computational resources on the action channel, the corresponding information processing rate (yellow bars) is quite low. The bounded-optimal perception adjusts accordingly and requires a much lower rate compared to the hand-crafted model of perception that does not adjust. The distributions $p_\lambda(a|x)$ and $p^*(a|x)$ are not shown in the figure but can easily be inspected in the Supplementary Jupyter Notebook "3-SerialHierarchy."

resources to reliably discriminate between large animals and all other animals. The overall behavioral policies $p(a|w)$ reflect the limited perceptual capacity in both cases, however, the bounded-optimal case scores a higher expected utility of ≈ 3 utils compared to the hand-crafted case. The overall objective $J_{ser}(p(x|w), p(a|x))$ is also higher for the bounded-optimal model, indicating that this model should be preferred because it finds a better trade-off between expected utility and information processing cost.

Note that in all three examples the optimal percept $p^*(x|w)$ often leads to a uniform mapping of an exclusive subset of world-states w to the same set of percepts x. Importantly, these percepts do not directly correspond to an observed animal size as in the case of the hand-crafted model of perception. Rather, the optimal percepts often encode more abstract concepts such as medium- or large-sized animal (as in **Figure 6**) or predator and prey animal (as in **Figures 7** and **8**). In a sense, abstractions similar to the ones shown in the recommender system example in the previous section (**Figure 3**) emerge in the predator-prey example as well but now they also manifest themselves in the form of abstract percepts. Crucially, the abstract percepts allow

for more efficient information processing further downstream in the decision-making part of the system. The formation of these abstract percepts is driven by the embodiment of the agent and reflects certain aspects of the utility function of the agent. For instance, unlike the actor in **Figure 6**, the actors in **Figures 7** and **8** would not "understand" the concept of medium-sized animals as it is of no use to them: with their very limited resources it is most important for them to have the two perceptual concepts of predator and prey. Note that the cardinality of X in the bounded-optimal model of perception is fixed in all examples in order to allow for easy comparison against the hand-crafted model, but it could be reduced further without any consequences (up to a certain point) – this can be explored in the Supplementary Jupyter Notebook "3-SerialHierarchy."

The solutions shown in this section were obtained by iterating the self-consistent equations until numerical convergence. Since there is no convergence-proof, it cannot be fully ruled out that the solutions are sub-optimal with respect to the objective. However, the point of the simulation results shown here is to allow for easier interpretation of the theoretical results and highlight

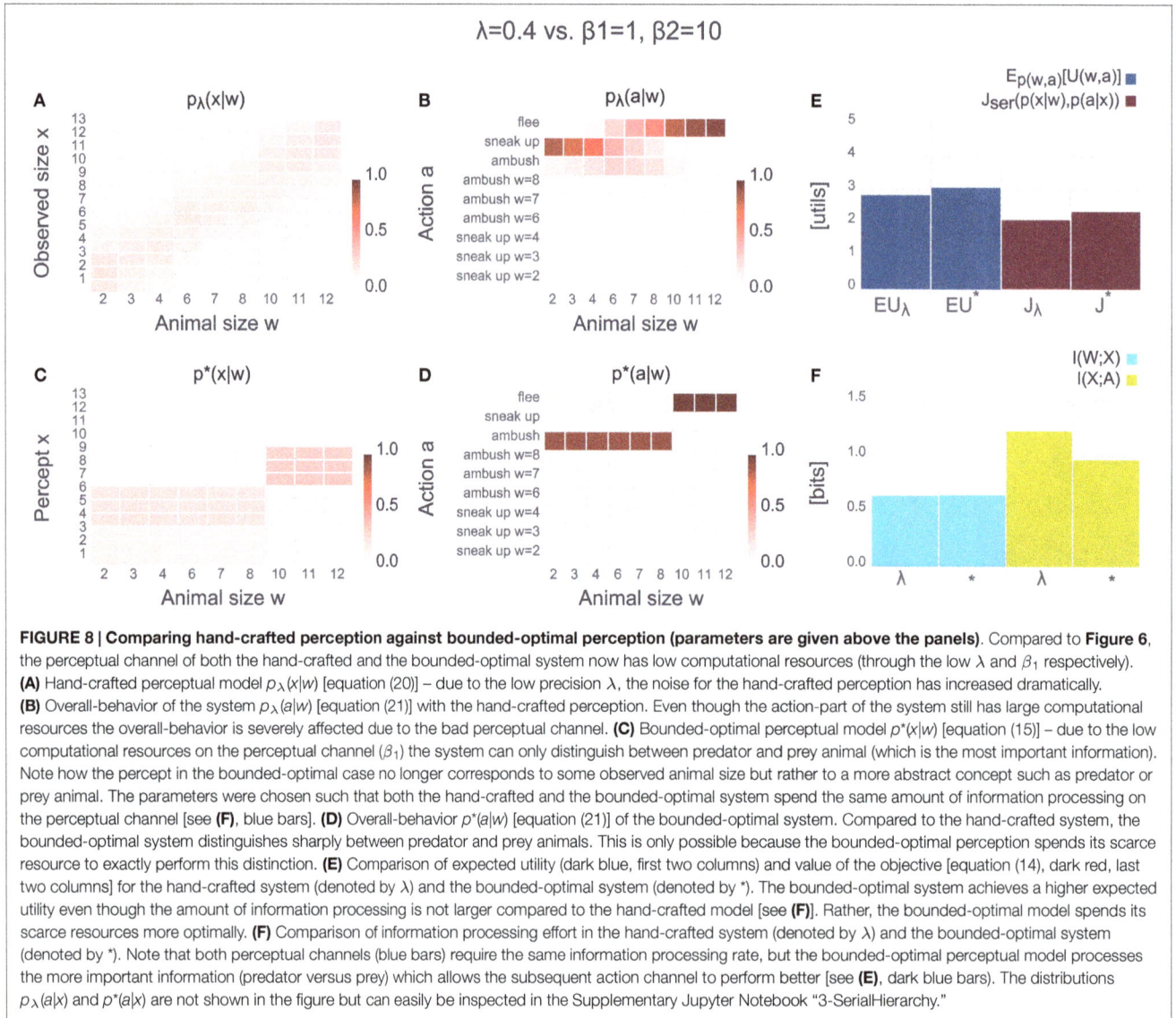

FIGURE 8 | Comparing hand-crafted perception against bounded-optimal perception (parameters are given above the panels). Compared to **Figure 6**, the perceptual channel of both the hand-crafted and the bounded-optimal system now has low computational resources (through the low λ and β_1 respectively). **(A)** Hand-crafted perceptual model $p_\lambda(x|w)$ [equation (20)] – due to the low precision λ, the noise for the hand-crafted perception has increased dramatically. **(B)** Overall-behavior of the system $p_\lambda(a|w)$ [equation (21)] with the hand-crafted perception. Even though the action-part of the system still has large computational resources the overall-behavior is severely affected due to the bad perceptual channel. **(C)** Bounded-optimal perceptual model $p^*(x|w)$ [equation (15)] – due to the low computational resources on the perceptual channel (β_1) the system can only distinguish between predator and prey animal (which is the most important information). Note how the percept in the bounded-optimal case no longer corresponds to some observed animal size but rather to a more abstract concept such as predator or prey animal. The parameters were chosen such that both the hand-crafted and the bounded-optimal system spend the same amount of information processing on the perceptual channel [see **(F)**, blue bars]. **(D)** Overall-behavior $p^*(a|w)$ [equation (21)] of the bounded-optimal system. Compared to the hand-crafted system, the bounded-optimal system distinguishes sharply between predator and prey animals. This is only possible because the bounded-optimal perception spends its scarce resource to exactly perform this distinction. **(E)** Comparison of expected utility (dark blue, first two columns) and value of the objective [equation (14), dark red, last two columns] for the hand-crafted system (denoted by λ) and the bounded-optimal system (denoted by *). The bounded-optimal system achieves a higher expected utility even though the amount of information processing is not larger compared to the hand-crafted model [see **(F)**]. Rather, the bounded-optimal model spends its scarce resources more optimally. **(F)** Comparison of information processing effort in the hand-crafted system (denoted by λ) and the bounded-optimal system (denoted by *). Note that both perceptual channels (blue bars) require the same information processing rate, but the bounded-optimal perceptual model processes the more important information (predator versus prey) which allows the subsequent action channel to perform better [see **(E)**, dark blue bars]. The distributions $p_\lambda(a|x)$ and $p^*(a|x)$ are not shown in the figure but can easily be inspected in the Supplementary Jupyter Notebook "3-SerialHierarchy."

certain aspects of the theoretical findings. We discuss this issue in Section 5.2.

4. PARALLEL INFORMATION-PROCESSING HIERARCHIES

Rational decision-making requires searching through a set of alternatives a and picking the option with the highest expected utility. Bounded rational decision-making replaces the "hard maximum" operation with a soft selection mechanism where the first action that satisfies a certain level of expected utility is picked. A parallel hierarchical architecture allows for a prior partitioning of the search space which reduces the effective size of the search space and thus speeds up the search process. For instance, consider a medical system that consists of general practitioners and specialist doctors. The general doctor can restrict the search space for a particular ailment of a patient by determining which specialist the patient should see. The specialist doctor in turn can determine the exact disease. This leads to a two-level decision-making hierarchy

consisting of a high-level partitioning that allows for making a subsequent low-level decision with reduced (search) effort. In statistics, the partitioning that is induced by the high-level decision is often referred to as a *model* and is commonly expressed as a probability distribution over the search space $p(a|m)$ (where m indicates the model) which also allows for a soft-partitioning. The advantage of hierarchical architectures is that the computation that leads to the high-level reduction of the search space can be stored in the model (or in a set of parameters in case of a parametric model). This computation can later be re-used by using the correct model (or set of parameters) in order to perform the low-level computation more efficiently. Interestingly, it should be most economic to put the most re-usable, and thus more abstract, information into the models $p(a|m)$ which leads to a hierarchy of abstractions. However, in order to make sure that the correct model is used, another deliberation process $p(m|w)$ is required (where w indicates the observed stimulus or data). Another problem is how to chose the partitioning to be most effective. In this section, we address both problems from a bounded rational point of view. We show that the bounded optimal solution $p^*(a|m)$ trades off the computational cost for choosing a model m against the reduction in computational cost for the low-level decision.

To keep the notation consistent across all sections of the paper we denote the model m in the rest of the paper with the variable x. This is in contrast to Section 3, where x played the role of a percept. The advantage of this notation is that it allows to easily see similarities and differences of the information terms and solution equations of the different cases. In particular, in Section 5 we present a unifying case that includes the serial and parallel case as special cases – by keeping the notation consistent this can easily be seen.

4.1. Optimal Partitioning of the Search Space

Constructing a two-level decision-making hierarchy requires the following three components: high-level models $p(a|x)$, a model selection mechanism $p(x|w)$ and a low-level decision maker $p(a|w, x)$ (w denotes the observed world-state, x indicates a particular model and a is an action). The first two distributions are free to be chosen by the designer of the system, for $p(a|w, x)$ a maximum expected utility decision-maker is the optimal choice if computational costs are neglected. Here, we take computational cost into account and replace the MEU decision-maker with a bounded rational decision-maker that includes MEU as a special case ($\beta_3 \to \infty$) – the bounded rational decision-maker optimizes equation (7) by implementing equation (9). In the following we show how all parts of the hierarchical architecture:

1. Selection of model (or expert): $\quad\quad\quad p(x|w)$ (22)

2. Prior knowledge of model (or expert): $\quad\quad p(a|x)$ (23)

3. Bounded rational decision of model (or expert):

$$p^*(a|w,x) = \frac{1}{Z(w,x)}p(a|x)e^{\beta_3 U(w,a)} \quad (24)$$

emerge from optimally trading off computational cost against gains in utility. Importantly, $p(a|x)$ plays the role of a prior distribution for the bounded rational decision-maker and reflects the high-level partitioning of the search space.

The optimization principle that leads to the bounded-optimal hierarchy trades off expected utility against the computational cost of model selection $I(W; X)$ and the cost of the low-level decision using the model as a prior $I(W; A|X)$:

$$\underset{p(x|w),p(a|w,x)}{\arg\max} \mathbf{E}_{p(w,x,a)}[U(w,a)] - \frac{1}{\beta_1} I(W;X) - \frac{1}{\beta_3} I(W;A|X)$$
$$= \underset{p(x|w),p(a|w,x)}{\arg\max} J_{\mathrm{par}}(p(x|w),p(a|w,x)). \quad (25)$$

The set of self-consistent solutions is given by

$$p^*(x|w) = \frac{1}{Z(w)}p(x)exp\left(\beta_1 \Delta F_{\mathrm{par}(w,x)}\right) \quad (26)$$

$$p(x) = \sum_w p(w)p^*(x|w) \quad (27)$$

$$p^*(a|w,x) = \frac{1}{Z(w,x)}p^*(a|x)exp\left(\beta_3 U(w,a)\right) \quad (28)$$

$$p^*(a|x) = \sum_w p(w|x)p^*(a|w,x), \quad (29)$$

where $Z(w)$ and $Z(w, x)$ denote the corresponding normalization constants or partition sums. $p(w|x)$ is given by Bayes' rule $p(w|x) = \frac{p(w)p^*(x|w)}{p(x)}$ and $\Delta F_{\mathrm{par}(w,x)}$ is the free energy difference of the low-level stage:

$$\Delta F_{\mathrm{par}}(w,x) := \mathbf{E}_{p^*(a|w,x)}[U(w,a)] - \frac{1}{\beta_3}D_{\mathrm{KL}}(p^*(a|w,x)||p^*(a|x)),$$
(30)

see equation (5). More details on the derivation of the solution equations can be found in the Supplementary Methods Section 3. By comparing the solution equations (26)–(29) with equations (22)–(24) the hierarchical structure of the bounded-optimal solution can be seen clearly. The bounded-optimal model selector in equation (26) maximizes the downstream free-energy trade-off $\Delta F_{\mathrm{par}(w,x)}$ in a bounded rational fashion and is similar to the optimal perceptual model of the serial case [equation (15)]. This means that the optimal model selection mechanism is shaped by the utility function as well as the computational process on the low-level stage of the hierarchy (governed by β_3) but also by the computational cost of model selection (governed by β_1). The optimal low-level decision-maker given by equation (28) turns out to be exactly a bounded rational decision-maker with $p(a|x)$ as a prior – identical to the low-level decision-maker that was motivated in equation (24). Importantly, the bounded-optimal solution provides a principled way of designing the models $p(a|x)$ [see equation (29)]. According to the equation, the optimal model $p^*(a|x)$ is given by a Bayesian mixture over optimal solutions $p^*(a|w, x)$ where w is known. The Bayesian mixture turns out to be the optimal compressor of actions for unknown w under the belief $p(w|x)$.

4.2. Illustrative Example

To illustrate the formation of bounded-optimal models, we designed the following example: in a simplified environment, only three diseases can occur – a heart disease or one of two possible lung diseases. Each of the diseases comes in two possible types (e.g., type 1 or type 2 diabetes). Depending on how much information is available on the symptoms of a patient, diseases can be treated according to the specific type (which is most effective) or with respect to the disease category (which is less effective but requires less information). See **Figure 9** for a plot of the utility function and a detailed description of the example. The goal is to design a medical analysis hierarchy that initiates the best possible treatment, given its limitations. The hierarchy consists of an automated medical system that can cheaply take standard measurements to partially assess a patient's disease category. Additionally, the patient is then sent to a specialist who can manually perform more elaborate measurements if necessary to further narrow down the patient's precise disease type and recommend a treatment. The automated system should be designed in a way that minimizes the additional measurements required by the specialists. More formally, the automated system delivers a first diagnosis x given the patient's precise disease type w according to $p(x|w)$. The first diagnosis narrows down the possible treatments a according to a model $p(a|x)$. For each x, a specialist can further reduce uncertainty about the correct treatment by performing more measurements $p(a|w, x)$. We compare the optimal design of the automated system and the corresponding optimal

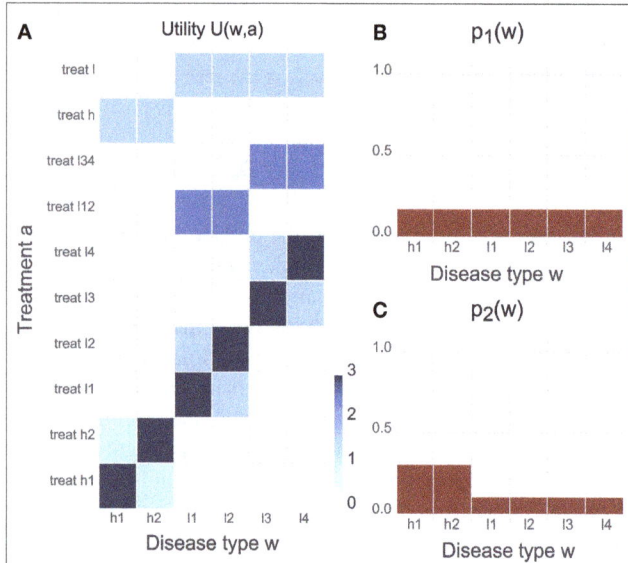

FIGURE 9 | Task setup of the medical system example. (A) Utility function. The disease category heart disease comes in the two types $h1$ or $h2$. There are two different categories of lung diseases that come in types $l1$ or $l2$ (lung disease A) or types $l3$ and $l4$ (lung disease B). The goal of the medical system is to gather information about the disease type and then initiate a treatment. Each of the disease types is best treated with a specific treatment. However, when the specific treatment is applied to the wrong disease type of the same disease category it is less effective. Additionally there are general treatments for the heart- and the two lung diseases that are slightly worse than the specific treatments but can be applied with less knowledge about the disease type. For lung diseases, there is a general treatment that works even if the specific lung disease is not known; however, it is less effective compared to the treatments where the disease but not the precise type is known. **(B)** Distribution of disease types in environment 1 – all diseases appear with equal probability. **(C)** Distribution of disease types in environment 2 where the heart diseases have an increased probability and the two corresponding disease types appear with higher chance.

treatment recommendations to the specialist $p^\star(a|x)$ according to equation (29) in two different environments: one, where all disease types occur with equal probability (**Figure 9B**) versus two, where heart diseases occur with increased chance (**Figure 9C**). For this example, the number of different high-level diagnoses X is set to $|\mathcal{X}| = 3$ which also means that there can be three different treatment recommendations $p(a|x)$. Since in the example the total budget for performing measurements is quite low (reflected by β_1, β_3 both being quite low), the whole system (automated plus specialists) can in general not gather enough information about the symptoms to treat every disease type with the correct specific treatment. Rather, the low budget has to be spent on gathering the most important information.

Figure 10 shows bounded-optimal hierarchies for the medical system in both environments. The top row in **Figure 10** shows the optimal hierarchy for the environment where all diseases appear with equal probability: the automated system $p^\star(x|w)$ (see **Figure 10A**) distinguishes between a heart disease, lung disease A and lung disease B, which means that there is one treatment recommendation for heart diseases and one treatment recommendation for each of the two possible lung diseases respectively (see the three columns of $p^\star(a|x)$ in **Figure 10B**). Since the general

treatment for the heart disease works less effective than the general treatments for the two lung diseases, the (very limited) budget of the specialists is completely spent on finding the correct specific heart treatment. Both lung diseases are treated with their respective general treatments since the two lung specialists have no budget for additional measurements. Since the automated system already distinguishes between the two lung diseases, it can narrow down the possible treatments to a delta over the correct general treatment, thus requiring no additional measurements by the lung specialists (shown by the two columns in $p^\star(a|x)$ that have a delta over the treatment).

The bottom row in **Figure 10** shows the optimal hierarchy for the environment where heart diseases appear with higher probability. In this case it is optimal to redesign the automated system to distinguish between the two types of the heart disease $h1$, $h2$, and lung diseases in general (see $p^\star(x|w)$ in **Figure 10D** of the figure). This means that there are now treatment recommendations $p^\star(a|x)$ for $h1$ and $h2$ that do not require any more measurements by the specialists (shown by the delta over a treatment in the first two columns of $p^\star(a|x)$ in **Figure 10E**) and there is another treatment recommendation for lung diseases. The corresponding specialist can use the limited budget to perform additional measurements to distinguish between the two categories of lung disease (but not between the four possible types as this would require more measurements than the budget allows). The example illustrates how the bounded-optimal decision-making hierarchy is shaped by the environment and emerges from optimizing the trade-off between expected utility and overall information processing cost. Readers can interactively explore the example in the Supplementary Jupyter Notebook "4-ParallelHierarchy" – in particular by changing the information processing costs of the specialists β_3 or changing the number of specialists by increasing or decreasing the cardinality of X.

The solutions shown in this section were obtained by iterating the self-consistent equations until numerical convergence. Since there is no convergence-proof, it cannot be fully ruled out that the solutions are sub-optimal with respect to the objective. However, the point of the simulation results shown here is to allow for easier interpretation of the theoretical results and highlight certain aspects of the theoretical findings. We discuss this issue in Section 5.2.

4.3. Comparing Parallel and Serial Information Processing

In order to achieve a certain expected utility, a certain overall rate $I(W; A)$ is needed. In the one-step rate-distortion case (Section 2) the channel from w to a must have a capacity larger or equal to that rate. In the serial case (Section 3) there is a channel from w to x and another channel from x to a. Both serial channels must at least have a capacity of $I(W; A)$ in order to achieve the same overall rate, as the following inequality always holds for the serial case

$$I(W; A) \leq \min \{I(W; X), I(X; A)\}.$$

In contrast, the parallel architecture allows for computing a certain overall rate $I(W; A)$ using channels with a lower capacity

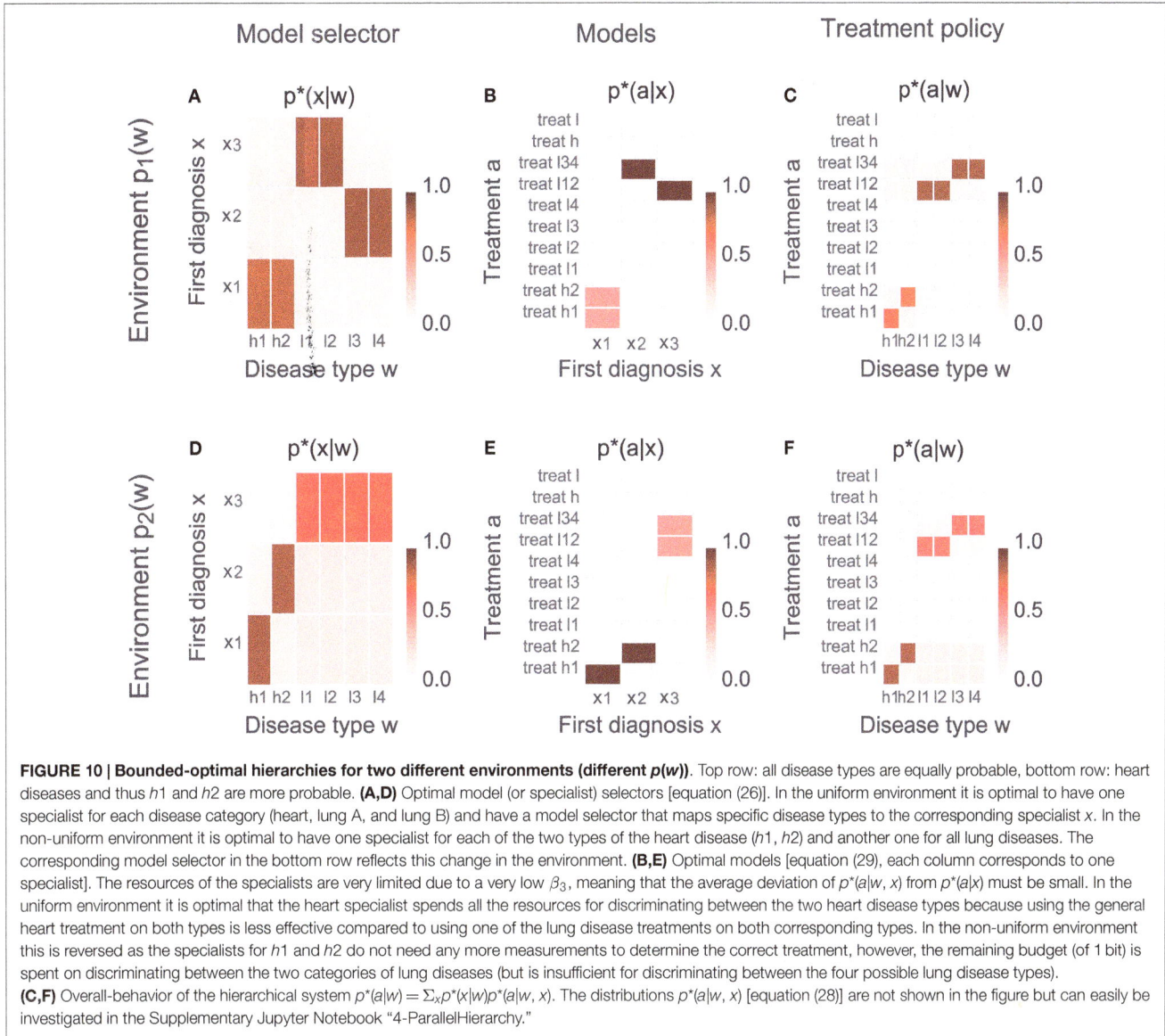

FIGURE 10 | Bounded-optimal hierarchies for two different environments (different *p(w)*). Top row: all disease types are equally probable, bottom row: heart diseases and thus *h*1 and *h*2 are more probable. **(A,D)** Optimal model (or specialist) selectors [equation (26)]. In the uniform environment it is optimal to have one specialist for each disease category (heart, lung A, and lung B) and have a model selector that maps specific disease types to the corresponding specialist *x*. In the non-uniform environment it is optimal to have one specialist for each of the two types of the heart disease (*h*1, *h*2) and another one for all lung diseases. The corresponding model selector in the bottom row reflects this change in the environment. **(B,E)** Optimal models [equation (29), each column corresponds to one specialist]. The resources of the specialists are very limited due to a very low β_3, meaning that the average deviation of $p^*(a|w, x)$ from $p^*(a|x)$ must be small. In the uniform environment it is optimal that the heart specialist spends all the resources for discriminating between the two heart disease types because using the general heart treatment on both types is less effective compared to using one of the lung disease treatments on both corresponding types. In the non-uniform environment this is reversed as the specialists for *h*1 and *h*2 do not need any more measurements to determine the correct treatment, however, the remaining budget (of 1 bit) is spent on discriminating between the two categories of lung diseases (but is insufficient for discriminating between the four possible lung disease types). **(C,F)** Overall-behavior of the hierarchical system $p^*(a|w) = \Sigma_x p^*(x|w) p^*(a|w, x)$. The distributions $p^*(a|w, x)$ [equation (28)] are not shown in the figure but can easily be investigated in the Supplementary Jupyter Notebook "4-ParallelHierarchy."

because the contribution in reducing uncertainty about *a* on each level of the hierarchy splits up as follows:

$$\underbrace{I(W, X; A)}_{\text{total reduction}} = \underbrace{I(X; A)}_{\text{high−level}} + \underbrace{I(W; A|X)}_{\text{low−level}},$$

which implies $I(W, X; A) \geq I(X; A)$. In particular, if the low-level step contributes information then $I(W; A|X) > 0$ and the previous inequality becomes strict. The same argument also holds when considering $I(W; A)$ (see Section 5.1).

In many scenarios the maximum capacity of a single processing element is limited and it is desirable to spread the total processing load on several elements that require a lower capacity. For instance, there could be technical reasons why processing elements with 5 bits of capacity can easily be manufactured but processing elements with a capacity of 10 bits cannot be manufactured or are disproportionally more costly to produce. In the one-step case and the serial case the only way to stay below a

certain capacity limit is by tuning β until the required rate is below the capacity – however, in both cases this also decreases the overall rate $I(W; A)$. In the parallel hierarchical case several building blocks with a limited capacity can be used to produce an overall rate $I(W; A)$ larger than the capacity of each processing block.

Splitting of information processing load onto several processing blocks is illustrated in **Figure 11**, where the one-step and parallel hierarchical solutions to the medical example are compared. In this example, the price of information processing in the one-step case is quite low ($\beta = 10$ bits per util) such that the corresponding solution leads to a deterministic mapping of each *w* to the best *a* (see **Figure 11A**). Doing so requires $I(W; A) \approx 2.6$ bits (see **Figure 11B**). Now assume for the sake of this example that processing elements, where information processing cost is reduced (15 bits per util), could be manufactured, but the maximum capacity of these elements is 1.58 bits. In the one-step case these processing elements can only be used if it is acceptable to reduce the rate $I(W; A)$ to 1.58 bits (by tuning β) which

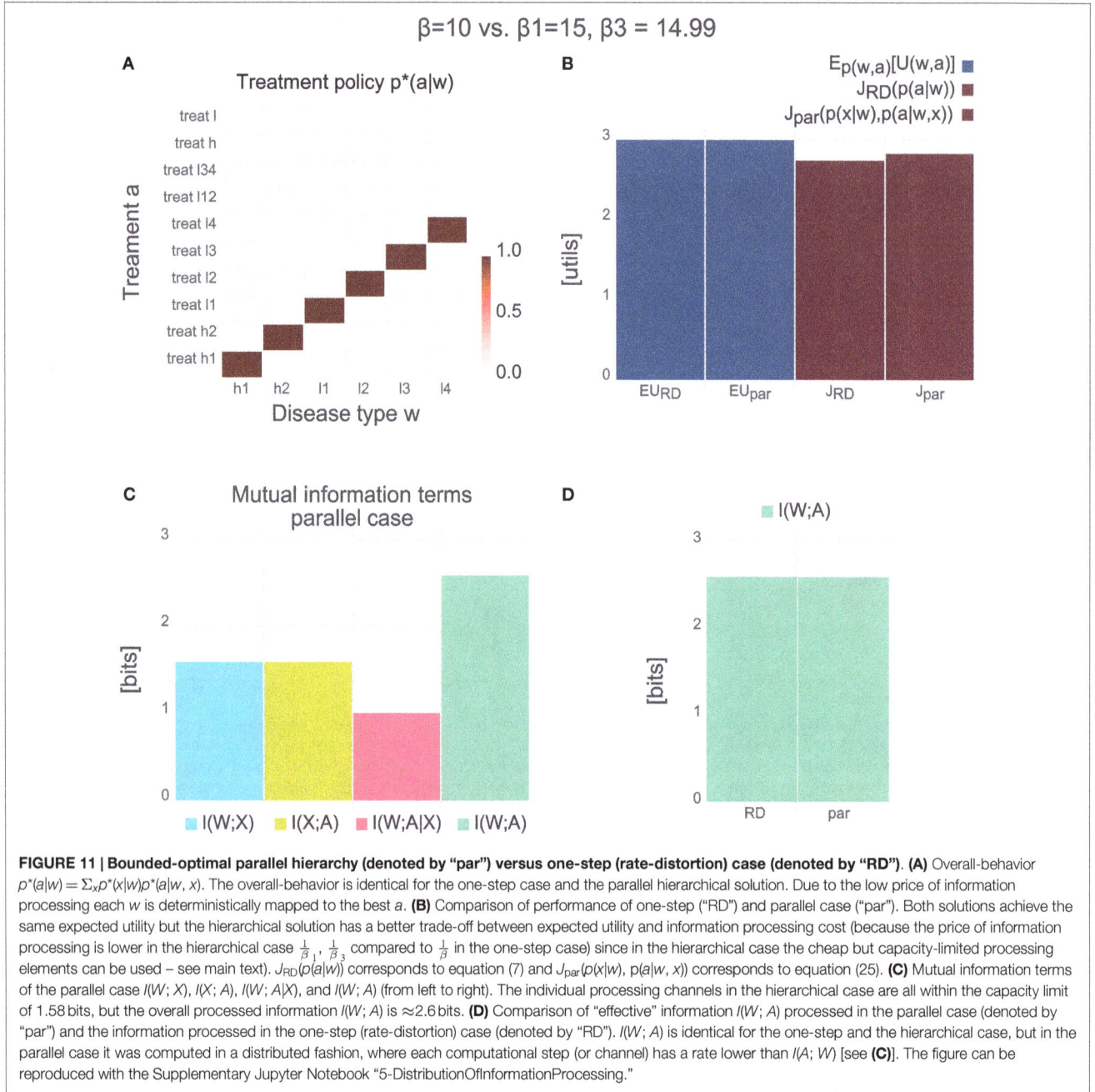

FIGURE 11 | Bounded-optimal parallel hierarchy (denoted by "par") versus one-step (rate-distortion) case (denoted by "RD"). **(A)** Overall-behavior $p^*(a|w) = \Sigma_x p^*(x|w) p^*(a|w, x)$. The overall-behavior is identical for the one-step case and the parallel hierarchical solution. Due to the low price of information processing each w is deterministically mapped to the best a. **(B)** Comparison of performance of one-step ("RD") and parallel case ("par"). Both solutions achieve the same expected utility but the hierarchical solution has a better trade-off between expected utility and information processing cost (because the price of information processing is lower in the hierarchical case $\frac{1}{\beta_1}, \frac{1}{\beta_3}$ compared to $\frac{1}{\beta}$ in the one-step case) since in the hierarchical case the cheap but capacity-limited processing elements can be used – see main text). $J_{RD}(p(a|w))$ corresponds to equation (7) and $J_{par}(p(x|w), p(a|w, x))$ corresponds to equation (25). **(C)** Mutual information terms of the parallel case $I(W; X)$, $I(X; A)$, $I(W; A|X)$, and $I(W; A)$ (from left to right). The individual processing channels in the hierarchical case are all within the capacity limit of 1.58 bits, but the overall processed information $I(W; A)$ is ≈ 2.6 bits. **(D)** Comparison of "effective" information $I(W; A)$ processed in the parallel case (denoted by "par") and the information processed in the one-step (rate-distortion) case (denoted by "RD"). $I(W; A)$ is identical for the one-step and the hierarchical case, but in the parallel case it was computed in a distributed fashion, where each computational step (or channel) has a rate lower than $I(A; W)$ [see **(C)**]. The figure can be reproduced with the Supplementary Jupyter Notebook "5-DistributionOfInformationProcessing."

would imply a lower expected utility. However, in the parallel hierarchical case the new processing elements can be used (see **Figure 11C**) which leads to a reduced price of information processing ($\beta_1 = 15$, $\beta_3 = 14.999$). In conjunction the new processing elements process the same effective information $I(W; A)$ (see **Figure 11D**) and achieve the same expected utility as the one-step case (see **Figure 11B**). However, since the price for information processing is lower on the more limited elements, the overall trade-off between expected utility and information processing cost is in favor of the parallel hierarchical architecture. Note that in this example it is important that the cardinality of X is limited (in this case $|\mathcal{X}| = 3$) and $\beta_1 > \beta_3$. We discuss this in the next paragraphs.

In the parallel hierarchical case, there are two possible pathways from w to a:

Two-stage serial pathway $I(W; X) \rightarrow I(X; A)$

Parallel pathway $I(W; A|X)$

Note that $I(X; A)$ does not appear in the objective [equation (25)]; however, it is crucial for distributing information processing on both levels of the hierarchy (see more analysis of the medical system example in **Figure 11**). $I(X; A)$ measures the average adaptation effort for going from $p(a)$ to $p(a|x)$. In the parallel hierarchical case it is a measure of how much the different models $p(a|x)$ narrow down the search space compared to the average

$p(a) = \Sigma_x p(x)p(a|x)$. If all models are equal $p(a|x) = p(a) \, \forall x$ the mutual information $I(X; A)$ is zero. Note, however, that a large $I(X; A)$ is rendered useless by a low $I(W; X)$ and vice versa – if the model selector is very bad, even the best models are not useful and vice versa.

Since there is no cost for having a large rate $I(X; A)$, the overall throughput of the serial pathway is effectively governed by β_1 as it affects the rate $I(W; X)$. Similarly, β_3 governs the rate on the parallel pathway $I(W; A|X)$. As a result, whenever one of the two inverse temperatures β_1 and β_3 is larger than the other, it becomes more economic to shift all the information processing to the cheaper pathway (either serial or parallel) thus rendering the other pathway obsolete. The only scenario where it can be advantageous to use both pathways (and distribute computation) is when the cheaper pathway has insufficient capacity and the more expensive pathway is used to take on additional computational load that cannot be handled by the cheap pathway alone. Effectively, this translates into the constraint that the serial pathway must be cheaper $\beta_1 > \beta_3$ and additionally the serial pathway must be limited in its capacity by limiting the cardinality $|\mathcal{X}|$ (see Supplementary Methods Section 5 for a detailed discussion).

Note the important difference between changing the cardinality of X which governs the channel capacity of the serial pathway (that is the maximally possible rates $I(W; X)$, $I(X; A)$) but has no influence on the price of information processing and changing β_1 which governs the price of processing $I(W; X)$ and hence affects the actual rate on the serial pathway but has no effect on the capacity of the channels of the serial pathway.

$$I(W; X) = H(X) - H(X|W) \leq \mathbf{H(X)} \tag{31}$$

$$I(X; A) = H(A) - H(A|X) = H(X) - H(X|A) \leq \mathbf{H(X)} \tag{32}$$

$$I(W; A|X) = H(A|X) - H(A|W, X) \leq H(A|X) \tag{33}$$

$H(X)$ is an upper bound for both $I(W; X)$ but also $I(X; A)$ and the upper bound of $H(X)$ itself is a function of $|\mathcal{X}|$. Note that since there is no cost associated with $I(X; A)$ it is generally desirable to maximize $I(X; A)$ at least such that $I(X; A) \geq I(W; X)$. To do so $H(A|X)$ must be pushed toward zero [equation (32)] – however, this simultaneously pushes the upper bound for $I(W; A|X)$ toward zero [equation (33)]. In case of a sufficiently limited $H(X)$ (through a low $|\mathcal{X}|$), $I(X; A)$ cannot be fully maximized, therefore leaving a non-zero upper bound for $I(W; A|X)$.

In the example shown in **Figure 11** information processing is performed on both the serial pathway ($I(W; X)$ and $I(X; A)$) but also on the parallel pathway ($I(W; A|X)$) because the constraints for distribution of information processing are fulfilled: $\beta_1 > \beta_3$ and the capacity (that is the maximum rate $I(W; X)$ and $I(X; A)$) of the serial pathway is limited by the (low) cardinality $|\mathcal{X}| = 3$. The cardinality of X for the example can easily be changed in the Supplementary Jupyter Notebook "4-ParallelHierarchy" – if it is for instance increased to $|\mathcal{X}| = 6$ while keeping all other parameters the same, the whole information processing load will be entirely on the serial pathway and $I(W; A|X) = 0$. Alternatively to limiting the cardinality of X, a cost for $I(X; A)$ could be introduced to limit the computational resources for computing $p(a|x)$ from $p(a)$. This is explored in Section 5.

5. TOWARD MORE GENERAL ARCHITECTURES

In the serial case in Section 3, information processing cost arises from adapting $p(x)$ to $p(x|w)$ and $p(a)$ to $p(a|x)$, and the average informational effort is measured by $I(W; X)$ and $I(X; A)$. In the parallel hierarchical case in Section 4 the two information processing terms considered are $I(W; X)$ and $I(W; A|X)$, where the latter measures the average informational effort for adapting from $p(a|x)$ to $p(a|w, x)$. In this section, we present a mathematically unifying case that considers all three mutual information terms and includes the serial and the parallel case as special cases. This unifying formulation might also be a starting point for generalizing toward more than three random variables as the corresponding objective function could easily be extended to include more variables.

The general case uses the same factorization of the three variables W, X, A as the parallel case: $p(w, x, a) = p(w)p(x|w)p(a|w, x)$. Given this factorization, the KL-divergence between the joint $p(w, x, a)$ and the product of all three marginals, also known as the total correlation $C(W, X, A)$, leads to:

$$\begin{aligned}
C(W, X, A) &= D_{\text{KL}}(p(w, x, a) \| p(w)p(x)p(a)) \\
&= H(W) + H(X) + H(A) - H(W, X, A) \tag{34} \\
&= \sum_{w,x,a} p(w, x, a) \log \frac{p(w, x, a)}{p(w)p(x)p(a)} \\
&= \sum_{w,x} p(w, x) \log \frac{p(x|w)}{p(x)} \\
&\quad + \sum_{w,a,x} p(w, x, a) \log \frac{p(a|w, x)}{p(a)} \\
&= I(W; X) + I(W, X; A) \tag{35} \\
&= I(W; X) + I(X; A) + I(W; A|X). \tag{36}
\end{aligned}$$

The total correlation (Watanabe, 1960), also called multivariate constraint (Garner, 1962) or multiinformation (Studený and Vejnarová, 1998), is the sum of the three information processing terms considered in the serial and parallel case. The general objective is formed by assigning different prices to each of the terms and trading off the resulting information processing cost against the expected utility:

$$\begin{aligned}
\arg\max_{p(x|w), p(a|w,x)} \; & \mathbf{E}_{p(w,x,a)}[U(w, a)] - \frac{1}{\beta_1} I(W; X) \\
& - \frac{1}{\beta_2} I(X; A) - \frac{1}{\beta_3} I(W; A|X) \\
= \; & \arg\max_{p(x|w), p(a|w,x)} \; J_{\text{gen}}(p(x|w), p(a|w, x)). \tag{37}
\end{aligned}$$

Identical to the parallel hierarchical case, the general case has two information processing pathways that allow for splitting up the total computational load: a serial pathway consisting of the two stages $I(W; X)$ and $I(X; A)$ and a parallel pathway $I(W; A|X)$. If any of the pathways is cheaper than the other one, it is more economical to shift all the computation to the cheaper pathway. However, the capacity of the serial pathway can be limited, for example by reducing the cardinality of X. In such a case the parallel pathway can take on additional computational load, leading to a parallel hierarchical information processing architecture.

The solution to the general objective is given by the following set of five self-consistent equations (the detailed derivation of the solutions is included in the Supplementary Methods Section 1):

$$p^*(x|w) = \frac{1}{Z(w)}p(x)\exp\left(\beta_1\Delta F_{\text{gen}}(w,x)\right.$$
$$\left. - \left(\frac{\beta_1}{\beta_3} - \frac{\beta_1}{\beta_2}\right)D_{\text{KL}}(p^*(a|w,x)||p^*(a|x))\right) \quad (38)$$

$$p(x) = \sum_w p(w)p^*(x|w) \quad (39)$$

$$p^*(a|w,x) = \frac{1}{Z(w,x)}p^*(a|x)\exp\left(\beta_3 U(w,a) - \frac{\beta_3}{\beta_2}\log\frac{p^*(a|x)}{p(a)}\right) \quad (40)$$

$$p^*(a|x) = \sum_w p(w|x)p^*(a|w,x) \quad (41)$$

$$p(a) = \sum_{w,x} p(w)p^*(x|w)p^*(a|w,x), \quad (42)$$

where $Z(w)$ and $Z(w,x)$ denote the corresponding normalization constants or partition sums. The conditional distribution $p(w|x)$ is given by Bayes' rule $p(w|x) = \frac{p(w)p^*(x|w)}{p(x)}$ and $\Delta F_{\text{gen}(w,x)}$ is the free energy difference

$$\Delta F_{\text{gen}}(w,x) := \mathbf{E}_{p^*(a|w,x)}[U(w,a)] - \frac{1}{\beta_2}D_{\text{KL}}(p^*(a|w,x)||p(a)).$$

For $\beta_3 < \beta_2$ the KL-term in equation (38) has a positive sign, implying that the KL-divergence is a utility instead of a cost which makes sense if computation on $I(W;A|X)$ is cheaper than computation on $I(A|X)$. For $\beta_3 > \beta_2$ the KL-term gets a negative sign, implying that the KL-divergence is a cost, as a result of computation on $I(W;A|X)$ being more expensive than computation on $I(A|X)$.

Equation 38 can also be rewritten as (see Supplementary Methods Section 1.2):

$$p^*(x|w) = \frac{1}{Z(w)}p(x)\exp\left(\beta_1\Delta F_{\text{par}}(w,x)\right.$$
$$\left. - \frac{\beta_1}{\beta_2}\sum_a p^*(a|w,x)\log\frac{p^*(a|x)}{p(a)}\right) \quad (43)$$

where $\Delta F_{\text{par}(w,x)}$ is the same free energy difference as in the parallel case

$$\Delta F_{\text{par}}(w,x) := \mathbf{E}_{p^*(a|w,x)}[U(w,a)] - \frac{1}{\beta_3}D_{\text{KL}}(p^*(a|w,x)||p^*(a|x)) \quad (44)$$

see equation (30).

Comparing the objective in equation (37) with the objective of the parallel case in equation (25), it can be seen that by setting $\beta_2 \to \infty$ the two objective functions become equal and the implicit assumption that in the parallel case there is no cost for going from $p(a)$ to $p(a|x)$ (as the latter is considered a prior) is made explicit. The solution equations of the general case also collapse

TABLE 2 | Recovery of special cases from the general, unifying case by specific settings of the inverse temperatures.

Case	β_1	β_2	β_3	(inverse) price per transformation
General	β_1	β_2	β_3	$\beta_1: p(x) \to p(x|w)$
				$\beta_2: p(a) \to p(a|x)$
				$\beta_3: p(a|x) \to p(a|w,x)$
Total correlation	β	β	β	$\beta: p(x) \to p(x|w)$
				$\beta: p(a) \to p(a|w,x)$
Degenerate TC	β_1	β	β	$\beta_1: p(x) \to p(x|w)$
				$\beta: p(a) \to p(a|w,x)$
Serial	β_1	β_2	$\to 0$	$\beta_1: p(x) \to p(x|w)$
				$\beta_2: p(a) \to p(a|x)$
				$p(a|w,x) = p(a|x) \forall w$
				$I(W;A|X) = 0$
Parallel	β_1	$\to \infty$	β_3	$\beta_1: p(x) \to p(x|w)$
				$\beta_3: p(a|x) \to p(a|w,x)$
Joint (x,a)	β	$\to \infty$	β	$\beta: p(x,a) \to p(x,a|w)$

The table shows how to set the inverse temperatures in the general case to recover particular special cases. The last column shows for all cases which probability-transformations are considered as computational effort and the corresponding (inverse) price. The case "degenerate total correlation" is not described in the main paper, but is outlined in the Supplementary Methods Section 4 – it could be relevant in a two-dimensional decision-making scenario, that is when x is considered one dimension of the decision and a is considered the other dimension. This implies that the utility function also depends on x: U(w,x,a). Similarly, the case "joint (x,a)" is only described in the Supplementary Methods Section 5 and describes how the one-step (rate-distortion) case is related to the general case.

to the solutions of the parallel case by letting $\beta_2 \to \infty$: compare equations (43) and (40) against equations (26) and (28). The general case thus also allows for designing more realistic hierarchical cases where there is a small cost for switching models $p^*(a|x)$ that arises, for instance, from loading a certain set of parameters or switching to a particular sampler or reading the model from memory. Similarly, the serial case can be recovered by $\beta_3 \to 0$. The special cases of the general objective are summarized in **Table 2**.

5.1. Effective Information Throughput $I(W; A)$

The amount of information processing that effectively contributes toward achieving a high expected utility is measured by $I(W;A)$ which does not directly appear in the objective of the general case (nor the serial and parallel case). However, the effective information throughput of the system is given by

$$I(W;A) = I(W;X;A) + I(W;A|X) \quad (45)$$
$$= I(W;X) + I(X;A) - I(X;W,A) + I(W;A|X) \quad (46)$$
$$= C(W,X,A) - I(X;W,A) \quad (47)$$

where $I(W;X;A)$ denotes the multivariate mutual information (MMI; Yeung, 1991). The first equation above is obtained by re-ordering the definition of the MMI $I(K;L;M) = I(K;M) - I(K;M|L)$. Note that in the serial hierarchical case $I(W;A|X) = 0$ always holds. The equations above also show how the total correlation $C(W,X,A)$ and the MMI $I(W;X;A)$ are related.

The multivariate mutual information is upper-bounded by $I(W;X;A) \leq \min\{I(W;X), I(W;A), I(X;A)\}$ (see (Yeung, 1991)). Using the bound in equation (45) leads to and upper bound for the effective information throughput:

$$I(W;A) \leq \min\{I(W;X), I(X;A)\} + I(W;A|X) \quad (48)$$

Equation 48 shows how information processing in the general case can be distributed between a two-stage serial pathway (consisting of $I(W;X)$ and $I(X;A)$) and a parallel pathway ($I(W;A|X)$). The general case forms a parallel hierarchy similar to Section 4, but it allows to associate a cost with $I(X;A)$ (which is a measure of how costly it is to switch models). Importantly the discussion on splitting up information processing between both levels of the parallel hierarchy as in Section 4.3 also holds for the general case.

5.2. Iterating the Self-Consistent Equations

For the simulation results shown in this paper the corresponding set of self-consistent equations was iterated until convergence (by checking that the total change in probability distributions between two iteration steps is below a certain threshold – see code underlying the Supplementary Notebooks for details). This is inspired by the Blahut-Arimoto scheme that is proven to converge to the global maximum in the rate-distortion case (Csiszar, 1974; Cover and Thomas, 1991) (Section 2). Unfortunately there is no such proof for iterating the sets of self-consistent equations of the general, serial or parallel case. It is not clear whether the optimization problems are still convex and have a global solution, nor is it clear that iterating the self-consistent equations would converge toward these global solutions. A convexity and convergence analysis is certainly among the most important steps for future investigations of the principles presented here. At this point, we can only report empirical observations and interested readers are encouraged to explore convergence behavior using the Supplementary Jupyter Notebooks (which include plots that show convergence behavior across iterations) but also the underlying code (published in the Supplementary Material).

6. DISCUSSION AND CONCLUSION

The overarching principle behind this paper is the consistent application of the trade-off of gains in expected utility against the computational cost that these gains require. Here, computational cost is defined as the average effort of computational adaptation (measured by the mutual information) multiplied by the price of information processing. This definition is motivated by first principles (Mattsson and Weibull, 2002; Ortega and Braun, 2010; Ortega and Braun, 2011) and is grounded in a thermodynamic framework for decision-making (Ortega and Braun, 2013). Mathematically, the basic principle is identical to the principle behind rate-distortion theory, the information-theoretic framework for lossy compression (Genewein and Braun, 2013; Still, 2014). This connection is no coincidence as bounded rational decision-making can be cast as a lossy compression problem in lossy compression the goal is to transmit the most relevant information (given the limited channel capacity) in order to minimize a distortion-function. In bounded rational decision-making the goal is to process the most relevant information in order maximize a utility function, given the limitations on information processing. In Section 2, we have shown how different levels of behavioral abstraction can be induced by different computational limitations. The authors in (van Dijk and Polani, 2013) use the *Relevant Information* method, which is a particular application

of rate-distortion theory and find a very similar emergence of "natural abstractions" and "ritualized behavior" when studying goal-directed behavior in the MDP case. We have shown how the basic principle can be extended to more complex cases and that analytic solutions can be obtained for these cases. Importantly, the solutions allow for interesting interpretations, highlighting how the same fundamental trade-off can lead to systems that elegantly solve more complex problems. For instance, when designing a perception-action system, the perceptual part of the system can easily be understood as a lossy compressor, but the corresponding distortion-function is not intuitively clear. We have shown in Section 3 how the extended lossy compression principle leads to a well-defined distortion-function for the perceptual part of the system that optimizes the downstream trade-off between expected utility and computational cost. In a similar fashion we have shown in Section 4 how the problem of designing bounded-optimal decision-making hierarchies is fundamentally equal to designing a distributed lossy compressor (that is spread over both levels of the hierarchy).

In the serial hierarchy in Section 3, we compared a perceptual channel that performs Bayesian inference against a bounded-optimal perceptual channel that optimizes the downstream free energy trade-off. We found that the difference between both models of perception was that in one case the likelihood model $p(x|w)$ was unspecified (Bayesian inference) whereas in the other case it was well defined (bounded-optimal solution). Perception is often conceptualized as (Bayesian) inference, however, given our findings there is a subtle but important difference. In our model of a perception-action system, the goal of the perceptual model $p(x|w)$ is to extract the most relevant information from w for choosing an action according to $p(a|x)$, given the computational limitations of the system. In plain inference, the goal is to predict w from x very well and the likelihood model is thus chosen to maximize predictive power. In many cases the two objectives coincide as achieving a large expected utility often requires precise knowledge about w. However, this must not always be the case and in particular for systems where computational limitations play a large role, the (limited) computational resources can often be spent more economically which allows for a higher expected utility at the cost of not being able to predict w from x that well. An interesting machine-learning application of the serial principle could be the design of optimal features for classification.

In Section 4, we showed how parallel bounded-optimal decision-making hierarchies can emerge from solving the trade-off between utility and cost of computation. We found that the condition for parallel hierarchies to being optimal solutions was that the price for model selection is lower than the price for processing information on the low level of the hierarchy ($\beta_1 > \beta_3$). At the same time, the upper level of the hierarchy must be limited in capacity (for instance through the cardinality $|\mathcal{X}|$). Intuitively this makes sense and fits with the general observation that often hardware that allows for cheap information processing is itself quite expensive to build (low signal to noise ratio, etc.). Therefore the amount of hardware that allows for cheap information processing is likely to be quite limited. It remains an open question whether this is a fundamental constraint for hierarchies being optimal solutions or whether there are other arguments in favor of hierarchical architectures. In changing environments, for example, the overall change required to adapt a system is smaller for a hierarchical system, compared to a flat system, because the more abstract levels of the hierarchy might require little or no change at all. It could also be that the upper levels of a hierarchical model based on our principle contain more transferable knowledge that can be applied to novel but similar tasks. Changing the task corresponds to changing the utility-function, which requires a non-equilibrium analysis (Grau-Moya and Braun, 2013) that we leave for future investigation.

In our simulations, we initialize $p(x|w)$ and either $p(a|w, x)$ (parallel hierarchies) or $p(a|x)$ (serial hierarchies) and iterate the equations until (numerical) convergence. We found sometimes that the solutions can be sensitive to the initialization. This hints at the problem being non-convex or the iteration-scheme being prone to get stuck in local optima or plateaus. In particular, we find that in the serial hierarchy with low cost of computation, a sparse, diagonal-like initialization of $p(x|w)$ works much better than a random initialization. For the parallel hierarchies, we found that a random or sparse initialization of $p(x|w)$ combined with a uniform initialization of $p(a|w, x)$ works most reliably. Additionally we found that in the hierarchical case if β_3 is slightly larger than β_1 the iterations converge to sub-optimal solutions where both pathways are used instead of shifting all the computation to the parallel pathway. The toy simulations presented here are illustrative examples only and numerically efficient implementations of the iteration-schemes are beyond the scope of the current paper. These problems might be addressed by other solution schemes like sampling-based or parametric model-based solutions. Nevertheless, these other solution schemes (that potentially do not even require the sets of analytical solutions) can benefit from the interpretations given by the analytic solution equations in this paper.

The ability to form abstractions is thought of as a hallmark of intelligence, both in cognitive tasks and in basic sensorimotor behaviors (Kemp et al., 2007; Braun et al., 2010a,b; Gershman and Niv, 2010; Tenenbaum et al., 2011; Genewein and Braun, 2012). Traditionally, the formation of abstractions is conceptualized as being computationally costly because particular entities have to be grouped together by neglecting irrelevant information. Recently, abstractions that arise from sensory evolution and hierarchical behaviors have been studied from an information-theoretic perspective (Salge and Polani, 2009; Van Dijk et al., 2011). Here, we study abstractions in the process of decision-making, where "similar" situations elicit the same behavior when partially ignoring the current situational context. Extending our principle to hierarchies with more than two levels might provide novel points of view on the formation of hierarchies in biological systems, such as the early visual system (DiCarlo et al., 2012). One fundamental prediction, based on our current work is that the formation of abstractions and concepts should be heavily shaped by the agent-environment structure (the utility function). Following the work of (Simon, 1972) decision-making with limited information-processing resources has been studied extensively in psychology, economics, political science, industrial organization, computer science, and artificial intelligence research. In

this paper, we use an information-theoretic model of decision-making under resource constraints (McKelvey and Palfrey, 1995; Kappen, 2005; Wolpert, 2006; Todorov, 2009; Peters et al., 2010; Theodorou et al., 2010; Rubin et al., 2012). In particular, Braun et al. (2011) and Ortega and Braun (2011, 2012, 2013) present a framework in which gain in expected utility is traded off against the adaptation cost of changing from an initial behavior to a posterior behavior. The variational problem that arises due to this trade-off has the same mathematical form as the minimization of a *free energy difference* functional in thermodynamics. Here, we discuss the close connection between the thermodynamic decision-making framework (Ortega and Braun, 2013) and rate-distortion theory which is an information-theoretic framework for lossy compression. The problem in lossy compression is essentially the problem of separating structure from noise and is thus highly related to finding abstractions (Tishby et al., 1999; Still and Crutchfield, 2007; Still et al., 2010). In the context of decision-making the rate-distortion framework can be applied by conceptualizing the decision-maker as a channel from observations to actions *with limited capacity*, which is known in economics as the framework of "Rational Inattention" (Sims, 2003).

The rate-distortion principle and all the extended principles presented in this paper measure computational cost with the mutual information which is an abstract measure that quantifies the average KL-divergence. The mutual information measures the actual transformation of probabilities and thus provides a lower bound for any possible implementation. In fact, different implementations could perform the same transformation more or less efficiently which should reflect in the price of information processing but not the amount of information processed. The advantage of using a generic measure is that the principle is universal and can be applied to any system. The downside of this is that it cannot be directly used to analyze specific implementations. In practice it can be hard to determine how difficult or "costly" it is to implement a certain transformation of probability distributions. Rather, the price for information processing is often set implicitly, for instance by certain computation-time constraints or by constraining the number of samples, etc. When applying the principle to a specific implementation it might be required to derive a novel, specific solution scheme for the corresponding optimization problem. In Leibfried and Braun (2015), for instance, the authors apply the rate-distortion principle for decision-making to a spiking neuron model by deriving a gradient-based update rule for tuning the parameters of the model (the weights of the neuron). In their case, the price of information processing β appears directly in the parameter update equations which leads to an interesting regularizer for the (online) parameter update rule.

The fundamental trade-off between large expected utility and low computational cost appears in many domains such as machine learning, AI, economics, computational biology or neuroscience, and many solutions, such as heuristics, sampling-based approaches, and model-based approximation schemes, exist (Gershman et al., 2015; Jordan and Mitchell, 2015; Parkes and Wellman, 2015). One of the exciting prospects of such an approach is that it might provide a common ground for research-questions from artificial intelligence and neuroscience, thus partially unifying the two fields that share common origins but have drifted apart over the last decades (Gershman et al., 2015). The main contribution of this paper is to advance a principled mathematical framework that formalizes the problem objective such that the trade-off between large expected utility and low computational cost and its solutions can be addressed in both a qualitative but also quantitative way. The main finding is that the consistent application of the principle beyond simple one-stage information processing systems leads to non-trivial solutions that address questions like optimal likelihood model design or the design of optimal decision-making hierarchies. Since the mathematics can easily be extended to more variables while the underlying principle remains the same, we believe that the formulation presented in this paper is a good candidate for a general underlying objective that is also applicable to biological organisms and evolutionary processes. We find the principle an interesting starting point for solving timely problems in machine learning, robotics, and AI but also for providing an interesting novel angle for research in computational neuroscience and biology. The principle also provides a promising basis for the design and analysis of guided self-organizing systems as most of the inner structure of systems following our principle is emergent (and thus self-organized) but ultimately aimed at solving particular tasks (through the utility function).

AUTHORS CONTRIBUTION

TG and DB conceived the project, TG and JG performed simulations, TG and FL did analysis and derivations, and TG, FL, JG, and DB wrote the paper.

FUNDING

This study was supported by the DFG, Emmy Noether grant BR4164/1-1.

SUPPLEMENTARY MATERIAL

The Supplementary Material for this article can be found online at http://journal.frontiersin.org/article/10.3389/frobt.2015.00027

A Supplementary Methods provides detailed steps to derive the solution to the general case (Section 5) and how to rewrite the solution equations of the general case. Additionally it outlines how to derive the solutions for the serial and the parallel case. It also provides the set of self-consistent equations for the "degenerate total correlation" and "total correlation" case that drop out mathematically from the general case but are not used in this paper (see **Table 2**). The Supplementary Methods provides details to the discussion on the different information processing pathways of the parallel case (Section 4.3). Finally, it contains the proof for the inequality based on equation (6).

The simulations underlying the results presented in this paper are published as supplementary material using Jupyter (http://jupyter.org/) notebooks. The notebooks are considered part of the results of this paper and readers are encouraged to use the notebooks to interactively explore the examples and concepts presented here. The underlying code is written in Julia (Bezanson et al., 2014) and uses the Gadfly package (http://gadflyjl.org/) for visualization. At the time of writing, the notebooks can be run with a local installation of Jupyter and Julia or without any installation in a web-browser through the JuliaBox project (https://www.juliabox.org/). The notebooks and code at the time of publication are provided in a supplementary.zip file but also under (Genewein, 2015). The notebooks and the code behind the notebooks as well as information on different methods to run the notebooks will be kept up-to-date in the accompanying GitHub repository: https://github.com/tgenewein/BoundedRationalityAbstraction AndHierarchicalDecisionMaking. If compatibility issues with future Julia versions are encountered, please refer to the GitHub repository and feel free to submit an issue. A readme-file on how to run the notebooks (with or without installation) is also provided in the supplementary data as well as in the GitHub repository.

The following notebooks are provided:

- "1-FreeEnergyForBoundedRationalDecisionMaking": Illustrates the results of Section 1 and reproduces **Figure 1**.
- "2-RateDistortionForDecisionMaking": Illustrates the results of Section 4 (the recommender system example) and reproduces **Figures 2–4**. The notebook can be used as a general template for setting up any of the examples presented in the paper and solving it using Blahut-Arimoto.
- "S1-SampleBasedBlahutArimoto": A simple proof-of-concept implementation of sample-based Blahut-Arimoto iterations. Due to space-constraints, this part has been omitted from the paper, but interested readers can find a short theoretical part on the sampling approach in the notebook. Additionally, the notebook shows an implementation of the sampling scheme and applies it to a toy example.
- "3-SerialHierarchy": Illustrates the comparison between handcrafted perception and bounded-optimal perception in the serial case (Section 3) using the predator-prey example. The notebook reproduces **Figures 5–8**. The notebook allows to easily modify the parameters (e.g., inverse temperatures) of the example or to switch to a different utility function. It can also be used to see how the parallel or general case solution for the predator-prey example would look like.
- "4-ParallelHierarchy": Illustrates the emergence of boundedoptimal hierarchies in two different environments of the medical system example as presented in Section 4 and reproduces **Figures 9** and **10**. The notebook can be used to easily explore the different information processing pathways in the parallel case but also to compare any two cases against each other (because it compares two general case solutions and they can be tuned to all of the special cases).
- "5-DistributionOfInformationProcessing": Compares the parallel hierarchical solution to the medical example to the one-step (rate-distortion) case as shown in **Figure 11**. Since it implements the parallel case through the general case, it also allows to compare any other case to the one-step solution.

REFERENCES

Arimoto, S. (1972). An algorithm for computing the capacity of arbitrary discrete memoryless channels. *IEEE Trans. Inf. Theory* 18, 14–20. doi:10.1109/TIT.1972.1054753

Ashby, W. R. (1956). *An Introduction to Cybcernetics*. London: Chapman & Hall.

Bezanson, J., Edelman, A., Karpinski, S., and Shah, V. B. (2014). Julia: a fresh approach to numerical computin. *arXiv preprint arXiv:1411.1607*.

Bishop, C. M. (2006). "Sampling methods," in *Pattern Recognition and Machine Learning, Number 4 in Information Science and Statistics*, Chap. 11 (New York: Springer).

Blahut, R. (1972). Computation of channel capacity and rate-distortion functions. *IEEE Trans. Inf. Theory* 18, 460–473. doi:10.1109/TIT.1972.1054855

Braun, D. A., Mehring, C., and Wolpert, D. M. (2010a). Structure learning in action. *Behav. Brain Res.* 206, 157–165. doi:10.1016/j.bbr.2009.08.031

Braun, D. A., Waldert, S., Aertsen, A., Wolpert, D. M., and Mehring, C. (2010b). Structure learning in a sensorimotor association task. *PLoS ONE* 5:e8973. doi:10.1371/journal.pone.0008973

Braun, D. A., and Ortega, P. A. (2014). Information-theoretic bounded rationality and epsilon-optimality. *Entropy* 16, 4662–4676. doi:10.3390/e16084662

Braun, D. A., Ortega, P. A., Theodorou, E., and Schaal, S. (2011). "Path integral control and bounded rationality," in *IEEE Symposium on Adaptive Dynamic Programming and Reinforcement Learning* (Piscataway: IEEE), 202–209.

Burns, E., Ruml, W., and Do, M. B. (2013). Heuristic search when time matters. *J. Artif. Intell. Res.* 47, 697–740. doi:10.1613/jair.4047

Camerer, C. (2003). *Behavioral Game Theory: Experiments in Strategic Interaction*. Princeton, NY: Princeton University Press.

Cover, T. M., and Thomas, J. A. (1991). *Elements of Information Theory*. Hoboken: John Wiley & Sons.

Csiszar, I. (1974). On the computation of rate-distortion functions. *IEEE Trans. Inf. Theory* 20, 122–124. doi:10.1109/TIT.1974.1055146

Csiszár, I., and Tusnády, G. (1984). Information geometry and alternating minimization procedures. *Stat. Decis.* 1, 205–237.

Daniel, C., Neumann, G., and Peters, J. (2012). "Hierarchical relative entropy policy search," in *International Conference on Artificial Intelligence and Statistics*. La Palma.

Daniel, C., Neumann, G., and Peters, J. (2013). "Autonomous reinforcement learning with hierarchical REPS," in *International Joint Conference on Neural Networks*. Dallas.

DiCarlo, J. J., Zoccolan, D., and Rust, N. C. (2012). How does the brain solve visual object recognition? *Neuron* 73, 415–434. doi:10.1016/j.neuron.2012.01.010

Fox, C. W., and Roberts, S. J. (2012). A tutorial on variational Bayesian inference. *Artif. Intell. Rev.* 38, 85–95. doi:10.1007/s10462-011-9236-8

Friston, K. (2010). The free-energy principle: a unified brain theory? *Nat. Rev. Neurosci.* 11, 127–138. doi:10.1038/nrn2787

Garner, W. R. (1962). *Uncertainty and Structure as Psychological Concepts*. New York: Wiley.

Genewein, T. (2015). Bounded rationality, abstraction and hierarchical decision-making: an information-theoretic optimality principle: supplementary code (v1.1.0). *Zenodo*. doi:10.5281/zenodo.32410

Genewein, T., and Braun, D. A. (2012). A sensorimotor paradigm for Bayesian model selection. *Front. Hum. Neurosci.* 6:291. doi:10.3389/fnhum.2012.00291

Genewein, T., and Braun, D. A. (2013). Abstraction in decision-makers with limited information processing capabilities. *arXiv preprint arXiv:1312.4353*.

Gershman, S. J., Horvitz, E. J., and Tenenbaum, J. B. (2015). Computational rationality: a converging paradigm for intelligence in brains, minds, and machines. *Science* 349, 273–278. doi:10.1126/science.aac6076

Gershman, S. J., and Niv, Y. (2010). Learning latent structure: carving nature at its joints. *Curr. Opin. Neurobiol.* 20, 251–256. doi:10.1016/j.conb.2010.02.008

Gigerenzer, G., and Brighton, H. (2009). Homo heuristicus: why biased minds make better inferences. *Top. Cogn. Sci.* 1, 107–143. doi:10.1111/j.1756-8765.2008.01006.x

Gigerenzer, G., and Todd, P. M. (1999). *Simple Heuristics That Make Us Smart*. Oxford: Oxford University Press.

Grau-Moya, J., and Braun, D. A. (2013). Bounded rational decision-making in changing environments. *arXiv preprint arXiv:1312.6726*.

Horvitz, E. (1988). "Reasoning under varying and uncertain resource constraints," in *AAAI*, Vol. 88 (Palo Alto: AAAI), 111–116.

Horvitz, E., and Zilberstein, S. (2001). Computational tradeoffs under bounded resources. *Artif. Intell.* 126, 1–4. doi:10.1016/S0004-3702(01)00051-0

Horvitz, E. J., Cooper, G. F., and Heckerman, D. E. (1989). "Reflection and action under scarce resources: theoretical principles and empirical study," in *Proceedings of the 11th International Joint Conference on Artificial Intelligence*, Vol. 2 (Detroit: Morgan Kaufmann Publishers, Inc.), 1121–1127.

Howes, A., Lewis, R. L., and Vera, A. (2009). Rational adaptation under task and processing constraints: implications for testing theories of cognition and action. *Psychol. Rev.* 116, 717–751. doi:10.1037/a0017187

Janssen, C. P., Brumby, D. P., Dowell, J., Chater, N., and Howes, A. (2011). Identifying optimum performance trade-offs using a cognitively bounded rational analysis model of discretionary task interleaving. *Top. Cogn. Sci.* 3, 123–139. doi:10.1111/j.1756-8765.2010.01125.x

Jones, B. D. (2003). Bounded rationality and political science: lessons from public administration and public policy. *J. Public Adm. Res. Theory* 13, 395–412. doi:10.1093/jopart/mug028

Jordan, M., and Mitchell, T. (2015). Machine learning: trends, perspectives, and prospects. *Science* 349, 255–260. doi:10.1126/science.aaa8415

Kahneman, D. (2003). Maps of bounded rationality: psychology for behavioral economics. *Am. Econ. Rev.* 93, 1449–1475. doi:10.1257/000282803322655392

Kappen, H. J. (2005). Linear theory for control of nonlinear stochastic systems. *Phys. Rev. Lett.* 95, 200–201. doi:10.1103/PhysRevLett.95.200201

Kappen, H. J., Gómez, V., and Opper, M. (2012). Optimal control as a graphical model inference problem. *Mach. Learn.* 87, 159–182. doi:10.1007/s10994-012-5278-7

Kemp, C., Perfors, A., and Tenenbaum, J. B. (2007). Learning overhypotheses with hierarchical Bayesian models. *Dev. Sci.* 10, 307–321. doi:10.1111/j.1467-7687.2007.00585.x

Leibfried, F., and Braun, D. A. (2015). A reward-maximizing spiking neuron as a bounded rational decision maker. *Neural Comput.* 27, 1686–1720. doi:10.1162/NECO_a_00758

Levy, R. P., Reali, F., and Griffiths, T. L. (2009). "Modeling the effects of memory on human online sentence processing with particle filters," in *Advances in Neural Information Processing Systems* (Vancouver: NIPS), 937–944.

Lewis, R. L., Howes, A., and Singh, S. (2014). Computational rationality: linking mechanism and behavior through bounded utility maximization. *Top. Cogn. Sci.* 6, 279–311. doi:10.1111/tops.12086

Lieder, F., Griffiths, T., and Goodman, N. (2012). "Burn-in, bias, and the rationality of anchoring," in *Advances in Neural Information Processing Systems* (Lake Tahoe: NIPS), 2690–2798.

Lipman, B. (1995). Information processing and bounded rationality: a survey. *Can. J. Econ.* 28, 42–67. doi:10.2307/136022

Mattsson, L. G., and Weibull, J. W. (2002). Probabilistic choice and procedurally bounded rationality. *Games Econ. Behav.* 41, 61–78. doi:10.1016/S0899-8256(02)00014-3

McKelvey, R. D., and Palfrey, T. R. (1995). Quantal response equilibria for normal-form games. *Games Econ. Behav.* 10, 6–38. doi:10.1006/game.1995.1023

Neal, R. M. (2003). Slice sampling. *Ann. Stat.* 31, 705–767. doi:10.1214/aos/1056562461

Neymotin, S. A., Chadderdon, G. L., Kerr, C. C., Francis, J. T., and Lytton, W. W. (2013). Reinforcement learning of two-joint virtual arm reaching in a computer model of sensorimotor cortex. *Neural Comput.* 25, 3263–3293. doi:10.1162/NECO_a_00521

Ortega, P., and Braun, D. (2010). "A conversion between utility and information," in *Third Conference on Artificial General Intelligence (AGI 2010)* (Lugano: Atlantis Press), 115–120.

Ortega, P. A., and Braun, D. A. (2014). Generalized Thompson sampling for sequential decision-making and causal inference. *Complex Adapt. Syst. Model.* 2, 269–274. doi:10.1186/2194-3206-2-2

Ortega, P. A., Braun, D. A., and Tishby, N. (2014). "Monte Carlo methods for exact & efficient solution of the generalized optimality equations," in *Proceedings of IEEE International Conference on Robotics and Automation*. Hong Kong.

Ortega, P. A., and Braun, D. A. (2011). "Information, utility and bounded rationality," in *Proceedings of the 4th International Conference on Artificial General Intelligence* (Mountain View: Springer-Verlag), 269–274.

Ortega, P. A., and Braun, D. A. (2012). "Free energy and the generalized optimality equations for sequential decision making," in *Journal of Machine Learning Research: Workshop and Conference Proceedings* (Edinburgh: JMLR W&C Proceedings), 1–10.

Ortega, P. A., and Braun, D. A. (2013). Thermodynamics as a theory of decision-making with information-processing costs. *Proc. R. Soc. A Math. Phys. Eng. Sci.* 469, 2153.

Palmer, S. E., Marre, O., Berry, M. J., and Bialek, W. (2015). Predictive information in a sensory population. *Proc. Natl. Acad. Sci. U.S.A.* 112(22), 6908–6913. doi: 10.1073/pnas.1506855112

Parkes, D. C., and Wellman, M. P. (2015). Economic reasoning and artificial intelligence. *Science* 349, 267–272. doi:10.1126/science.aaa8403

Peters, J., Mülling, K., and Altun, Y. (2010). "Relative entropy policy search," in *AAAI*. Atlanta.

Ramsey, F. P. (1931). "Truth and probability," in *The Foundations of Mathematics and Other Logical Essays*, ed. R. B. Braithwaite (New York, NY: Harcourt, Brace and Co), 156–198.

Rawlik, K., Toussaint, M., and Vijayakumar, S. (2012). "On stochastic optimal control and reinforcement learning by approximate inference," in *Proceedings Robotics: Science and Systems*. Sydney.

Rubin, J., Shamir, O., and Tishby, N. (2012). "Trading value and information in mdps," in *Decision Making with Imperfect Decision Makers* (Springer), 57–74.

Rubinstein, A. (1998). *Modeling Bounded Rationality*. Cambridge: MIT Press.

Russell, S. (1995). "Rationality and intelligence," in *Proceedings of the Fourteenth International Joint Conference on Artificial Intelligence*, ed. C. Mellish (San Francisco, CA: Morgan Kaufmann), 950–957.

Russell, S. J., and Norvig, P. (2002). *Artificial Intelligence: A Modern Approach*. Upper Saddle River: Prentice Hall.

Russell, S. J., and Subramanian, D. (1995). Provably bounded-optimal agents. *J. Artif. Intell. Res.* 2, 575–609.

Salge, C., and Polani, D. (2009). Information-driven organization of visual receptive fields. *Adv. Complex Syst.* 12, 311–326. doi:10.1142/S0219525909002234

Sanborn, A. N., Griffiths, T. L., and Navarro, D. J. (2010). Rational approximations to rational models: alternative algorithms for category learning. *Psychol. Rev.* 117, 1144. doi:10.1037/a0020511

Savage, L. J. (1954). *The Foundations of Statistics*. New York: Wiley.

Shannon, C. E. (1948). A mathematical theory of communication. *Bell Syst. Tech. J.* 27, 379–423, 623–656. doi:10.1002/j.1538-7305.1948.tb00917.x

Simon, H. A. (1955). A behavioral model of rational choice. *Q. J. Econ.* 69, 99–118. doi:10.2307/1884852

Simon, H. A. (1972). Theories of bounded rationality. *Decis. Organ.* 1, 161–176.

Sims, C. A. (2003). Implications of rational inattention. *J. Monet. Econ.* 50, 665–690. doi:10.1016/S0304-3932(03)00029-1

Sims, C. A. (2005). "Rational inattention: a research agenda," in *Deutsche Bundesbank Spring Conference, Number 4*. Berlin.

Sims, C. A. (2006). Rational inattention: beyond the linear-quadratic case. *Am. Econ. Rev.* 96, 158–163. doi:10.1257/000282806777212431

Sims, C. A. (2010). "Rational inattention and monetary economics," in *Handbook of Monetary Economics*, Vol. 3, Chap. 4 (Elsevier), 155–181.

Spiegler, R. (2011). *Bounded Rationality and Industrial Organization*. Oxford: Oxford University Press.

Still, S. (2009). Information-theoretic approach to interactive learning. *Europhys. Lett.* 85, 28005. doi:10.1209/0295-5075/85/28005

Still, S. (2014). "Lossy is lazy," in *Workshop on Information Theoretic Methods in Science and Engineering* (Helsinki: University of Helsinki), 17–21.

Still, S., and Crutchfield, J. P. (2007). Structure or noise? *arXiv preprint arXiv:0708.0654*.

Still, S., Crutchfield, J. P., and Ellison, C. J. (2010). Optimal causal inference: estimating stored information and approximating causal architecture. *Chaos* 20, 037111. doi:10.1063/1.3489885

Studený, M., and Vejnarová, J. (1998). "The multiinformation function as a tool for measuring stochastic dependence," in *Learning in Graphical Models* (New York: Springer), 261–297.

Tenenbaum, J. B., Kemp, C., Griffiths, T. L., and Goodman, N. D. (2011). How to grow a mind: statistics, structure, and abstraction. *Science* 331, 1279–1285. doi:10.1126/science.1192788

Theodorou, E., Buchli, J., and Schaal, S. (2010). A generalized path integral control approach to reinforcement learning. *J. Mach. Learn. Res.* 11, 3137–3181.

Tishby, N., Pereira, F. C., and Bialek, W. (1999). "The information bottleneck method," in *The 37th Annual Allerton Conference on Communication, Control, and Computing*.

Tishby, N., and Polani, D. (2011). "Information theory of decisions and actions," in *Perception-Action Cycle*, Chap. 19 (New York: Springer), 601–636.

Tkačik, G., and Bialek, W. (2014). Information processing in living systems. *arXiv preprint arXiv:1412.8752*.

Todorov, E. (2007). "Linearly-solvable Markov decision problems," in *Advances in Neural Information Processing Systems* (Vancouver: NIPS), 1369–1376.

Todorov, E. (2009). Efficient computation of optimal actions. *Proc. Natl. Acad. Sci. U.S.A.* 106, 11478–11483. doi:10.1073/pnas.0710743106

Tversky, A., and Kahneman, D. (1974). Judgment under uncertainty: heuristics and biases. *Science* 185, 1124–1131. doi:10.1126/science.185.4157.1124

van Dijk, S. G., and Polani, D. (2013). Informational constraints-driven organization in goal-directed behavior. *Adv. Complex Syst.* 16:1350016. doi:10.1142/S0219525913500161

Van Dijk, S. G., Polani, D., and Nehaniv, C. L. (2011). "Hierarchical behaviours: getting the most bang for your bit," in *Advances in Artificial Life: Darwin Meets von Neumann*, eds. R. Goebel, J. Siekmann, and W. Wahlster (New York: Springer), 342–349.

Von Neumann, J., and Morgenstern, O. (1944). *Theory of Games and Economic Behavior*. Princeton: Princeton University Press.

Vul, E., Alvarez, G., Tenenbaum, J. B., and Black, M. J. (2009). "Explaining human multiple object tracking as resource-constrained approximate inference in a dynamic probabilistic model," in *Advances in Neural Information Processing Systems* (New York: Wiley), 1955–1963.

Vul, E., Goodman, N., Griffiths, T. L., and Tenenbaum, J. B. (2014). One and done? Optimal decisions from very few samples. *Cogn. Sci.* 38, 599–637. doi:10.1111/cogs.12101

Watanabe, S. (1960). Information theoretical analysis of multivariate correlation. *IBM J. Res. Dev.* 4, 66–82. doi:10.1147/rd.41.0066

Wiener, N. (1961). *Cybernetics or Control and Communication in the Animal and the Machine*, Vol. 25. Cambridge: MIT press.

Wolpert, D. H. (2006). "Information theory-the bridge connecting bounded rational game theory and statistical physics," in *Complex Engineered Systems*, eds D. Braha, A. A. Minai, and Y. Bar-Yam (New York: Springer), 262–290.

Yeung, R. W. (1991). A new outlook on Shannon's information measures. *IEEE Trans. Inf. Theory* 37, 466–474. doi:10.1109/18.79902

Yeung, R. W. (2008). *Information Theory and Network Coding*. New York: Springer.

Conflict of Interest Statement: The authors declare that the research was conducted in the absence of any commercial or financial relationships that could be construed as a potential conflict of interest.

8

Encoded and crossmodal thermal stimulation through a fingertip-sized haptic display

Simon Gallo[1*†], Giulio Rognini[1,2,3*†], Laura Santos-Carreras[1], Tristan Vouga[1],
Olaf Blanke[2,3,4] and Hannes Bleuler[1]

[1] Laboratory of Robotic Systems, School of Engineering, Ecole Polytechnique Fédérale de Lausanne, Lausanne, Switzerland, [2] Center for Neuroprosthetics, Ecole Polytechnique Fédérale de Lausanne, Geneva, Switzerland, [3] Laboratory of Cognitive Neuroscience, Brain Mind Institute, Ecole Polytechnique Fédérale de Lausanne, Geneva, Switzerland, [4] Department of Neurology, University Hospital of Geneva, Geneva, Switzerland

Edited by:
Rob Lindeman,
Worcester Polytechnic Institute, USA

Reviewed by:
Ryan Patrick McMahan,
University of Texas at Dallas, USA
Gustavo A. Patow,
Universitat de Girona, Spain

*Correspondence:
Simon Gallo
simon.gallo@epfl.ch;
Giulio Rognini
giulio.rognini@epfl.ch

[†]Simon Gallo and Giulio Rognini have contributed equally to this work.

Haptic displays aim at artificially creating tactile sensations by applying tactile features to the user's skin. Although thermal perception is a haptic modality, it has received scant attention possibly because humans process thermal properties of objects slower than other tactile properties. Yet, thermal feedback is important for material discrimination and has been used to convey thermally encoded information in environments in which vibrotactile feedback might be masked by noise and/or movements. Moreover, the well-reported influence of temperature over tactile processing makes thermal displays good candidates for the development of crossmodal haptic interfaces, in which temperature is used to manipulate other sensations. Here, we present a thermal display able to render four individually controlled temperatures at the user's fingertip along with its technical characterization and psychophysical evaluation. Device performance was assessed in terms of accuracy and repeatability. In the psychophysical evaluation, we first show that the device can render perceivable temperature gradients at the level of the fingertip, thereby extending the concept of thermally encoded information to fingertip-sized thermal displays. Second, we show that increasing temperature improves stiffness precision. Results show that neglected features of thermal feedback, i.e., encoded and crossmodal thermal stimulation, can be provided by fingertip-sized thermal displays to improve haptic manipulations.

Keywords: multimodal haptics, stiffness perception, thermal feedback, tele-manipulation, human–machine interaction

INTRODUCTION

Haptic technology aims at artificially creating tactile sensations by applying different features of touch (e.g., vibration, texture, tapping) to the user's skin. Recent advances in the field have led to the development of tactile displays able to render features such as shapes (Ottermo et al., 2008), lumps with different stiffness, and sizes (Gwilliam et al., 2012) or force feedback due to contact with an object (McMahan et al., 2011). Thermal perception is also often included within these haptic modalities. In fact, as with the sense of touch, temperature is felt through receptors in the skin and the heat exchange between the skin and the environment is bidirectional: when our skin touches an object, the heat exchanged alters the temperature of the skin as well as the object temperature.

To date, many research groups have investigated the role of temperature or thermal flow sensation in the tactile identification of materials or tissues. For this reason, the grand majority of the current temperature displays have been exclusively used to simulate different thermal properties of materials in order to improve object identification in virtual and tele-operated environments (Ino et al., 1993; Yamamoto et al., 2004; Guiatni et al., 2008). This focus on virtual rendering of material properties has limited the design of thermal displays and bounded the research on thermal perception. For example, current thermal displays cannot provide multiple thermal stimuli simultaneously, and, respectively, we do not know whether humans can differentiate multiple temperatures applied on a small skin surface, like the fingertip. Yet, this matter is of particular interest for novel applications of thermal interfaces such as using the display to present scalar, non-thermal information to the user under the form of encoded thermal stimuli (Zerkus et al., 1994; MacLean and Roderick, 1999). For example, Wilson et al. (2013) successfully used encoded thermal feedback to convey scalar information in environments in which audio or vibrotactile feedback might be masked by noise or movements.

Most of today's tactile displays are able to render only one tactile feature (e.g., vibration, friction, tapping, or temperature) at a time. However, to perceive our surrounding environment, we rely on several sensory modalities that we integrate into a coherent percept (Ernst and Bülthoff, 2004). Thus, to increase realism it might be beneficial to display several features of tactile stimuli as well as feedback from other sensory modalities (e.g., vision and audition). Going in this direction, in the last decade, several tactile multi-feature devices, including miniaturized thermal displays, have been developed (Kammermeier et al., 2004; Yang et al., 2007; Kim et al., 2010). However, while successfully rendering different haptic features (e.g., temperature, pressure or force), these devices neglect crucial interactions among different sensory modalities or different haptic features. This is especially surprising for the case of thermal feedback as a large number of studies have reported an influence of skin temperature on tactile perception (Russ et al., 1987; Markand et al., 1990; Phillips and Matthews, 1993). This characteristic influence of temperature upon different features of tactile processing makes thermal feedback a good candidate for the development of crossmodal haptic displays in which one modality is used to influence another.

In summary, current thermal displays are not able to provide more than one temperature or a temperature gradient at the level of the fingertip and they are often designed to only simulate different thermal properties of materials in virtual environments.

In a previous study, we have presented a proof-of-concept thermal display capable of rendering multiple distinct temperatures at the level of the fingertip (Gallo et al., 2012). However, the presentation of the device lacked of both an in-depth technical and psychophysical evaluation (and the related applications). Here, we present an upgraded version of the fingertip-sized thermal display and perform its technical characterization. Then, in the psychophysical evaluation, we show the possibility to use it in two important but often neglected applications of thermal feedback: conveying thermally encoded information (experiment 1)

and using thermal feedback to manipulate tactile processing (experiment 2: crossmodal stimulation).

The design of the thermal display is presented first. It has several innovative features such as a high spatial resolution due to four adjacent small and powerful thermal units fitting under the fingertip. The characterization of the device is performed in terms of thermal accuracy and repeatability. In the psychophysical evaluation, we first assess the capability of the user to identify multiple thermal stimuli presented simultaneously under the fingertip. This is done to validate the use of a high resolution thermal display under the fingertip to render combinations of thermal stimuli in order to convey encoded information (experiment 1: thermally encoded information). Finally, the thermal display is integrated into a commercial haptic interface to test the possibility to manipulate stiffness perception through thermal feedback (experiment 2: crossmodal stimulation). Given the well-reported influence of tactile cues on stiffness discrimination (Srinivasan and LaMotte, 1995; Bergmann Tiest and Kappers, 2009), and of temperature on mechanoreceptor's sensitivity (Stevens and Green, 1978; Green et al., 1979; Stevens, 1982, 1989; Stevens and Hooper, 1982; Srinivasan and LaMotte, 1995; Lowrey et al., 2013), we hypothesized that temperature affects stiffness perception. More specifically, as the sensitivity of mechanoreceptors to mechanical stimulation was found to increase/decrease with increasing/decreasing skin temperature, we expected the precision of stiffness perception to improve with increasing skin temperature.

THE THERMAL DISPLAY

To date, most thermal displays are composed of a Peltier element (PE), a heatsink, and a temperature sensor (Ino et al., 1993; Yamamoto et al., 2004; Guiatni et al., 2008). The temperature of the top side of the PE is controlled using the sensor input in a feedback loop while the heatsink, placed against the bottom side of the PE, is used to evacuate excess heat from the system. These devices are not meant to be integrated in multimodal haptic stations or worn; they are voluminous with large Peltier elements (PEs) and large heatsinks to evacuate the heat on the hot side. Furthermore, current thermal displays typically provide a single thermal channel, stimulating the fingertip in its entirety.

The thermal display employed in this study is fingertip-sized and can provide individually controlled thermal stimulations at four distinct locations under the fingertip.

Design

This compact display, presented in **Figure 1**, measures 28.5 mm × 21.6 mm × 10 mm. It features four thermal units each consisting of a PE (3.8 mm × 4.8 mm × 1 mm, KSAH018, Komatsu Electronics KELK Ltd., Japan), a K-Type Nickelchromium/Nickel-aluminum thermocouple (LABFACILITY, UK), and a copper plate. A thermocouple is glued over each PE with a thermal adhesive paste (Arctic Silver™) and surmounted by a copper plate used as a heat diffuser and as the interface with the finger. The four thermal units are aligned and glued along a water-cooled copper heat sink. Each thermal unit measures 14 mm × 4.2 mm × 1.9 mm. The 0.4-mm gap between adjacent units is filled with a hot-melt adhesive to obtain a smooth continuous contact surface without undesired tactile cues, while thermally insulating the copper plates from one another.

Cooling System

Peltier elements transfer heat from one surface to the other. However, the PE efficiency decreases with the increase of the temperature difference between its two surfaces (ΔT_{pelt}). Furthermore, when one surface is being cooled, the other can easily overheat due to the Joule effect adding up to the heat transferred by the Peltier effect as described in Eq. 1. The generated heat (H) is:

$$H = R \times I_{max}^2 + Q_{max} \tag{1}$$

where Q_{max} is the maximum heat transferred, R is the electrical resistance of the PE, and I_{max} is the maximum current applied.

To successfully control one surface of the PE and boost the performance, the heat generated on the unused surface must be removed. Natural convection does not require any actuation. However, to dissipate the 26 W of heat generated by four PEs (from Eq. 1, with Q_{max} = 2.8 W, R = 0.7 Ω, I_{max} = 2.3 A) a large heatsink is required. Thus, forced convection was favored to increase the heat exchange and reduce the size of the heatsink. The Nusselt Number (Nu) represents the ratio between convective heat transfer and conductive heat transfer. Therefore, the higher the Nu the stronger the convective heat transfer. With the custom heatsink and pump of the presented thermal display, Nu of 20.1 (heat transfer coefficient h of 1895 W/m²K) is obtained. This water cooling system is able to maintain the temperature of the heatsink (thus of the non-exposed side of the PE) at a constant temperature for any input current applied on the four PEs simultaneously [see Gallo et al. (2012) for details].

Control

The previous version of the display was driven using a Proportional-Integral-Derivative (PID) voltage control, a common control strategy for thermal displays (Ho and Jones, 2007;

Guiatni et al., 2008). While voltage control using pulse width modulation (PWM) of the DC input voltage is simple to implement, it has some drawbacks compared to direct current control. In fact, while the average current (determining the cooling power of the PE) is the same in both cases, the peak current used with PWM control generates larger Joule heating that, in turn, decreases the overall PE efficiency. For this reason and because PEs are non-ohmic devices, which resistance varies with temperature, current control is preferable (Yamamoto et al., 2004). We upgraded from voltage to direct current control of the PEs by using MAX1978 (Maxim Integrated™) single-chip temperature controllers. Built-in thermal control-loop circuitry and ripple cancelation circuitry avoid current surges and non-linearities usually encountered with voltage control. Overall, this control provides an excellent temperature stability and dynamic response (see **Figure 2**).

TECHNICAL CHARACTERIZATION

A considerable number of tactile displays have been developed in the last decade (Yamamoto et al., 2004; Yang et al., 2007; Guiatni et al., 2008). However, metrics to assess the thermal performance of such displays are still lacking. Commonly, the maximum temperature change rate (i.e., how fast the temperature on the device can change during both cooling and heating) that can be achieved by the display is used as a measurement of performance. This measure is dependent on many factors, such as room temperature, heat evacuation on the other side of the PE, and is especially affected by the contact with a thermal load such as the user's finger. Furthermore, depending on the application, the importance of specific characteristics of the device can change. In the following characterization, we will first discuss the heat dissipation capability of the cooling system, which is essential to achieve an optimal PE control in any thermal application. Then, the temperature stability (i.e., the capability of the device to maintain the temperature close to the desired temperature) and rejection of thermal perturbations (i.e., maintain stability despite an external perturbation) due to the user's finger are discussed. These characteristics are especially important when providing encoded thermal information. Indeed, strong perturbations could result in the user perceiving undesired cooling or heating cues. When touching an object, the initial temperature change rate induces most of our thermal sensation (Yamamoto et al., 2004; Tiest and Kappers, 2009). Thus, we further characterize the dynamic response (i.e., the time-varying behavior of the system) of the system in terms of fall and rise times (i.e., the time taken by the display to decrease, fall, or increase, rise, from a specified temperature to another specified temperature) and measure its repeatability (i.e., how variable these times are for multiple repetitions of the same temperature step). These dynamic characteristics of the system are important both for providing thermal encoded information through rapid temperature variations and for simulating the thermal interaction during contact with a virtual object (although the latter application was not investigated in this study). Finally, to show that this device can simulate most of the objects found in our environment, we compare the cooling rate induced on the finger by a copper slab with the one induced by the device.

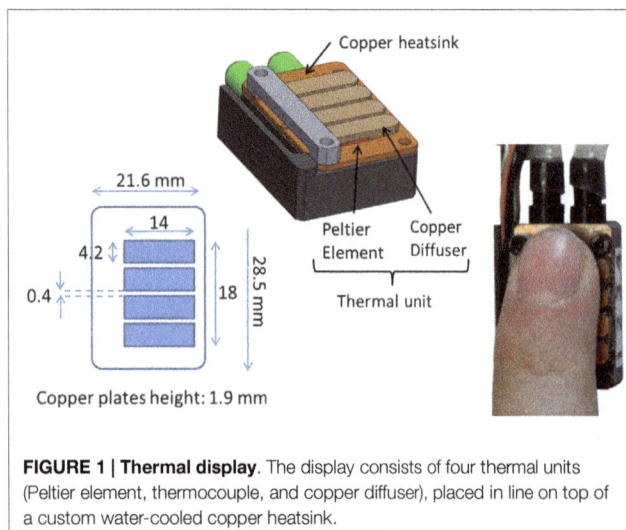

FIGURE 1 | **Thermal display**. The display consists of four thermal units (Peltier element, thermocouple, and copper diffuser), placed in line on top of a custom water-cooled copper heatsink.

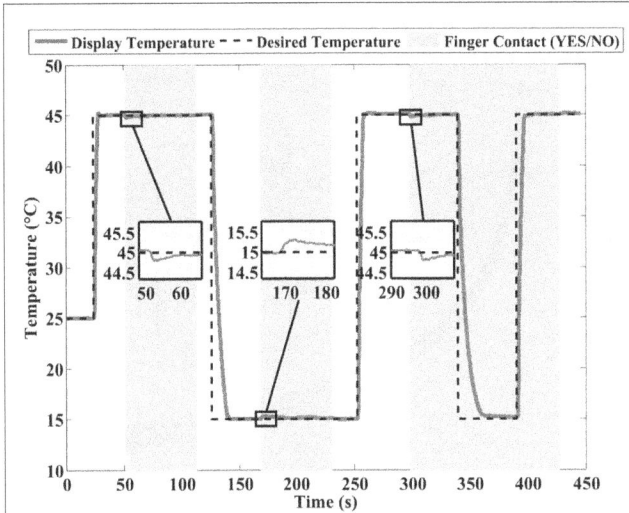

FIGURE 2 | Single Peltier behavior. The behavior of a single Peltier unit with and without finger contact for temperature steps from 15 to 45°C (and vice versa) is illustrated. Temperature stability is within ±0.15°C of the desired temperature and response to the external perturbation is within ±0.5°C.

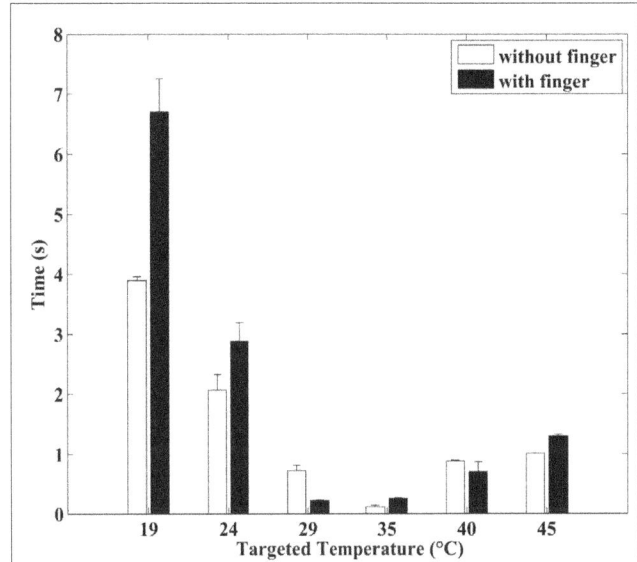

FIGURE 3 | Effect of thermal load and temperature step magnitude on the temperature change rate. Rise and fall time (from 10 to 90% of the desired value and 90 to 10%, respectively) were measured (six different steps, all starting from 32°C) with and without finger contact. Smaller time indicates faster response of the device (higher bandwidth). Error bars represent the SD of the mean.

Commercial water heatsinks are efficient but too bulky for portable applications. Thus, in order to miniaturize thermal displays, custom heatsinks must be produced and their performance assessed. The custom heatsink mounted on the proposed device was found to maintain a constant temperature for any input current of the PE, as shown earlier (Gallo et al., 2012), and can thus evacuate the heat generated by the PE.

Figure 2 illustrates the behavior of a single PE with and without finger for command steps from 15 to 45°C [not in the thermal pain range (Jones and Ho, 2008)] and vice versa. Temperature stability within ±0.15°C of the desired temperature and within ±0.5°C with the finger contact was obtained. From the same figure one can also notice the influence of the finger on the step response.

In order to quantitatively determine the influence of the thermal load as well as the temperature step magnitude on the temperature change rate, the rise and fall time (from 10 to 90% of the desired value and 90–10%, respectively) were measured for six different step heights with and without finger contact. **Figure 3** presents the mean fall and rise times and final temperature for each step in each condition (with and without finger). Before each step, the finger–PE contact temperature was stabilized at a 32°C initial temperature. During the experiment, room temperature varied between 24 and 27°C. The finger thermal load has little influence on the performance of the device for most steps, especially between room temperature and finger temperature (typical range for object discrimination). Fall/rise times increase with cooling/heating step size. This is expected as the PE cooling power decreases with increasing ΔT_{pelt}. Small SDs despite variations in room temperature suggest a strong robustness of the device design.

When a finger comes in contact with an object at room temperature, at first the skin temperature drops rapidly, then the temperature change rate decreases as the finger and object temperatures tend toward equilibrium. This first phase, called early time sensation in (Yamamoto et al., 2004), is responsible for most of the thermal sensation. Therefore, during object manipulation, the perceived thermal sensation strongly depends on the initial temperature change rate determined by the thermal diffusivity of the object (Tiest and Kappers, 2009). The rise and fall times do not assess whether these parameters are precisely reproduced from one step to another. To evaluate their repeatability, a different metric is required. To do so, we tested the device by performing temperature steps from 32 to 24°C with and without index finger contact. Each condition was repeated 10 times for a total of 20 measurements per subject and 60 measurements in total (three male subjects, age range 26–29 years). The selected temperature step covers the transient temperature drop for most materials. The mean temperature profile and its SD were then derived for both conditions (with and without finger). Thus a mean and a SD were obtained for each discrete sample. In order to estimate the repeatability of the cooling profile, we considered the maximum SD (out of all the discrete points SDs) for both conditions, i.e., the largest spread point along the temperature profile. The maximum SDs (over the entire temperature profile), 0.42°C with finger and 0.47°C without finger, were below 6% of the step value. Considering that changes of 10% in the skin thermal properties (variance between subjects) can result in changes up to 8% in the skin temperature drop (Ho and Jones, 2008), the repeatability of this device is clearly satisfactory.

Finally, to compare the presented device with the state of the art devices, mainly used in object discrimination tasks, we determined whether each individual thermal unit had sufficient cooling power to simulate the temperature drop occurring when the finger contacts objects at room temperature. Due to its high thermal conductivity and density, copper is one of the materials in our environment with the highest thermal diffusivity thus generating the fastest skin temperature drops. Hence, if the thermal unit is able to cool the skin faster than a copper slab, it should have sufficient cooling power to simulate most materials in our environment. Hence, the temperature profile of a finger contacting a copper slab at room temperature (21°C) was measured. As the geometry of the copper slab will influence the initial heat transfer (Tiest and Kappers, 2008), the finger was pressed against the smaller surface of a 17 mm × 17 mm × 145 mm slab granting a sufficiently low Fourier number (~0.005) and ensuring the validity of the semi-infinite model used in many discrimination studies (Yamamoto et al., 2004; Ho and Jones, 2008). The finger was placed on one of the four PE and its temperature stabilized to the average initial finger temperature recorded just before contacting the copper slab in the previous experiment. The surface/finger was then cooled at the PE maximum rate down to the arbitrary temperature of 23°C. **Figure 4** shows the average and SD of the finger temperature drop when touching copper out of 15 measures as well as the average PE-induced temperature drop for a typical subject. The finger cooling rate is clearly higher with the PE than with the real copper slab. Therefore, it is safe to assume that each of the four thermal units (PE with water cooling) can be used to realistically simulate the interaction of our skin with most objects found in our environment.

PSYCHOPHYSICAL EVALUATION: ENCODED AND CROSSMODAL THERMAL STIMULATION

Experiment 1: Gradient Sensitivity and Spatial Modulation (Thermally Encoded Information)

No thermal display is currently able to provide several thermal stimuli simultaneously. Although this might be related to previous research showing that two distinct materials cannot be discriminated under one single fingertip based on thermal cues alone (Yang et al., 2009), no study has yet investigated whether several distinct and constant temperatures can be differentiated under the fingertip. This may be of great importance for novel applications of thermal interfaces such as using the display to present scalar, non-thermal information to the user under the form of encoded thermal stimuli (Zerkus et al., 1994; MacLean and Roderick, 1999). For example, Wilson et al. (2013) successfully used encoded thermal feedback to convey information in environments in which audio or vibrotactile feedback might be masked by noise or movements. In their studies, thermal feedback was used instead of vibration to provide warnings about incoming cellphone messages. By varying thermal parameters such as subjective intensity (moderate warm, intensive cold, etc.) or direction of the temperature change, it was possible to inform the user on the urgency of the message (very warm for urgent, mild warm for not urgent) or the identity of the sender (warm for family, cold for work, for example). The limited number of

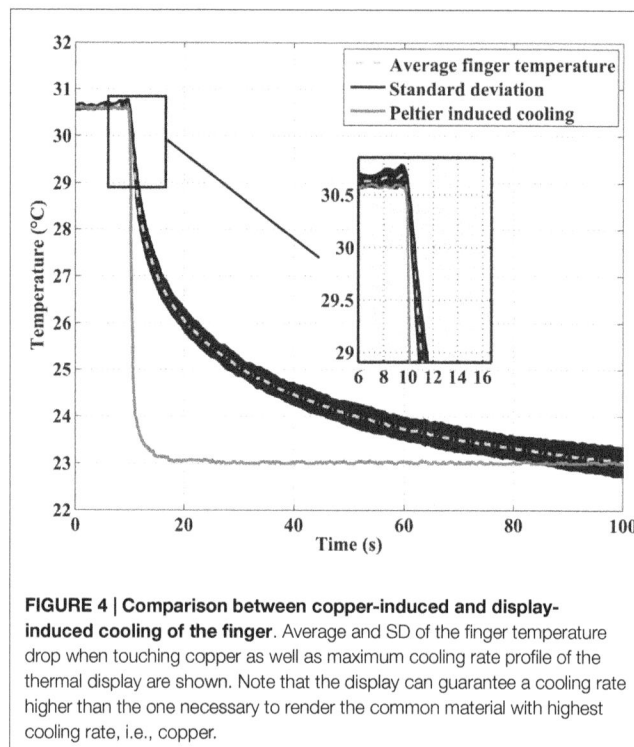

FIGURE 4 | Comparison between copper-induced and display-induced cooling of the finger. Average and SD of the finger temperature drop when touching copper as well as maximum cooling rate profile of the thermal display are shown. Note that the display can guarantee a cooling rate higher than the one necessary to render the common material with highest cooling rate, i.e., copper.

thermal parameters used to convey information simplifies the design and control of the thermal display. In comparison, thermal discrimination of virtual objects requires an accurate modeling of the heat transferred between the skin and the object (Bergamasco and Alessi, 1997; Yamamoto et al., 2004; Ho and Jones, 2008). This complicates the control scheme as thermal properties of the object and the skin, such as their thermal diffusivities (Tiest and Kappers, 2009) need to be taken into account. A display providing multiple thermal stimuli on small skin surfaces has the potential to provide a larger amount of information to the user, e.g., compared to a vibrator, while still being compact, discrete, and wearable.

Participants

A total of eight healthy right-handed participants took part in experiment 1 (two females; age range 23–30 years). All participants had neither history of neurological disorders nor any known tactile or thermal systems abnormalities. All participants gave written informed consent, and were compensated for their participation. The study protocol was approved by the local ethics research committee – La Commission d'Ethique de la Recherche Clinique de la Faculté de Biologie et de Médecine – at the University of Lausanne, Switzerland, and was performed in accordance with the ethical standards laid down in the Declaration of Helsinki.

Material, Methods, and Procedure

The aim of experiment 1 was to investigate the possibility to render encoded thermal information (perceive at least two different temperatures) using compact displays at the level of the fingertip. The minimum distance required to differentiate two distinct thermal stimuli will determinate the minimum size of the display. Although differentiating more than two simultaneous stimuli under the fingertip would increase the amount of encoded information, spatial summation (Kenshalo et al., 1967) and low spatial resolution (Yang et al., 2009) of the skin suggests that the differentiation of more than two neighboring stimuli on the fingertip will be difficult. Thus, we studied the minimal perceivable temperature difference between two stimuli (just noticeable difference, JND) at two distinct locations. We tested four different spatial configurations, 1–4, 2–4, 1–3, and 2–3 as shown in **Figure 5**. These configurations correspond to three distances between the centers of the two copper plates providing the thermal stimuli: 4.6, 9.2 (administrated twice at two different locations of the fingertip: 2–4 and 1–3), and 13.8 mm. The contact area on the four plates was estimated from images of the finger taken after it was colored with ink and pressed on a paper footprint of the device. The index finger was positioned on the device in order to obtain four similar contact surfaces of ~57 mm² for the central plates, and 54 mm² for the outer plates.

To calculate the JND, we used a yes/no choice paradigm with a two-down one-up staircase method (Wetherill and Levitt, 1965; Levitt, 1971), which estimates the JND with a 70.7% correctness. Thus, in this experiment, the JND is calculated as the minimum interval for which two different stimuli are recognized as such 70.7% of the times (Grassi and Soranzo, 2009). During the threshold tracking, two different step sizes were used for the online adjustment of the stimuli levels; one large (6°C) for the first four reversals and one small (2°C) for the next eight. The procedure stopped after 12 reversals (varying number of trials between subjects) and the final JND was calculated as an average over the last eight reversals (Grassi and Soranzo, 2009). These are based on typical practice in psychophysics and are due to an empirical trade-off between accuracy (the larger number of reversal the better) and experimental time (Wetherill and Levitt, 1965; Ernst and Banks, 2002; Hillis et al., 2002). The participants were asked to place their finger on an additional PE for 5 s in between each trial to reset the initial finger temperature to the neutral temperature measured before the tests.

Results

Globally, the estimated JND ranged from a minimum of 5.1°C to a maximum (poorest sensitivity) of 19.4°C across all subjects and all tested distances. Single subject data shows that the large variability is due to inter-individual rather than between condition differences. Importantly, all the reported thresholds were therefore within the range of investigation (i.e., a difference of 20°C centered at the body temperature of each participant). Repeated measures one-way ANOVA (four spatial conditions) did not show any significant difference between the four configurations ($P > 0.25$). More specifically, the JNDs (mean °C ± SD) were 12.6 ± 6°C for condition 1–4, 14.1 ± 5°C for condition 1–3, 13.3 ± 4.6°C for condition 2–3, and 12.1 ± 5°C for condition 2–4 (see **Figure 5**). Overall, when pooling all data together the average JND was 13.0 ± 4.2°C. These results provide no evidence for an effect of spatial configuration (contact distances along the finger) on the JND.

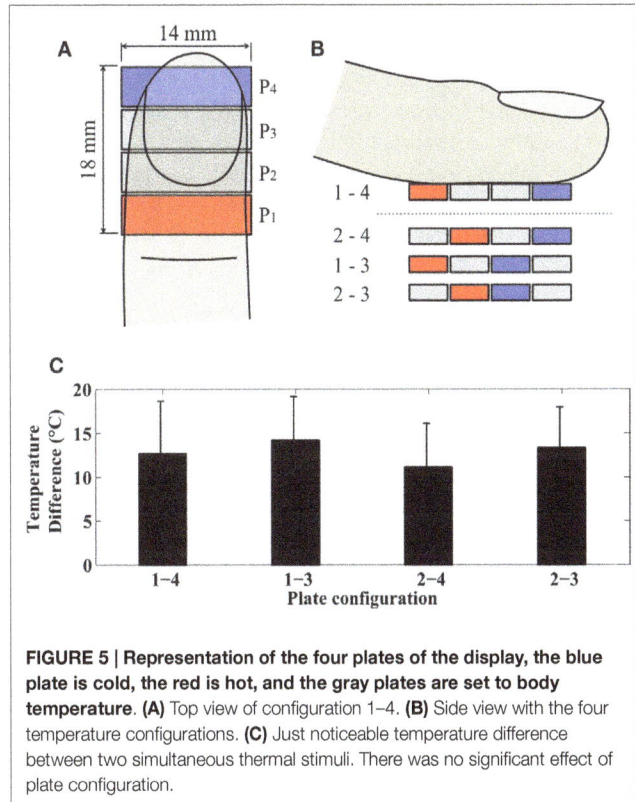

FIGURE 5 | Representation of the four plates of the display, the blue plate is cold, the red is hot, and the gray plates are set to body temperature. (A) Top view of configuration 1–4. **(B)** Side view with the four temperature configurations. **(C)** Just noticeable temperature difference between two simultaneous thermal stimuli. There was no significant effect of plate configuration.

Experiment 2: Manipulating Stiffness Perception Through Thermal Feedback (Crossmodal Thermal Stimulation)

Currently, haptic devices capable of rendering multiple haptic features exist (e.g., temperature, pressure, or force), but neglect crucial interactions among these features. This paucity is especially surprising for the case of skin temperature and tactile perception. Indeed, several studies highlighted the effect of skin temperature on the sensitivity of most mechanoreceptors, as reflected by changes in their activation threshold (Bolanowski and Verrillo, 1982; Verrillo and Bolanowski, 1986; Kunesch et al., 1987; Harazin and Harazin-Lechowska, 2007; Lowrey et al., 2013). This effect of temperature on mechanoreceptors has been shown to have a direct influence on tactile acuity (Stevens, 1982), the perception of roughness (Green et al., 1979), or two-point discrimination thresholds (Stevens, 1989). Interestingly, combined coding of mechanical and thermal stimulations could account for reported effects of skin temperature on object recognition by touch (Stevens and Hooper, 1982). In particular, the silver Thaler illusion, first described by Weber in 1846 [see Stevens and Green (1978)] involving cold objects to feel heavier than warm objects of equal weight and dimensions, appears to derive from the effects of thermal gradients on slow adaptive mechanoreceptors of both type I or II (SAI and/or SAII) (Cahusac and Noyce, 2007). Furthermore, for the perception of complex features resulting from a multimodal haptic integration, the influence of skin temperature is more intricate. For example, a strong implication of temperature in the perception of wetness was observed with effects different to the ones observed for vibrotactile perception (Filingeri et al., 2014). In fact, warm and wet stimuli have been found to suppress the perception of skin wetness (Filingeri et al., 2015) while on the contrary cool but dry objects were observed to evoke the perception of wetness (Filingeri et al., 2013). Given both the influence of tactile cues on stiffness discrimination (Srinivasan and LaMotte, 1995; Bergmann Tiest and Kappers, 2009) and of temperature on mechanoreceptor's sensitivity, our hypothesis was that temperature affects stiffness perception. More specifically, we expected the precision of stiffness perception to improve with increasing skin temperature.

Participants

A total of five healthy right-handed participants took part in experiment 2 (one female; mean age 25.8 years). All participants were free from neurological disorders and had no known history of somatic or thermoesthesia abnormalities. All participants gave written informed consent and were compensated for their participation. The study protocol was approved by the local ethics research committee – La Commission d'Ethique de la Recherche Clinique de la Faculté de Biologie et de Médecine – at the University of Lausanne, Switzerland, and was performed in accordance with the ethical standards laid down in the Declaration of Helsinki.

Material and Methods

In this experiment, the temperature of the entire fingertip had to be controlled. To provide a homogenous temperature, we covered the four PEs with one large copper plate (instead of the four separate plates), and placed a thermocouple on top of this copper plate, as close as possible to the finger contact surface. In order to display several controlled levels of stiffness, we then mounted the thermal display on an Omega 3 (Force Dimension, Switzerland) force feedback device. The three arms of this device, acting in parallel, are connected to a vertical end-plate on which different types of end-effectors can be mounted (**Figure 6A**). We manufactured a custom end-effector with two superposed horizontal plates, the bottom one supporting a force sensor and the top one housing the thermal display (**Figure 6A** and inset). The force sensor (CentoNewton; range: 0–10 N; Size: 16 mm square; Span: ±3%; Response time <10 ms) is taped to the top side of the bottom plate. The thermal display slides horizontally into a slot on the superior face of the upper plate and is secured with a screw. The finger force exerted on the thermal display is transmitted to the force sensor through a steal sphere, thus creating a one point contact removing undesired torques. This design, in combination with the blade in the upper plate, was implemented to measure only vertical forces and ensures a standardized range of forces exertion for stiffness judgments across participants. Indeed, previous studies (Yamamoto et al., 2004; Ho and Jones, 2008) have emphasized the role of the contact force on the quality of the heat exchange, but also on physiological factors such as the reduction of the blood flow in the fingertip. For forces above 2 N, the contact surface and the corresponding thermal resistance increase very slowly and can be considered constant (Ho and Jones, 2008). The mechanical interface can be schematized as two parallel springs as illustrated by **Figure 6** inset. This system is described by the following equation:

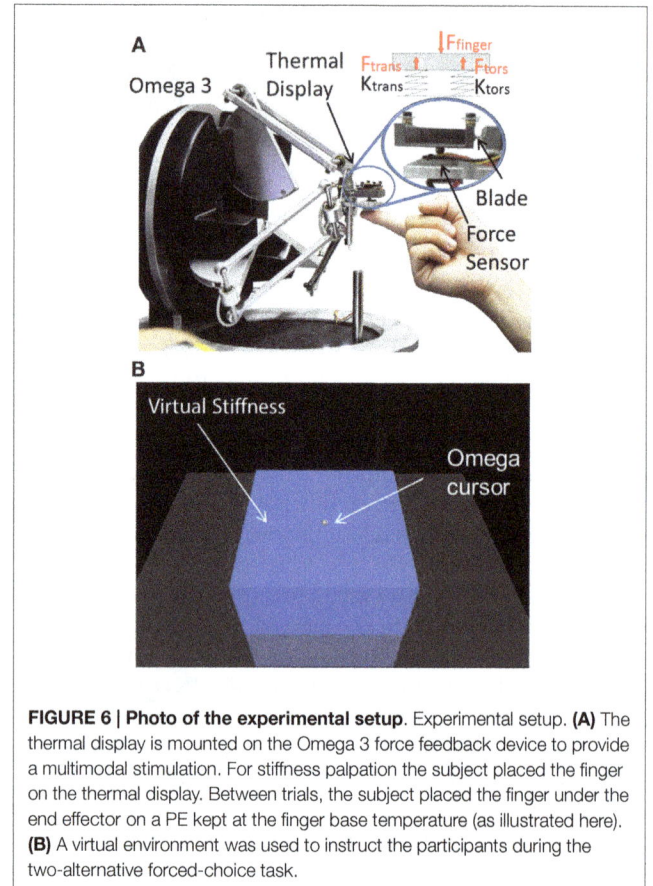

FIGURE 6 | Photo of the experimental setup. Experimental setup. **(A)** The thermal display is mounted on the Omega 3 force feedback device to provide a multimodal stimulation. For stiffness palpation the subject placed the finger on the thermal display. Between trials, the subject placed the finger under the end effector on a PE kept at the finger base temperature (as illustrated here). **(B)** A virtual environment was used to instruct the participants during the two-alternative forced-choice task.

$$F_{finger} = K_{tors} \times \delta_1 + K_{trans} \times \delta_2 \qquad (2)$$

where F_{finger} is the force applied by the user, K_{tors} and K_{trans} are the stiffnesses of the two springs representing the stiffness of the blade and of the force sensor, respectively, and δ_1 and δ_2 are the displacements of the springs. If K_{trans} is much larger than K_{tors}, the second spring can be neglected, which implies that the force measured by the force sensor is equal to the vertical component of the force applied by the user.

Procedure

To test our hypothesis that changing the temperature of mechanoreceptors at the fingertip would influence sensibility to stiffness, participants judged the relative stiffness of two virtual surfaces (the standard and the test) for three contact temperatures ($T1$: 15°C, $T2$: 25°C or $T3$: 40°C) in a two-alternative forced-choice task. The stiffness of the reference was 150 N/m. It was compared with any of seven equally spaced test stiffnesses in the interval 30–270 N/m (i.e., 30, 70, 110, 150, 190, 230, and 270 N/m) using the method of constant stimuli. The sensation of hardness can be perceived starting from 10,000 N/m (Lawrence and Chapel, 1994); the rendered objects are thus soft. These values were determined from stiffness specifications of the Omega 3 device and preliminary tests conducted on two subjects. The set of test stiffnesses is bounded by values corresponding to each end of the psychometric curve where discrimination could be achieved with nearly 100% accuracy.

In a typical trial, participants were asked to place the tip of their right index finger on the thermal display mounted on the Omega device and to push down until the contact with the surface of a virtual cube was perceived haptically (**Figure 6B**). An open-source platform (CHAI 3D)[1] was used for modeling and simulating the haptics (1 kHz refresh rate) and for visualization. The virtual scenario, displayed on a desktop screen, was only used to guide the subjects though the experimental procedure by signaling the beginning or end of a trial (break). The displayed cube was there to ensure that the subjects would push at the same approximate location (the center of the top face of the cube), but no visual cues related to the applied force, such as deformation or displacement of the cube, were provided. A "go" signal appeared on the screen if the finger force exerted on the surface was between 2 and 6 N and the contact lasted longer than 5 s. This was done to instruct participants about the amount of force they would need to apply during the task. Small movements of the index finger during stiffness perception were allowed (the role of kinesthetic cues in this experiment and in stiffness perception in general is discussed in the Section "Discussion and Conclusion"). The two stiffness stimuli, reference and test, were always presented with the same contact temperature in each trial. Subjects were given 5 s to palpate each stimulus. The stimuli were separated by an interval of 5 s. The entire trial duration was 20 s (2 s× 5 s force adjustment + 2 s× 5 s palpation). The time interval between trials, which varied from 5 to 30 s, depended on the temperature difference between consecutive trials (the larger the difference, the longer the time for the device to display the desired temperature). During this time interval, participants were asked to place the index finger on a PE kept at their base skin temperature measured at the beginning of the experiment. At the beginning of this time interval participants had 6 s to respond, via a button press, whether the second stiffness was greater than the first.

The standard stiffness was randomly assigned to the first or second presentation, but measured responses were always comparing the test with the standard. Trials in which the participants fail to give an answer or answered after 6 s were discarded (this accounted for only 2.0% of all trials). Each comparison was repeated 16 times for each contact temperature. This yielded a total of 336 trials per participant, recorded in four 2-h sessions performed on different days. Before the experiment, participants underwent a short training to familiarize themselves with the task.

Statistical Analysis

For each stiffness test at each selected temperature, individual responses were pooled across all participants to obtain a probabilistic measure of the response and yield a sufficient sample to fit a psychometric function and perform the statistical analysis (Wichmann and Hill, 2001a,b). This consisted in calculating the proportion of "stiffer" responses for every stiffness test for each

given temperature. Then, a cumulative Gaussian function of this type was used to fit the data:

$$y = \frac{1}{2} \times \mathrm{erfc}\left(\frac{\mu - x}{\sqrt{2}\sigma}\right) \qquad (3)$$

where, y is the fraction of trials perceived as stiffer, x represents the stiffness levels, μ is its mean and σ its SD. Psychophysically, μ represents the level of stiffness that subjects perceived 50% of trials as stiffer than the reference stiffness (i.e., the point of subjective equality, PSE), whereas σ corresponds to the 0.84 level of the curve and represents the precision of the stiffness judgment. Smaller σ corresponds to greater precision. Thus, the fitting procedure allows the extraction of measures of mean and SD for each temperature condition (i.e., three μ: $\mu T1$, $\mu T2$, $\mu T3$; and three SDs: $\sigma T1$, $\sigma T2$, $\sigma T3$). A bootstrap bias-corrected accelerated (BCa) analysis (9999 resamples) provided a 68.2% confidence interval for each measure [i.e., μ and σ; (Wichmann and Hill, 2001a,b; Hesterberg et al., 2005)]. This was chosen as it corresponds to ±1 SD when measures are normally distributed.

Permutation tests were used for the statistical analyses on the extracted parameters [μ and σ (Hesterberg et al., 2005)]. This approach was preferred to parametric testing as it uses a direct computation of the cumulative distribution of a test rather than an asymptotic approximation.

The permutation test re-samples N times the total number of observations, in a population sample, to build an empirical estimate of the null distribution from which the test statistic has been drawn (Belmonte and Yurgelun-Todd, 2001). In the case of this study, for each test (e.g., $\mu T1$ vs. $\mu T2$, $\sigma T1$ vs. $\sigma T2$) the null distribution was built by resampling without replacement 1000 times (N) from the two experiment matrices [80 (16 trials × 5 subjects) × 7 (stiffness test levels)] of the tested temperatures. For each repetition, a permuted experimental matrix was first created in which each row was randomly picked from one of the two original experimental matrices. Then, a cumulative Gaussian distribution was fit and the statistical parameters (μ, σ) of each permuted experimental matrix were computed.

P values were finally calculated by counting the times (M), the statistic value obtained in the original data set was smaller than the norm (two-tailed comparison) of or the statistic value itself (one-tailed comparison) obtained from the permuted data sets (null distribution), and dividing that value by the number of random permutations, i.e., $M/1000$. All the fits were performed using the open source library for MATLAB "psignifit"[2] (Wichmann and Hill, 2001a,b). Due to the initial hypothesis assuming that SD decreases with increasing temperature, a one-tailed permutation test was used when investigating the difference among SDs at different temperatures. Because of the absence of any prediction for possible temperature-related changes in the PSE, a two-tailed comparison was used in these cases. Significant effects were reported for $P < 0.05$.

---

[1]http://www.chai3d.org/

[2]http://bootstrap-software.org/psignifit/

Results

The hypothesis that warming the mechanoreceptors at the fingertip would increase stiffness precision entails that σ should decrease (increased precision) with increasing temperature (15, 25 or 40°C). As predicted, the data shows a progressive decrease of σ with increasing temperature:

$$\sigma T1 = 63.9\text{N}/\text{m} \left(\text{Confidence Interval, CI}\left[\text{N}/\text{m}\right]:\ 59.5,\ 68.6\right),$$

$$\sigma T2 = 56.6\text{N}/\text{m} \left(\text{CI}\left[\text{N}/\text{m}\right]:\ 53.6,\ 61.8\right),$$

$$\sigma T3 = 49.6\text{N}/\text{m} \left(\text{CI}\left[\text{N}/\text{m}\right]:\ 46.8,\ 53.9\right).$$

Crucially, $\sigma T3$ was significantly smaller than $\sigma T1$ [(one tailed permutation test; $P < 0.02$; (difference = 14.3 N/m, 22.3% change)]. However, neither the difference between $\sigma T1$ and $\sigma T2$ or $\sigma T2$ and $\sigma T3$ reached significance ($P > 0.05$, see **Figure 7**).

The PSE (mean of the different fit cumulative Gaussian) was not affected by temperature ($P > 0.05$): $\mu T1 = 162.2$ N/m (CI [N/m]: 157.4, 170.2), $\mu T2 = 165.9$ N/m (CI [N/m]: 162.2, 169.8), $\mu T3 = 157.8$ N/m (CI [N/m]: 153.7, 161.1). As mentioned in the method section for experiment 2, we decided to pool together all the trials from different subjects to have a better fit of the psychometric function (Wichmann and Hill, 2001a). However, fittings for individual participants showed a very similar pattern of results, with all subjects having a larger σ for $T1$ than $T3$. The decrease of the σ from $T1$ to $T3$ was of 14.6 N/m (25%) for subject 1, 0.7 N/m (1%) for subject 2, 36.4 N/m (38%) for subject 3, 10.0 N/m (17%) for subject 4, and 9.7 N/m (18%) for subject 5.

DISCUSSION AND CONCLUSION

The present study introduces a new high-spatial resolution finger-sized thermal display. The performance of the thermal device was assessed by means of new metrics accounting for both accuracy and repeatability. In the psychophysical evaluation, we first show that our device can render a perceivable temperature gradient (two different temperatures) at the level of the fingertip (experiment 1: thermally encoded information), thus extending the concept of temperature feedback as vector of information to fingertip-sized thermal displays. Second, we show that our display can be used to manipulate stiffness judgments through thermal feedback (experiment 2: crossmodal stimulation).

Device Performance

The finger-sized thermal display presented here can produce four independent in-line thermal stimuli under the fingertip within 18 mm of length. Thanks to its small size, the thermal display can be integrated on a commercial force feedback device, thus providing multimodal feedback. The PEs are current controlled, which grants a better dynamic response, stability, and repeatability compared to voltage-controlled PEs (as in our previous study; Gallo et al., 2012). The system response error is below 0.5°C with finger contact (<0.15°C without finger). The fall and rise times (which give an estimate of the dynamic performance of the device) calculated for several steps are unaffected by finger contact and the low SD shows a good repeatability. These time

constants are measured between 10 and 90% of the desired step; however, the initial transient response accounts for most of the thermal sensation (Yamamoto et al., 2004) needed to discriminate materials. To assess whether the transient response is repeatable, we calculated the mean and SD of the system cooling response to a step going from 32 to 24°C derived from multiple measures. The maximum SDs, 0.42°C with finger and 0.47°C without finger, are within 6% of the step value, which confirms the repeatability of the system. Finally, the display was shown to drop the finger temperature faster than any material in our environment, which confirms its usability also for thermal discrimination tasks.

Thermal Feedback to Convey Encoded Information at the Fingertip

In experiment 1, the possibility to convey encoded information using thermal feedback was tested. This feature of thermal feedback has been recently introduced (Wilson et al., 2013) and could allow the usage of thermal cues as detectable/alert signals in environment in which audio or vibrotactile cues might be masked by noise or movements. If each PE can provide n distinguishable temperature levels, then the thermal display featuring four PEs can potentially provide n^4 levels of information. However, the poor spatial resolution of the skin (Jones and Ho, 2008) is known to be a limiting factor in the perception of several thermal stimuli on a small skin surface. For this reason, the just noticeable difference was only investigated between two thermal stimuli applied on the fingertip, but for four different spatial configurations.

Results showed that a constant thermal gradient can be perceived under the fingertip for a minimal temperature difference of 13.0°C. Although our data provide no evidence for a dependency of the thermal gradient on the four tested configurations, all subjects were able to perceive a thermal gradient (two distinct temperatures) at the level of the fingertip. This finding is novel as previous research reported no capability to discriminate two different, adjacent, materials through thermal stimulation only (Yang et al., 2009). The divergence between the results might be due to the difference of the thermal stimulations. Indeed, Yang and colleagues simulated real objects, thus the two areas of the finger are cooled down at different rates. Conversely, in this study one area of the fingertip is warmed while the other is cooled down to predefined temperatures, thus the two temperatures change in opposite directions. In addition, the thermal stimuli in this study are presented along the finger while Yang and colleagues presented them transversally. In any case, our results are not conclusive and thereby further research is needed to understand the spatial modulation of thermal gradient perception. The capability to detect temperature differences between two adjacent stimulation sites suggests that compact displays (down to 10 mm × 10 mm stimulation area, i.e., the smallest tested distance in this experiment) could be used to convey up to three levels of thermally encoded information (warm, cool or different) under the fingertip. The proposed design can be easily adapted to provide two thermal stimulations with a very small heatsink facilitating integration and wearability. This design however, suffers from drawbacks including high-power consumption of

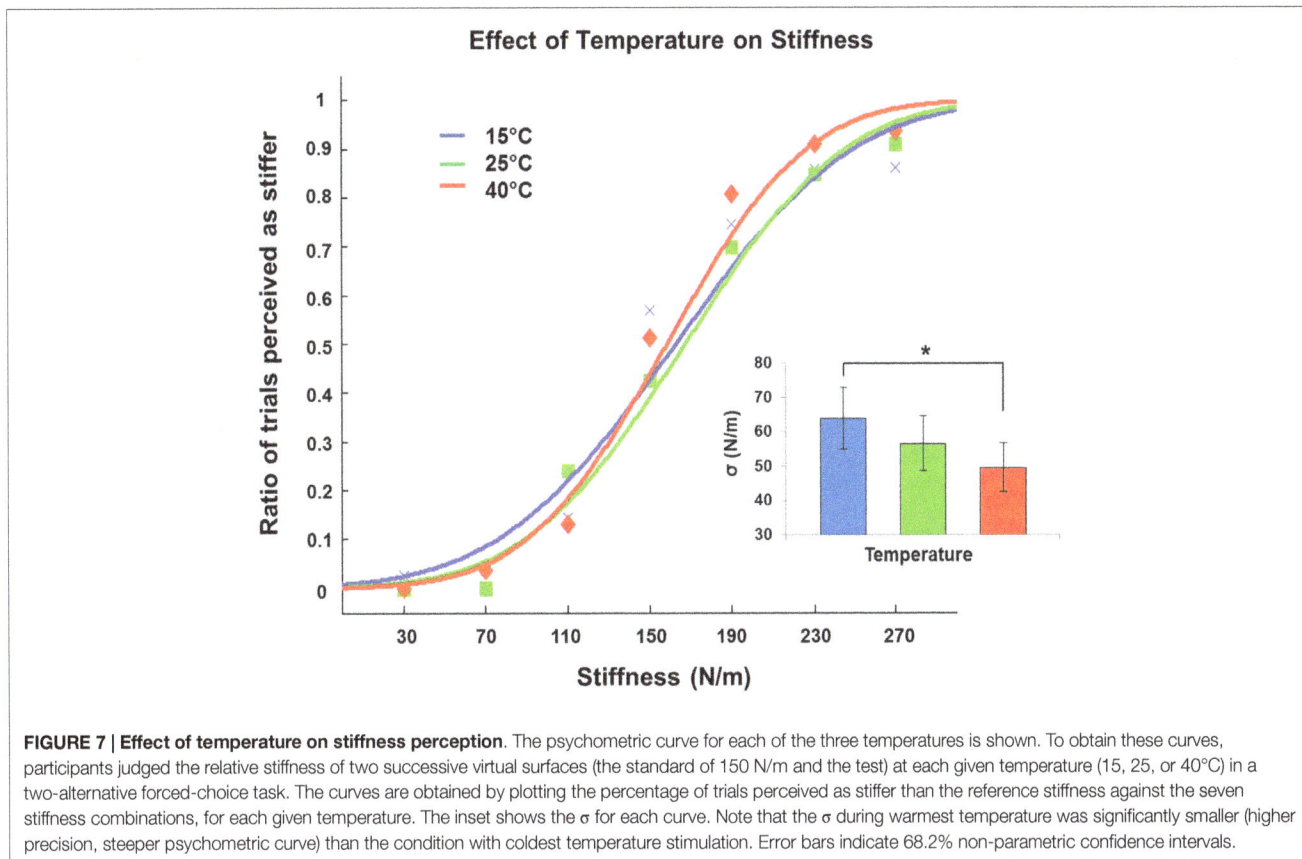

FIGURE 7 | Effect of temperature on stiffness perception. The psychometric curve for each of the three temperatures is shown. To obtain these curves, participants judged the relative stiffness of two successive virtual surfaces (the standard of 150 N/m and the test) at each given temperature (15, 25, or 40°C) in a two-alternative forced-choice task. The curves are obtained by plotting the percentage of trials perceived as stiffer than the reference stiffness against the seven stiffness combinations, for each given temperature. The inset shows the σ for each curve. Note that the σ during warmest temperature was significantly smaller (higher precision, steeper psychometric curve) than the condition with coldest temperature stimulation. Error bars indicate 68.2% non-parametric confidence intervals.

the PEs requiring a water-cooling system involving additional components such as a pump, a tank and a larger battery, thus decreasing wearability. On the other hand, providing thermally encoded information requires considerably less PE cooling power than simulating the contact with objects. Hence, for this application, low power PEs can be used and their excessive heat can be controlled using a compact air-cooling system, thus enabling the use of highly portable, multi-channel thermal displays to present encoded information on the user's skin.

Crossmodal Aspects of Thermal Feedback: Effect of Temperature on Stiffness Judgments

In experiment 2, we showed that increasing temperature increased the precision of stiffness judgments. Indeed, perceptual precision was 22.3% greater for the highest (40°C) than lowest (15°C) temperature tested.

For soft objects, tactile cues were found to be both necessary and sufficient to discriminate between different compliances, while for hard surfaces both tactile and kinesthetic cues were found to be necessary (Srinivasan and LaMotte, 1995). Tactile surface deformation cues were estimated to account for 90% of the information used to perceive compliance while force/displacement cues accounted for the remaining 10% (Bergmann Tiest and Kappers, 2009). These findings highlight the predominance of tactile cues (over kinesthetic cues), more specifically

surface deformation cues, for both the perception of hard and soft surfaces. Since the reported effect was found by using a hard surface while applying the thermal stimulation at the fingertip (thus affecting only the perception of tactile cues), we assume that the same or stronger effect will be present for soft surface objects in which compliance can be identified using tactile cues alone.

Possible Physiological Mechanisms

Several electrophysiological studies speak against a role of SAII and FAII in the coding of pressure distribution on the fingertip (Bolanowski and Verrillo, 1982; Kunesch et al., 1987; Harazin and Harazin-Lechowska, 2007), which – as mentioned above – has been argued to be the crucial component accounting for stiffness detection. Thus, the involvement of both receptors in this study is unlikely.

On the contrary, rate of change of pressure distribution and force are likely coded by SAIs and/or FAIs on the finger pad (Bergmann Tiest and Kappers, 2009); and both receptors have been shown to be influenced by temperature changes. Psychophysical studies based on vibration perception thresholds (VPT) have reported that the SAIs response to mechanical indentation is reduced in response to cooling (Kunesch et al., 1987). SAIs were also found to respond to cold gradients (Hensel and Zotterman, 1951; Duclaux and Kenshalo, 1972; Cahusac and Noyce, 2007) and warm gradients (Cahusac and Noyce, 2007) without a mechanical stimulation. This response appears as an increase of spontaneous firing for both cold and warm gradients.

These spontaneous responses could suggest that SAI are not responsible for the studied effect. However, warm gradients generate significantly larger spontaneous responses than cold gradients, with firing rates similar to the response evoked by mechanical stimuli (Cahusac and Noyce, 2007). In addition, the evoked response of SAI is reduced by cooling (Lowrey et al., 2013) and increased by heating of the skin (Cahusac and Noyce, 2007). Recent neurophysiological studies in humans have reported a reduction of the mechanically evoked response of FAI receptors resulting from the cooling of the subjects' skin (Lowrey et al., 2013). Altogether, these findings suggest that the hereby reported increase in stiffness precision due to warming of the skin is likely to be caused by the increase in sensitivity to static pressure and skin curvature during indentation of the SAIs and/or FAIs.

Finally, we note that, although our data show the possibility to use our device to provide thermal feedback to influence stiffness perception, more studies are needed to assess the robustness of our findings (e.g., using larger sample size) and to further investigate the perceptual mechanisms that connect temperature and stiffness perception.

Applications

Our results show that two crucial but often neglected features of thermal feedback – conveying encoded information and influencing tactile processing – can be provided by a fingertip-sized thermal display.

REFERENCES

Belmonte, M., and Yurgelun-Todd, D. (2001). Permutation testing made practical for functional magnetic resonance image analysis. *IEEE Trans. Med. Imaging* 20, 243–248. doi:10.1109/42.918475

Bergamasco, M., and Alessi, A. (1997). Thermal feedback in virtual environments. *Presence* 6, 617–629.

Bergmann Tiest, W. M., and Kappers, A. M. L. (2009). Cues for haptic perception of compliance. *IEEE Trans. Haptics* 2, 189–199. doi:10.1109/TOH.2009.16

Bolanowski, S., and Verrillo, R. (1982). Temperature and criterion effects in a somatosensory subsystem: a neurophysiological and psychophysical study. *J. Neurophysiol.* 48, 836–855.

Cahusac, P. M. B., and Noyce, R. (2007). A pharmacological study of slowly adapting mechanoreceptors responsive to cold thermal stimulation. *Neuroscience* 148, 489–500. doi:10.1016/j.neuroscience.2007.06.018

Duclaux, R., and Kenshalo, D. R. (1972). The temperature sensitivity of the type I slowly adapting mechanoreceptors in cats and monkeys. *J. Physiol.* 224, 647–664. doi:10.1113/jphysiol.1972.sp009917

Ernst, M. O., and Banks, M. S. (2002). Humans integrate visual and haptic information in a statistically optimal fashion. *Nature* 415, 429–433. doi:10.1038/415429a

Ernst, M. O., and Bülthoff, H. H. (2004). Merging the senses into a robust percept. *Trends Cogn. Sci.* 8, 162–169. doi:10.1016/j.tics.2004.02.002

Filingeri, D., Fournet, D., Hodder, S., and Havenith, G. (2014). Why wet feels wet? A neurophysiological model of human cutaneous wetness sensitivity. *J. Neurophysiol.* 112, 1457–1469. doi:10.1152/jn.00120.2014

Filingeri, D., Redortier, B., Hodder, S., and Havenith, G. (2013). The role of decreasing contact temperatures and skin cooling in the perception of skin wetness. *Neurosci. Lett.* 551, 65–69. doi:10.1016/j.neulet.2013.07.015

Filingeri, D., Redortier, B., Hodder, S., and Havenith, G. (2015). Warm temperature stimulus suppresses the perception of skin wetness during initial contact with a wet surface. *Skin Res. Technol.* 21, 9–14. doi:10.1111/srt.12148

Gallo, S., Santos-Carreras, L., Rognini, G., Hara, M., Yamamoto, A., and Higuchi, T. (2012). "Towards multimodal haptics for teleoperation: design of a tactile thermal display," in *Proc. 12th IEEE International Workshop on Advanced Motion Control (AMC)* (Sarajevo: IEEE), 1–5.

With wearable technology quickly expanding, the usage of a finger-worn thermal display has potential, especially as an alternative to vibrotactile feedback. In fact, thermal feedback can provide a greater amount of information per actuator compared to vibrotactile feedback. Moreover, while providing useful information, vibrotactile feedback also has a negative impact as it can generate undesired proprioceptive cues (illusion of skin being stretched) and can induce non-volitional movements (Lee et al., 2012) going against its original assistive purpose. In this case, a thermal feedback could be a better solution as it provides a sensible feedback without the undesired side effects of vibrotactile stimulation.

The reported effect of temperature on stiffness judgments opens up new scenarios in which thermal feedback could be used to influence stiffness perception precision (crossmodal feedback). We propose that crossmodal effects should be taken into account, or even exploited, when designing multimodal haptic displays.

ACKNOWLEDGMENTS

The authors wish to thank Bernard Martin and Roger Gassert for their proofreading and advice. This work was supported by a grant from the Swiss National Science Foundation (NCCR Robotics, Grant: 51AU40_125773).

Grassi, M., and Soranzo, A. (2009). MLP: A MATLAB toolbox for rapid and reliable auditory threshold estimation. *Behav. Res. Methods* 41, 20–28. doi:10.3758/BRM.41.1.20

Green, B. G., Lederman, S. J., and Stevens, J. C. (1979). The effect of skin temperature on the perception of roughness. *Sens. Processes* 3, 327–333.

Guiatni, M., Benallegue, A., and Kheddar, A. (2008). "Learning-based thermal rendering in telepresence," in *Haptics: Perception, Devices and Scenarios* ed M. Ferre (Madrid: Springer), 820–825.

Gwilliam, J. C., Degirmenci, A., Bianchi, M., and Okamura, A. M. (2012). "Design and control of an air-jet lump display," in *Proc. IEEE Haptics Symposium (IEEE)* (Vancouver: IEEE), 45–49.

Harazin, B., and Harazin-Lechowska, A. (2007). Effect of changes in finger skin temperature on vibrotactile perception threshold. *Int. J. Occup. Med. Environ. Health* 20, 223–227. doi:10.2478/v10001-007-0027-z

Hensel, H., and Zotterman, Y. (1951). The response of mechanoreceptors to thermal stimulation. *J. Physiol.* 115, 16–24. doi:10.1113/jphysiol.1951.sp004649

Hesterberg, T., Moore, D. S., Monaghan, S., Clipson, A., and Epstein, R. (2005). *Bootstrap methods and permutation tests*, 2nd Edn. New York: W. H. Freeman and Company.

Hillis, J. M., Ernst, M. O., Banks, M. S., and Landy, M. S. (2002). Combining sensory information: mandatory fusion within, but not between, senses. *Science* 298, 1627–1630. doi:10.1126/science.1075396

Ho, H.-N., and Jones, L. A. (2007). Development and evaluation of a thermal display for material identification and discrimination. *ACM Trans. Appl. Percept.* 4, 13. doi:10.1145/1265957.1265962

Ho, H.-N., and Jones, L. A. (2008). Modeling the thermal responses of the skin surface during hand-object interactions. *J. Biomech. Eng.* 130, 021005. doi:10.1115/1.2899574

Ino, S., Shimizu, S., Odagawa, T., Sato, M., Takahashi, M., Izumi, T., et al. (1993). "A tactile display for presenting quality of materials by changing the temperature of skin surface," in *Proc. 2nd IEEE International Workshop on Robot and Human Communication (IEEE)* (Tokyo: IEEE), 220–224.

Jones, L. A., and Ho, H.-N. (2008). Warm or cool, large or small? The challenge of thermal displays. *IEEE Trans. Haptics* 1, 53–70. doi:10.1109/TOH.2008.2

Kammermeier, P., Kron, A., Hoogen, J., and Schmidt, G. (2004). Display of holistic haptic sensations by combined tactile and kinesthetic feedback. *Presence* 13, 1–15. doi:10.1162/105474604774048199

Kenshalo, D. R., Decker, T., and Hamilton, A. (1967). Spatial summation on the forehead, forearm, and back produced by radiant and conducted heat. *J. Comp. Physiol. Psychol.* 63, 510. doi:10.1037/h0024610

Kim, K., Colgate, J. E., Santos-Munné, J. J., Makhlin, A., and Peshkin, M. A. (2010). On the design of miniature haptic devices for upper extremity prosthetics. *IEEE/ASME Trans. Mechatron.* 15, 27–39. doi:10.1109/TMECH.2009.2013944

Kunesch, E., Schmidt, R., Nordin, M., Wallin, U., and Hagbarth, K. (1987). Peripheral neural correlates of cutaneous anaesthesia induced by skin cooling in man. *Acta Physiol. Scand.* 129, 247–257. doi:10.1111/j.1748-1716.1987.tb08065.x

Lawrence, D., and Chapel, J. D. (1994). "Performance trade-offs for hand controller design," in *Proc. IEEE International Conference on Robotics and Automation (IEEE)* (San Diego: IEEE), 3211–3216.

Lee, B.-C., Martin, B. J., and Sienko, K. H. (2012). Directional postural responses induced by vibrotactile stimulations applied to the torso. *Exp. Brain Res.* 222, 471–482. doi:10.1007/s00221-012-3233-2

Levitt, H. (1971). Transformed up-down methods in psychoacoustics. *J. Acoust. Soc. Am.* 49, 467–477. doi:10.1121/1.1912375

Lowrey, C. R., Strzalkowski, N. D., and Bent, L. R. (2013). Cooling reduces the cutaneous afferent firing response to vibratory stimuli in glabrous skin of the human foot sole. *J. Neurophysiol.* 109, 839–850. doi:10.1152/jn.00381.2012

MacLean, K. E., and Roderick, J. B. (1999). "Smart tangible displays in the everyday world: a haptic door knob," in *Proc. IEEE/ASME International Conference on Advanced Intelligent Mechatronics (IEEE)* (Atlanta: IEEE), 203–208.

Markand, O. N., Warren, C., Mallik, G. S., King, R. D., Brown, J. W., and Mahomed, Y. (1990). Effects of hypothermia on short latency somatosensory evoked potentials in humans. *Electroencephalogr. Clin. Neurophysiol.* 77, 416–424. doi:10.1016/0168-5597(90)90002-U

McMahan, W., Gewirtz, J., Standish, D., Martin, P., Kunkel, J. A., Lilavois, M., et al. (2011). Tool contact acceleration feedback for telerobotic surgery. *IEEE Trans. Haptics* 4, 210–220. doi:10.1109/TOH.2011.31

Ottermo, M. V., Stavdahl, Ø, and Johansen, T. A. (2008). Design and performance of a prototype tactile shape display for minimally invasive surgery. *Haptics-e*, 4.

Phillips, J. R., and Matthews, P. B. (1993). Texture perception and afferent coding distorted by cooling the human ulnar nerve. *J. Neurosci.* 13, 2332–2341.

Russ, W., Sticher, J., Scheld, H., and Hempelmann, G. (1987). Effects of hypothermia on somatosensory evoked responses in man. *Br. J. Anaesth.* 59, 1484–1491. doi:10.1093/bja/59.12.1484

Srinivasan, M., and LaMotte, R. (1995). Tactual discrimination of softness. *J. Neurophysiol.* 73, 88–101.

Stevens, J. C. (1982). Temperature can sharpen tactile acuity. *Percept. Psychophys.* 31, 577–580. doi:10.3758/BF03204192

Stevens, J. C. (1989). Temperature and the two-point threshold. *Somatosens. Mot. Res.* 6, 275–284. doi:10.3109/08990228909144677

Stevens, J. C., and Green, B. G. (1978). Temperature–touch interaction: Weber's phenomenon revisited. *Sens. Processes* 2, 206–209.

Stevens, J. C., and Hooper, J. E. (1982). How skin and object temperature influence touch sensation. *Percept. Psychophys.* 32, 282–285. doi:10.3758/BF03206232

Tiest, W. M. B., and Kappers, A. M. (2008). Thermosensory reversal effect quantified. *Acta Psychol.* 127, 46–50. doi:10.1016/j.actpsy.2006.12.006

Tiest, W. M. B., and Kappers, A. M. (2009). Tactile perception of thermal diffusivity. *Atten. Percept. Psychophys.* 71, 481–489. doi:10.3758/APP.71.3.481

Verrillo, R. T., and Bolanowski, S. J. Jr. (1986). The effects of skin temperature on the psychophysical responses to vibration on glabrous and hairy skin. *J. Acoust. Soc. Am.* 80, 528. doi:10.1121/1.394047

Wetherill, G. B., and Levitt, H. (1965). Sequential estimation of points on a psychometric function. *Br. J. Math. Stat. Psychol.* 18, 1–10. doi:10.1111/j.2044-8317.1965.tb00689.x

Wichmann, F. A., and Hill, N. J. (2001a). The psychometric function: I. Fitting, sampling, and goodness of fit. *Percept. Psychophys.* 63, 1293–1313. doi:10.3758/BF03194544

Wichmann, F. A., and Hill, N. J. (2001b). The psychometric function: II. Bootstrap-based confidence intervals and sampling. *Percept. Psychophys.* 63, 1314–1329. doi:10.3758/BF03194545

Wilson, G., Brewster, S., Halvey, M., and Hughes, S. (2013). "Thermal feedback identification in a mobile environment," in *Proc. 8th International Workshop on Haptic and Audio Interaction Design (HAID)* (Daejeon: Springer Berlin Heidelberg), Vol. 7989, 10–19.

Yamamoto, A., Cros, B., Hashimoto, H., and Higuchi, T. (2004). "Control of thermal tactile display based on prediction of contact temperature," in *Proc. IEEE International Conference on Robotics and Automation (IEEE)* (New Orleans: IEEE), 1536–1541.

Yang, G.-H., Kwon, D.-S., and Jones, L. A. (2009). Spatial acuity and summation on the hand: The role of thermal cues in material discrimination. *Percept. Psychophys.* 71, 156–163. doi:10.3758/APP.71.1.156

Yang, G.-H., Yang, T.-H., Kim, S.-C., Kwon, D.-S., and Kang, S.-C. (2007). "Compact tactile display for fingertips with multiple vibrotactile actuator and thermoelectric module," in *Proc. IEEE International Conference on Robotics and Automation (IEEE)* (Roma: IEEE), 491–496.

Zerkus, M., Becker, B., Ward, J., and Halvorsen, L. (1994). "Thermal feedback in virtual reality and telerobotic systems," in *Proc. of the Fourth International Symposium on Measurement and Control in Robotics*, (Houston: NASA. Johnson Space Center) 107–113.

Conflict of Interest Statement: The authors declare that the research was conducted in the absence of any commercial or financial relationships that could be construed as a potential conflict of interest.

9

Robot Control for Task Performance and Enhanced Safety under Impact

Yiannis Karayiannidis[1,2], Leonidas Droukas[3], Dimitrios Papageorgiou[3] and Zoe Doulgeri[3]*

[1] *Department of Signals and Systems, Chalmers University of Technology, Gothenburg, Sweden,* [2] *Center for Autonomous Systems, Royal Institute of Technology (KTH), Stockholm, Sweden,* [3] *Department of Electrical and Computer Engineering, Aristotle University of Thessaloniki, Thessaloniki, Greece*

A control law combining motion performance quality and low stiffness reaction to unintended contacts is proposed in this work. It achieves prescribed performance evolution of the position error under disturbances up to a level related to model uncertainties and responds compliantly and with low stiffness to significant disturbances arising from impact forces. The controller employs a velocity reference signal in a model-based control law utilizing a non-linear time-dependent term, which embeds prescribed performance specifications and vanishes in case of significant disturbances. Simulation results with a three degrees of freedom (DOF) robot illustrate the motion performance and self-regulation of the output stiffness achieved by this controller under an external force, and highlights its advantages with respect to constant and switched impedance schemes. Experiments with a KUKA LWR4+ demonstrate its performance under impact with a human while following a desired trajectory.

Keywords: motion performance, safety, unintentional contact, control, variable stiffness

Edited by:
Nikola Miskovic,
University of Zagreb, Croatia

Reviewed by:
Ahmed Chemori,
CNRS, France
Arnaud Leleve,
Université de Lyon, France
Ivana Palunko,
University of Dubrovnik, Croatia

*Correspondence:
Zoe Doulgeri
doulgeri@eng.auth.gr

1. INTRODUCTION

A key challenge for the successful introduction of robots in human centered environments, as domestic assistants or co-workers, is the concurrent resolution of the issues of task performance and safety for the coexisting human (De Santis et al., 2010). Despite the need for the existence of collision avoidance mechanisms, to date, and in the context of service robots, there exists no sensor system that can guarantee collision avoidance with sufficient reliability. Hence, the possibility of collision with a human should still be accounted by minimizing the harm of such collisions.

Humans cope superbly with collisions and contact uncertainty by flexibly modulating their arm/hand compliance. Compliance protects the human from excessive forces during impact and can be achieved in robots either passively by using flexible components in the robot's structure or actively by the controller. Passive compliance is very important for the reduction of the initial collision force, which is responsible for the so-called pre-collision safety (Heinzmann and Zelinsky, 2003), and may be achieved by using deformable material to cover the robot or by building robots with compliant joints (Choi et al., 2008; Wolf and Hirzinger, 2008; Tsagarakis et al., 2011; Albu-Schaeffer et al., 2012). Variable stiffness actuation allows the regulation of the joint stiffness to values set by a higher level controller. When joint mechanical compliance is absent, then all the responsibility of keeping collisions harmless is being transferred to the controller. Active compliance or joint stiffness regulation enhances comfort and gives the impression that the human is in control in intentional contacts but at unintentional contacts, it may fail to reduce the initial collision force due to the delays introduced in the control system by the contact detection and reaction mechanisms. The delay introduced by the proposed detection methods vary depending on sensing and/or the general method's computational requirements (Golz et al., 2015). The residual torque method is a robot model-based contact detection utilizing proprioceptive sensors and one or more external RGB-D sensors to localize the contact point (De Luca and Mattone, 2005; De Luca et al., 2006; Magrini et al., 2015). Alternatively, variations in control effort can be utilized for estimating human contact without force sensing (Erden and Tomiyama, 2010). Recently, the utilization of a disturbance observer is proposed in a frequency-shaped impedance control scheme, which, however, is mainly addressed to intentional physical human–robot interaction (Oh et al., 2014). The problem of discriminating contacts to intentional and unintentional is also examined in Golz et al. (2015) where a machine learning method combined with features of physical contact models is proposed. Once a collision is detected, the robot switches from the control law associated to its nominal task to that of a reaction control law. Switching may be another source of delay and may in general adversely affect the stability of the overall switched system (Haddadin et al., 2008). Moreover, in the case of variable stiffness actuators, the bandwidth of the stiffness actuating system is crucial for responding promptly to unexpected impacts. On the other hand, controllers achieving an output impedance at safety level are characterized by poor performance. Traditionally, performing robot tasks with performance quality is associated with stiffness and hence a potentially unsafe contact (Bicchi and Tonietti, 2004). Thus, non-linear stiffness terms are introduced in impedance controllers for physical human–robot interaction control purposes setting different stiffness values in relation to deviation sizes around a nominal trajectory (Lee and Ott, 2011).

As service robots have to perform useful tasks for humans in a dynamic and uncertain environment, quality of performance is desired. The prescribed performance control methodology introduced in Bechlioulis and Rovithakis (2008) has been utilized for designing robot motion and/or force controllers guaranteeing prescribed performance for the output error (Bechlioulis et al., 2012; Karayiannidis and Doulgeri, 2012). In fact, prescribed performance controllers do not allow the output error to escape the performance region guaranteeing prescribed performance; they are robust to any external disturbance by utilizing a transformed error, which is approaching infinity at the performance boundaries and is not defined outside the performance region. Consequently, the control effort generated by a prescribed performance controller is increasing as the error approaches the boundary under the effect of a disturbance. The considerable stiffness induced by the prescribed performance control action may be undesired or even dangerous, if humans share the robot's workspace. Moreover, in practice, the output error may be forced outside the performance region due to the inability of the physical actuator to provide the demanded control effort or by not employing sufficiently high sampling rates. In such instances the system becomes uncontrollable.

The aim of this work is to concurrently address the competing requirements of motion performance and compliance under impact by designing a control scheme that achieves prescribed performance in nominal operation (high stiffness), a compliant reaction at impact (low stiffness) and smooth transition between the two modes. The system self-regulates the output stiffness

according to the disturbance level without explicit collision detection and control switching which are subjective to delays and jeopardize performance and stability. The feedback controller is model-based assuming knowledge of the robot's model and measurements of joint positions and velocities. The paper is organized as follows: In Section 2, we consider a simple first-order integrator system in order to define the *nominal performance operation* and *impact reaction* modes and introduce the basic control idea. Section 3 proposes a passivity-based motion controller utilizing the robot model, with an outer loop based on the control idea introduced in Section 2 achieving prescribed performance of a robot's task position tracking error and low stiffness compliance under impact. Section 4 presents simulation and experimental results illustrating the motion performance and self output stiffness regulation of the proposed controller under impact while conclusions are drawn in Section 5.

2. OPERATION MODES AND CONTROL PRELIMINARIES

Consider a first-order integrator scalar system of a tracking error e under disturbance $d(t)$:

$$\dot{e} = u + d(t), \text{ with } |d(t)| \leq \Delta, \forall t , \qquad (1)$$

where u is the control input; such type of system may model a kinematically or velocity controlled, robotic degree of freedom. We shall utilize this system to define operation modes and introduce the basic control idea.

2.1. Operation Modes

For system [equation (1)], we define two modes of system operation: the *nominal performance operation* and *impact reaction*. The definition of the nominal performance system operation is motivated by the prescribed performance concept (Bechlioulis and Rovithakis, 2008). A system is under its nominal performance operation if the tracking error $e(t)$ evolves strictly within a predefined region that is bounded by a decaying function of time constructed by the designer. Otherwise, the system is operating in the impact reaction mode. The following is the mathematical expression of the nominal performance operation:

$$-\rho(t) < e(t) < \rho(t) , \; \forall t \geq 0 \qquad (2)$$

where $\rho(t)$ is a bounded, smooth, strictly positive, and decreasing function satisfying $\lim_{t \to \infty} \rho(t) = \rho_\infty > 0$ called performance function. A candidate performance function is the exponential

$$\rho(t) = (\rho_0 - \rho_\infty) \exp(-lt) + \rho_\infty \qquad (3)$$

with ρ_0, ρ_∞, l strictly positive constants expressing nominal performance specifications. Constant $\rho_0 = \rho(0) > |e(0)|$ and is selected as described in Remark 2. Constant ρ_∞ represents the maximum allowable size of the output error $e(t)$ at steady state. Furthermore, constant l, which is related to the decreasing rate of $\rho(t)$, introduces a lower bound on the required speed of convergence of $e(t)$. An illustration of the nominal performance error

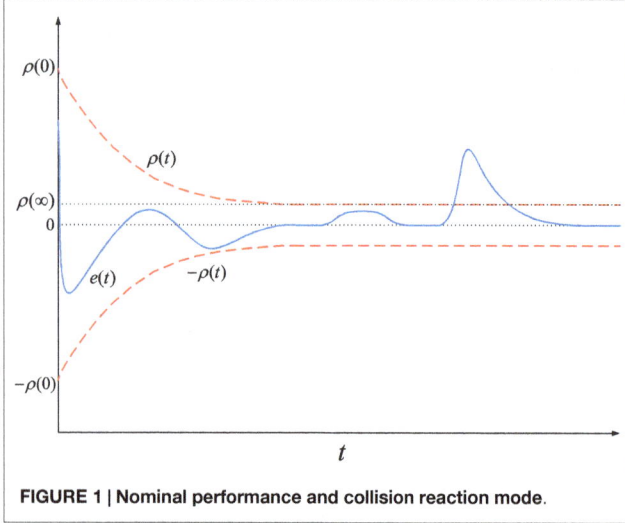

FIGURE 1 | Nominal performance and collision reaction mode.

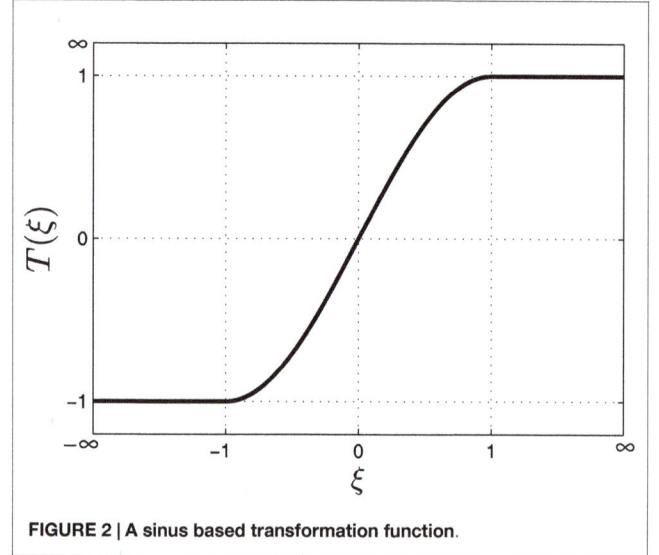

FIGURE 2 | A sinus based transformation function.

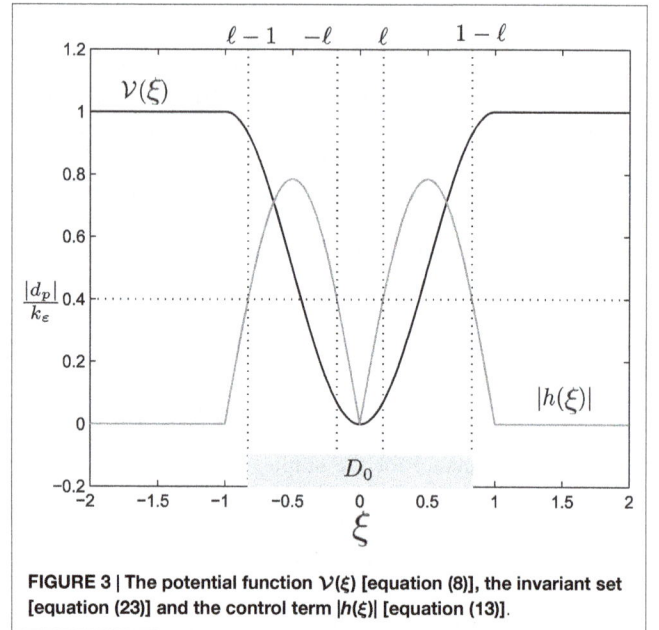

FIGURE 3 | The potential function $\mathcal{V}(\xi)$ [equation (8)], the invariant set [equation (23)] and the control term $|h(\xi)|$ [equation (13)].

evolution and of the error evolution in the impact reaction mode is shown in **Figure 1**.

By considering the modulated error $\xi = \frac{e(t)}{\rho(t)}$, we can define the system in its nominal performance operation if $|\xi(t)| < 1$, and in impact reaction mode if $|\xi(t)| \geq 1$, i.e., when the output error evolves outside the performance bounds (**Figure 1**). Let us denote with D the region of nominal performance operation, i.e., $D \triangleq (-1,1)$ while $D^c = \Re \setminus D$ is the complement set of D.

2.2. Basic Control Idea

In this work, we combine both requirements of nominal performance and enhanced safety operation under impact by proposing a new controller design based on a transformation, which defines the following smooth, non-decreasing, non-linear, surjective mapping of the modulated error domain:

$$
\begin{aligned}
T : (-\infty, -1) &\to -1, && \text{for } \xi(t) \leq -1 \,, \\
T : [-1, 1] &\to [-1, 1], && \text{for } -1 \leq \xi(t) \leq 1 \,, \quad (4)\\
T : (1, \infty) &\to 1, && \text{for } \xi(t) \geq 1 \,,
\end{aligned}
$$

further satisfying the following properties:

$$
\begin{aligned}
&T(0) = 0 \,, \\
&\frac{\partial T}{\partial \xi} > 0, && \forall \xi \in D \,, \\
&\frac{\partial^2 T}{\partial \xi^2} < 0, && \forall \xi : 0 < \xi < 1 \,, \quad (5)\\
&\frac{\partial^2 T}{\partial \xi^2} > 0, && \forall \xi : -1 < \xi < 0 \,.
\end{aligned}
$$

Hence, the transformation is strictly increasing in ξ for $\xi \in D$ and saturated on and beyond the prescribed performance boundaries, i.e., for $\xi \in D^c$. Moreover, the transformation is concave in the first quadrant and convex in the third. The following is a candidate transformation function illustrated in **Figure 2**:

$$
T(\xi) = \begin{cases} \sin\left(\frac{\pi}{2}\xi\right) & \text{for } -1 \leq \xi(t) \leq 1 \\ 1 & \text{for } \xi(t) > 1 \\ -1 & \text{for } \xi(t) < -1 \end{cases} . \quad (6)
$$

Notice that in place of equation (6), standard polynomials satisfying equation (4), and (5) may be utilized.

The artificial potential induced by such saturated transformation:

$$
\mathcal{V}(\xi) = T^2(\xi) : \Re \to [0, 1] \quad (7)
$$

is continuously differentiable, positive definite, i.e., $\mathcal{V}(\xi) > 0$ for $\xi \in \Re - \{0\}$, but it is not radially unbounded since regions $\mathcal{V}(\xi) \leq \beta$ are only closed for values of $\beta < 1$. Such potentials may allow a solution to escape the nominal performance region as opposed to the potentials induced by the transformations utilized in the prescribed performance controllers (Karayiannidis and Doulgeri, 2012). **Figure 3** gives an illustration of the potential function induced by equation (6):

$$
\mathcal{V}(\xi) = \begin{cases} \frac{1}{2}(1 - \cos(\pi\xi)) & \text{for } \xi \in D \\ 1 & \text{for } \xi \notin D \end{cases} . \quad (8)
$$

Let us further define:

$$h(\xi) \triangleq \frac{1}{2} \frac{\partial \mathcal{V}(\xi)}{\partial \xi} = \frac{\partial T}{\partial \xi} T(\xi) . \tag{9}$$

Notice that $h(\xi)$ satisfies the following properties:

$$h(\xi) = 0 \text{ for } \xi \in D^c , \tag{10}$$

$$h(\xi) \le c_h \xi \text{ for some } c_h > 0 , \tag{11}$$

$$|h(\xi)| \le h_M . \tag{12}$$

Property [equation (10)] is enabled by the bounded potential $\mathcal{V}(\xi)$, which is not typical in control design. This type of potential is unsuitable for global asymptotic stabilization and robustness analysis but allows different control actions without switching. For the transformation function given by equation (6):

$$h(\xi) = \begin{cases} \frac{\pi}{4} \sin(\pi \xi), & \text{for } \xi \in D \\ 0, & \text{for } \xi \notin D \end{cases} . \tag{13}$$

Function $h(\xi)$ lies in the first and third quadrant satisfying equation (11); its absolute value is illustrated in **Figure 3** yielding $h_M = \frac{\pi}{4}$.

Using $h(\xi)$ [equation (9)], we can design a simple control input u for equation (1) as follows:

$$u = - \left[\alpha(t) + k_s \right] e - k h(\xi) \tag{14}$$

where k_s, k are positive control constants, and $\alpha(t) \triangleq \frac{-\dot{\rho}(t)}{\rho(t)}$ is non-negative and bounded; for the exponential performance function, $\alpha(t)$ is, further, strictly decreasing with $0 < \alpha(t) \le \alpha(0) < l$, $\lim_{t \to \infty} \alpha(t) = 0$.

REMARK 1. (control philosophy): In case, the error is forced outside the nominal performance region [equation (2)] by a significant disturbance owing to a collision, the control term involving $h(\xi)$ vanishes due to property [equation (10)], while the remaining terms can be viewed as a proportional control action with a small decreasing gain in the case of the exponential performance. Hence, the proposed control law works independently of a detection or observation of the interaction force and has the advantage of avoiding switching between two controllers as opposed to other strategies.

Substituting control input [equation (14)] to the system [equation (1)] yields:

$$\dot{e} = - \left[\alpha(t) + k_s \right] e - k h(\xi) + d(t) . \tag{15}$$

Differentiating $\xi = \frac{e(t)}{\rho(t)}$ with respect to time yields $\dot{\xi} = \frac{\dot{e} + \alpha(t) e}{\rho(t)}$ and substituting $\dot{e}$ from equation (15), we get the closed-loop system expressed with respect to the modulated error:

$$\dot{\xi} = \frac{1}{\rho(t)} \left[-k h(\xi) + d(t) \right] - k_s \xi . \tag{16}$$

For the unforced non-autonomous system [equation (16)], i.e., $d(t) = 0$, it is easy to establish that the origin $\xi = 0$ is a uniformly

asymptotically stable equilibrium in $D \oplus D^c$. Notice that if the system operates in impact reaction mode then the unforced closed-loop system [equation (16)] becomes $\dot{\xi} = -k_s \xi$; hence ξ is drawn to ± 1 with a time constant $1/k_s$ that is, the error e is reaching the boundary of the prescribed performance region $\pm \rho(t)$. Given that no disturbance is acting at the system, ξ will return to the nominal operation mode ($|\xi| < 1$); hence, e will cross the boundary converging uniformly and asymptotically to $e = 0$ ($\xi = 0$).

In the presence of a bounded input $d(t)$, the following theorem establishes the range of disturbances guaranteeing system operation in nominal mode:

THEOREM 1. Consider a bounded disturbance input $|d(t)| < \Delta$ for the non-linear system [equation (16)] such that:

$$\Delta \le h_M k, \tag{17}$$

Then, there exists an invariant set $D_0 \subset D$ for the system state ξ; that is, initializing within D_0 guarantees a nominal performance error evolution in the sense of equation (2).

Proof: Using equation (7) for equation (16) the following can be satisfied in D: $\alpha_1(|\xi|) \le \mathcal{V}(\xi) \le \alpha_2(|\xi|)$ where α_1, α_2 are class $\mathcal{K}$ functions, and

$$\dot{\mathcal{V}}(\xi) = -\frac{2k}{\rho(t)} h^2(\xi) + \frac{2h(\xi) d(t)}{\rho(t)} - k_s \frac{\partial \mathcal{V}(\xi)}{\partial \xi} \xi , \tag{18}$$

$$\dot{\mathcal{V}}(\xi) \le \frac{1}{\rho(t)} \left[-k|h(\xi)|^2 + \frac{|d(t)|^2}{k} \right] - k_s \frac{\partial \mathcal{V}(\xi)}{\partial \xi} \xi , \tag{19}$$

If $|h(\xi)| \ge \frac{\Delta}{k}$ holds, then the quantity inside the brackets in equation (19) i.e., $-k|h(\xi)|^2 + \frac{|d(t)|^2}{k}$ is negative. Thus, we may write:

$$\dot{\mathcal{V}}(\xi) \le -k_s \frac{\partial \mathcal{V}(\xi)}{\partial \xi} \xi, \text{ for } |h(\xi)| \ge \frac{\Delta}{k} . \tag{20}$$

Next, we simplify the analysis by considering odd $h(\xi)$ functions although the analysis can be easily extended for the case of non-symmetric functions. If $h(\xi)$ is odd then, $|h(\xi)| = h(|\xi|)$ and it is now easier to calculate the domain of ξ wherein $h(|\xi|) \ge \frac{\Delta}{k}$. The equation $h(|\xi|) = \frac{\Delta}{k}$ can be solved with respect to $|\xi|$ if $\Delta \ne 0$ and equation (17) holds; the solution ζ_1, ζ_2 satisfy $0 < \zeta_1 < \zeta_2 < 1$ as shown in **Figure 3**. We can then write

$$\dot{\mathcal{V}}(\xi) \le -k_s \frac{\partial \mathcal{V}(\xi)}{\partial \xi} \xi, \text{ for } \varsigma_1 \le |\xi| \le \varsigma_2 . \tag{21}$$

Hence, defining

$$D_0 \triangleq \{ \xi \in D : |\xi| \le \varsigma_2 \}, \tag{22}$$

if $\xi(0) \in D_0$ then $\xi(t) \in D_0 \subset D$, $\forall t \in R^+$, which implies that D_0 is invariant and the system remains in nominal performance operation.

For the specific case of the candidate transformation function [equation (6)], the invariant set D_0 illustrated in **Figure 3** exists if $\Delta \le \frac{\pi k}{4}$ and is given by:

$$D_0 = \{ \xi \in D : |\xi| \le 1 - \ell \} \tag{23}$$

with

$$\ell = \frac{1}{\pi} \arcsin\left(\frac{4\Delta}{\pi k}\right) , \tag{24}$$

deriving from solving equation $h(|\xi|) = \frac{\Delta}{k}$ with respect to $|\xi|$, while taking into consideration [equation (13)] for $\xi \in D$.

REMARK 2. The maximum disturbance allowing a nominal performance operation mode [equation (17)] can be regulated by the control design constant k. Moreover, for nominal performance operation, constant ρ_0 of the performance function should be selected such that equation (22) is satisfied at $t = 0$, i.e., $\rho_0 \geq \frac{e(0)}{\varsigma_2}$.

REMARK 3. When the system operates in impact reaction mode, the closed-loop system $\dot{e} = -[\alpha(t) + k_s]e + d(t)$ or $\dot{\xi} = -k_s\xi + \frac{d(t)}{\rho(t)}$ is ISS (input-to-state stable) for the disturbance input $d(t)$ since $\alpha(t) + k_s \geq k_s > 0$. If t_e is the time instant the disturbance vanishes, the system will return to the nominal operation in $\frac{\ln \xi(t_e)}{k_s}$ s.

Since our objective is a robot-control design, which complies with large disturbances, there is no need of choosing high values for k_s in order to shrink the ultimate bound of the system given by $\frac{d(t)}{k_s}$ (since $\lim_{t \to \infty} \alpha(t) = 0$).

3. THE PROPOSED ROBOT CONTROLLER

Consider a n_q DOF robotic manipulator with $q \in \Re^{n_q}$ denoting its joint position vector and $p_e \in \Re^3$, $R_e \in SO(3)$ describing the position and the orientation of the end-effector with respect to the inertial frame, respectively. Let $v \triangleq [\dot{p}_e^T \ \omega_e^T]^T \in \Re^6$ denotes the end-effector generalized velocity with ω_e being the rotational velocity of the end-effector expressed at the inertial frame. Then, joint velocities are related to the generalized velocity with the robot Jacobian $J(q) \in \Re^{6 \times n_q}$ as follows:

$$v = J(q)\dot{q} . \tag{25}$$

The robot dynamic model can be written as follows:

$$M(q)\ddot{q} + C(q, \dot{q})\dot{q} + G(q) + \tau_d(t) = u , \tag{26}$$

where $M(q) \in \Re^{n_q \times n_q}$ is the positive definite robot inertia matrix, $C(q, \dot{q})\dot{q} \in \Re^{n_q}$ is the vector of Coriolis and centripetal forces, $G(q) \in \Re^{n_q}$ is the gravity vector, $u \in \Re^{n_q}$ is the control input joint torques and $\tau_d(t) \in \Re^{n_q}$ is a bounded joint disturbance typically arising by unforeseen collisions of the arm with a human or the environment. Let $p_d(t) \in \Re^3$, $R_d(t) \in SO(3)$ denotes the smooth and bounded desired end-effector position and orientation trajectories. Operational space tracking control employs both the position and orientation error. The position error is given by $e = p_e(t) - p_d(t)$ and for the orientation error, the outer product formulation, the Euler angle representation or quaternions may be used (Siciliano et al., 2010).

For simplicity and without loss of generality, we proceed by considering the position tracking problem. Thus, our objective is to design a state feedback control law, in order to force the robot's end-effector position $p_e(t)$ to track a given desired trajectory $p_d(t)$ with prescribed performance under its nominal operation mode in the sense of confining the evolution of each position error

coordinate $e_i(t)$ within a predefined region that is bounded by $\pm\rho_i(t)$ under small disturbances and to enable a smooth compliant reaction outside the performance region when the impact force is greater than an allowed level, returning to the nominal mode after the disturbance vanishes. We shall call this problem *Prescribed Motion Performance and Compliant Reaction(PMPCR)*.

THEOREM 2. Consider the model of a robotic manipulator equations (25) and (26), the desired trajectory $p_d(t) \in \Re^3$ and performance functions $\rho_i(t)$, $i = 1, 2, 3$ as defined in equation (3) that incorporate the desired performance bounds of the task position tracking error elements $e_i = p_{ei}(t) - p_{di}(t)$ in the nominal operation mode as well as transformations $T_i(\xi_i)$ as in equation (6) for the modulated error elements $\xi_i = \frac{e_i}{\rho_i}$. Moreover, define the intermediate control signals:

$$v_{ri} = \dot{p}_{di} - [\alpha_i(t) + k_{si}]e_i(t) - k_i h_i(\xi_i) , \tag{27}$$

where k_i, k_{si} are positive control constants, $h_i(\xi_i)$ is defined as in equation (9) and $\alpha_i(t) = \frac{-\dot{\rho}_i(t)}{\rho_i(t)}$. Assuming a robot motion away from singular positions, the passivity model-based control law:

$$u = -K_v s_q + M(q)\ddot{q}_r + C(q, \dot{q})\dot{q}_r + G(q) , \tag{28}$$

with K_v being a diagonal matrix of positive control constants, $s_q = \dot{q} - \dot{q}_r$ where $\dot{q}_r = J^+(q)v_r$, $J^+(q)$ with being a generalized pseudo-inverse of the Jacobian ($J^+(q) = J^{-1}(q)$ for the non-redundant case,) and $v_r \in \Re^3$ having v_{ri} entries given in equation (27), solves the PMPCR problem.

Proof: Substituting equation (28) in equation (26), we obtain the closed-loop system (Slotine and Li, 1991):

$$M(q)\dot{s}_q + C(q, \dot{q})s_q + K_v s_q + \tau_d(t) = 0 . \tag{29}$$

Consider now the positive definite radially unbounded function:

$$\mathcal{L} = \frac{1}{2}s_q^T M(q)s_q , \tag{30}$$

which satisfies the following inequality

$$\frac{\lambda_m}{2}\|s_q\|^2 \leq \mathcal{L} \leq \frac{\lambda_M}{2}\|s_q\|^2 , \tag{31}$$

where λ_m, λ_M are positive constants related to the robot's minimum and maximum eigenvalue of $M(q) \ \forall q$. Differentiating equation (30) with respect to time and substituting $\dot{s}_q$ from equation (29), while taking into account the skew symmetry of $\dot{M}(q) - C(q, \dot{q})$, we obtain:

$$\dot{\mathcal{L}} = -s_q^T K_v s_q - s_q^T \tau_d(t) . \tag{32}$$

Let k_v be the minimum entry of K_v; then, $\dot{\mathcal{L}}$ can be upper bounded as follows:

$$\dot{\mathcal{L}} \leq -k_v\|s_q\|^2 + \|s_q\|\|\tau_d(t)\| , \tag{33}$$

$$\dot{\mathcal{L}} \leq -\frac{1}{2}k_v\|s_q\|^2 - \frac{1}{2}k_v\|s_q\|\left(\|s_q\| - \frac{2\|\tau_d(t)\|}{k_v}\right) . \tag{34}$$

Defining the region $B = \{s_q \in \Re^n : \|s_q\| \leq \frac{2\|\tau_d(t)\|}{k_v}\}$, it is clear that:

$$\dot{\mathcal{L}} \leq -\frac{1}{2}k_v\|s_q\|^2, \quad \text{for } s_q \notin B, \tag{35}$$

which proves the uniform ultimately boundedness of s_q. In fact, using equations (31) and (35), it can be shown that

$$\|s_q\| \leq \sqrt{\frac{\lambda_M}{\lambda_m}}\|s_q(0)\|e^{-(\frac{k_v}{2\lambda_M})t}, \quad \text{for } s_q \notin B, \tag{36}$$

$$\|s_q\| \leq \sqrt{\frac{\lambda_M}{\lambda_m}}\frac{2\|\tau_d(t)\|}{k_v}, \quad \text{for } s_q \in B, \tag{37}$$

which can be combined in the following:

$$\|s_q\| \leq \max\sqrt{\frac{\lambda_M}{\lambda_m}}\left(\|s_q(0)\|e^{-(\frac{k_v}{2\lambda_M})t}, \frac{2\|\tau_d(t)\|}{k_v}\right), \tag{38}$$

demonstrating an input-to-state stability (Marquez, 2003), for the pair $\tau_d(t)$, s_q of equation (29).

Given equation (38), s_q is bounded for a bounded disturbance $\tau_d(t)$ and there exists a function of time $d_p(t)$ satisfying equation (38) such that $s_q = d_p(t)$ and substituting $s_q = \dot{q} - \dot{q}_r$ yields:

$$\dot{q} = J^+(q)v_r + d_p(t). \tag{39}$$

Multiplying with the robot Jacobian $J(q)$ and substituting equation (27) we obtain:

$$\dot{e}(t) = -[A(t) + K_s]e(t) - KH(\xi) + J(q)d_p(t), \tag{40}$$

where $A(t)$, K_s, K are diagonal matrices with entries $\alpha_i(t)$ and k_{si}, $k_i > 0$, respectively, and $H(\xi)$ is a vector with elements $h_i(\xi_i)$. Each element of equation (40) is related to the error scalar system [equation (15)] having as disturbance input the ith element of $J(q)d_p(t)$ in place of the $d(t)$ of equation (15). Let us take the example of an impact force F_{ext} applied to the robot's end-effector, which is mapped to the joint space as a disturbance torque $\tau_d(t) = J^T F_{ext}$. Let the system [equation (29)] be operating at the steady state, i.e., $e^{-(\frac{k_v}{2\lambda_M})t} \simeq 0$. From equation (38), we observe that $\|d_p(t)\| \sim \|\tau_d\|/k_v$ remains the main source of disturbance at the velocity control level. As Theorem 1 implies, the controller guarantees that the system is in nominal performance operation mode when the disturbance force is less than a tunable threshold (reflecting modeling errors) but allows the system to escape this mode for higher disturbances as are those arising from collision impacts. In this case, the robot reaction is stable and compliant, returning to the nominal operation mode after the disturbance vanishes thus solving the PMPSC problem.

REMARK 4. Notice that when the system operates in the impact reaction mode where $h_i(\xi_i) = 0$ and $\alpha_i(t) \simeq 0$ (in the steady state region of the performance function), the reference velocity [equation (27)] becomes $v_{ri} = \dot{p}_{di} - k_{si}e_i$ and the model-based controller [equation (28)] guarantees system stability.

REMARK 5. A brief unexpected contact is so far assumed in the presentation of the proposed control law. However, if the contact persists contact forces will keep increasing even with low output stiffness since the reference position advances. In that case, a post-impact strategy that abandons the desired trajectory is mandatory so that tracking errors cease to build up.

4. RESULTS AND DISCUSSION

4.1. Simulation Results

We consider a three DOF rotational joint spatial robotic manipulator with link masses $m_1 = m_2 = m_3 = 1$ kg, link lengths $l_2 = l_3 = 0.5$ m, and link inertias $I_{x2} = I_{x3} = I_{z1} = 4.15 \times 10^{-4}$ kg m^2, $I_{y2} = I_{z2} = 2.1 \times 10^{-2}$ kg m^2, and $I_{y3} = I_{z3} = 0.39 \times 10^{-2}$ kg m^2. The robot is initially at rest at the position $p_e(0) = [0.55\ 0.55\ 0.55]^T$ (m) and configuration $q(0) = [45\ 53\ -35.4]^T$ (deg) and is desired to move to the target location $p_{df} = [0.249\ 0.249\ 0.249]^T$ (m) for a duration of $T = 3$ s, following a fifth-order polynomial trajectory for each position coordinate: $p_d(t) = p_{do} + (p_{df} - p_{do})(10(\frac{t}{T})^3 - 15(\frac{t}{T})^4 + 6(\frac{t}{T})^5)$ where $p_{do} = [0.549\ 0.549\ 0.549]^T$ (m) is the desired trajectory's initial position resulting an initial position error of $e(0) = [0.001\ 0.001\ 0.001]^T$ (m). The performance function is defined as in equation (3) and considered to be the same for all position errors $e_i(t)$ $i = 1, 2, 3$, with $\rho_{i0} = 0.02$ set high enough to ensure initialization within the invariant set D_0 for a range of disturbance magnitudes, $\rho_{i\infty} = 10^{-3}$ corresponding to an accuracy of 1 mm and $l_i = 20$ for a fast transient response. Control constants from equations (28) to (27) are set to: $K_v = 15I_3$, with I_3 being the identity matrix of dimension 3, $k_{si} = 5$ and $k_i = 0.3$, $i = 1, 2, 3$.

We initially consider an impact force $F_{ext}(t)$ applied to the robot's end-effector along the x Cartesian direction in the form of a smooth pulse simulated by the function: $F_{ext}(t) = \frac{F_E}{2}(\tanh(100(t - 0.95)) - \tanh(100(t - 1.05)))$ N, where F_E is the pulse amplitude in order to evaluate the robot's reaction to a disturbance from the point of view of an apparent output stiffness (K_{stiff}) via a series of simulation runs with impact forces of various amplitudes in the range of 5–60 N with a step of 1 N. The stiffness values are calculated by the ratio of the pulse amplitude F_E to the maximum error displacement. Results from two simulation cases in nominal and impact reaction modes with $F_E = 10$ N and $F_E = 40$, respectively, are shown in **Figures 4A,B** depicting positions error responses and the respective impulse as well as the associated control efforts (**Figures 5A,B**). Notice the sudden control input increase when crossing the performance boundaries. **Figure 6** is an interpolation of the calculated output stiffness, revealing two distinct areas of stiffness values (K_{stiff}) corresponding to the nominal performance with high stiffness values and impact reaction modes with low stiffness values (**Figure 6** subplot). With the specific gain selection, it is clear that a nominal performance is achieved for disturbances up to approximately 12 N. This level can be regulated by changing the value of k_i. An impedance scheme with stiffness values in the range of the safe region would have resulted in a comparable compliant reaction under impact but in higher tracking errors during the nominal operation. In fact, simulating an impedance control scheme that does not require the measurement of the interaction force, which yields the following closed-loop $M_d\ddot{e} + K_d\dot{e} + K_pe = M_dJ(q)M(q)^{-1}\tau_d(t)$ (Siciliano et al., 2010) with Cartesian stiffness K_p at 660 N/m in all directions, results in tracking errors higher for at least one order of magnitude to those achieved by the proposed controller ($<10^{-3}$ m as prescribed).

Next, we consider joint disturbances arising by an uncertain gravity vector model. Joint disturbances arising from a partially

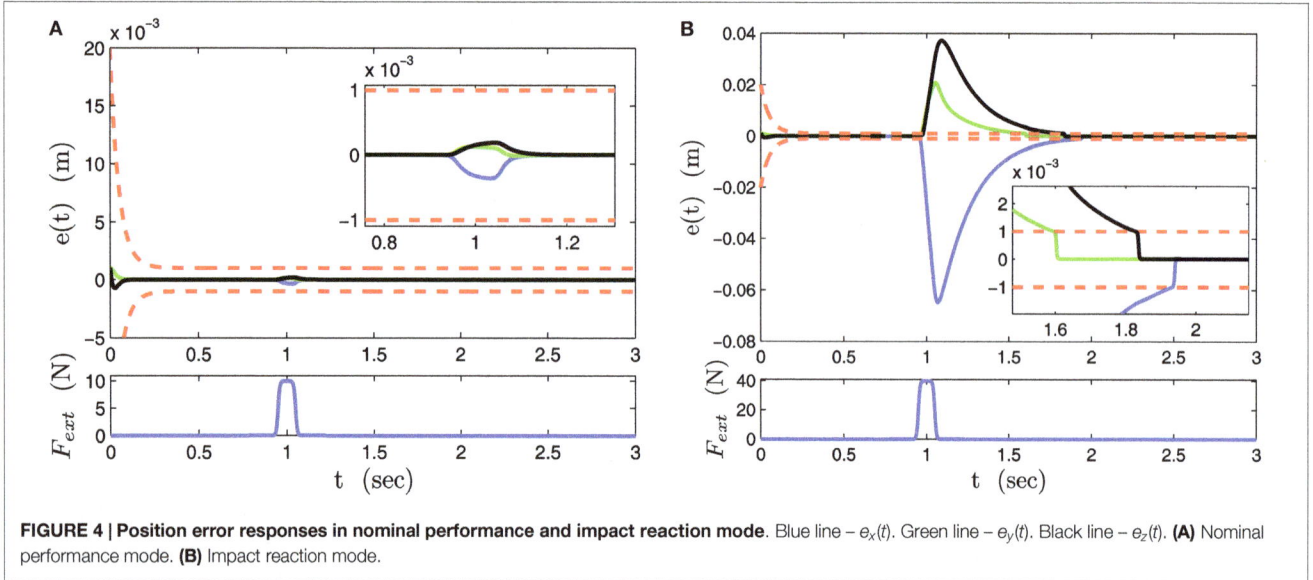

FIGURE 4 | Position error responses in nominal performance and impact reaction mode. Blue line – $e_x(t)$. Green line – $e_y(t)$. Black line – $e_z(t)$. **(A)** Nominal performance mode. **(B)** Impact reaction mode.

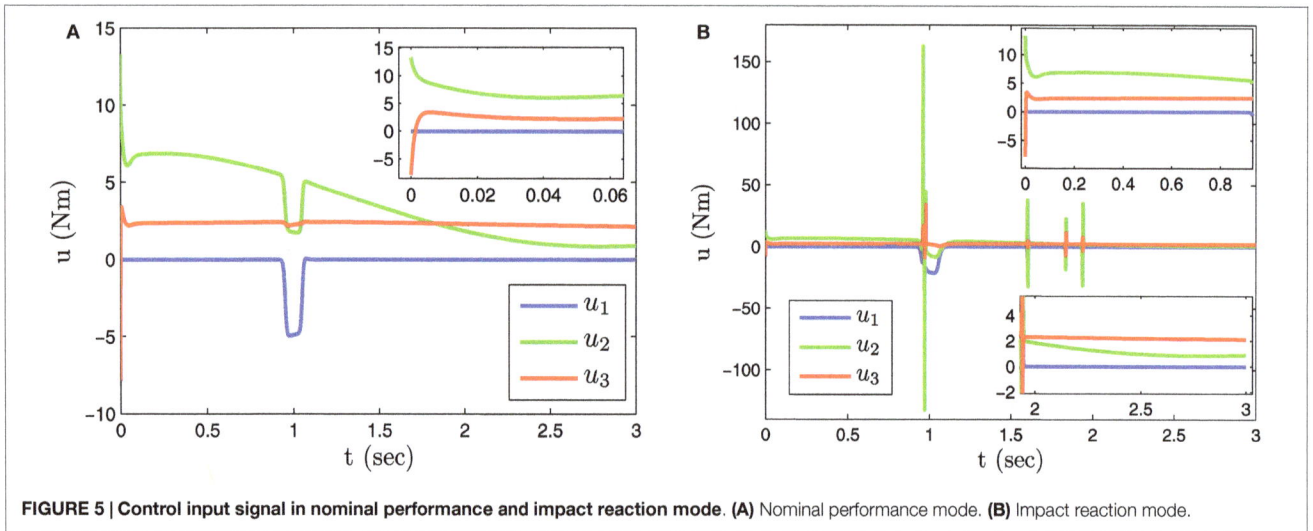

FIGURE 5 | Control input signal in nominal performance and impact reaction mode. (A) Nominal performance mode. **(B)** Impact reaction mode.

compensated gravity $0.6G(q)$ result in Cartesian disturbance forces shown in **Figure 7B**, while error responses together with the response of the system being fully compensated for gravity are shown in **Figure 7A**. Notice how the system stays in nominal operation respecting preselected performance boundaries despite the presence of disturbances.

Last, we have simulated the case of an impact with an environment modeled as a spring with stiffness of 1000 N/m, obstructing the motion of the arm for 0.5 s. For comparison purposes, we have simulated the case of the robot being under the impedance control scheme of high targeted stiffness as well as a switched impedance between the high and a low stiffness with a delay of 0.001 s (an ideal case examined for comparison purposes) and 0.2 s from the moment of impact in order to account for the time needed for the impact detection and reaction response (a practical switched impedance case). Stiffness values were selected from those appearing in the two modes of operation for the proposed controller and were 28,000 and 600 N/m, respectively. **Figure 8** displays the interaction forces developed during the impact. The proposed

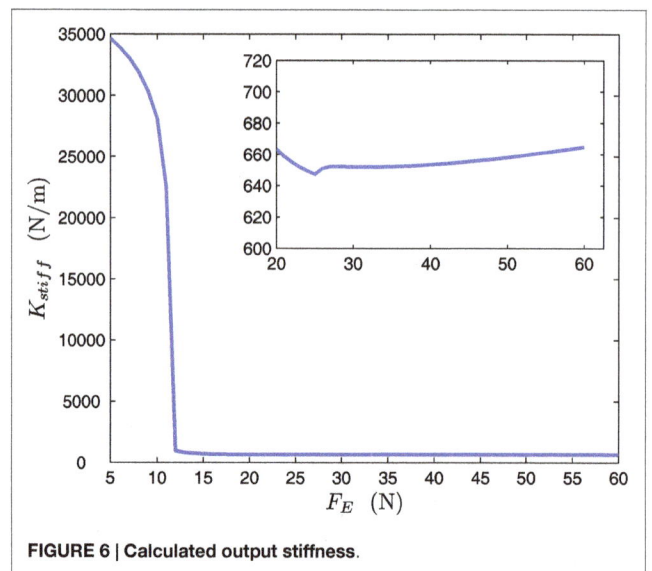

FIGURE 6 | Calculated output stiffness.

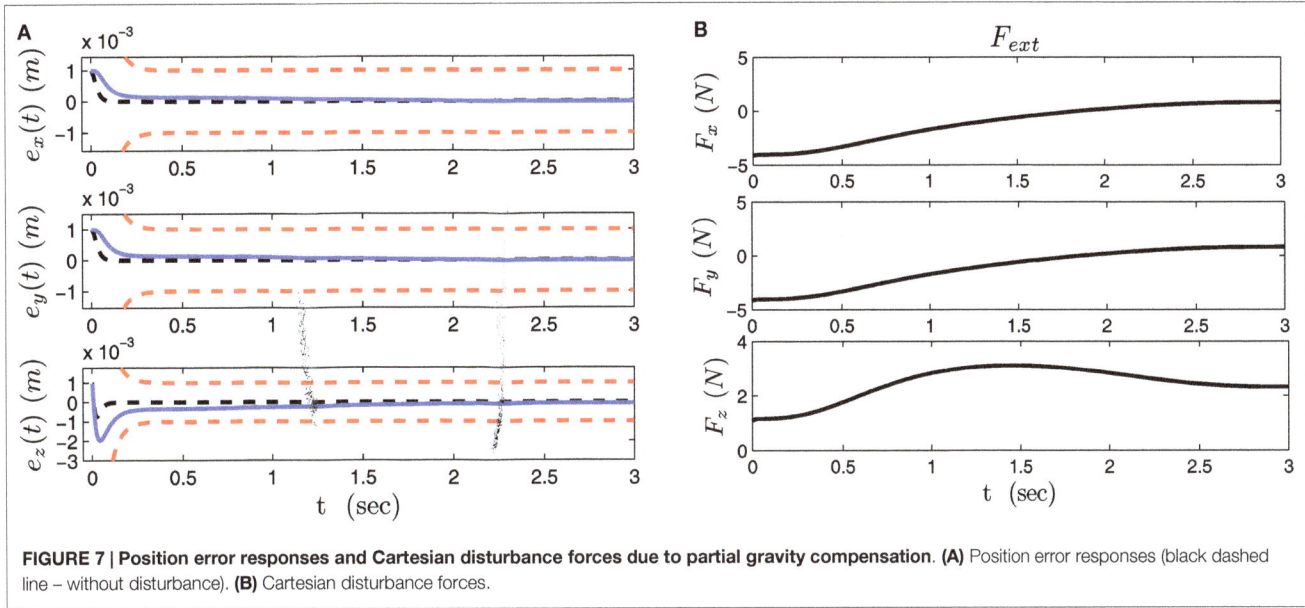

FIGURE 7 | Position error responses and Cartesian disturbance forces due to partial gravity compensation. (A) Position error responses (black dashed line – without disturbance). **(B)** Cartesian disturbance forces.

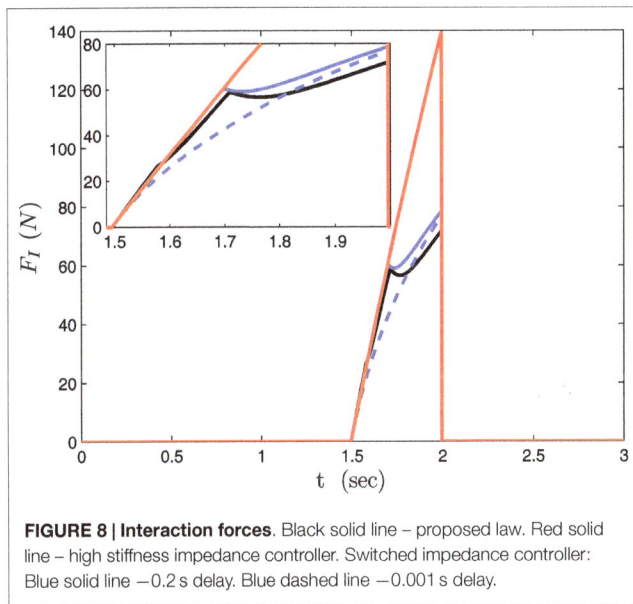

FIGURE 8 | Interaction forces. Black solid line – proposed law. Red solid line – high stiffness impedance controller. Switched impedance controller: Blue solid line −0.2 s delay. Blue dashed line −0.001 s delay.

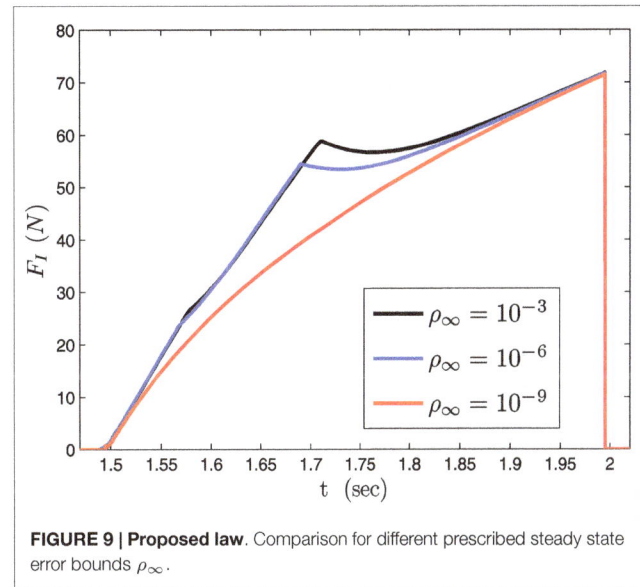

FIGURE 9 | Proposed law. Comparison for different prescribed steady state error bounds ρ_∞.

controller achieves enhanced safety like the practical switched impedance but by smoothly traversing the two stiffness areas. In the proposed controller, the instant the error traverses the boundary is the time of the first interaction force peak and can be regulated by prescribing a different performance bound at steady state. As shown in **Figure 9**, a lower ρ_∞ results in a lower force peak appearing earlier; with a very small value ($\rho_\infty = 10^{-9}$) the interaction force behaves like the ideal switch case.

4.2. Experimental Results

Experiments are conducted with a KUKA LWR4+ 7 DOF robotic manipulator. The control law of equation (27) is utilized, ignoring the inertia and Coriolis terms in order to demonstrate the system's robustness to disturbances due to model uncertainties in the nominal performance mode. The control parameter values are

selected as follows: $\rho_{i0} = 0.01$, $\rho_{i\infty} = 0.005$, and $l_i = 20$, $k_{si} = 3$, $k_i = 0.4$, $i = 1, 2, 3$, and $K_v = \text{diag}\,(25, 50, 25, 25, 2.5, 0.25, 0.025)$. In order to demonstrate the apparent Cartesian stiffness of the arm in the impact reaction and nominal mode of operation, two pushing forces of relatively high and low magnitude are exerted by a human to the robot's end-effector being stationary at position $p_d = [-0.431\ 0.6\ 0.5]^T$ m having initial configuration $q = [-7.03\ 44.45\ 7.789\ -113.26\ -8.54\ 21.27\ -8.99]^T$ deg. **Figure 10** shows the two pushing forces, at $t = 11.6$ s and $t = 21.6$ s, along the y direction, and the respective error displacements. For estimating the apparent stiffness, we have utilized the maximum error displacement with the respective force magnitude (the spike appearing in the first pulse **Figure 10** is excluded from this calculation). The apparent Cartesian stiffness is calculated as approximately 204.37 N/m in the impact reaction mode, and approximately 10,073 N/m in nominal operation mode.

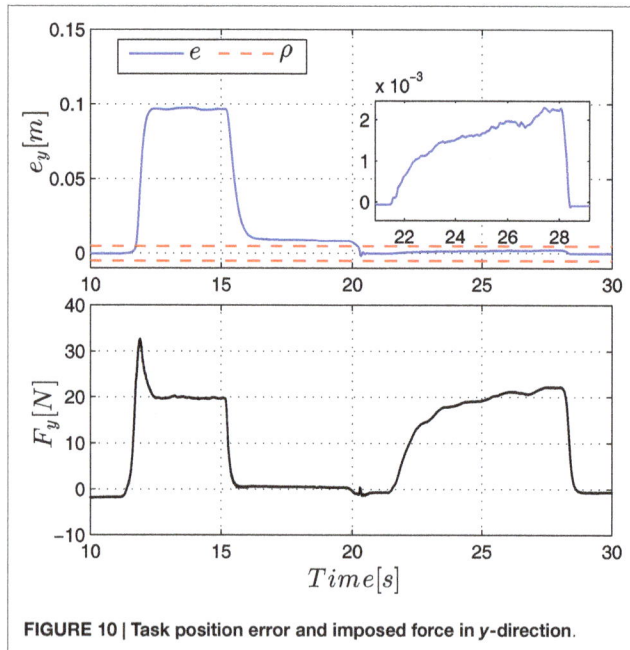

FIGURE 10 | Task position error and imposed force in *y*-direction.

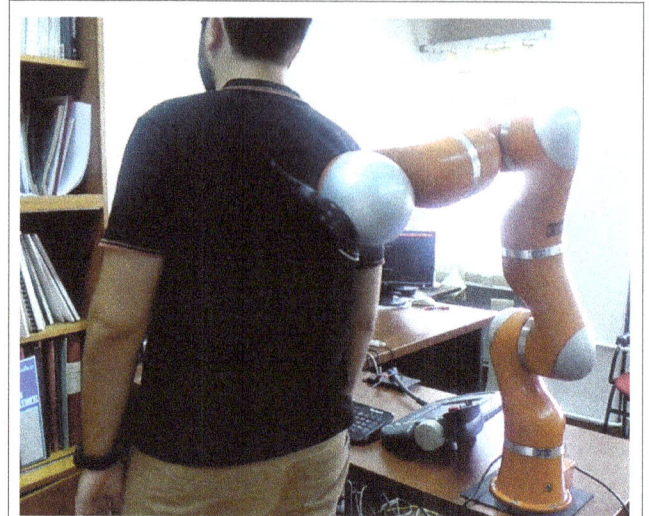

FIGURE 11 | Experiments with KUKA LWR4+.

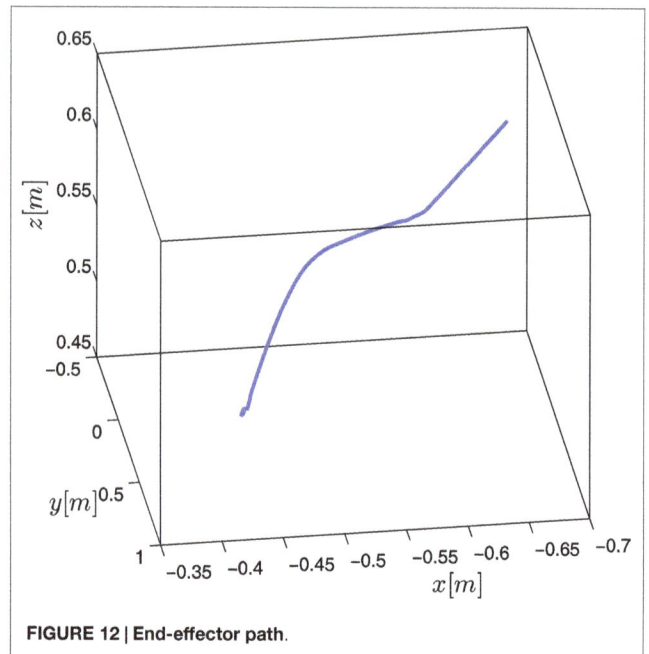

FIGURE 12 | End-effector path.

Next, we consider the case of a human standing in the robot's way causing an unintentional contact on his back (**Figure 11**); the human was instructed to move away as soon as he realized the collision. The arm is executing a desired 5*th* order polynomial trajectory starting from $p(0) = [-0.668 \ -0.058 \ 0.626]^T$ m with configuration $q(0) = [0 \ 30 \ 9 \ -60 \ -14 \ 22 \ -9]^T$ deg and moving on a linear path toward the target location $p_{df} = [-0.431 \ 0.6 \ 0.5]^T$ m in 2 s (the maximum velocity is 0.66 m/s). The end-effector path and the position response are shown in **Figures 12** and **13** demonstrating the system's compliance when operating in the impact reaction mode. As KUKA LWR4+ is equipped with torque sensors at each robot joint, joint torque measurements are used by the KUKA software to estimate external forces at the end-effector. Force readings are not utilized in the proposed controller, but they are depicted in **Figure 14** together with error deviations in order to demonstrate the drop of stiffness that cannot be analytically obtained. Notice how the force exerted in the *x*-direction is just below the value causing the transition from the nominal to the reaction mode, as compared to the case in *y* and *z* directions. Notice the respective error deviations in the latter case owing to the drop of stiffness. In all cases, impact forces stay relatively low. As the human moves away after impact the robot follows the desired motion with errors returning within the high stiffness prescribed performance region. **Figure 15** shows the control effort associated with this case excluding gravity which is provided by the KUKA/FRI (Fast Research Interface) default torque control method.

4.3. Discussion

It is well-known that setting a low-desired stiffness in a conventional impedance controller for safety reasons adversely affects performance. On the other hand, high-targeted stiffness in impedance control can achieve a certain tracking quality and robustness but adversely affects human and robot safety. By contrast, the proposed controller addresses both objectives of

motion performance and enhanced safety in one scheme. With regards to motion performance quality, it achieves prescribed performance tracking both in transient and steady state respecting performance bounds under disturbances up to a tunable level conceptually separating the nominal operation and the impact reaction modes. Performance in the nominal operation mode has been demonstrated under disturbances arising by partially compensating gravity in simulations and by ignoring the feed-forward control terms related to inertial and Coriolis forces in experiments. Thus, the model-based structure in the inner loop does not jeopardize performance. The threshold of the disturbances allowing the operation in the nominal performance mode can be regulated by changing the value of k_i. Furthermore, the prescribed performance property of the proposed law facilitates

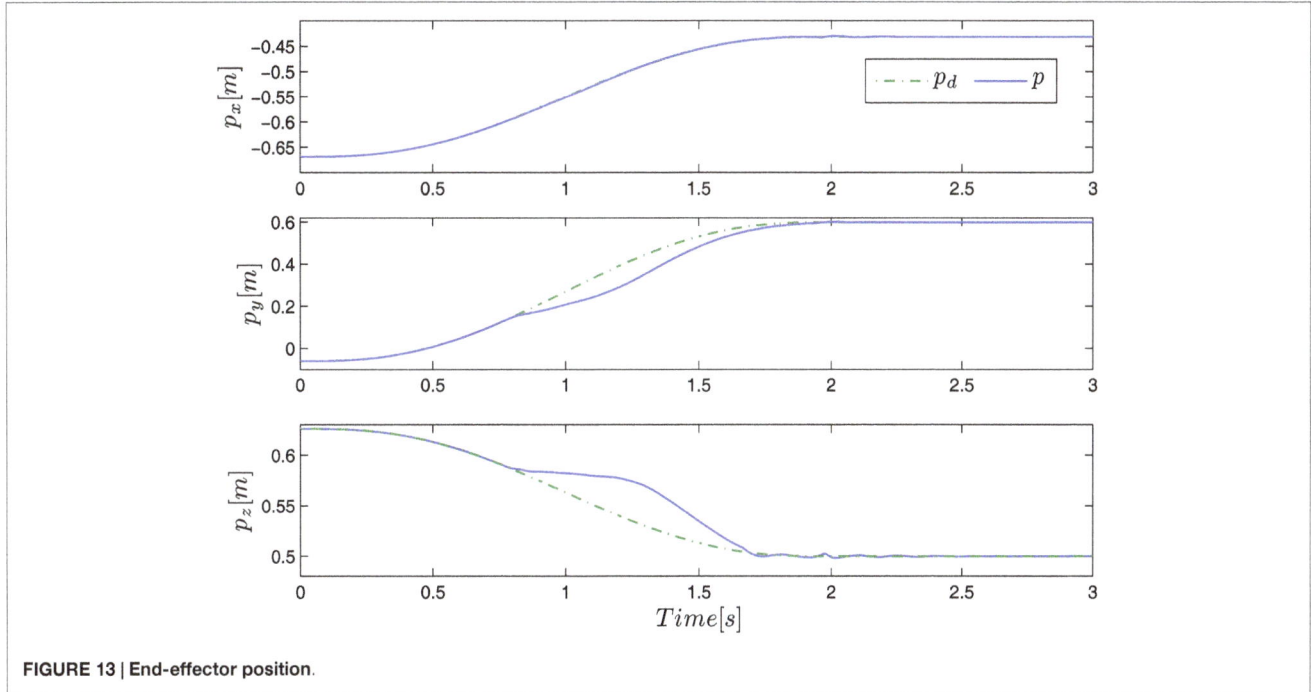

FIGURE 13 | End-effector position.

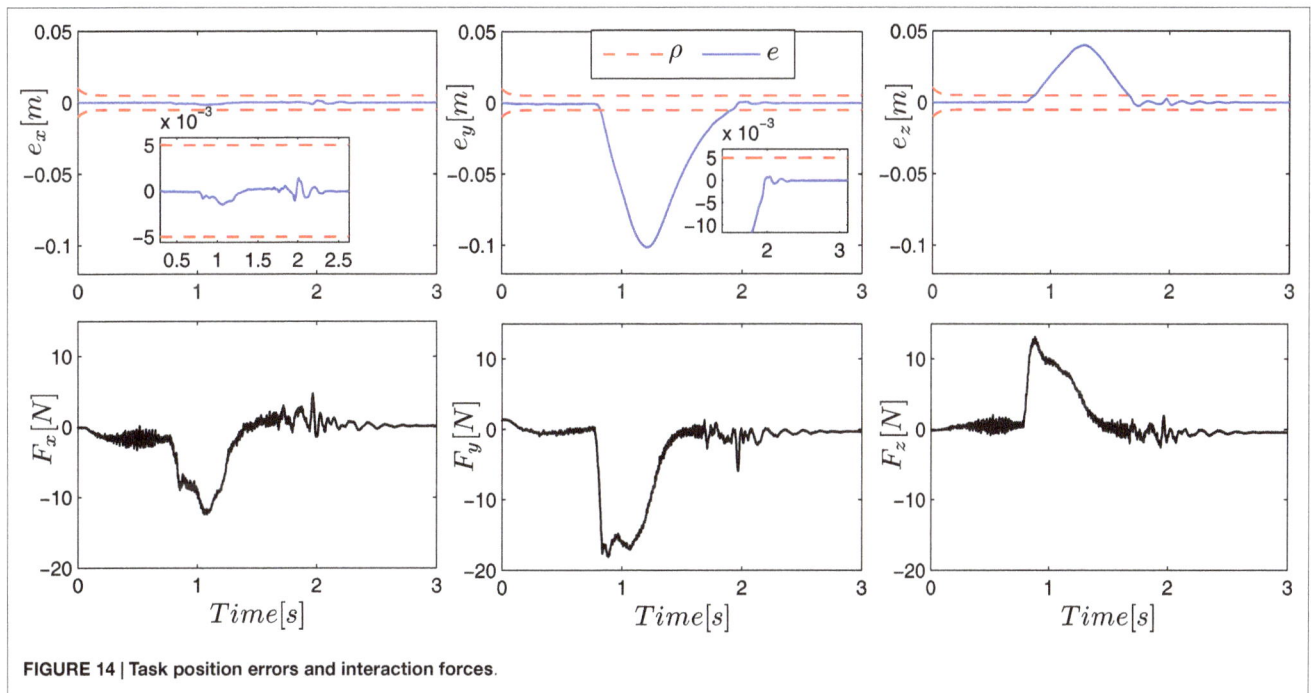

FIGURE 14 | Task position errors and interaction forces.

control parameter tuning as this is relaxed into adopting those values that lead to reasonable control effort with respect to magnitude and slew-rate.

Enhanced human and robot safety is on the other hand achieved by operating in the impact reaction mode where the apparent output stiffness is characterized by low values as compared to the high stiffness values characterizing the nominal operation. Stiffness drop was demonstrated in both simulations and experiments where force values shown in Subsections 4.1 and

4.2 were obtained either via a simulated contact or via the force estimation provided by KUKA LWR4+ that is based on torque readings during the impact with a human; they also show how impact forces stay relatively low. The proposed scheme furthermore realizes smooth traversing between the two stiffness areas without control switching or force detection mechanisms that rely on the existence of force measurements or estimates and thus may incur considerable delays. As it has been also demonstrated in simulations, the force peak observed even in the case of a practical

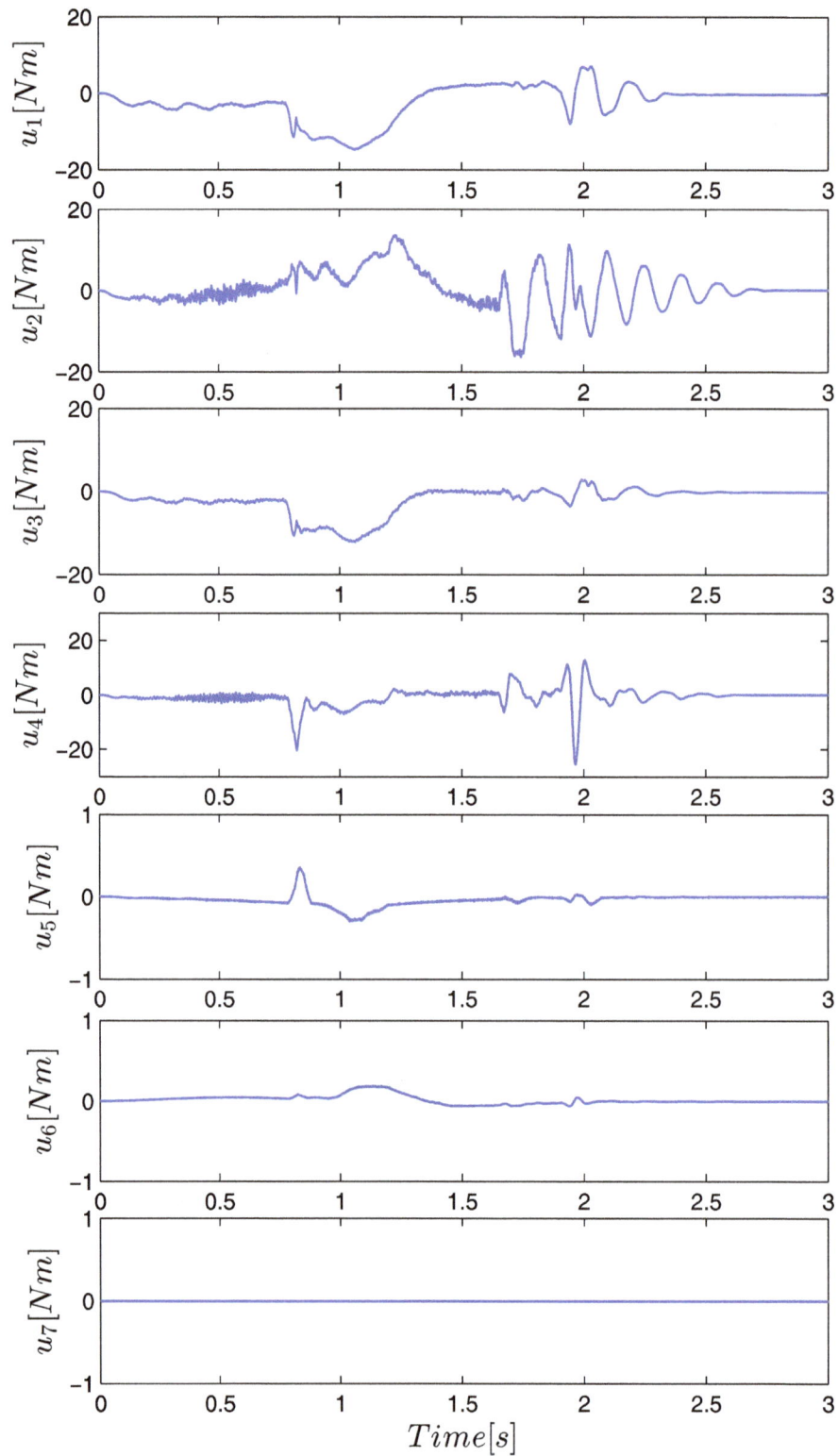

FIGURE 15 | Control input signal not including gravity.

switched impedance can be reduced with the proposed controller by reducing the performance bound at steady state, ρ_∞. Notice, however, that ρ_∞ values cannot be chosen arbitrarily small since they should reflect the accuracy achieved by the measurement device, and ensure that the steady state performance zone is wide enough to accommodate the measurement noise.

5. CONCLUSION

This work proposes a controller in which the robot output stiffness is self-regulated according to the disturbance magnitude. Moreover, the error is shown to evolve within a predefined performance region in nominal operation mode exhibiting robustness to disturbances up to a tunable threshold. The system reacts stably by reducing its output stiffness under bigger disturbances like those arising from impact, returning to the nominal operation after the disturbance vanishes. The controller achieves a continuous and smooth transition between the two modes without switching, eliminating the need for separate collision detection, and reaction strategies. Simulations and experimental results demonstrate the enhanced safety achieved by the proposed controller under impact with initial impact force magnitudes connected to the prescribed steady state error bounds of the nominal operation mode. Comparison with a practical switched impedance control scheme shows that the proposed control law achieves a slightly better performance without making use of any switching. Future work will further investigate safety under unintentional contacts by taking into account disturbance frequency content.

AUTHOR CONTRIBUTIONS

I have developed the basic idea for the proposed controller together with my co-author YK, and I have supervised and coordinated the simulated and experimental work, which was performed by my PhD students and co-authors LD and DP.

ACKNOWLEDGMENTS

This research is co-financed by the EU-ESF and Greek national funds through the operational program "Education and Lifelong Learning" of the National Strategic Reference Framework (NSRF) – Research Funding Program ARISTEIA I under Grant PIROS/506. This work is also partially funded by the Swedish Research Council (VR).

REFERENCES

Albu-Schaeffer, A., Ott, C., and Petit, F. (2012). "Energy shaping control for a class of underactuated Euler-Langrange systems," in *Proc. 10th Intern. IFAC Symposium on Robot Control* (Dubrovnik: International Federation of Automatic Control), 567–574.

Bechlioulis, C. P., Doulgeri, Z., and Rovithakis, G. A. (2012). Guaranteeing prescribed performance and contact maintenance via an approximation free robot force/position controller. *Automatica* 48, 360–365. doi:10.1016/j.automatica.2011.07.009

Bechlioulis, C. P., and Rovithakis, G. A. (2008). Robust adaptive control of feedback linearizable MIMO nonlinear systems with prescribed performance. *IEEE Trans. Automat. Contr.* 53, 2090–2099. doi:10.1109/TAC.2008.929402

Bicchi, A., and Tonietti, G. (2004). Fast and soft arm tactics: dealing with the safety-performance trade-off in robot arms design and control. *IEEE Robot. Autom. Mag.* 11, 22–33. doi:10.1109/MRA.2004.1310939

Choi, J., Park, S., Lee, W., and Kang, S. C. (2008). "Design of a robot joint with variable stiffness," in *Proceedings – IEEE International Conference on Robotics and Automation* (Pasadena, CA: IEEE), 1760–1765.

De Luca, A., Albu-Schaffer, A., Haddadin, S., and Hirzinger, G. (2006). "Collision detection and safe reaction with the DLR-III lightweight manipulator arm," in *Proceedings of the 2006 IEEE/RSJ International Conference on Intelligent Robots and Systems* (Beijing: IEEE), 1623–1630.

De Luca, A., and Mattone, R. (2005). "Sensorless robot collision detection and hybrid force/motion control," in *Proceedings of the 2005 IEEE International Conference on Robotics and Automation* (Barcelona: IEEE), 999–1004.

De Santis, A., Siciliano, B., De Luca, A., and Bicchi, A. (2010). An atlas of physical humanrobot interaction. *Mech. Mach. Theory* 43, 253–270. doi:10.1016/j.mechmachtheory.2007.03.003

Erden, M. S., and Tomiyama, T. (2010). Human-intent detection and physically interactive control of a robot without force sensors. *IEEE Trans. Robot.* 26, 370–382. doi:10.1109/TRO.2010.2040202

Golz, S., Osendorfer, C., and Haddadin, S. (2015). "Using tactile sensation for learning contact knowledge: discriminate collision from physical interaction," in *IEEE International Conference on Robotics and Automation (ICRA)* (Seattle, WA: IEEE), 3788–3794.

Haddadin, S., Albu-Schaeffer, A., De Luca, A., and Hirzinger, G. (2008). "Collision detection and reaction: a contribution to safe physical human-robot interaction," in *IEEE International Conference on Intelligent Robotics and System (IROS)* (Nice: IEEE), 3356–3363.

Heinzmann, J., and Zelinsky, A. (2003). Quantitative safety guarantees for physical human-robot interaction. *Int. J. Rob. Res.* 22, 479–504. doi:10.1177/02783649030227004

Karayiannidis, Y., and Doulgeri, Z. (2012). Model-free robot joint position regulation and tracking with prescribed performance guarantees. *Rob. Auton. Syst.* 60, 214–226. doi:10.1016/j.robot.2011.10.007

Lee, D., and Ott, C. (2011). Incremental kinesthetic teaching of motion primitives using the motion refinement tube. *Auton. Robots* 31, 115–131. doi:10.1007/s10514-011-9234-3

Magrini, E., Flacco, F., and De Luca, A. (2015). "Control of generalized contact motion and force in physical human-robot interaction," in *IEEE International Conference on Robotics and Automation (ICRA)* (Seattle, WA: IEEE), 2298–2304.

Marquez, H. (2003). *Nonlinear Control Systems.* New Jersey: Wiley-Interscience.

Oh, S., Woo, H., and Kong, K. (2014). Frequency-shaped impedance control for safe human-robot interaction in reference tracking application. *IEEE/ASME Trans. Mechatron.* 19, 1907–1916. doi:10.1109/TMECH.2014.2309118

Siciliano, B., Sciavicco, L., Villani, L., and Oriolo, G. (2010). *Robotics: Modelling, Planning and Control. Advanced Textbooks in Control and Signal Processing.* London: Springer.

Slotine, J.-J. E., and Li, W. (1991). *Applied Nonlinear Control.* New Jersey: Prentice Hall.

Tsagarakis, N. G., Sardellitti, I., and Caldwell, D. G. (2011). "A new variable stiffness actuator (CompAct-VSA): design and modelling," in *IEEE International Conference on Intelligent Robots and Systems* (San Francisco, CA: IEEE), 378–383.

Wolf, S., and Hirzinger, G. (2008). "A new variable stiffness design: matching requirements of the next robot generation," in *Proceedings – IEEE International Conference on Robotics and Automation* (Pasadena, CA: IEEE), 1741–1746.

Conflict of Interest Statement: The authors declare that the research was conducted in the absence of any commercial or financial relationships that could be construed as a potential conflict of interest.

10

The Sensorimotor Loop as a Dynamical System: How Regular Motion Primitives May Emerge from Self-Organized Limit Cycles

Bulcsú Sándor*, Tim Jahn, Laura Martin and Claudius Gros

Institute for Theoretical Physics, Goethe University Frankfurt, Frankfurt am Main, Germany

We investigate the sensorimotor loop of simple robots simulated within the LPZRobots environment from the point of view of dynamical systems theory. For a robot with a cylindrical shaped body and an actuator controlled by a single proprioceptual neuron, we find various types of periodic motions in terms of stable limit cycles. These are self-organized in the sense that the dynamics of the actuator kicks in only, for a certain range of parameters, when the barrel is already rolling, stopping otherwise. The stability of the resulting rolling motions terminates generally, as a function of the control parameters, at points where fold bifurcations of limit cycles occur. We find that several branches of motion types exist for the same parameters, in terms of the relative frequencies of the barrel and of the actuator, having each their respective basins of attractions in terms of initial conditions. For low drivings stable limit cycles describing periodic and drifting back-and-forth motions are found additionally. These modes allow to generate symmetry breaking explorative behavior purely by the timing of an otherwise neutral signal with respect to the cyclic back-and-forth motion of the robot.

Keywords: sensorimotor loop, adaptive behavior, self-organization, limit cycles, period tripling, embodiment, explorative behavior, symmetry breaking

Edited by:
Joschka Boedecker,
University of Freiburg, Germany

Reviewed by:
Sakyasingha Dasgupta,
RIKEN Brain Science Institute, Japan
Carlos Gershenson,
Universidad Nacional Autónoma de
México, Mexico

*Correspondence:
Bulcsú Sándor
sandor@itp.uni-frankfurt.de

1. INTRODUCTION

Robots moving through an environment need to take the physical laws into account. This can be achieved either via classical control theory (de Wit et al., 2012), or by considering the full sensorimotor loop as an overarching dynamical system (Ay et al., 2012). This distinction could be cast, alternatively, into open-loop control, e.g., via central pattern generators (Ijspeert, 2008), and closed-loop schemes using feedback to control the states of an internal dynamical system (Dorf and Bishop, 1998). The presence of such feedback mechanisms capable of amplifying local instabilities are key components leading to the emergence of self-organization (Der and Martius, 2012). A closely related notion is that of embodiment (Ziemke, 2003), for which no need arises for an explicit modeling of the interactions between the robot and its surroundings. The agent situated in a given environment can be treated, in an embodied approach, as an overarching dynamical system, incorporating both the external dynamics (body–environment interaction) and the internal (controller body) processes. Thus, combining the closed-loop control with the embodied approach leads to movements generated through self-organizing processes. These may in turn be guided by generic, e.g., information-theoretical objective functions (Martius et al., 2013), such as predictive information (Ay et al., 2008), resulting in explorative or even playful behavior (Der and Martius, 2012).

Similar objective functions, such as the free energy (Friston, 2010), can also be considered for the brain as a whole (Baddeley et al., 2008) and in the context of adaptive behavior (Friston and Ao, 2012). Distinct control mechanisms for neural networks can also be derived from other information-theoretical generating functionals, such as the relative information entropy (Triesch, 2007), the mutual information (Toyoizumi et al., 2005), the Fisher information (Echeveste and Gros, 2014), and the recently introduced active information storage measure (Lizier et al., 2012; Dasgupta et al., 2013). Starting from first principles Hebbian learning rules have also been derived (Echeveste et al., 2015).

A parallel approach for studying the power of embodiment is provided by evolutionary robotics. Robots, selected through evolutionary processes (Nolfi and Floreano, 2000) take environmental feedback naturally into account, as they would otherwise not be positively selected. The notion of an acting agent in a reacting environment becomes blurry, to a certain extent, when the full sensorimotor loop is considered, with the motion coming to a standstill without a fully functional feedback cycle. Within other approaches to embodiment, the physical constraints acting on compliant real-world robots are studied (Pfeifer et al., 2007), or the flow of information, e.g., in terms of transfer entropy, through the sensorimotor loop (Schmidt et al., 2013). A related question is how to ground actions generically, i.e., without a priori knowledge, in sensorimotor perceptions (Olsson et al., 2006), or how to select actions from universal and agent-centric measures of control (Klyubin et al., 2005).

Abstracting from the sensorimotor loop, one may regard, from the point of view of dynamical system theory (Beer, 2000), motions as organized sequences of movement primitives in terms of attractor dynamics (Schaal et al., 2000), which the agent needs first to acquire by learning attractor landscapes (Ijspeert et al., 2002, 2013). These may be used later on for encoding the transients leading to periodic motions (Ernesti et al., 2012) or may furthermore self-organize into complex behaviors (Tani and Ito, 2003). In this context, the fully embodied approach may serve as an algorithmic first step to generate a palette of motion primitives. One may also observe that all regular motions are, per definition, attractors in terms of stable limit cycles in the overarching sensorimotor loop, which may be controlled either actively (Laszlo et al., 1996) or passively in terms of limit-cycle walking (Hobbelen, 2008). As an alternative approach for creating and controlling limit cycles, one could use prototype dynamical systems, a concept recently proposed for the study of complex bifurcation scenarios (Sándor and Gros, 2015).

In the present study, we examine in detail the notion of periodic movements as stable limit cycles, using the LPZRobots package (Der and Martius, 2012; Martius et al., 2013) for simulating robots (current development version), which are geometrically simple enough to allow for an at least partial modeling in terms of dynamical system theory (Gros, 2015). Our robots, see **Figures 1** and **2**, are controlled by a single proprioceptual neuron with a time-dependent threshold $b = b(t)$. We find a region of parameters in which the motion is fully embodied, and where the movement $v_b = v_b(t)$ of the robot and the threshold dynamics are mutually fully interdependent, vanishing when one of them, either $b(t)$ or $v_b(t)$, is clamped. In engineering terms, the engine db/dt powering

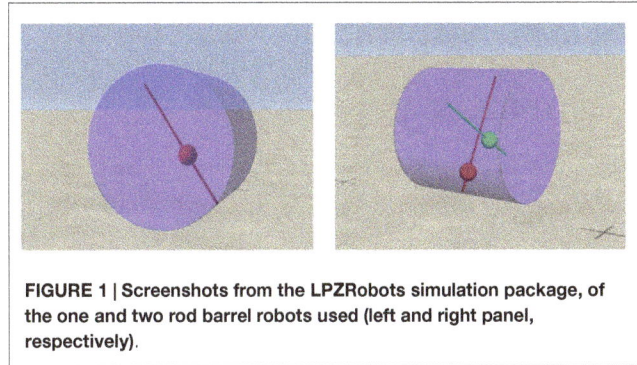

FIGURE 1 | Screenshots from the LPZRobots simulation package, of the one and two rod barrel robots used (left and right panel, respectively).

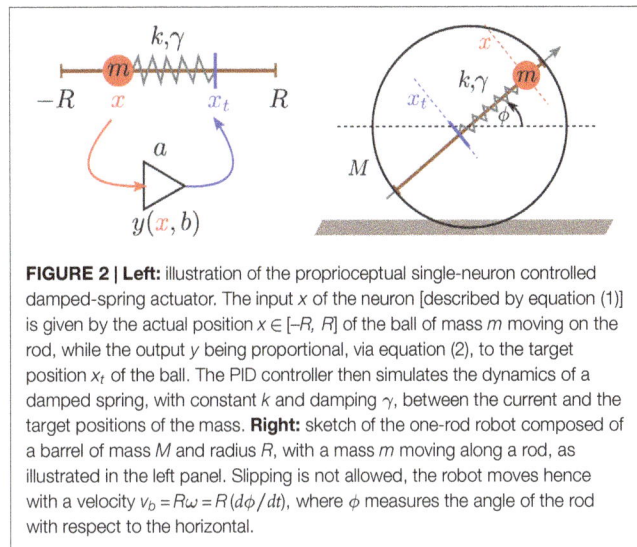

FIGURE 2 | Left: illustration of the proprioceptual single-neuron controlled damped-spring actuator. The input x of the neuron [described by equation (1)] is given by the actual position $x \in [-R, R]$ of the ball of mass m moving on the rod, while the output y being proportional, via equation (2), to the target position x_t of the ball. The PID controller then simulates the dynamics of a damped spring, with constant k and damping γ, between the current and the target positions of the mass. **Right:** sketch of the one-rod robot composed of a barrel of mass M and radius R, with a mass m moving along a rod, as illustrated in the left panel. Slipping is not allowed, the robot moves hence with a velocity $v_b = R\omega = R(d\phi/dt)$, where ϕ measures the angle of the rod with respect to the horizontal.

the motion of the robot is turned on dynamically through the feedback of its very motion.

We also find that a set of qualitatively distinct movements can arise for identical settings of the parameters in terms of stable limit cycles, having their own distinct basins of attraction in phase space. Control signals may hence switch between different motion primitives without the need to interfere with the parameter setting of the sensorimotor loop. Most modes found lead to regular motions with finite average velocities. We discovered, however, also a particular mode corresponding to a cyclic back-and-forth movement, without an average translational motion of the robot. When the parameter settings are changed in this mode, the robot will enter a rolling motion, either to the left or to the right, depending on the timing of the signal with respect to the phase of the cycle, allowing, as a matter of principle, for a truly explorative behavior.

A central result of the present study is that even very simple controller dynamics (a single differential equation, in our case) may lead via the sensorimotor loop to surprisingly rich repertoires of regular motion primitives, which may be selected in turn through higher order decision processes. This is due to the self-stabilization of motion patterns within the sensorimotor loop. Goal-oriented behavior would in this context be achieved not by optimizing motion directly, but by selecting from the many attracting states generated by an embodied controller within the overall sensorimotor loop.

2. MATERIALS AND METHODS

We start by describing the one-neuron controller used together with the actuator in terms of a damped spring, and the actual setup of the robot.

2.1. Rate Encoding Neurons with Internal Adaption

In this paper, we consider actuators controlled by simple rate encoding neurons, characterized by a sigmoidal transfer function

$$y(x, b) = \frac{1}{1 + e^{a(b-x)}}, \qquad \dot{b} = \varepsilon a(2y - 1) \qquad (1)$$

between the membrane potential x and the firing rate y, where a is the gain, taken to be fixed, and $b = b(t)$ a time-dependent threshold. The dynamics $\dot{b}$ for the threshold in equation (1) would lead to $b \to x$ and $y \to 1/2$ for any constant input $x(t) = x$, with a relaxation time being inversely proportional to the adaption rate ε. This adaption rate can also be motivated by information-theoretical considerations for the distribution of the firing rates (Triesch, 2005; Marković and Gros, 2010).

2.2. Damped-Spring Actuators

Our robots are controlled by actuators regulating the motion of the ball of mass m on a rod, as illustrated in **Figure 2**, from its actual position x on the rod, to its target position

$$x_t = 2R\left(y(x, b) - \frac{1}{2}\right), \qquad (2)$$

where R is the radius of the barrel containing the rod and where $y(x, b)$ is the sigmoidal equation (1). We note that the input and the output of the neuron are, via equation (2), of the same dimensionality, namely positions. The force $F = m\ddot{x}$ moving the ball is evaluated by the PID controller

$$F = g_P(x_t - x) + g_I \int_0^t (x_t - x)dt + g_D\frac{d(x_t - x)}{dt}, \qquad (3)$$

provided by the LPZRobots simulation environment (Der and Martius, 2012), characterized by the standard PID-control parameters g_P, g_I, and g_D.

For our simulations, we considered the case $g_I = 0$, for which the PID controller reduces to a damped spring, see **Figure 2**,

$$m\ddot{x} = -k(x - x_t) - \gamma\frac{d(x - x_t)}{dt}, \qquad (4)$$

with $k = g_P$ and $\gamma = g_D$.

- Equation (4) represents only the contribution of the actuator to the force moving the ball along the rod. The gravitational pull acting on the mass m, and the centrifugal force resulting from the rolling motion of the barrel on the ground are to be added to the RHS of equation (4).

- The target position $x_t = x_t(t)$ is time-dependent through equations (2) and (1).

- Equation (4) is strictly dissipative, due to the damping $\gamma > 0$. The same holds for the rolling motion of the barrel on the ground, which is also characterized by a finite rolling friction. Thus, the dynamics db/dt of the threshold in equation (1) can be considered as an engine, providing, by adjusting continuously the target position x_t of the ball, and hence the length of the spring, the energy dissipated by the physical motions.

2.3. Motion of a Mass on a Fixed Rod

As an example we consider a robot, for which we keep the angle ϕ between the rod and the horizontal fixed, $\phi = \phi_0$, by preventing it from rolling. We are then left with a self-coupled motion of a ball along a rod, as illustrated in the left panel of **Figure 2**, resulting in a dynamics similar to the one of a self coupled neuron (Marković and Gros, 2012; Gros et al., 2014). Using $\Omega^2 = k/m$ and $\Gamma = \gamma/m$, we find in this case

$$\begin{aligned} \dot{x} &= v & \dot{x}_t &= 2Ray(1 - y)(v - \dot{b}) \\ \dot{v} &= -\Omega^2(x - x_t) & \dot{b} &= 2\varepsilon a(y - 1/2) \\ &\quad -\Gamma(v - \dot{x}_t) - g\sin(\phi_0) \end{aligned}, \qquad (5)$$

when combining equations (1), (2), and (4). The gravitational term $-g\sin(\phi_0)$ can be transformed away via

$$x \to x - g/\Omega^2 \sin\phi_o, \qquad b \to b - g/\Omega^2 \sin\phi_o, \qquad (6)$$

and does hence not influence the phase diagram, which is shown in **Figure 3** for $\Omega^2 = 200$, $\Gamma = 2\Omega$, and $g = 9.81$. We have used standard numerical methods (Clewley, 2012).

We find a Hopf bifurcation line separating the stability regions for the trivial fixpoint and for a limit cycle, denoted, respectively, as off and on modes. This behavior is similar to the one observed for a self coupled neuron with intrinsic adaption (Marković and Gros, 2012; Gros et al., 2014).

3. RESULTS

In **Figure 1**, the screenshots of the one- and two-rod robots simulated with the LPZRobots package (current development version) (Der and Martius, 2012; Martius et al., 2013) are presented. Throughout the simulations the control parameters $\Gamma = 2\Omega$ and $\Omega^2 = 200$ for the actuator, $\Lambda = 1$ for the mass ratio m/M (ball to barrel), $R = 1$ for the radius for the barrel, and $\Psi = 0.3$ for the coefficient of the rolling friction have been held constant, varying only the adaption rate ε for the threshold of the neuron, and the gain a. For the simulations, a step size of 0.001 was used. In the figures (and in the rest of the paper), the parameters will be presented in dimensionless units, with SI units being implied: seconds/meter for the time and length, respectively, and $g = 9.81$ m/s^2 for the gravitational acceleration. Our barrel has a radius of 1 m and a moving mass of 1 kg, rolling typically at speeds of (1–4) m/s $\approx$ (3–12) km/h. A table of the parameters is given in the Supplementary Material.

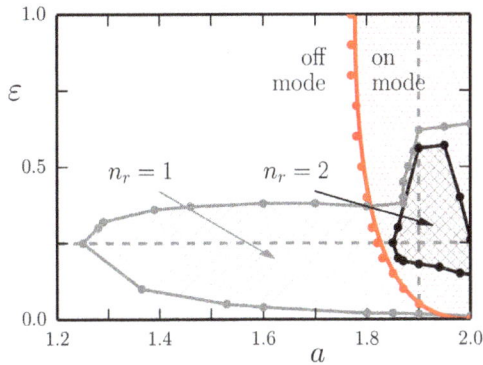

FIGURE 3 | The phase diagram of the one-rod barrel, as a function of the gain a and of the adaption rate ε. The results are obtained using the LPZRobots package, apart from the red solid line separating the off- and the on-mode, which follows from equation (5), for a fixed but otherwise arbitrary angle $\phi = \phi_0$. The dashed vertical and horizontal gray lines indicate the cuts used for the phase diagrams presented in **Figure 5**. The number of stable limit cycles found in the respective parameter regions is denoted by n_r. *Non-rolling modes:* the red dots/lines indicate the locus of a Hopf bifurcation, where a stable non-rolling limit cycle (on mode) emerges from the trivial non-rolling fixpoint (off mode). In the off mode the "engine" $db(t)/dt$, see equation (1), kicks in only when the barrel is already moving. *Rolling modes:* shown are the regions containing $n_r = 1$ (enclosed by the solid gray line) and $n_r = 2$ (enclosed by the solid black line) attracting limit-cycles corresponding to a barrel moving with a finite velocity $<v_b>$. Note that the robot is able to move also in the off mode (of the engine). The stationary and the drifting back-and-forth modes, discussed in **Figure 6**, have been omitted, in order to avoid overcrowding.

3.1. One-Rod Barrel

The overall phase diagram of the one-rod barrel shown in **Figure 3** contains regions of non-rolling fixpoints or limit cycles, and regions where one or more attracting limit cycles corresponding to a continuously rolling barrel are present, in part additionally. Depending on the initial conditions the system will eventually settle into one of the attracting states.

3.1.1. Coexisting Modes as Behavioral Primitives

Standard robot control aims at achieving a predefined outcome, and for this purpose it is indispensable that identical robot actions lead also to identical movements. This is not necessarily the case for robots controlled by self-organized processes, as investigated here.

In **Figure 4**, we illustrate the time series and the corresponding phase-space plots of the dominant modes of the one-rod barrel shown in **Figure 1**. The simulation parameters $a = 1.9$ for the gain, and the $\varepsilon = 0.25$ adaption rate are close to the Hopf bifurcation line shown in **Figure 3**, but in the on mode, which means that the ball moves both for fixed horizontal and vertical rods.

The first of the three coexisting stable limit cycles, illustrated in **Figure 4**, corresponds to the non-moving barrel with the ball oscillating vertically along the rod (first column). For the second, 1:1 mode, the average rolling frequency of the barrel and of the oscillation of the ball along the rod match (second column). For the 1:3 mode, the corresponding ratio of frequencies is, however, 1:3 (third column).

The occurrence of several distinct limit cycles for identical parameters can be interpreted in terms of behavioral primitives, potentially allowing an agent to switch rapidly between different types of motions, by shortly destabilizing the currently active limit cycle.

Note that the self-coupled neuron, controlling the dynamics of the ball along the horizontally fixed rod, has only two possible stable states (a fixpoint and a limit cycle). Considering, however, the fully embodied rolling robot, coexisting states are arising, which can lead to different behavioral patterns purely as a result of the environmental context. An external force applied to the robot can qualitatively change its behavior, indicating the sign of multifunctionality (Williams and Beer, 2013).

3.1.2. Embodiment as Self-Organized Motion

Most robots are autonomously active in the sense that the motion is not essentially dependent on the feedback of the environment. For the case of self-organized motion, as considered here, there would be, on the other side, no motion when the sensorimotor loop would be interrupted.

We present in the left plot of **Figure 5**, the evolution of the self-sustained rolling modes, in terms of the averaged measured velocity, for $a = 1.9$ and as a function of adaption rate ε. The dashed black line indicates, as a guide to the eye, that the velocity increases roughly $\propto \sqrt{\varepsilon}$ for the 1:1 mode. The two branches are stable for $\varepsilon \in [0.018, 0.55]$ and $\varepsilon \in [0.19, 0.61]$, respectively, for the 1:1 and the 1:3 mode, and terminate (presumably) through saddle node bifurcations of limit cycles. We have indicated this scenario by adding by hand in **Figure 5**, as guides to the eye, the respective unstable branches.

The locus of the Hopf bifurcation shown in **Figure 3**, at $\varepsilon \approx 0.05$, is indicated in (the left panel of) **Figure 5** by the dashed vertical line, separating the off from the on mode. In the off and on modes, the non-rolling attractors are a fixpoint and a limit cycle, respectively. Note that self-sustained rolling modes exist in the off mode as well, where the "engine" $db(t)/dt$ of the barrel only kicks in, through amplifying local fluctuations (damped oscillations around the fixpoint), when the barrel is already moving. This underlines the embodied nature of the motion, which arises in a truly self-organized fashion [in terms of dynamical systems theory (Gros, 2015)] through the bidirectional feedback between environment and both the body and the controller of the robot.

However, in the absence of feedback mechanisms (such as centrifugal- and Coriolis-forces), the neuron controlled actuator could only generate a single regular rolling motion, similar to the ones achieved by sending motor signals generated by some central pattern generators (Der and Martius, 2012). This is not the case for our robot, which exhibits, as shown in **Figure 5** (and in **Figure 6**, see Discussion below) a wide spectrum of possible rolling modes.

3.1.3. Avoided Pitchfork Bifurcations of Limit Cycles

In the right panel of **Figure 5**, we present the measured mean velocity $<v_b>$ of the ball for $\varepsilon = 0.25$, as a function of the gain a. The Hopf bifurcation between the off- and on- non-rolling modes occurs at $a_H \approx 1.83$, compare **Figure 3**.

For $1.23 < a < 1.83$, the ball hence is moving in the off mode, with the engine kicking in only through the feedback from the

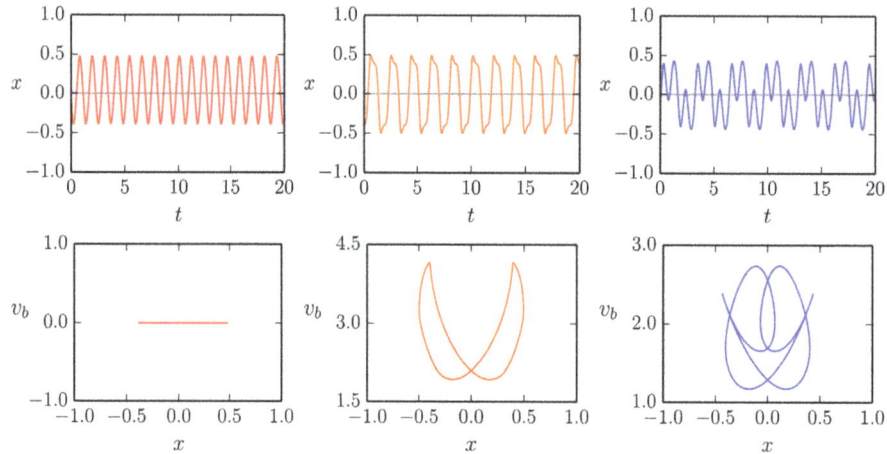

FIGURE 4 | The motion $x(t)$ of the mass along the rod of the one-rod barrel (top row), together with the corresponding phase-plane trajectories ($x(t)$, $v_b(t)$)) (bottom row), compare Figure 2. The gain and the adaption rate are $a = 1.9$ and $\varepsilon = 0.25$, respectively. Shown are the 0:1, 1:1, and 1:3 modes (left/middle/right column). Note that the velocity $v_b(t)$ of the barrel vanishes for the 0:1 mode, oscillating but remaining otherwise positive for the 1:1 and the 1:3 mode. For the corresponding videos, see the Supplementary Material.

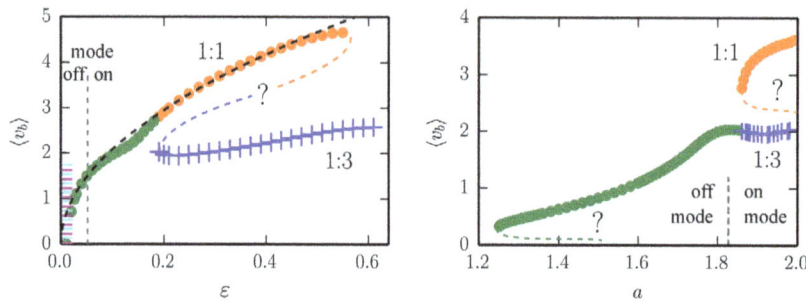

FIGURE 5 | The average speed $<v_b>$ of the one-rod barrel for the 1:1 (green/orange dots) and for 1:3 (blue crosses) mode. The vertical dashed line denotes the locus of the Hopf bifurcation line shown in **Figure 3**. In the off mode (on mode), the attracting state for the non-rolling mode is a stable fixpoint (limit cycle), respectively. Presumably existing unstable limit cycles are indicated by dashed lines (labeled with question marks). *Left:* for a gain $a = 1.9$. The colored region for very small adaption rates ε indicates a region with both stable and drifting back-and-forth modes, further described in **Figure 6**. *Right:* for an adaption rate $\varepsilon = 0.25$.

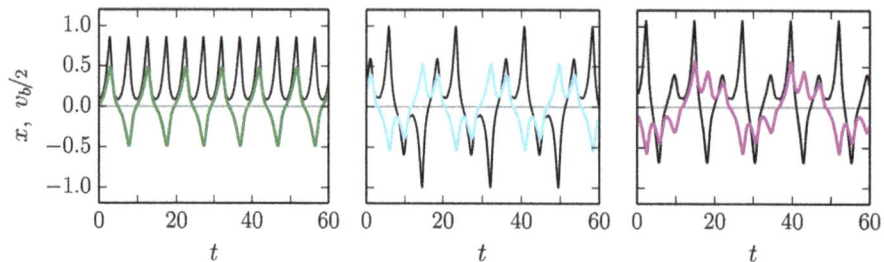

FIGURE 6 | The time evolution of the position x (colored lines) of the mass along the rod, and of the (rescaled) speed v_b of the barrel (black lines), for $a = 1.9$ and $\varepsilon = 0.019/0.017/0.015$ (left/center/right), all in the off mode (compare Figure 5). The respective average velocities are $<v_b> = 0.63/0.00/0.25$ for the 1:1 mode (left), the stationary back-and-forth mode (middle), and the drifting back-and-forth mode (right). For the corresponding videos, see the Supplementary Material.

environment, which we interpret as self-organized embodied motion, with the environment being an essential component of the overarching dynamical system.

Comparing both panels of **Figure 5**, one can notice that the low-velocity mode (green dots) connects either to the 1:1 mode (as in the left panel) or to the 1:3 mode (as in the right panel). The reason for the apparent discrepancy lies in the fact that the respective bifurcation line is oblique in the phase space plane (a, ε). The evolution of these modes suggests in any case that the low-velocity mode connects to the two higher-velocity modes via an avoided pitchfork transition of limit cycles (Gros, 2015).

3.1.4. Explorative Motion via Noise-Induced Directional Switching

Our robot contains a single dynamical variable, the threshold $b(t)$, generating self-stabilizing motions via the sensorimotor loop. The palette of modes generated is, despite this apparent simplicity, surprisingly large and may be used to generate higher order behavior.

There are three dominant branches, the 0:1, 1:1, and 1:3 modes (in terms of the ratios of the respective barrel and mass frequencies), compare **Figures 4** and **5**, which are stable for a wide range of parameters. We found in addition also a parameter region for which different types of motions arise from minute changes of control parameters, such as the adaption rate ε.

In **Figure 6**, the motion $x(t)$ of the ball along the rod and the velocity $v_b(t)$ of the barrel are given for three closely spaced adaption rates $\varepsilon = 0.019, 0.017$, and 0.015, for which three qualitatively different types of motions are found (which have partially, but not completely overlapping stability regions).

- For $\varepsilon = 0.019$, the standard 1:1 rolling motion is recovered, with an average velocity $<v_b> = 0.63$.
- For $\varepsilon = 0.017$, a new mode arises, for which the ball rolls back-and-forth forever. The motion is exactly symmetric with respect to the left and to the right, and the average velocity $<v_b> = 0.0$ of the barrel hence vanishes exactly.
- For $\varepsilon = 0.015$, the ball also rolls back-and-forth, but asymmetrically, giving rise to a drifting motion with small but finite average velocity of $<v_b> = 0.25$.

The occurrence of a limit cycle corresponding to a symmetric back-and-forth rolling motion, sandwiched between symmetry breaking modes, gives rise to an interesting venue for the generation of explorative behaviors, as the robot will be sensitive to finite but otherwise very small perturbations influencing its internal control parameters. This behavior is illustrated in **Figure 7**. Depending on the timing of the perturbation with respect to the back-and-forth rolling cycle, the robot will settle into a left- or into a right-moving motion (in the 1:1 or in the back-and-forth drifting mode, respectively, for increasing/decreasing ε). It is

hence possible to break spatial symmetries, in general, purely via the timing of a perturbation. The perturbation itself, here acting on the adaption rate ε, does not need to carry any information about the direction of motion.

3.2. Two-Rod Barrel

Adding a second actuator perpendicular to the first one, a neuron controlled ball moving along a rod, one can increase the complexity of the robot (see the right picture of **Figure 1**). Both actuators work, in our setup, independently, with the crosstalk being provided exclusively by the environmental feedback. Both actuators are identical to the rod used for the single-rod barrel, with each rod having its own adapting threshold $b_\alpha(t)$ and membrane potential $x_\alpha(t)$, with $\alpha = 1,2$. The adaption rate ε, the gain a, and all other parameters are identical for the two rods.

In **Figure 8**, we show in the right panel the stability range, for $a = 1.9$ and as a function of the adaption rate ε, of the three most dominant rolling modes (1:1, 1:3, and 1:5) of the two-rod barrel. A large variety of higher order 1:M modes (with M being an integer) is found in addition. We did not carry out a systematic search of their stability range, which becomes progressively smaller with increasing M, and present here only exemplary parameter settings for which the respective modes have been found by trial-and-error (by randomly kicking the barrel). A blow-up is given in the right panel of **Figure 8**. Most values of M found are odd, but not exclusively. We cannot exclude, at this stage that an infinite cascade M $\rightarrow \infty$ of higher order limit cycles may possibly occur.

The time series and the respective phase space trajectories $(x_1(t), x_2(t))$ of the 1:1, 1:3, and of the 1:5 modes are presented in **Figure 9**. As one can see in the time series plots, the two independent actuators, being only coupled through the dynamics of the barrel, self-organize themselves in a constant phase-shift, necessary for a consistent rolling. In the reduced phase space (x_1, x_2), the trajectories exactly close on themselves, needing, respectively, 1, 3, and 5 revolutions around the origin $(0,0)$ to close, for, respectively, the 1:1, 1:3, and for the 1:5 limit cycles. In **Figure 10**, we show the corresponding phase-space trajectories of the M = 9, 13, and 21 limit cycles. These modes have progressively slower average velocities $\langle v_b \rangle$, compare **Figure 8**, and smaller basins of attractions, being otherwise regular stable limit cycles. Whether

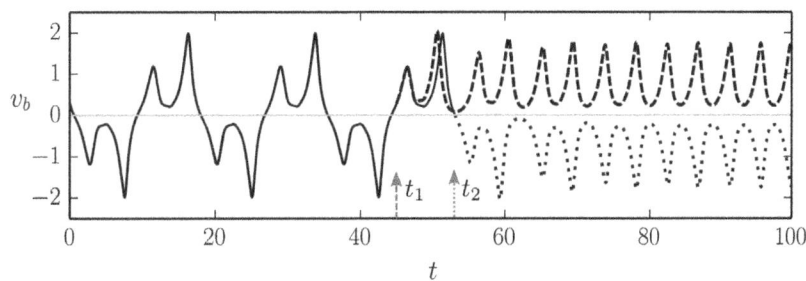

FIGURE 7 | Two superimposed runs for the time evolution of the speed v_b of the barrel (black lines), for $a = 1.9$. In the first run, the adaption rate ε is changed discontinuously at time $t_1 = 45$ from $\varepsilon = 0.017$ (corresponding to the stationary back-and-forth mode, see **Figure 6**) to $\varepsilon = 0.02$ (corresponding to the 1:1 rolling mode). In the second run, identical initial conditions have been used and an identical change is made to the adaption rated ε, but now at time $t_2 = 53$. In both runs (dashed and dotted lines, respectively), the barrel settles into the 1:1 rolling motion, albeit in opposite directions (to the left/right with $\langle v_b \rangle > 0$ and $\langle v_b \rangle < 0$, respectively). For the corresponding videos, see the Supplementary Material.

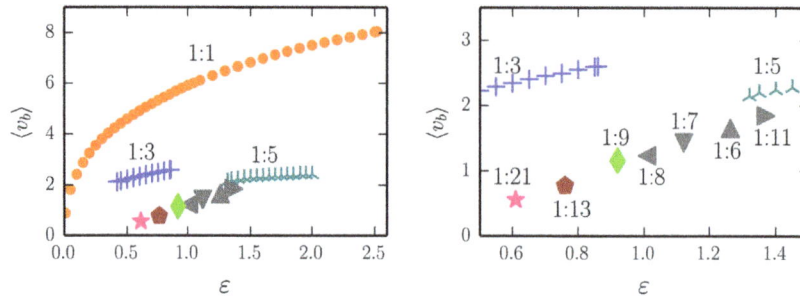

FIGURE 8 | *Left:* **the average speed <v_b> of the two-rod barrel for the 1:1/1:3/1:5 (orange dots, blue crosses, dark-cyan stars) modes**. The gain is $a = 1.9$, all other parameters are identical to the ones used for the one-rod barrel. The respective time series and phase-space plots are presented in **Figure 9**. The filled symbols denote examples of additional higher order modes, of which the 1:21, 1:13, and 1:9 (pink star, maroon pentagon, and green rhombus) are illustrated in **Figure 10**. *Right:* a blow-up, showing the relative location of the 1:8, 1:7, 1:6, and 1:11 modes found at $\varepsilon = 1.009$, 1.122, 1.263, and 1.370, respectively.

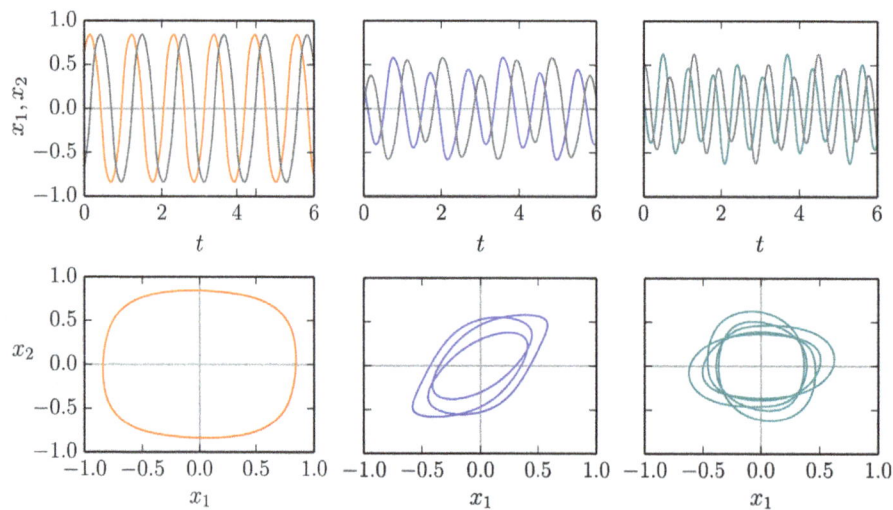

FIGURE 9 | **Time series $x_1(t)$ and $x_2(t)$ of the balls along the two rods of the two-rod barrel (top row), and the respective phase plots ($x_1(t)$, $x_2(t)$)**. Shown are the 1:1/1:3/1:5 modes (left/middle/right column) for $\varepsilon = 1.0/0.5/1.5$, compare **Figure 8**, needing, respectively, 1/3/5 revolutions around the origin (x_1, x_2) = (0,0) in order to close. For the corresponding videos, see the Supplementary Material.

they arise through a bifurcation cascade of limit cycles (Sándor and Gros, 2015), or via some other mechanism, is, however, beyond the scope of the present study.

4. DISCUSSION

It is, in a certain sense, a trivial statement, that the environment is part of the dynamical system a biological or artificial agent lives in. Little of the environmental dynamics is, however, in general accessible, or known, from the perspective of a robot, and it is hence often more suitable, as in closed-loop control (Dorf and Bishop, 1998), to consider the sensorimotor loop as a sequence of stimulus–response reactions of the agent, eliciting at every step the subsequent environmental signal. Here, we have considered simple barrel-shaped robots in a simulated environment, for which the sensorimotor loop constitutes truly a dynamical system, capable of generating, even in a simple setup, a very rich palette of dynamical modes and hence a wide range of qualitatively different types of motions.

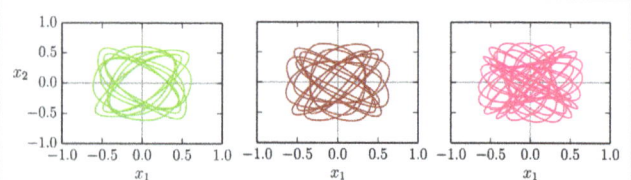

FIGURE 10 | **Examples of higher order limit cycles found for the two-rod barrel, closing (within numerical accuracy, viz the thickness of the lines) after 9/13/21 revolutions around the origin (x_1, x_2) = (0,0) (left/middle/right)**. The gain is $a = 1.9$ and the respective adaption rates are $\varepsilon = 0.92$, 0.76, and 0.61, compare **Figure 8**.

The dominant rolling modes found are 1:M attractors, where the actuators cycle M = 1, 3, 5, ... times during one revolution $\phi \rightarrow \phi + 2\pi$ of the barrel. These modes coexist with non-rolling modes, having their own respective basins of attractions, emerging from the mutual feedback of robot and environment. There exist, in addition, regions of phase space with stationary rolling modes (rolling periodically back-and-forth), and drifting back-and-forth modes. We have also found preliminary indications of rolling modes living on two or higher dimensional tori, with incommensurate revolution frequencies, which we did, however, not investigate in detail in the present study. There may additionally exist further attracting states, yet not discovered when performing numerical simulations within the LPZRobots environment.

All modes found are attracting dynamical states and hence robust against noise. This robustness varies, however, with the dominant 1:1 being the most stable, and higher order modes, like the 1:3 or the 1:21 limit cycles, being relatively less stable. There is, in addition, the need to overcome the dissipation, which is present in the simulated environment, by an appropriate energy intake of the actuator. As for all robots the question then arises, whether the observed behavior can be considered as dominantly driven, in the sense of actuator overpowering, or as self-organized, via an inherent and essential feedback loop through the environment [in this context, see Egbert et al. (2010) for an analogous discussion in the context of bacterial sensorimotor system involving chemotaxis].

Actuator-controlled behavior would generally lead, in our perspective, to rather stereotypical movements modes. The fact that our robots show a very large variety of modes upon changing the adaption rate ε, viz, the reaction time $1/\varepsilon$ of the actuator, indicates self-organization. These modes are also partially overlapping with several rolling modes possibly coexisting for the same settings. It is then a question of starting conditions, into which behavior the robot then settles.

We have also investigated the dynamics of the actuators employed, a damped-spring ball moving along a rod, when the rolling motion $d\phi/dt \to 0$ of the barrel is turned off. In this setting, the environmental feedback from the rolling motion is not present. We find parameter regions where the engine is autonomously active and parameter regions, where the engine shuts itself off. In the later region, the engine may be kicked in again, when the barrel is given a kick, and allowed to roll normally.

In this case, the environmental feedback is hence essential, and the motion of the robot is a consequence of self-organizing processes in the combined phase space of the internal degrees of freedom of the robot and of the physical environment.

Thus, the behavior of the robot can not be attributed to merely one of the subsystems, but it is a property of the coupled brain–body–environment system, a result also found in the context of minimally cognitive agents (Beer, 2003; Beer and Williams, 2015). Since we are not aiming here for the presence of higher level cognitive processes, our work can be seen as a purely dynamical systems approach for understanding embodiment directly within the sensorimotor loop.

Our work has been performed with the LPZRobots simulation package, which has been used extensively to investigate the emergence of "playful" behavior and sensorimotor intelligence in terms of intermittent chaotic motion patterns (Der and Martius, 2012; Martius et al., 2013). In this context, our investigation is embedded in the long-standing effort (Taga et al., 1991; Kelso, 1994; Pfeifer et al., 2007; Der and Martius, 2015) to reduce the demanding problem of programing robots by investigating the emergence of self-organized motions within the sensorimotor loop.

AUTHOR CONTRIBUTIONS

Most data and figures where produced by BS, the paper written by CG and BS, with TJ and LM adding data and material.

ACKNOWLEDGMENTS

We thank Georg Martius for extensive discussions and for helping setting up the LPZRobots simulation environment.

REFERENCES

Ay, N., Bernigau, H., Der, R., and Prokopenko, M. (2012). Information-driven self-organization: the dynamical system approach to autonomous robot behavior. *Theory Biosci.* 131, 161–179. doi:10.1007/s12064-011-0137-9

Ay, N., Bertschinger, N., Der, R., Güttler, F., and Olbrich, E. (2008). Predictive information and explorative behavior of autonomous robots. *Eur. Phys. J. B* 63, 329–339. doi:10.1140/epjb/e2008-00175-0

Baddeley, R., Hancock, P., and Földiák, P. (2008). *Information Theory and the Brain*. New York: Cambridge University Press. p. 1.

Beer, R. D. (2000). Dynamical approaches to cognitive science. *Trends Cogn. Sci.* 4, 91–99. doi:10.1016/S1364-6613(99)01440-0

Beer, R. D. (2003). The dynamics of active categorical perception in an evolved model agent. *Adaptive Behav.* 11, 209–243. doi:10.1177/1059712303114001

Beer, R. D., and Williams, P. L. (2015). Information processing and dynamics in minimally cognitive agents. *Cogn. Sci.* 39, 1–38. doi:10.1111/cogs.12142

Clewley, R. (2012). Hybrid models and biological model reduction with pydstool. *PLoS Comput. Biol.* 8:e1002628. doi:10.1371/journal.pcbi.1002628

Dasgupta, S., Wörgötter, F., and Manoonpong, P. (2013). Information dynamics based self-adaptive reservoir for delay temporal memory tasks. *Evolving Syst.* 4, 235–249. doi:10.1007/s12530-013-9080-y

de Wit, C. C., Siciliano, B., and Bastin, G. (2012). *Theory of Robot Control*. London: Springer Science & Business Media.

Der, R., and Martius, G. (2012). *The Playful Machine: Theoretical Foundation and Practical Realization of Self-Organizing Robots, Volume 15*. Berlin: Springer Science & Business Media.

Der, R., and Martius, G. (2015). A novel plasticity rule can explain the development of sensorimotor intelligence. arXiv preprint arXiv 1505.00835.

Dorf, R. C., and Bishop, R. H. (1998). *Modern Control Systems*. Menlo Park, CA: Pearson (Addison-Wesley).

Echeveste, R., Eckmann, S., and Gros, C. (2015). The fisher information as a neural guiding principle for independent component analysis. *Entropy* 17, 3838–3856. doi:10.3390/e17063838

Echeveste, R., and Gros, C. (2014). Generating functionals for computational intelligence: the fisher information as an objective function for self-limiting Hebbian learning rules. *Front. Rob. AI* 1. doi:10.3389/frobt.2014.00001

Egbert, M. D., Barandiaran, X. E., and Di Paolo, E. A. (2010). A minimal model of metabolism-based chemotaxis. *PLoS Comput. Biol.* 6:e1001004. doi:10.1371/journal.pcbi.1001004

Ernesti, J., Righetti, L., Do, M., Asfour, T., and Schaal, S. (2012). "Encoding of periodic and their transient motions by a single dynamic movement primitive," in *12th IEEE-RAS International Conference on Humanoid Robots (Humanoids 2012)* (Piscataway, NJ: IEEE), 57–64.

Friston, K. (2010). The free-energy principle: a unified brain theory? *Nat. Rev. Neurosci.* 11, 127–138. doi:10.1038/nrn2787

Friston, K., and Ao, P. (2012). Free energy, value, and attractors. *Comput. Math. Methods Med.* 2012:937860. doi:10.1155/2012/937860

Gros, C. (2015). *Complex and Adaptive Dynamical Systems: A Primer.* Heidelberg: Springer.

Gros, C., Linkerhand, M., and Walther, V. (2014). "Attractor metadynamics in adapting neural networks," in *Artificial Neural Networks and Machine Learning–ICANN 2014* (Heidelberg: Springer-Verlag), 65–72.

Hobbelen, D. G. (2008). *Limit Cycle Walking.* Delft: TU Delft, Delft University of Technology.

Ijspeert, A. J. (2008). Central pattern generators for locomotion control in animals and robots: a review. *Neural Networks* 21, 642–653. doi:10.1016/j.neunet.2008.03.014

Ijspeert, A. J., Nakanishi, J., Hoffmann, H., Pastor, P., and Schaal, S. (2013). Dynamical movement primitives: learning attractor models for motor behaviors. *Neural Comput.* 25, 328–373. doi:10.1162/NECO_a_00393

Ijspeert, A. J., Nakanishi, J., and Schaal, S. (2002). "Learning attractor landscapes for learning motor primitives," in *Advances in Neural Information Processing Systems* (Cambridge, MA: MIT Press), 1547–1554.

Kelso, J. (1994). The informational character of self-organized coordination dynamics. *Hum. Mov. Sci.* 13, 393–413. doi:10.1016/0167-9457(94)90047-7

Klyubin, A. S., Polani, D., and Nehaniv, C. L. (2005). "Empowerment: a universal agent-centric measure of control," in *The 2005 IEEE Congress on Evolutionary Computation, 2005,* Vol. 1 (Piscataway, NJ: IEEE), 128–135.

Laszlo, J., van de Panne, M., and Fiume, E. (1996). "Limit cycle control and its application to the animation of balancing and walking," in *Proceedings of the 23rd Annual Conference on Computer Graphics and Interactive Techniques* (New York: ACM), 155–162.

Lizier, J. T., Prokopenko, M., and Zomaya, A. Y. (2012). Local measures of information storage in complex distributed computation. *Inf. Sci.* 208, 39–54. doi:10.1016/j.ins.2012.04.016

Marković, D., and Gros, C. (2010). Self-organized chaos through polyhomeostatic optimization. *Phys. Rev. Lett.* 105, 068702. doi:10.1103/PhysRevLett.105.068702

Marković, D., and Gros, C. (2012). Intrinsic adaptation in autonomous recurrent neural networks. *Neural Comput.* 24, 523–540. doi:10.1162/NECO_a_00232

Martius, G., Der, R., and Ay, N. (2013). Information driven self-organization of complex robotic behaviors. *PLoS ONE* 8:e63400. doi:10.1371/journal.pone.0063400

Nolfi, S., and Floreano, D. (2000). *Evolutionary Robotics: The Biology, Intelligence, and Technology of Self-Organizing Machines.* Cambridge, MA: MIT Press.

Olsson, L. A., Nehaniv, C. L., and Polani, D. (2006). From unknown sensors and actuators to actions grounded in sensorimotor perceptions. *Connection Sci.* 18, 121–144. doi:10.1080/09540090600768542

Pfeifer, R., Lungarella, M., and Iida, F. (2007). Self-organization, embodiment, and biologically inspired robotics. *Science* 318, 1088–1093. doi:10.1126/science.1145803

Sándor, B., and Gros, C. (2015). A versatile class of prototype dynamical systems for complex bifurcation cascades of limit cycles. *Sci. Rep.* 5, 12316. doi:10.1038/srep12316

Schaal, S., Kotosaka, S., and Sternad, D. (2000). "Nonlinear dynamical systems as movement primitives," in *IEEE International Conference on Humanoid Robotics* (Cambridge, MA: CD-Proceedings), 1–11.

Schmidt, N. M., Hoffmann, M., Nakajima, K., and Pfeifer, R. (2013). Bootstrapping perception using information theory: case studies in a quadruped robot running on different grounds. *Adv. Complex Syst.* 16, 1250078. doi:10.1142/S0219525912500786

Taga, G., Yamaguchi, Y., and Shimizu, H. (1991). Self-organized control of bipedal locomotion by neural oscillators in unpredictable environment. *Biol. Cybern.* 65, 147–159. doi:10.1007/BF00198086

Tani, J., and Ito, M. (2003). Self-organization of behavioral primitives as multiple attractor dynamics: a robot experiment. *IEEE Trans. Syst. Man Cybern. Part A Syst. Humans* 33, 481–488. doi:10.1109/TSMCA.2003.809171

Toyoizumi, T., Pfister, J.-P., Aihara, K., and Gerstner, W. (2005). Generalized Bienenstock-Cooper-Munro rule for spiking neurons that maximizes information transmission. *Proc. Natl. Acad. Sci. U.S.A.* 102, 5239–5244. doi:10.1073/pnas.0500495102

Triesch, J. (2005). "A gradient rule for the plasticity of a neurons intrinsic excitability," in *Artificial Neural Networks: Biological Inspirations ICANN 2005* (Berlin: Springer), 65–70.

Triesch, J. (2007). Synergies between intrinsic and synaptic plasticity mechanisms. *Neural Comput.* 19, 885–909. doi:10.1162/neco.2007.19.4.885

Williams, P., and Beer, R. (2013). "Environmental feedback drives multiple behaviors from the same neural circuit," in *Advances in Artificial Life, ECAL 2013,* Vol. 12 (Cambridge, MA: MIT Press), 268–275.

Ziemke, T. (2003). "Whats that thing called embodiment," in *Proceedings of the 25th Annual Meeting of the Cognitive Science Society* (Mahwah, NJ: Lawrence Erlbaum), 1305–1310.

Conflict of Interest Statement: The authors declare that the research was conducted in the absence of any commercial or financial relationships that could be construed as a potential conflict of interest.

11

Cryptobotics: why robots need cyber safety

Santiago Morante*, Juan G. Victores and Carlos Balaguer

Robotics Lab, Automation and Engineering Systems Department, Universidad Carlos III de Madrid, Madrid, Spain

Keywords: cryptography, robotics, cyber safety, communications, cyber-physical, cryptobotics, cyber security

Introduction

With the expected introduction of robots into our daily lives, providing mechanisms to avoid undesired attacks and exploits in robot communication software is becoming increasingly required. Just as during the beginnings of the computer age (Pfleeger and Pfleeger, 2002), robotics is established in a "happy naivety," where security rules against external attacks are not adopted, assuming that robotics knowledgeable people are well intended. While this may have been true in the past, the mass adoption of robots will increase the possibilities of attacks. This fact is especially relevant in defense, medical and other critical fields involving humans, where tampering can result in serious bodily harm and/or privacy invasions. For these reasons, we consider that researchers and industry should deploy efforts in cyber safety and acquire good practices when developing and distributing robot software. We propose the term *Cryptobotics* as a unifying term for research and applications of computer and microcontrollers' security measures in robotics.

Stating the Problem

Edited by:
Lorenzo Natale,
Istituto Italiano di Tecnologia, Italy

Reviewed by:
Fulvio Mastrogiovanni,
University of Genoa, Italy
Emanuele Ruffaldi,
Scuola Superiore Sant'Anna, Italy
Ali Paikan,
Istituto Italiano di Tecnologia, Italy

*Correspondence:
Santiago Morante
smorante@ing.uc3m.es

The problems that the field of robotics will face are similar to those the computer revolution faced with the widespread of the Internet 30 years ago. Among the common attacks computers may suffer, there are: denial-of-service, eavesdropping, spoofing, tampering, privilege escalation, or information disclosure for instance. To these problems, robots add the additional factor of physical interaction. While taking the control of a desktop computer or a server may result in loss of information (with its associated costs), taking the control of a robot may endanger whatever or whoever is near.

As robots become more integrated on the communications networks, it seems appropriate to reuse the tools designed for web applications in order to controls the robots. However, the authors consider there are differences between regular computers communicating through the network, and robots performing the same actions. Mohanarajah et al. (2015) states differences between web and robotic applications: "Web applications are typically stateless, single processes that use a request-response model to talk to the client. Meanwhile, robotic applications are stateful, multiprocessed, and require a bidirectional communication with the client. These fundamental differences may lead to different tradeoffs and design choices and may ultimately result in different software solutions for web and robotics applications." To these differences, we could also add the real-time constraints that characterize robotics applications. Despite other sources of issues, like software bugs or vulnerabilities [buffer overflow, command injection, etc. (Tanenbaum and Bos, 2014)], we consider that communications currently are one of the main vulnerabilities in robotics.

A number of fields in robotics where security and privacy are particularly relevant can be addressed.

- Defense and Space: The military field should be very aware of the best practices in cyber security to be followed regarding its robots. Unmanned aerial vehicles, commonly called "drones," are being destined to surveillance and also to combat missions. Common sense dictates that any communications with these vehicles should be encrypted (Javaid et al., 2012), but reality shows us differently. For example, in the year 2012 it was reported that only between 30 and 50 percent of America's Predators and Reapers (two of the most used drones in US) were using fully encrypted transmissions.[1]

> *Situation: a non-authorized entity eavesdrops surveillance images of drones, takes its control, exploiting a non-encrypted connection, and crashes it into a populated area.*
>
> *Situation: a non-authorized entity takes control of a robot inside International Space Station and sabotages an ongoing experiment.*

- Telemedicine and Remote surgery: This exciting field can make remote surgery become an everyday reality, where experts can operate patients from the other side of the world. While this is beneficial to society, we must consider the potential dangers. In 2009, the Interoperable Telesurgery Protocol (ITP) (King et al., 2009) was proposed as a preliminary specification for interoperability among robotic telesurgery systems. Recently, the fact that ITP does not use any form of encryption or authentication was discovered.[2] This is an obvious system exposure to exploits using a man-in-the-middle attack for taking control of the robot (Bonaci et al., 2015).

> *Situation: a non-authorized entity takes control of a surgery robot during an operation, endangering the life of the patient.*

- Household robots: This market is growing both in research and commercially available robots. Robots will be used as assistants at home. For instance, one of these projects is Care-O-bot (Hans et al., 2002), a robotic assistant in homes. In one of the available versions, this robot is equipped with microphones, cameras and 3D sensors. This set of sensors can collect a huge amount of information, which must be protected (Denning et al., 2009). Service robots may one day also collect data about the health status of a person; law regulations require that this data is handled with extra care.

> *Situation: a non-authorized entity takes control of a household robot and obtains streams of images with private data.*

- Disaster robots: Since the Fukushima Daiichi nuclear disaster in 2011, the robotics community has increased its efforts to build and deploy robots for disaster scenarios. One of the expected tasks these robots will have to face in a disaster scenario is related to accessing and repairing/disconnecting dangerous systems. Due to the potential danger that may arise in these situations (Vuong et al., 2014), robots should not be able to be externally modified by an external attack.

> *Situation: a non-authorized entity takes control of a robot deployed to disconnect a nuclear platform that*

may suffer a partial meltdown, and can thwart the disconnection operation.

Current State of Security in Mainstream Robotic Software

Robots are a combination of mechanical structures, sensors, actuators, and computer software that manages and controls these devices. Mainstream practices in robotics involve component-based software engineering. Each component is designed as an individual computer program (e.g., a motor moving program) which communicates with other components using predefined protocols. While a large quantity of software libraries for communication already exist, the robotics community has developed a number of "software architectures." Currently, one of the most popular robotics-oriented architecture is ROS (Robot Operating System) (Quigley et al., 2009). Another co-existing architecture is YARP (Yet Another Robot Platform) (Metta et al., 2006). Both systems work similarly: a system built using ROS or YARP consists of a number of programs (nodes or modules), potentially on several different hosts, connected in a peer-to-peer topology.

According to ROS documentation[3]: "Topics are named buses over which nodes exchange messages. Topics have anonymous publish/subscribe semantics (.) In general, nodes are not aware of who they are communicating with." From the point of view of security, this anonymous communication scheme is a welcome sign toward exploits (McClean et al., 2013). Messages are sent unencrypted through TCP/IP or UDP/IP. The default check performed is an initial MD5 sum of the message structure, a mechanism used to assure the parties agree on the layout of the message. Some researchers have developed an authentication mechanism for achieving secure authentication for remote, non-native clients in ROS (Toris et al., 2014). While it can increase the security of the overall system, without data encryption, an eavesdropper could acquire non-encrypted information.

Part of the ROS community is dedicating efforts to integrating OMG's DDS (Data Distribution Service) as a transport layer for ROS 2.0.[4] A preliminary alpha version has just been released. DDS is a standard specification followed by several vendors for a middleware providing publish-subscribed communications for real-time and embedded systems. RTI provides plugins which comply with the DDS Security specification including authentication, access control and cryptography. It would be a big step forward for securing our robots if ROS 2.0 aimed to comply with the DDS Security specification as well.

YARP states among its documentation[5]: "A [default] new connection to a YARP port is established via handshaking on a TCP port. So everyone who can access this TCP port can connect to your YARP port. So if you are not behind a firewall, you are exposing your YARP infrastructure to the world (.) And if your application is vulnerable to corrupted data, it is a security

---

[1] Most U.S. Drones Openly Broadcast Secret Video Feeds: http://www.wired.com/2012/10/hack-proof-drone/

[2] Interoperable Telesurgery Protocol (ITP) Plaintext Unauthenticated MitM Hijacking: http://osvdb.org/121842

[3] http://wiki.ros.org/Topics

[4] http://design.ros2.org

[5] http://wiki.icub.org/yarpdoc/yarp_port_auth.html

leak." Other YARP documentation reads clearly[6]: "If you expose machines running YARP to the Internet, expect your robot to 1 day be commanded to make a crude gesture at your funders by a script kiddie in New Zealand." However, an authentication mechanism can be activated in YARP, which adds a key exchange to the initial handshaking in order to authenticate any connection request. It has been enabled by default so it is always compiled. However, to preserve backward compatibility, the feature is skipped at runtime if the user does not configure it by providing a file that contains the authentication key.

Additionally, a new port monitoring and arbitration (Paikan et al., 2014) functionality inside YARP has been used to implement a LUA encoder/decoder of data.[7] Data are passed through a Base64 encoder before being sent, and decoded upon reception at the target port. A similar mechanism could potentially be used to encrypt and decrypt the data.

A limited amount of other works has also focused on securing robot communications. In Groza and Dragomir (2008), they implement an authentication protocol to assure the authenticity of the information when controlling a robot via TCP/IP. However, they do not implement encrypted communications. In Coble et al. (2010), they implemented a hardware system that verifies integrity and health of the system software (to avoid tampering) in telesurgical robots. Regarding the previously mention ITP protocol, some researchers are working on security enhancements (Lee and Thuraisingham, 2012). One commercially available robot that does take cyber security into account is BeamPro, a telepresence robot[8] where secure protocols, symmetric encryption, and data authentication are used, thus providing security and privacy.

Secure communications are even more important in new trends in robotics which aim at outsourcing computation, namely *Cloud Robotics*. In this paradigm, robots use their sensors to collect data, and then upload the information to a remote computation center, where the information is processed, and may be shared with other robots. Rapyuta (Mohanarajah et al., 2015) is an example of this paradigm where the technologies used (e.g., WebSockets) allow to secure the information.

Another usual way of communications between robot's devices is through communication buses (CAN, EtherCAT, etc.). Currently, none of the traditional field buses offers security features against intentional attacks (Dzung et al., 2005). However, those based on ethernet could potentially make use of the security measures included in TCP/UDP/IP. For instance, secure routers (e.g., EDR-G903), include firewalls and VPNs, and support EtherCAT.

Discussion

A big market of opportunities for research regarding cyber safety in robotics exists. Most robots are not yet prepared, from a security point of view, to be deployed in daily life. The software is not prepared to protect against attacks, because communications are usually unencrypted.

Regarding the dates of the exploits presented, and the current hype in deployment of daily robotics (vacuum cleaners, amateur drones, etc.), *Cryptobotics*, understood as a mix of cyber security and robotics, comes just in time to prepare these systems to be safely used.

An important issue to be discussed is whether the implementation of encrypted communications may affect the performance, especially in real-time systems. The question about performance is highly dependant on the hardware, the software and the network used. Encrypted communications on the Internet (https, ssh) show us that it is possible to perform secure communications and offer remote services. For instance, Adam Langley (Google Senior Staff Software Engineer) has stated: "when Google changed Gmail from http to https (.) we had to deploy no additional machines and no special hardware. On our production front-end machines, SSL/TLS accounts for less than 1 of the CPU load.[9]" From our experience in humanoid robotics, a 1% overhead (while respecting determinism in time) can be acceptable if it means our devices can be less vulnerable to cyber attacks. Could an 8 MHz microcontroller perform real-time encryption? Is it reasonable to implement authentication mechanisms along field buses in time-constrained scenarios? This article intends to raise awareness for developers to determine whether it is viable to integrate these mechanisms depending on each specific use case.

Some may ask why these problems have not been addressed previously. In recent years, intrinsically safe industrial robots, the rise of domestic robots, and the use of mobile robots in public spaces, have arisen issues that the robotics community did not have to face in its previous 60 years of existence. Researchers are now focused on developing applications to make robots useful, which may have made cyber safety a low priority.

Author Contributions

SM discovered the potential issue in mainstream robotics software and wrote part of the paper. JV found the technical support behind the security issues and wrote and improved the text. CB provided context for the topic, reviewed the text, defined the criteria to evaluate the work, and contributed to the text.

References

Bonaci, T., Herron, J., Yusuf, T., Yan, J., Kohno, T., and Chizeck, H. J. (2015). "To make a robot secure: an experimental analysis of cyber security threats against teleoperated surgical robots," in *arXiv Preprint arXiv:1504.04339*.

Coble, K., Wang, W., Chu, B., and Li, Z. (2010). Secure software attestation for military telesurgical robot systems" in *Military Communications Conference, 2010 – Milcom 2010*, 965–970. doi:10.1109/MILCOM.2010.5679580

Denning, T., Matuszek, C., Koscher, K., Smith, J. R., and Kohno, T. (2009). "A spotlight on security and privacy risks with future household robots: attacks

---

[6]http://wiki.icub.org/yarpdoc/what_is_yarp.html
[7]http://wiki.icub.org/yarpdoc/coder_decoder.html
[8]http://www.suitabletech.com/
[9]http://www.imperialviolet.org/2010/06/25/overclocking-ssl.html

and lessons," in *Proceedings of the 11th International Conference on Ubiquitous Computing* (New York, NY: ACM), 105–114. doi:10.1145/1620545.1620564

Dzung, D., Naedele, M., Von Hoff, T. P., and Crevatin, M. (2005). Security for industrial communication systems. *Proc IEEE* 93, 1152–1177. doi:10.1109/JPROC.2005.849714

Groza, B., and Dragomir, T.-L. (2008). "Using a cryptographic authentication protocol for the secure control of a robot over TCP/IP," in *IEEE International Conference on Automation, Quality and Testing, Robotics, 2008. AQTR 2008*, Vol. 1, 184–189. doi:10.1109/AQTR.2008.4588731

Hans, M., Graf, B., and Schraft, R. (2002). "Robotic home assistant care-o-bot: past-present-future," in *Proceedings of the 11th IEEE International Workshop on Robot and Human Interactive Communication, 2002*, 380–385. doi:10.1109/ROMAN.2002.1045652

Javaid, A. Y., Sun, W., Devabhaktuni, V. K., and Alam, M. (2012). "Cyber security threat analysis and modeling of an unmanned aerial vehicle system," in *2012 IEEE Conference on Technologies for Homeland Security (HST)*, 585–590. doi:10.1109/THS.2012.6459914

King, H. H., Tadano, K., Donlin, R., Friedman, D., Lum, M. J., Asch, V., et al. (2009). "Preliminary protocol for interoperable telesurgery," in *International Conference on Advanced Robotics, 2009. ICAR 2009*, 1–6. Available at: http://ieeexplore.ieee.org/stamp/stamp.jsp?tp=&arnumber=5174711&isnumber=5174665

Lee, G. S., and Thuraisingham, B. (2012). Cyberphysical systems security applied to telesurgical robotics. *Comput Stand Interfaces* 34, 225–229. doi:10.1016/j.csi.2011.09.001

McClean, J., Stull, C., Farrar, C., and Mascareñas, D. (2013). "A preliminary cyber-physical security assessment of the robot operating system (ROS)," in *Proceedings of SPIE 8741, Unmanned Systems Technology XV*, 874110. doi:10.1117/12.2016189

Metta, G., Fitzpatrick, P., and Natale, L. (2006). Yarp: yet another robot platform. *Int J Adv Rob Syst* 3, 43–48. doi:10.5772/5761

Mohanarajah, G., Hunziker, D., D'Andrea, R., and Waibel, M. (2015). Rapyuta: a cloud robotics platform. *IEEE Trans Autom Sci Eng* 12, 481–493. doi:10.1109/TASE.2014.2329556

Paikan, A., Fitzpatrick, P., Metta, G., and Natale, L. (2014). Data flow ports monitoring and arbitration. *J Software Eng Rob* 5, 80–88.

Pfleeger, C. P., and Pfleeger, S. L. (2002). *Security in Computing*. Prentice Hall Professional Technical Reference.

Quigley, M., Conley, K., Gerkey, B., Faust, J., Foote, T., Leibs, J., et al. (2009). "ROS: an open-source robot operating system," in *ICRA Workshop on Open Source Software*, 3, 5.

Tanenbaum, A. S., and Bos, H. (2014). *Modern Operating Systems*. Prentice Hall Press.

Toris, R., Shue, C., and Chernova, S. (2014). "Message authentication codes for secure remote non-native client connections to ros enabled robots," in *2014 IEEE International Conference on Technologies for Practical Robot Applications (TePRA)*, 1–6. doi:10.1109/TePRA.2014.6869141

Vuong, T., Filippoupolitis, A., Loukas, G., and Gan, D. (2014). "Physical indicators of cyber attacks against a rescue robot," in *2014 IEEE International Conference on Pervasive Computing and Communications Workshops (PERCOM Workshops)*, 338–343. doi:10.1109/PerComW.2014.6815228

Conflict of Interest Statement: The authors declare that the research was conducted in the absence of any commercial or financial relationships that could be construed as a potential conflict of interest.

12

Grammars for Games: A Gradient-Based, Game-Theoretic Framework for Optimization in Deep Learning

David Balduzzi*

School of Mathematics and Statistics, Victoria University Wellington, Wellington, New Zealand

Deep learning is currently the subject of intensive study. However, fundamental concepts such as representations are not formally defined – researchers "know them when they see them" – and there is no common language for describing and analyzing algorithms. This essay proposes an abstract framework that identifies the essential features of current practice and may provide a foundation for future developments. The backbone of almost all deep learning algorithms is backpropagation, which is simply a gradient computation distributed over a neural network. The main ingredients of the framework are, thus, unsurprisingly: (i) game theory, to formalize distributed optimization; and (ii) communication protocols, to track the flow of zeroth and first-order information. The framework allows natural definitions of semantics (as the meaning encoded in functions), representations (as functions whose semantics is chosen to optimized a criterion), and grammars (as communication protocols equipped with first-order convergence guarantees). Much of the essay is spent discussing examples taken from the literature. The ultimate aim is to develop a graphical language for describing the structure of deep learning algorithms that backgrounds the details of the optimization procedure and foregrounds how the components interact. Inspiration is taken from probabilistic graphical models and factor graphs, which capture the essential structural features of multivariate distributions.

Keywords: deep learning, representation learning, optimization, game theory, neural networks

Edited by:
Fabrizio Riguzzi,
Università degli Studi di Ferrara, Italy

Reviewed by:
Raul Vicente,
Max-Planck Institute for Brain
Research, Germany
Kate Cerqueira Revoredo,
Federal University of the State of Rio
de Janeiro, Brazil

***Correspondence:**
David Balduzzi
david.balduzzi@vuw.ac.nz

1. INTRODUCTION

Deep learning has achieved remarkable successes in object and voice recognition, machine translation, reinforcement learning, and other tasks (Hinton et al., 2012; Krizhevsky et al., 2012; Sutskever et al., 2014; LeCun et al., 2015; Mnih et al., 2015). From a practical standpoint, the problem of supervised learning is well-understood and has largely been solved – at least in the regime where both labeled data and computational power are abundant. The workhorse underlying most deep learning algorithms is error backpropagation (Werbos, 1974; Rumelhart et al., 1986a,b; Schmidhuber, 2015), which is simply gradient descent distributed across a neural network via the chain rule.

Gradient descent and its variants are well-understood when applied to convex or nearly convex objectives (Robbins and Monro, 1951; Nemirovski and Yudin, 1978; Nemirovski, 1979; Nemirovski et al., 2009). In particular, they have strong performance guarantees in the stochastic and adversarial settings (Zinkevich, 2003; Cesa-Bianchi and Lugosi, 2006; Bousquet and Bottou, 2008; Shalev-Shwartz, 2011). The reasons for the success of gradient descent in non-convex settings are less clear, although recent work has provided evidence that most local minima are good enough (Choromanska et al., 2015a,b); that modern convolutional networks are close enough to convex for many results on rates of convergence apply (Balduzzi, 2015); and that the rate of convergence of gradient descent can control generalization performance, even in non-convex settings (Hardt et al., 2015).

Taking a step back, gradient-based optimization provides a well-established set of computational primitives (Gordon, 2007), with theoretical backing in simple cases and empirical backing in others. First-order optimization, thus, falls in broadly the same category as computing an eigenvector or inverting a matrix: given sufficient data and computational resources, we have algorithms that reliably find good enough solutions for a wide range of problems.

This essay proposes to abstract out the optimization algorithms used for weight updates and focus on how the components of deep learning algorithms interact. Treating optimization as a computational primitive encourages a shift from low-level algorithm design to higher-level mechanism design: we can shift attention to designing architectures that are guaranteed to learn distributed representations suited to specific objectives. The goal is to introduce a language at a level of abstraction where designers can focus on formal specifications (grammars) that specify how plug-and-play optimization modules combine into larger learning systems.

1.1. What Is a Representation?

Let us recall how representation learning is commonly understood. Bengio et al. (2013) describe representation learning as "learning transformations of the data that make it easier to extract useful information when building classifiers or other predictors." More specifically, "a deep learning algorithm is a particular kind of representation learning procedure that discovers multiple levels of representation, with higher-level features representing more abstract aspects of the data" (Bengio, 2013). Finally, LeCun et al. (2015) state that multiple levels of representations are obtained "by composing simple but non-linear modules that each transform the representation at one level (starting with the raw input) into a representation at a higher, slightly more abstract level. With the composition of enough such transformations, very complex functions can be learned. For classification tasks, higher layers of representation amplify aspects of the input that are important for discrimination and suppress irrelevant variations."

The quotes describe the operation of a successful deep learning algorithm. What is lacking is a characterization of what makes a deep learning algorithm work in the first place. What properties must an algorithm have to learn layered representations? What does it mean for the representation learned by one layer to be useful to another? What, exactly, is a representation?

In practice, almost all deep learning algorithms rely on error backpropagation to "align" the representations learned by different layers of a network. This suggests that the answers to the above questions are tightly bound up in first-order (that is, gradient-based) optimization methods. It is, therefore, unsurprisingly that the bulk of the paper is concerned with tracking the flow of first-order information. The framework is intended to facilitate the design of more general first-order algorithms than backpropagation.

1.1.1. Semantics

To get started, we need a theory of the meaning or semantics encoded in neural networks. Since there is nothing special about neural networks, the approach taken is inclusive and minimalistic. Definition 1 states that the meaning of *any* function is how it implicitly categorizes inputs by assigning them to outputs. The next step is to characterize those functions whose semantics encode knowledge, and for this we turn to optimization (Sra et al., 2012).

1.1.2. Representations from Optimizations

Nemirovski and Yudin (1983) developed the black-box computational model to analyze the computational complexity of first-order optimization methods (Agarwal et al., 2009; Raginsky and Rakhlin, 2011; Arjevani et al., 2016). The black-box model is a more abstract view on optimization than the Turing machine model: it specifies a *communication protocol* that tracks how often an algorithm makes *queries* about the objective. It is useful to refine Nemirovski and Yudin's terminology by distinguishing between black-boxes, which *respond* with zeroth-order information (the value of a function at the query-point), and gray-boxes,[1] which respond with zeroth- and first-order information (the gradient or subgradient).

With these preliminaries in hand, Definition 4 proposes that a *representation* is a function that is a *local* solution to an optimization problem. Since we do not restrict to convex problems, finding global solutions is not feasible. Indeed, recent experience shows that global solutions are often not necessary practice (Hinton et al., 2012; Krizhevsky et al., 2012; Sutskever et al., 2014; LeCun et al., 2015; Mnih et al., 2015). The local solution has similar semantics to – that is, it represents – the ideal solution. The ideal solution usually cannot be found: due to computational limitations, since the problem is non-convex, because we only have access to a finite sample from an unknown distribution, etc.

To see how Definition 4 connects with representation learning as commonly understood, it is necessary to take a detour through distributed optimization and game theory.

1.2. Distributed Representations

Game theory provides tools for analyzing distributed optimization problems where a set of players aim to minimizes losses that depend not only on their actions but also the actions of all other players in the game (von Neumann and Morgenstern, 1944; Nisan et al., 2007). Game theory has traditionally focused on convex losses since they are more theoretically amenable. Here, the only restriction imposed on losses is that they are differentiable almost everywhere.

Allowing non-convex losses means that error backpropagation can be reformulated as a game. Interestingly, there is enormous freedom in choosing the players. They can correspond to individual units, layers, entire neural networks, and a variety of other, intermediate choices. An advantage of the game-theoretic formulation is, thus, that it applies at many different scales.

Non-convex losses and local optima are essential to developing a *scale-free* formalism. Even when it turns out that particular units

---

[1] Gray for gradient.

or a particular layer of a neural network are solving a convex problem, convexity is destroyed as soon as those units or layers are combined to form larger learning systems. Convexity is not a property that is preserved in general when units are combined into layers or layers into networks. It is, therefore, convenient to introduce the computational primitive arglocopt to denote the output of a first-order optimization procedure, see Definition 4.

1.2.1. A Concern about Excessive Generality

A potential criticism is that the formulation is too broad. Very little can be said about non-convex optimization in general; introducing games where many players jointly optimize a set of arbitrary non-convex functions only compounds the problem.

Additional structure is required. A successful case study can be found in Balduzzi (2015), which presents a detailed game-theoretic analysis of rectifier neural networks. The key to the analysis is that rectifier units are almost convex. The main result is that the rate of convergence of a neural network to a local optimum is controlled by the (waking-)regret of the algorithms applied to compute weight updates in the network.

Whereas Balduzzi (2015) relied heavily on specific properties of rectifier non-linearities, this paper considers a wide-range of deep learning architectures. Nevertheless, it is possible to carve out an interesting subclass of non-convex games by identifying the composition of simple functions as an essential feature common to deep learning architectures. Compositionality is formalized via distributed communication protocols and grammars.

1.2.2. Grammars for Games

Neural networks are constructed by composing a series of elementary operations. The resulting feedforward computation is captured via as a computation graph (Griewank and Walther, 2008; Bergstra et al., 2010; Bastien et al., 2012; Baydin and Pearlmutter, 2014; Schulman et al., 2015; van Merriënboer et al., 2015). Backpropagation traverses the graph in reverse and recursively computes the gradient with respect to the parameters at each node.

Section 3 maps the feedforward and feedback computations onto the queries and responses that arise in Nemirovski and Yudin's model of optimization. However, queries and responses are now highly structured. In the query phase, players feed parameters into a computation graph (the Query graph Q) that performs the feedforward sweep. In the response phase, oracles reveal first-order information that is fed into a second computation graph (the Response graph R).

In most cases, the Response graph simply implements backpropagation. However, there are examples where it does not. Three are highlighted here, see Section 3.5, and especially Sections 3.6 and 3.7. Other algorithms where the Response graphs do not simply implement backprop include difference target propagation (Lee et al., 2015) and feedback alignment (Lillicrap et al., 2014) [both discussed briefly in Section 3.7] and truncated backpropagation through time (Elman, 1990; Williams and Peng, 1990; Williams and Zipser, 1995), where a choice is made about where to cut backprop short. Examples where the query and response graph differ are of particular interest, since they point toward more general classes of deep learning algorithms.

A *distributed communication protocol* is a game with additional structure: the Query and Response graphs, see Definition 7. The graphs capture the compositional structure of the functions learned by a neural network and the compositional structure of the learning procedure, respectively. It is important for our purposes that (i) the feedforward and feedback sweeps correspond to two distinct graphs and (ii) the communication protocol is kept distinct from the optimization procedure. That is, the communication protocol specifies how information flows through the networks without specifying how players make use of it. Players can be treated as plug-and-play rational agents that are provided with carefully constructed and coordinated first-order information to optimize as they see fit (Russell and Norvig, 2009; Gershman et al., 2015).

Finally, a *grammar* is a distributed communication protocol equipped with a guarantee that the response graph encodes sufficient information for the players to jointly find a local optimum of an objective function. The paradigmatic example of a grammar is backpropagation. A grammar is a, thus, a game designed to perform a task. A representation learned by one (p)layer is useful to another if the game is guaranteed to converge on a local solution to an objective – that is, if the players interact though a grammar. It follows that the players build representations that jointly encode knowledge about the task.

1.3. Contribution

The content of the paper is sketched above. In summary, the main contributions are as follows:

1. A characterization of *representations* as local solutions to functional optimization problems, see Definition 4.
2. An extension of Nemirovski and Yudin's first-order (Query–Response) protocol to deep learning, see Definition 7.
3. *Grammars*, Definition 8, which generalize the first-order guarantees provided by the error backpropagation to Response graphs that do not implement the chain rule.
4. Examples of grammars that do not reduce to the chain rule, see Sections 3.5, 3.6, and 3.7.

The essay presents a provisional framework; see Balduzzi (2015), Balduzzi and Ghifary (2015), Balduzzi et al. (2015) for applications of the ideas presented here. The essay is not intended to be comprehensive. Many details are left out and many important aspects are not covered: most notably, probabilistic and Bayesian formulations, and various methods for unsupervised pre-training.

1.3.1. A Series of Worked Examples

In line with its provisional nature, much of the essay is spent applying the framework to worked examples: error backpropagation as a supervised model (Rumelhart et al., 1986a); variational autoencoders (Kingma and Welling, 2014) and generative adversarial networks (Goodfellow et al., 2014) for unsupervised learning; the deviator-actor-critic (DAC) model for deep reinforcement learning (Balduzzi and Ghifary, 2015); and kickback, a biologically plausible variant of backpropagation (Balduzzi et al., 2015). The examples were chosen, in part, to maximize variety and, in part, based on familiarity. The discussions are short; the interested reader is encouraged to consult the original papers to fill in the gaps.

The last two examples are particularly interesting since their Response graphs differ substantially from backpropagation. The DAC model constructs a zeroth-order black-box to estimate gradients rather than querying a first-order gray-box. Kickback prunes backprop's Response graph by replacing most of its gray-boxes with black-boxes and approximating the chain rule with (primarily) local computations.

1.3.2. Related Work

Bottou and Gallinari (1991) proposed to decompose neural networks into cooperating modules (Bottou, 2014). Decomposing more general algorithms or models into collections of interacting agents dates back to the shrieking demons that comprised of Selfridge's Pandemonium (Selfridge, 1958) and a long line of related work (Klopf, 1982; Barto, 1985; Minsky, 1986; Baum, 1999; Kwee et al., 2001; von Bartheld et al., 2001; Seung, 2003; Lewis and Harris, 2014). The focus on components of neural networks as players, or rational agents, in their own right developed here derives from work aimed at modeling biological neurons game-theoretically, see Balduzzi and Besserve (2012), Balduzzi (2013, 2014), Balduzzi and Tononi (2013), and Balduzzi et al. (2013).

A related approach to semantics based on general value functions can be found in Sutton et al. (2011), see Remark 1. Computation graphs applied to backprop are the basis of the deep learning library Theano (Bergstra et al., 2010; Bastien et al., 2012; van Merriënboer et al., 2015) among others and provide the backbone for algorithmic differentiation (Griewank and Walther, 2008; Baydin and Pearlmutter, 2014).

Grammars are a technical term in the theory of formal languages relating to the Chomsky hierarchy (Hopcroft and Ullman, 1979). There is no apparent relation between that notion of grammar and the one presented here, aside from both relating to structural rules governing composition. Formal languages and deep learning are sufficiently disparate fields that there is little risk of terminological confusion. Similarly, the notion of semantics introduced here is distinct from semantics in the theory of programing languages.

Although game theory was developed to model human interactions (von Neumann and Morgenstern, 1944), it has been pointed out that it may be more directly applicable to interacting populations of algorithms, the so-called *machina economicus* (Lay and Barbu, 2010; Abernethy and Frongillo, 2011; Storkey, 2011; Frongillo and Reid, 2015; Parkes and Wellman, 2015; Syrgkanis et al., 2015). This paper goes one step further to propose that games played over first-order communication protocols are a key component of the foundations of deep learning.

A source of inspiration for the essay is Bayesian networks and Markov random fields. Probabilistic graphical models and factor graphs provide simple, powerful ways to encode a multivariate distribution's independencies into a diagram (Pearl, 1988; Kschischang et al., 2001; Wainwright and Jordan, 2008). They have greatly facilitated the design and analysis of algorithms for probabilistic inference. However, there is no comparable framework for distributed optimization and deep learning. The essay is intended as a first step in this direction.

2. SEMANTICS AND REPRESENTATIONS

This section defines semantics and representations. In short, the semantics of a function is how it categorizes its inputs; a function is a representation if it is selected to optimize an objective. The connection between the definition of representation below and "representation learning" is clarified in Section 3.3.

Possible world semantics was introduced by Lewis (1986) to formalize the meaning of sentences in terms of counterfactuals. Let $\mathcal{P}$ be a proposition about the world. Its truth depends on its content and the state of the world. Rather than allowing the state of the world to vary, it is convenient to introduce the set W of all possible worlds.

Let us denote proposition $\mathcal{P}$ applied in world $w \in W$ by $\mathcal{P}(w)$. The meaning of $\mathcal{P}$ is then the mapping $v_{\mathcal{P}}: W \to \{0,1\}$ which assigns 1 or 0 to each $w \in W$ according to whether or not proposition $\mathcal{P}(w)$ is true. Equivalently, the meaning of the proposition is the ordered pair consisting of: all worlds, and the subset of worlds where it is true:

$$\underbrace{W}_{\text{set of possible worlds}} \supset \underbrace{v_{\mathcal{P}}^{-1}(1)}_{\text{subset of worlds where } \mathcal{P} \text{ is true}} \ .$$

For example, the meaning of $\mathcal{P}_{blue}(that) = $ "that is blue" is the subset $v_{\mathcal{P}_{blue}}^{-1}(1)$ of possible worlds where I am pointing at a blue object. The concept of blue is rendered explicit in an exhaustive list of possible examples.

A simple extension of possible world semantics from propositions to arbitrary functions is as follows (Balduzzi, 2011):

DEFINITION 1 (*semantics*).

Given function $f: X \to Y$, the **semantics** or **meaning** of output $y \in Y$ is the ordered pair of sets:

$$\underbrace{X}_{\text{set of possible inputs}} \supset \underbrace{f^{-1}(y)}_{\text{subset causing } f \text{ to output } y} \ .$$

Functions implicitly categorize inputs by assigning outputs to them; the meaning of an output is the category.

Whereas propositions are true or false, the output of a function is neither. However, if two functions both optimize a criterion, then one can refer to how *accurately* one function *represents* the other. Before we can define representations, we therefore need to take a quick detour through optimization:

DEFINITION 2 (*optimization problem*).

An **optimization problem** is a pair $(\Theta, \mathbf{R})$ consisting of parameter-space $\Theta \subset \mathbb{R}^d$ and objective $\mathbf{R}: \Theta \to \mathbb{R}$ that is differentiable almost everywhere.

The **solution** to the global optimization problem is:

$$\theta^* = \underset{\theta \in \Theta}{\arg \text{opt}} \ \mathbf{R}(\theta),$$

which is either a maximum or minimum according to the nature of the objective.

The solution may not be unique; it also may not exist unless further restrictions are imposed. Such details are ignored here.

Next recall the black-box optimization framework introduced by Nemirovski and Yudin (1983) (Agarwal et al., 2009; Raginsky and Rakhlin, 2011; Arjevani et al., 2016).

Definition 3 (*communication protocol*).

A **communication protocol** for optimizing an unknown objective $\mathbf{R} \colon \Theta \to \mathbb{R}$ consists in a User (or Player) and an Oracle. On each round, User presents a **query** $\theta \in \Theta$. Oracle can **respond** in one of two ways, depending on the nature of the protocol:

- Black-box (zeroth-order) protocol.
  Oracle responds with the value $\mathbf{R}(\theta)$.

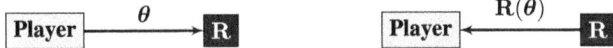

- Gray-box (first-order) protocol.
  Oracle responds with either the gradient $\triangledown \mathbf{R}\ (\theta)$ or with the gradient together with the value.

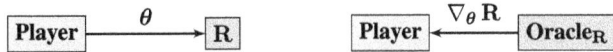

The protocol specifies how Player and Oracle interact without specifying the algorithm used by Player to decide which points to query. The next section introduces *distributed communication protocols* as a general framework that includes a variety of deep learning architectures as special cases – again without specifying the precise algorithms used to perform weight updates.

Unlike Nemirovski and Yudin (1983), Raginsky and Rakhlin (2011), we do not restrict to convex problems. Finding a global optimum is not always feasible, and in practice often unnecessary.

Definition 4 (*representation*).

Let $\mathcal{F} \subset \{f \colon X \to Y\}$ be a function space and:

$$f \colon \Theta \to \mathcal{F} \colon \theta \mapsto f_\theta(\bullet)$$

be a map from parameter space to functions. Further suppose that objective function $\mathbf{R} \colon \mathcal{F} \to \mathbb{R}$ is given.

A **representation** is a local solution to the optimization problem:

$$f_{\hat{\theta}} \quad where \quad \hat{\theta} \in \arg \operatorname*{locopt}_{\theta \in \Theta} \mathbf{R}(f_\theta),$$

corresponding to a *local* maximum or minimum according to whether the objective is minimized or maximized.

Intuitively, the objective quantifies the extent to which functions in $\mathcal{F}$ categorize their inputs similarly. The operation arglocopt applies a first-order method to find a function whose semantics resembles the optimal solution $f_{\theta^\star}$ where $\theta^* = \operatorname{argopt}_{\theta \in \Theta} \mathbf{R}(f_\theta)$.

In short, representations are functions with useful semantics, where usefulness is quantified using a specific objective: the lower the loss or higher the reward associated with a function, the more useful it is. The relation between Definition 4 and representations as commonly understood in the deep learning literature is discussed in Section 3.3 below.

Remark 1 (*value function semantics*).

In related work, Sutton et al. (2011) proposed that semantics – i.e., knowledge about the world – can be encoded in general value functions that provide answers to specific questions about expected rewards. Definition 1 is more general than their approach since it associates a semantics to *any* function. However, the function must arise from optimizing an objective for its semantics to accurately represent a phenomenon of interest.

2.1. Supervised Learning

The main example of a representation arises under supervised learning.

Representation 1 (*supervised learning*).

Let X and Y be an input space and a set of labels and $\ell \colon Y \times Y \to \mathbb{R}$ be a loss function. Suppose that $\{f_\theta \colon X \to Y | \theta \in \Theta\}$ is a parameterized family of functions.

- *Nature* which samples labeled pairs (x, y) i.i.d. from distribution $\mathbb{P}_{XY}$, singly or in batches.
- *Predictor* chooses parameters $\theta \in \Theta$.
- Objective is

$$\mathbf{R}(\theta) = \mathop{\mathbf{E}}_{(x,y) \sim \mathbb{P}_{XY}} [\ell(f_\theta(x), y)].$$

The query and responses phases can be depicted graphically as

The predictor $f_{\hat{\theta}} = \operatorname{arglocmin}_{\theta \in \Theta} \mathbf{R}(\theta)$ *is then a representation of the optimal predictor* $f_{\theta^\star} = \operatorname{argmin}_{\theta \in \Theta} \mathbf{R}(\theta)$.

A commonly used mapping from parameters to functions is

$$f \colon \Theta \to \mathcal{F} \colon \theta \mapsto f_\theta(\bullet) := \langle \phi(\bullet), \theta \rangle$$

where a feature map $\phi \colon X \to \mathbb{R}^d$ is fixed.

The setup admits a variety of complications in practice. First, it is typically infeasible even to find a local optimum. Instead, a solution that is within some small $\epsilon > 0$ of the local optimum suffices. Second, the distribution $\mathbb{P}_{XY}$ is unknown, so the expectation is replaced by a sum over a finite sample. The quality of the resulting representation has been extensively studied in statistical learning theory (Vapnik, 1995). Finally, it is often convenient to modify the objective, for example, by incorporating a regularizer. Thus, a more detailed presentation would conclude that

$$\hat{\theta} \approx \operatorname*{arg\,locmin}_{\theta \in \Theta} \sum_{i=1}^{n} \ell(f_\theta(x_i), y_i) + \Omega(\theta)$$

yields a representation $f_{\hat{\theta}}$ of the solution to $\arg \min_\theta \mathbf{E}_{\mathbb{P}_{XY}} [\ell(f_\theta(x), y)]$. To keep the discussion and notation simple, we do not consider any of these important details.

It is instructive to unpack the protocol, by observing that the objective $\mathbf{R}$ is a composite function involving $f(\theta, x)$, $\ell(f, y)$, and $\mathbf{E}[\bullet]$:

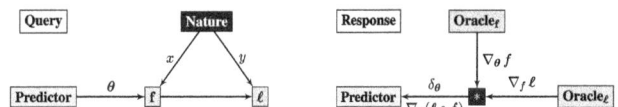

The notation δ_θ is borrowed from backpropagation. It is shorthand for the derivative of the objective with respect to parameters θ.

Nature is not a deterministic black-box since it is not queried directly: nature produces (x,y) pairs stochastically, rather than in response to specific inputs. Our notion of black-box can be extended to stochastic black-boxes, see Schulman et al. (2015). However, once again we prefer to keep the exposition as simple as possible.

2.2. Unsupervised Learning

The second example concerns fitting a probabilistic or generative model to data. A natural approach is to find the distribution under which the observed data is most likely:

REPRESENTATION 2 (*maximum likelihood estimation*).
Let X be a data space.

- *Nature* samples points from distribution $\mathbb{P}_X$.
- *Estimator* chooses parameters $\theta \in \Theta$.
- *Operator* $\mathbb{Q}(x; \theta) = \mathbb{Q}_\theta(x)$ computes a probability density on X that depends on parameter θ.
- *Operator* $-\log(\cdot)$ acts as a loss. The objective is to minimize

$$\mathbf{R}(\theta) := -\underset{x \sim \mathbb{P}_X}{\mathbf{E}} \left[\log \mathbb{Q}_\theta(x) \right].$$

The estimate $\mathbb{Q}(x; \hat{\theta})$, where $\hat{\theta} \in \operatorname{arglocmin}_{\theta \in \Theta} \mathbf{R}(\theta)$, is a representation of the optimal solution, and can also be considered a representation of $\mathbb{P}_X$. The setup extends easily to maximum *a posteriori* estimation.

As for supervised learning, the protocol can be unpacked by observing that the objective has a compositional structure:

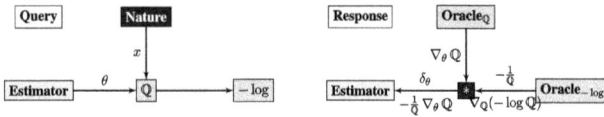

2.3. Reinforcement Learning

The third example is taken from reinforcement learning (Sutton and Barto, 1998). We will return to reinforcement learning in Section 3.6, so the example is presented in some detail. In reinforcement learning, an agent interacts with its environment, which is often modeled as a Markov decision process consisting of state space $\mathcal{S} \subset \mathbb{R}^m$, action space $\mathcal{A} \subset \mathbb{R}^d$, initial distribution $\mathbb{P}_1(s)$ on states, stationary transition distribution $\mathbb{P}(s_{t+1}|s_t, a_t)$ and reward function $r: \mathcal{S} \times \mathcal{A} \to \mathbb{R}$. The agent chooses actions based on a *policy*: a function $\mu_\theta: \mathcal{S} \to \mathcal{A}$ from states to actions. The goal is to find the optimal policy.

Actor-critic methods break up the problem into two pieces (Barto et al., 1983). The critic estimates the expected value of state-action pairs given the current policy, and the actor attempts to find the optimal policy using the estimates provided by the critic. The critic is typically trained via temporal difference methods (Sutton, 1988; Dann et al., 2014).

Let $\mathbb{P}_t$ $(s \to s', \mu)$ denote the distribution on states s' at time t given policy μ and initial state s at $t=0$ and let

$\rho^\mu(s') = \int_\mathcal{S} \sum_{t=0}^\infty \gamma^t \mathbb{P}_1(s) \mathbb{P}_t(s \to s', \mu) ds$. Let $r_t^\gamma = \sum_{\tau=t}^\infty \gamma^{\tau-t} r(s_\tau, a_\tau)$ be the discounted future reward. Define the value of a state-action pair as

$$Q^\mu(s, a) = \mathbf{E}[r_1^\gamma | S_1 = s, A_1 = a; \mu].$$

Unfortunately, the value–function $Q^\mu(s,a)$ cannot be queried. Instead, temporal difference methods take a bootstrapped approach by minimizing the Bellman error:

$$\ell_{BE}(v) = \underset{(s,a) \sim (\rho^\mu, \mu)}{\mathbf{E}} \left[(r(s, a) + \gamma Q^v(s', \mu(s')) - Q^v(s, a))^2 \right]$$

where s' is the state subsequent to s.

REPRESENTATION 3 (*temporal difference learning*).
Critic interacts with black-boxes Actor and Nature.[2]

- *Critic* plays parameters $\mathbf{v}$.
- *Operator* Q and ℓ_{BE} estimates the value function and compute the Bellman error. In practice, it turns out to *clone* the value-estimate periodically and compute a slightly modified Bellman error:

$$\ell_{BE}(v) = \underset{(s,a) \sim (\rho^\mu, \mu)}{\mathbf{E}} \left[(r(s, a) + \gamma Q^{\tilde{v}}(s', \mu(s')) - Q^v(s, a))^2 \right]$$

where $Q^{\tilde{v}}$ is the cloned estimate. Cloning improves the stability of TD-learning (Mnih et al., 2015). A nice conceptual side-effect of cloning is that TD-learning reduces to gradient descent.

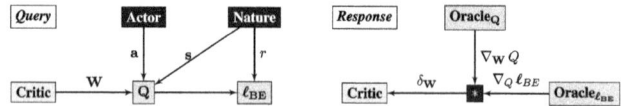

The estimate is a representation of the true value function.

REMARK 2 (*on temporal difference learning as first-order method*).

Temporal difference learning is not strictly speaking a gradient-based method (Dann et al., 2014). The residual gradient method performs gradient descent on the Bellman error, but suffers from double sampling (Baird, 1995). Projected fix point methods minimize the *projected* Bellman error via gradient descent and have nice convergence properties (Sutton et al., 2009a,b; Maei et al., 2010). An interesting recent proposal is implicit TD-learning (Tamar et al., 2014), which is based on implicit gradient descent (Toulis et al., 2014).

Section 3.6 presents the Deviator-Actor-Critic model, which simultaneously learns a value–function estimate and a locally optimal policy.

3. PROTOCOLS AND GRAMMARS

It is often useful to decompose complex problems into simpler subtasks that can handled by specialized modules. Examples

---

[2]Nature's outputs depend on Actor's actions, so the Query graph should technically have an additional arrow from Actor to Nature.

include variational autoencoders, generative adversarial networks, and actor-critic models. Neural networks are particularly well-adapted to modular designs, since units, layers, and even entire networks can easily be combined analogously to bricks of lego (Bottou and Gallinari, 1991).

However, not all configurations are viable models. A methodology is required to distinguish good designs from bad. This section provides a basic language to describe how bricks are glued together that may be a useful design tool. The idea is to extend the definitions of optimization problems, protocols, and representations from Section 2 from single to multi-player optimization problems.

DEFINITION 5 (*game*).

A **distributed optimization problem** or **game** $([N], \Theta, \boldsymbol{\ell})$ is a set $[N] = \{1, \ldots N\}$ of players, a parameter space $\Theta = \prod_{i=1}^{N} \Theta_i$, and loss vector $\boldsymbol{\ell} = (\ell_1, \ldots, \ell_N) \colon \Theta \to \mathbb{R}^N$. Player i picks moves from $\Theta_i \subset \mathbb{R}^{d_i}$ and incurs loss determined by $\ell_i \colon \Theta \to \mathbb{R}$. The goal of each player is to minimize its loss, which depends on the moves of the other players.

The classic example is a *finite game* (von Neumann and Morgenstern, 1944), where player i has a menu of d_i-actions and chooses a distribution over actions, $\boldsymbol{\theta}_i \in \Theta_i = \Delta_{d_i} = \{(\theta_1, \ldots, \theta_{d_i}) \colon \sum_{j=1}^{d_i} \theta_j = 1 \text{ and } \theta_j \geq 0\}$ on each round. Losses are specified for individual actions, and extended linearly to distributions over actions. A natural generalization of finite games is *convex games* where the parameter spaces are compact convex sets and each loss ℓ_i is a convex function in its ith-argument (Stoltz and Lugosi, 2007). It has been shown that players implementing no-regret algorithms are guaranteed to converge to a correlated equilibrium in convex games (Foster and Vohra, 1997; Blum and Mansour, 2007; Stoltz and Lugosi, 2007).

The notion of game in Definition 5 is too general for our purposes. Additional structure is required.

DEFINITION 6 (*computation graph*).

A **computation graph** is a directed acyclic graph with two kinds of nodes:

- *Inputs* are set externally (in practice by Players or Oracles).
- *Operators* produce outputs that are a fixed function of their parents' outputs.

Computation graphs are a useful tool for calculating derivatives (Griewank and Walther, 2008; Bergstra et al., 2010; Bastien et al., 2012; Baydin and Pearlmutter, 2014; van Merriënboer et al., 2015). For simplicity, we restrict to deterministic computation graphs. More general stochastic computation graphs are studied in Schulman et al. (2015).

A *distributed* communication protocol extends the communication protocol in Definition 3 to multiplayer games using two computation graphs.

DEFINITION 7 (*distributed communication protocol*).

A **distributed communication protocol** is a game where each round has two phases, determined by two computation graphs:

- *Query phase.* Players provide inputs to the Query graph (Q) that Operators transform into outputs.

- *Response phase.* Operators in Q act as Oracles in the Response graph (R): they input subgradients that are transformed and communicated to the Players.

The moves chosen by Players depend only on their prior moves and the information communicated to them by the Response graph.

The protocol specifies how Players and Oracles communicate without specifying the optimization algorithms used by the Players. The addition of a Response graph allows more general computations than simply backpropagating the gradients of the Query phase. The additional flexibility allows the design of new algorithms, see Sections 3.6 and 3.7 below. It is also sometimes necessary for computational reasons. For example, backpropagation through time on recurrent networks typically runs over a truncated Response graph (Elman, 1990; Williams and Peng, 1990; Williams and Zipser, 1995).

Suppose that we wish to optimize an objective function $\mathbf{R} \colon \Theta \to \mathbb{R}$ that depends on all the moves of all the players. Finding a global optimum is clearly not feasible. However, we may be able to construct a protocol such that the players are jointly able to find local optima of the objective. In such cases, we refer to the protocol as a grammar:

DEFINITION 8 (*grammar*).

A **grammar** for objective $\mathbf{R} \colon \Theta \to \mathbb{R}$ is a distributed communication protocol where the Response graph provides *sufficient* first-order information to find a local optimum of $(\mathbf{R}, \Theta)$.

The guarantee ensures that the representations constructed by Players in a grammar can be combined into a coherent distributed representation. That is, it ensures that the representations constructed by the Players transform data in a way that is useful for optimizing the shared objective $\mathbf{R}$.

The Players' losses need not be explicitly computed. All that is necessary is that the Response phase communicates the gradient information needed for Players to locally minimize their losses – and that doing so yields a local optimum of the objective.

3.1. Basic Building Blocks: Function Composition (Q) and the Chain Rule (R)

Functions can be inserted into grammars as lego-like building blocks via function composition during queries and the chain rule during responses. Let $G(\boldsymbol{\theta}, F)$ be a function that takes inputs $\boldsymbol{\theta}$ and F, provided by a Player and by upstream computations, respectively. The output of G is communicated downstream in the Query phase:

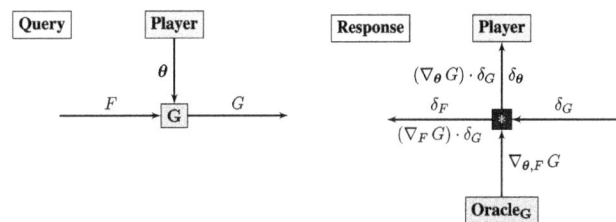

The chain rule is implemented in the Response phase as follows. Oracle_G reports the gradient $\nabla_{\boldsymbol{\theta}, F} G \colon = (\nabla_{\boldsymbol{\theta}} G, \nabla_F G)$ in the

Response phase. Operator "∗" computes the products ($\nabla_\theta\, G \cdot \delta_G$, $\nabla_F\, G \cdot \delta_G$) via matrix multiplication. The projection of the product onto the first and second components[3] is reported to Player and upstream, respectively.

3.2. Summary of Guarantees

A selection of examples is presented below. Guarantees fall under the following broad categories:

1. Exact gradients: under error backpropagation the Response graph implements the chain rule, which guarantees that Players receive the gradients of their loss functions; see Section 3.3.
2. Surrogate objectives: the variational autoencoder uses a surrogate objective: the variational lower bound. Maximizing the surrogate is guaranteed to also maximize the true objective, which is computational intractable; see Section 3.4.
3. Learned objectives: in the case of generative adversarial network and the DAC-model, some of the players learn a loss that is guaranteed to align with the true objective, which is unknown; see Sections 3.5 and 3.6.
4. Estimated gradient: in the DAC-model and kickback, gradient estimates are substituted for the true gradient; see Sections 3.5 and 3.6. Guarantees are provided on the estimates.

Remark 3 (*fine- and coarse-graining*).

There is considerable freedom regarding the choice of players. In the examples below, players are typically chosen to be layers or entire neural networks to keep the diagrams simple. It is worth noting that zooming in, such that players correspond to individual units, has proven to be a useful tool when analyzing neural networks (Balduzzi, 2015; Balduzzi and Ghifary, 2015; Balduzzi et al., 2015).

The game-theoretic formulation is thus scale-free and can be coarse or fine grained as required. A mathematical language for tracking the structure of hierarchical systems at different scales is provided by operads, see Spivak (2013) and the references therein, which are the natural setting to study the composition of operators that receive multiple inputs.

3.3. Error Backpropagation

The main example of a grammar is a neural network using error backpropagation to perform supervised learning. Layers in the network can be modeled as players in a game. Setting each (p)layer's objective as the network's loss, which it minimizes using gradient ascent, yields backpropagation.

Syzygy 1 (backpropagation).

An L-layer neural network can be reformulated as a game played between $L+1$ players, corresponding to *Nature* and the *Layers* of the network. The query graph for a 3-layer network is:

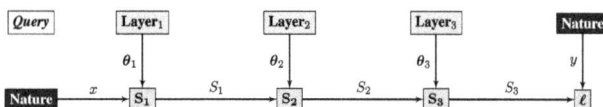

- *Nature* plays samples data points (x, y) i.i.d. from $\mathbb{P}_{X \times Y}$ and acts as the zeroth player.
- *Layer$_i$* plays weight matrices θ_i.
- *Operators* compute $S_i(\theta_i, S_{i-1}) := S_i(\theta_i \cdot S_{i-1})$ for each layer, along with loss $\ell(S_L, y)$.

The response graph performs error backpropagation:

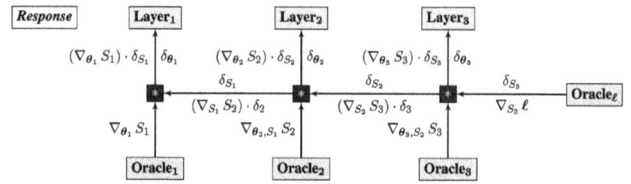

The protocol can be extended to convolutional networks by replacing the matrix multiplications performed by each operator, $S_i(\theta_i \cdot S_{i-1})$, with convolutions and adding parameterless max-pooling operators (LeCun et al., 1998).

Guarantee. The loss of every (p)layer is

$$\ell(\theta, x, y) = \ell_y \circ S_{\theta_L} \circ \cdots \circ S_{\theta_1}(x) \quad \text{where} \quad \ell_y(\bullet) := \ell(\bullet, y)$$
$$\text{where} \quad S_{\theta_i}(\bullet) := S_i(\theta_i \cdot \bullet).$$

It follows by the chain rule that R communicates $\nabla_{\theta_i} \ell$ to player i. □

3.3.1. Representation Learning

We are now in a position to relate the notion of representation in Definition 4 with the standard notion of representation learning in neural networks. In the terminology of Section 2, each player learns a representation. The representations learned by the different players form a coherent distributed representation because they jointly optimize a single objective function.

Abstractly, the objective can be written as

$$\mathbf{R}(\theta_1, \ldots, \theta_L) = \mathop{\mathbf{E}}_{(x,y)\sim\mathbb{P}_{XY}} [\ell(S(\theta_1, \ldots, \theta_L, x), y)],$$

where $S(\theta_1, \ldots, \theta_L, x) = S_{\theta_L} \circ \cdots \circ S_{\theta_1}(x)$. The goal is to minimize the composite objective.

If we set $\hat{\theta}_{1:L} \in \text{arglocmin}_{(\theta_1,\ldots,\theta_L)\in\Theta} \mathbf{R}(\theta_1, \ldots, \theta_L)$ then the function $S_{\hat{\theta}_{1:L}} : X \to Y$ fits the definition of representation above. Moreover, the compositional structure of the network implies that $S_{\hat{\theta}_{1:L}}$ is composed of subrepresentations corresponding to the optimizations performed by the different players in the grammar: each function $S_{\hat{\theta}_j}(\bullet)$ is a local optimum – where $\hat{\theta}_j \in$ arglocmin$_{\theta_j \in \Theta_j} \mathbf{R}(\hat{\theta}_1, \ldots, \theta_j, \ldots, \hat{\theta}_L)$ is optimized to transform its inputs into a form that is useful to network as a whole.

3.3.2. Detailed Analysis of Convergence Rates

Little can be said in general about the rate of converge of the layers in a neural network since the loss is not convex. However, neural networks can be decomposed further by treating the individual units as players. When the units are linear or rectilinear, it turns out that the network is a *circadian game*. The circadian structure provides a way to convert results about the convergence of convex optimization methods into results about the global convergence a rectifier network to a local optimum, see Balduzzi (2015).

---

[3] Alternatively, to avoid having "∗" produce two outputs, the entire vector can be reported in both directions with the irrelevant components ignored.

3.4. Variational Autoencoders

The next example extends the unsupervised setting described in Section 2.2. Suppose that observations $\{\mathbf{x}^{(i)}\}_{i=1}^{N}$ are sampled i.i.d. from a two-step stochastic process: a latent value $\mathbf{z}^{(i)}$ is sampled from $\mathbb{P}(\mathbf{z})$, after which $\mathbf{x}^{(i)}$ is sampled from $\mathbb{P}(\mathbf{x}|\mathbf{z}^{(i)})$.

The goal is to (i) find the maximum likelihood estimator for the observed data and (ii) estimate the posterior distribution on $\mathbf{z}$ conditioned on an observation $\mathbf{x}$. A straightforward approach is to maximize the marginal likelihood

$$\boldsymbol{\theta}^{*} := \arg\max_{\boldsymbol{\theta}} \prod_{i=1}^{N} \mathbb{Q}_{\boldsymbol{\theta}}(\mathbf{x}^{(i)}), \;\text{ where}$$

$$\mathbb{Q}_{\boldsymbol{\theta}}(\mathbf{x}) = \int \mathbb{Q}_{\boldsymbol{\theta}}(\mathbf{x}|\mathbf{z})\mathbb{Q}_{\boldsymbol{\theta}}(\mathbf{z})d\mathbf{z}, \tag{1}$$

and then compute the posterior

$$\mathbb{Q}_{\boldsymbol{\theta}^{*}}(\mathbf{z}|\mathbf{x}) = \frac{\mathbb{Q}_{\boldsymbol{\theta}^{*}}(\mathbf{x}|\mathbf{z})\mathbb{Q}_{\boldsymbol{\theta}^{*}}(\mathbf{z})}{\mathbb{Q}_{\boldsymbol{\theta}^{*}}(\mathbf{x})}.$$

However, the integral in equation (1) is typically untractable, so a more roundabout tactic is required. The approach proposed in Kingma and Welling (2014) is to construct two neural networks, a decoder $\mathbb{D}_{\boldsymbol{\theta}}(\mathbf{x}|\mathbf{z})$ that learns a generative model approximating $\mathbb{P}(\mathbf{x}|\mathbf{z})$, and an encoder $\mathbb{E}_{\boldsymbol{\phi}}(\mathbf{z}|\mathbf{x})$ that learns a recognition model or posterior approximating $\mathbb{P}(\mathbf{z}|\mathbf{x})$.

It turns out to be useful to replace the encoder with a deterministic function, $G_{\boldsymbol{\phi}}(\boldsymbol{\epsilon},\mathbf{x})$, and a noise source, $\mathbb{P}_{noise}(\boldsymbol{\epsilon})$ that are compatible. Here, compatible means that sampling $\tilde{\mathbf{z}} \sim \mathbb{E}_{\boldsymbol{\phi}}(\mathbf{z}|\mathbf{x})$ is equivalent to sampling $\boldsymbol{\epsilon} \sim \mathbb{P}_{noise}(\boldsymbol{\epsilon})$ and computing $\tilde{\mathbf{z}} := G_{\boldsymbol{\phi}}(\boldsymbol{\epsilon},\mathbf{x})$.

SYZYGY 2 (variational autoencoder).
A variational autoencoder is a game played between Encoder, Decoder, Noise, and Environment. The query graph is

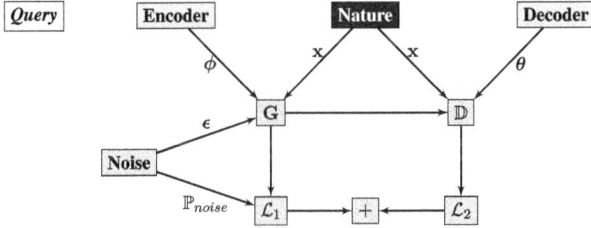

- *Environment* plays i.i.d. samples from $\mathbb{P}(\mathbf{x})$.
- *Noise* plays i.i.d. samples from $\mathbb{P}_{noise}(\boldsymbol{\epsilon})$. It also communicates its density function $\mathcal{P}_{noise}(\boldsymbol{\epsilon})$, which is analogous to a gradient – and the reason that *Noise* is gray rather than black-box.
- *Encoder* and *Decoder* play parameters $\boldsymbol{\phi}$ and $\boldsymbol{\theta}$, respectively.
- *Operator* $\mathbf{z} = G_{\boldsymbol{\phi}}(\boldsymbol{\epsilon},\mathbf{x})$ is a neural network that encodes samples into latent variables.
- *Operator* $\mathbb{D}_{\boldsymbol{\theta}}(\mathbf{z},\mathbf{x})$ is a neural network that estimates the probability of $\mathbf{x}$ conditioned on $\mathbf{z}$.
- The remaining operators compute the (negative) variational lower bound

$$\mathcal{L}(\boldsymbol{\theta},\boldsymbol{\phi};\mathbf{x}) = \underbrace{\int \mathbb{P}_{noise}(\boldsymbol{\epsilon}) \log \frac{\mathbb{P}_{noise}(\boldsymbol{\epsilon})}{\mathbb{P}_{prior}(G_{\boldsymbol{\phi}}(\boldsymbol{\epsilon},\mathbf{x}))}}_{\mathcal{L}_1}$$

$$+ \underbrace{\mathop{\mathbb{E}}_{\boldsymbol{\epsilon} \sim \mathbb{P}_{noise}(\boldsymbol{\epsilon})} \left[-\log \mathbb{D}_{\boldsymbol{\theta}}(G_{\boldsymbol{\phi}}(\boldsymbol{\epsilon},\mathbf{x}),\mathbf{x})\right]}_{\mathcal{L}_2}.$$

The response graph implements backpropagation:

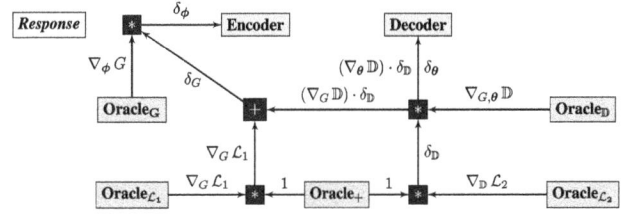

Guarantee. The guarantee has two components:

1. Maximizing the variational lower bound yields (i) a maximum likelihood estimator and (ii) an estimate of the posterior on the latent variable (Kingma and Welling, 2014).
2. The chain rule ensures that the correct gradients are communicated to Encoder and Decoder.

The first guarantee is that the surrogate objective computed by the query graph yields good solutions. The second guarantee is that the response graph communicates the correct gradients.   □

3.5. Generative-Adversarial Networks

A recent approach to designing generative models is to construct an adversarial game between Forger and Curator (Goodfellow et al., 2014). Forger generates samples; Curator aims to discriminate the samples produced by Forger from those produced by Nature. Forger aims to create samples realistic enough to fool Curator.

If Forger plays parameters $\boldsymbol{\theta}$ and Curator plays $\boldsymbol{\phi}$ then the game is described succinctly via

$$\underset{\boldsymbol{\theta}}{\text{arglocmin}} \; \underset{\boldsymbol{\phi}}{\text{arglocmax}}$$

$$\times \left[\mathop{\mathbb{E}}_{\mathbf{x}\sim\mathbb{P}(\mathbf{x})} \left[\log D_{\boldsymbol{\phi}}(\mathbf{x})\right] + \mathop{\mathbb{E}}_{\boldsymbol{\epsilon}\sim\mathbb{P}_{noise}(\boldsymbol{\epsilon})} \left[\log(1 - D_{\boldsymbol{\phi}}(G_{\boldsymbol{\theta}}(\boldsymbol{\epsilon})))\right]\right],$$

where $G_{\boldsymbol{\theta}}(\boldsymbol{\epsilon})$ is a neural network that converts noise in samples and $D_{\boldsymbol{\phi}}(\mathbf{x})$ classifies samples as fake or not.

SYZYGY 3 (generative adversarial networks).
Construct a game played between Forger and Curator, with ancillary players Noise and Environment:

- *Environment* samples images i.i.d. from $\mathbb{P}(\mathbf{x})$.
- *Noise* samples i.i.d. from $\mathbb{P}(\boldsymbol{\epsilon})$.
- *Forger* and *Curator* play parameters $\boldsymbol{\theta}$ and $\boldsymbol{\phi}$, respectively.
- *Operator* $G_{\boldsymbol{\theta}}(\boldsymbol{\epsilon})$ is a neural network that produces fake image $\tilde{\mathbf{x}} = G_{\boldsymbol{\theta}}(\boldsymbol{\epsilon})$.
- *Operator* $D_{\boldsymbol{\phi}}(\tilde{\mathbf{x}})$ is a neural network that estimates the probability that an image is fake.
- The remaining operators compute a loss that *Curator* minimizes and *Forger* maximizes

$$\mathcal{L}(\boldsymbol{\theta},\boldsymbol{\phi}) = \underbrace{\mathop{\mathbb{E}}_{\mathbf{x}\sim\mathbb{P}(\mathbf{x})} \left[\log D_{\boldsymbol{\phi}}(\mathbf{x})\right]}_{\ell_{disc}}$$

$$+ \underbrace{\mathop{\mathbb{E}}_{\boldsymbol{\epsilon}\sim\mathbb{P}(\boldsymbol{\epsilon})} \left[\log(1 - D_{\boldsymbol{\phi}}(G_{\boldsymbol{\theta}}(\boldsymbol{\epsilon})))\right]}_{\ell_{gen}}$$

Note there are two copies of Operator D in the Query graph. The response graph implements the chain rule, with a tweak that multiplies the gradient communicated to *Forger* by (-1) to ensure that *Forger* maximizes the loss that *Curator* is minimizing.

Guarantee. For a fixed Forger that produces images with probability $\mathcal{P}_{\text{Forger}}(\mathbf{x})$, the optimal Curator would assign

$$D^*_{\mathbb{P}_{\text{Forger}},\mathbb{P}_{\text{Nature}}}(\mathbf{x}) = \frac{\mathbb{P}_{\text{Nature}}(\mathbf{x})}{\mathbb{P}_{\text{Nature}}(\mathbf{x}) + \mathbb{P}_{\text{Forger}}(\mathbf{x})} \qquad (2)$$

The guarantee has two components:

1. For fixed Forger, the Curator in equation (2) is the global optimum for $\mathcal{L}$.
2. The chain rule ensures the correct gradients are communicated to Curator and Forger.

It follows that the network converges to a local optimum where Curator represents [equation (2)] and Forger represents the "ideal Forger" that would best fool Curator.   □

The generative-adversarial network is the first example where the Response graph does not simply backpropagate gradients: the arrow labeled δ_G is computed as $-(\nabla_G D) \cdot \delta_D$, whereas backpropagation would use $(\nabla_G D) \cdot \delta_D$. The minus sign arises due to the adversarial relationship between Forger and Curator – they do not optimize the same objective.

3.6. Deviator-Actor-Critic (DAC) Model

As discussed in Section 2.3, actor-critic algorithms decompose the reinforcement learning problem into two components: the critic, which learns an approximate value function that predicts the total discounted future reward associated with state-action pairs, and the actor, which searches for a policy that maximizes the value approximation provided by the critic. When the action-space is continuous, a natural approach is to follow the gradient (Sutton et al., 2000; Deisenroth et al., 2013; Silver et al., 2014). In Sutton et al. (2000), it was shown how to compute the policy gradient given the true value function. Furthermore, sufficient conditions were provided for an approximate value function learned by the critic to yield an unbiased estimator of the policy gradient.

More recently, Silver et al. (2014) provided analogous results for deterministic policies.

The next example of a grammar is taken from Balduzzi and Ghifary (2015), which builds on the above work by introducing a third algorithm, Deviator, that directly estimates the gradient of the value function estimated by Critic.

SYZYGY 4 (DAC model).

Construct a game played by Actor, Critic, Deviator, Noise, and Environment:

- *Nature* samples states from $\mathbb{P}(\mathbf{s}_{t+1}|\mathbf{s}_t,\mathbf{a}_t)$ and announces rewards $r(\mathbf{s}_t,\mathbf{a}_t)$ that are a function of the prior state and action; *Noise* samples $\epsilon \sim N(0, \sigma^2 \cdot \mathbf{I}_d)$.
- *Actor, Critic,* and *Deviator* play parameters θ, $\mathbf{V}$, and $\mathbf{W}$, respectively.
- *Operator* μ is a neural network that computes actions $\mathbf{a} = \mu_\theta(\mathbf{s})$.
- *Operator* $Q^\mathbf{V}(\mathbf{s}, \mu_\theta(\mathbf{s}))$ is a neural network that estimates the value of state-action pairs.
- *Operator* $\mathbf{G}^\mathbf{W}(\mathbf{s}, \mu_\theta(\mathbf{s}))$ is a neural network that estimates the gradient of the value function.
- The remaining *Operator* computes the Bellman gradient error (BGE) which *Critic* and *Deviator* minimize

$$\ell_{BGE}(r_t, Q, \tilde{Q}, \mathbf{G}, \epsilon) = \left(r_t + \gamma\tilde{Q} - Q - \langle \mathbf{G}, \epsilon \rangle\right)^2.$$

The response graph backpropagates the gradient of ℓ_{BGE} to *Critic* and *Deviator*, and communicates the output of Operator $\mathbf{G}$, which is a *gradient estimate*, to *Actor*:

Note that instead of backpropagating first-order information in the form of gradient $\nabla_\mu \mathbf{G}$, the Response graph instead backpropagates zeroth-order information in the form of *gradient-estimate* $\mathbf{G}$, which is computed by the Query graph during the feedforward sweep. We therefore write $\hat{\delta}_\mu$ and $\hat{\delta}_\theta$ (instead of δ_μ and δ_θ) to emphasize that the gradients communicated to Actor are estimates.

As in Section 2.3, an arrow from Actor to Nature is omitted from the Query graph for simplicity.

Guarantee. The guarantee has the following components:

1. *Critic* estimates the value function via TD-learning (Sutton and Barto, 1998) with cloning for improved stability (Mnih et al., 2015).
2. *Deviator* estimates the value gradient via TD-learning and the gradient perturbation trick (Balduzzi and Ghifary, 2015).

3. *Actor* follows the correct gradient by the policy gradient theorem (Sutton et al., 2000; Silver et al., 2014).

4. The internal workings of each neural network are guaranteed correct by the chain rule.

It follows that Critic and Deviator represent the value function and its gradient; and that Actor represents the optimal policy. □

Two appealing features of the algorithm are that (i) Actor is insulated from Critic, and only interacts with Deviator and (ii) Critic and Deviator learn different features adapted to representing the value function and its gradient, respectively. Previous work used the derivative of the value–function estimate, which is not guaranteed to have compatible function approximation, and can lead to problems when the value-function is estimated using functions such as rectifiers that are not smooth (Prokhorov and Wunsch, 1997; Hafner and Riedmiller, 2011; Heess et al., 2015; Lillicrap et al., 2015).

3.7. Kickback (Truncated Backpropagation)

Finally, we consider Kickback, a biologically motivated variant of Backprop with reduced communication requirements (Balduzzi et al., 2015). The problem that kickback solves is that backprop requires two distinct kinds of signals to be communicated between units – feedforward and feedback – whereas only one signal type – spikes – are produced by cortical neurons. Kickback computes an estimate of the backpropagated gradient using the signals generated during the feedforward sweep. Kickback also requires the gradient of the loss with respect to the (one-dimensional) output to be broadcast to all units, which is analogous to the role played by diffuse chemical neuromodulators (Schultz et al., 1997; Pawlak et al., 2010; Dayan, 2012).

SYZYGY 5 (*kickback*).
The query graph is the same as for backpropagation, except that the Operator for each layer produces the additional output $\tau_{i-1} := \boldsymbol{\theta}_{i+1}^{\top} \bullet S_{i+1}$:

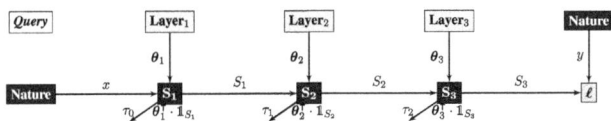

- *Nature* samples labeled data (x, y) from $\mathbb{P}_{X \times Y}$.
- *Layers* by weight matrices $\boldsymbol{\theta}_i$. The output of the neural network is required to be one-dimensional.
- *Operators* for each layer compute two outputs: $S_i = \max(0, \boldsymbol{\theta}_i \cdot S_{i-1})$ and $\tau_{i-1} = \boldsymbol{\theta}_i^{\top} \bullet S_i$ where $\bullet_a = 1$ if $a \geq 0$ and 0 otherwise.
- The task is regression or binary classification with loss given by the mean-squared or logistic error. It follows that the derivative of the loss with respect to the network's output $\beta = \nabla_{S_3} \ell$ is a *scalar*.

The response graph contains a single Oracle that broadcasts the gradient of the loss with respect to the network's output (which is a scalar). Gradient *estimates* for each *Layer* are computed using a mixture of Oracle and local zeroth-order information referred to as *Kicks*:

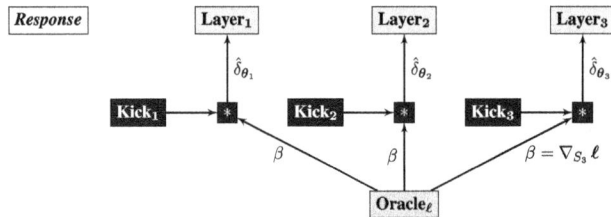

Kick$_i$ is computed using locally available zeroth-order information as follows

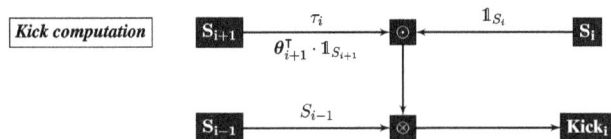

where $\odot$ is coordinatewise multiplication and $\otimes$ is the outer product. If $i = 1$ then *Nature* is substituted for S_{i-1}. If $i = L$ then S_{i+1} is replaced with the scalar value 1.

The loss functions for the layers are not computed in the query graph. Nevertheless, the gradients communicated to the layers by the response graph are exact with respect to the layers' losses, see Balduzzi et al. (2015). For our purposes, it is more convenient to focus on the global objective of the neural network and treat the gradients communicated to the layers as *estimates* of the gradient of the global objective with respect to the layers' weights.

Guarantee. Define unit j to be coherent if $\tau_j > 0$. A network is coherent if all its units are coherent. A sufficient condition for a rectifier to be coherent is that its weights are positive.

The guarantee for Kickback is that, if the network is coherent, then the gradient estimate $\hat{\delta}_{\boldsymbol{\theta}_i}$ computed using the zeroth-order Kicks has the same sign as the backpropagated error $\delta_{\boldsymbol{\theta}_i}$ computed using gradients, see Balduzzi et al. (2015) for details. As a result, small steps in the direction of the gradient estimates are guaranteed to decrease the network's loss. □

REMARK 4 (*biological plausibility of kickback*).
Kickback uses a single oracle, analogous to a neuromodulatory signal, in contrast to Backprop which requires an oracle per layer. The rest of the oracles are replaced by kicks – zeroth-order information from which gradient estimates are constructed. Importantly, the kick computation for layer i only requires locally available information produced by its neighboring layers $i-1$ and $i+1$ during the feedforward sweep. The feedback signals τ_i are analogous to the signals transmitted by NMDA synapses.

Finally, rectifier units with non-negative weights (for which coherence holds) can be considered a simple model of excitatory neurons (Glorot et al., 2011; Balduzzi and Besserve, 2012; Balduzzi, 2014).

Two recent alternatives to backprop that also do not rely on backpropagating exact gradients are target propagation (Lee et al., 2015) and feedback alignment (Lillicrap et al., 2014). Target propagation makes do without gradients by implementing autoencoders at each layer. Unfortunately, optimization problems force the authors to introduce a correction term involving *differences* of targets. As a consequence, and in contrast to Kickback, the information required by layers in difference target propagation cannot be computed locally but instead requires recursively backpropagating differences from the output layer.

Feedback alignment solves a different problem: that feedback and forward weights are required to be equal in backprop (and also in kickback). The authors observe that using random feedback weights can suffice. Unfortunately, as for difference target propagation, feedback alignment still requires separate feedforward and recursively backpropagated training signals, so weight updates are not local.

Unfortunately, at a conceptual level kickback, target propagation and feedback alignment all tackle the wrong problem. The cortex performs reinforcement learning: mammals are not provided with labels, and there is no clearly defined output layer from which signals could backpropagate. A biologically plausible deep learning algorithm should take advantage of the particularities of the reinforcement learning setting.

4. CONCLUSION

Backpropagation was proposed by Rumelhart et al. (1986a) as a method for learning representations in neural networks. Grammars are a framework for distributed optimization that includes backprop as a special case. Grammars abstract two essential features of backprop:

- distributing first-order information about the objective to nodes in a graph (generalizing the backpropagation algorithm itself) such that,

- the first-order information is sufficient to find a local optimum of the objective (generalizing the guarantee that follows from the chain-rule).

Generative-adversarial networks, the deviator-actor-critic model, and kickback are examples of grammars that do not straightforwardly implement backprop, but nevertheless perform well since they communicate the necessary gradient information.

Grammars enlarge the design space for deep learning. A potential application of the framework is to connect deep learning with cortical learning. Thirty years after backpropagation's discovery, no evidence for backpropagated error signals has been found in cortex (Crick, 1989; Lamme and Roelfsema, 2000; Roelfsema and van Ooyen, 2005). Nevertheless, backpropagation is an essential ingredient in essentially all state-of-the-art algorithms for supervised, unsupervised, and reinforcement learning. This suggests investigating algorithms with similar guarantees to backprop that do not directly implement the chain rule.

AUTHOR CONTRIBUTIONS

DB wrote the article.

ACKNOWLEDGMENTS

I am grateful to Marcus Frean, J. P. Lewis, and Brian McWilliams for useful comments and discussions.

FUNDING

Research funding in part by VUW Research Establishment Grant.

REFERENCES

Abernethy, J. D., and Frongillo, R. M. (2011). "A collaborative mechanism for crowdsourcing prediction problems," in *Advances in Neural Information Processing Systems 24*, eds J. Shawe-Taylor, R. S. Zemel, P. L. Bartlett, F. Pereira, and K. Q. Weinberger (Curran Associates, Inc.), 2600–2608.

Agarwal, A., Bartlett, P. L., Ravikumar, P. K., and Wainwright, M. J. (2009). "Information-theoretic lower bounds on the oracle complexity of convex optimization," in *Advances in Neural Information Processing Systems 22*, eds Y. Bengio, D. Schuurmans, J. D. Lafferty, C. K. I. Williams, and A. Culotta (Curran Associates, Inc.), 1–9.

Arjevani, Y., Shalev-Shwartz, S., and Shamir, O. (2016). *On Lower and Upper Bounds for Smooth and Strongly Convex Optimization Problems. J. Mach. Learn. Res.* arXiv:1503.06833.

Baird, L. C. III. (1995). "Residual algorithms: reinforcement learning with function approximation," in *Machine Learning, Proceedings of the Twelfth International Conference on Machine Learning, Tahoe City, California, USA, July 9–12, 1995*, 30–37.

Balduzzi, D. (2011). "Falsification and future performance," in *Algorithmic Probability and Friends. Bayesian Prediction and Artificial Intelligence – Papers from the Ray Solomonoff 85th Memorial Conference, Melbourne, VIC, Australia, November 30 – December 2, 2011*, Vol. 7070, ed. D. Dowe (Springer), 65–78. doi:10.1007/978-3-642-44958-1_5

Balduzzi, D. (2013). "Randomized co-training: from cortical neurons to machine learning and back again," in *Randomized Methods for Machine Learning Workshop, Neural Inf Proc Systems (NIPS)*.

Balduzzi, D. (2014). "Cortical prediction markets," in *International conference on Autonomous Agents and Multi-Agent Systems, AAMAS '14, Paris, France, May 5–9, 2014*, 1265–1272.

Balduzzi, D. (2015). *Deep Online Convex Optimization by Putting Forecaster to Sleep.* arXiv:1509.01851.

Balduzzi, D., and Besserve, M. (2012). "Towards a learning-theoretic analysis of spike-timing dependent plasticity," in *Advances in Neural Information Processing Systems 25*, eds F. Pereira, C. J. C. Burges, L. Bottou, and K. Q. Weinberger (Curran Associates, Inc.), 2456–2464.

Balduzzi, D., and Ghifary, M. (2015). *Compatible Value Gradients for Reinforcement Learning of Continuous Deep Policies.* arXiv:1509.03005.

Balduzzi, D., Ortega, P. A., and Besserve, M. (2013). Metabolic cost as an organizing principle for cooperative learning. *Adv. Complex Syst.* 16, 1350012. doi:10.1142/S0219525913500124

Balduzzi, D., and Tononi, G. (2013). What can neurons do for their brain? Communicate selectivity with spikes. *Theory Biosci.* 132, 27–39. doi:10.1007/s12064-012-0165-0

Balduzzi, D., Vanchinathan, H., and Buhmann, J. M. (2015). "Kickback cuts Backprop's red-tape: biologically plausible credit assignment in neural networks," in *Proceedings of the Twenty-Ninth AAAI Conference on Artificial Intelligence, January 25–30, 2015, Austin, Texas, USA*, 485–491.

Barto, A. G. (1985). Learning by statistical cooperation of self-interested neuron-like computing elements. *Hum. Neurobiol.* 4, 229–256.

Barto, A. G., Sutton, R. S., and Anderson, C. W. (1983). Neuronlike adaptive elements that can solve difficult learning control problems. *IEEE Trans. Syst. Man. Cybern.* 13, 834–846. doi:10.1109/TSMC.1983.6313077

Bastien, F., Lamblin, P., Pascanu, R., Bergstra, J., Goodfellow, I. J., Bergeron, A., et al. (2012). Theano: new features and speed improvements. *CoRR*.

Baum, E. B. (1999). Toward a model of intelligence as an economy of agents. *Mach. Learn.* 35, 155–185. doi:10.1023/A:1007593124513

Baydin, A. G., and Pearlmutter, B. A. (2014). "Automatic differentiation of algorithms for machine learning," in *Journal of Machine Learning Research: Workshop and Conference Proceedings*, 1–7.

Bengio, Y. (2013). "Deep learning of representations: looking forward," in *Statistical Language and Speech Processing – First International Conference, SLSP 2013, Tarragona, Spain, July 29–31, 2013. Proceedings*, eds A.-H. Dediu, C. Martín-Vide, R. Mitkov, and B. Truthe (Springer) 1–37. doi:10.1007/978-3-642-39593-2_1

Bengio, Y., Courville, A., and Vincent, P. (2013). Representation learning: a review and new perspectives. *IEEE Trans. Pattern Anal. Mach. Intell.* 35, 1798–1828. doi:10.1109/TPAMI.2013.50

Bergstra, J., Breuleux, O., Bastien, F., Lamblin, P., Pascanu, R., Desjardins, G., et al. (2010). "Theano: a CPU and GPU math expression compiler," in *Proceedings of the Python for Scientific Computing Conference (SciPy)*, Austin.

Blum, A., and Mansour, Y. (2007). From external to internal regret. *J. Mach. Learn. Res.* 8, 1307–1324.

Bottou, L. (2014). From machine learning to machine reasoning: an essay. *Mach. Learn.* 94, 133–149. doi:10.1007/s10994-013-5335-x

Bousquet, O., and Bottou, L. (2008). "The tradeoffs of large scale learning," in *Advances in Neural Information Processing Systems 20*, eds J. C. Platt, D. Koller, Y. Singer, and S. T. Roweis (Curran Associates, Inc.) 161–168.

Bottou, L., and Gallinari, P. (1991). "A framework for the cooperation of learning algorithms," in *Advances in Neural Information Processing Systems 3*, eds R. P. Lippmann, J. E. Moody, and D. S. Touretzky (Morgan-Kaufmann) 781–788.

Cesa-Bianchi, N., and Lugosi, G. (2006). *Prediction, Learning and Games*. Cambridge, UK: Cambridge University Press.

Choromanska, A., Henaff, M., Mathieu, M., Arous, G. B., and LeCun, Y. (2015a). "The loss surface of multilayer networks," in *Proceedings of the Eighteenth International Conference on Artificial Intelligence and Statistics, AISTATS 2015, San Diego, California, USA, May 9–12, 2015*.

Choromanska, A., LeCun, Y., and Arous, G. B. (2015b). "Open problem: the landscape of the loss surfaces of multilayer networks," in *Proceedings of The 28th Conference on Learning Theory, COLT 2015, Paris, France, July 3–6, 2015*, 1756–1760.

Crick, F. (1989). The recent excitement about neural networks. *Nature* 337, 129–132. doi:10.1038/337129a0

Dann, C., Neumann, G., and Peters, J. (2014). Policy evaluation with temporal differences: a survey and comparison. *J. Mach. Learn. Res.* 15, 809–883.

Dayan, P. (2012). Twenty-five lessons from computational neuromodulation. *Neuron* 76, 240–256. doi:10.1016/j.neuron.2012.09.027

Deisenroth, M. P., Neumann, G., and Peters, J. (2013). A survey on policy search for robotics. *Found. Trends in Robotics* 2, 1–142. doi:10.1561/2300000021

Elman, J. (1990). Finding structure in time. *Cogn. Sci.* 14, 179–211. doi:10.1207/s15516709cog1402_1

Foster, D. P., and Vohra, R. V. (1997). Calibrated learning and correlated equilibrium. *Games Econ. Behav.* 21, 40–55. doi:10.1006/game.1997.0595

Frongillo, R., and Reid, M. (2015). "Convergence analysis of prediction markets via randomized subspace descent," in *NIPS*.

Gershman, S. J., Horvitz, E. J., and Tenenbaum, J. (2015). Computational rationality: a converging paradigm for intelligence in brains, minds, and machines. *Science* 349, 273–278. doi:10.1126/science.aac6076

Glorot, X., Bordes, A., and Bengio, Y. (2011). "Deep sparse rectifier neural networks," in *Proceedings of the Fourteenth International Conference on Artificial Intelligence and Statistics, AISTATS 2011, Fort Lauderdale, USA, April 11–13, 2011*, 315–323.

Goodfellow, I., Pouget-Abadie, J., Mirza, M., Xu, B., Warde-Farley, D., Ozair, S., et al. (2014). "Generative adversarial nets," in *Advances in Neural Information Processing Systems 27*, eds Z. Ghahramani, M. Welling, C. Cortes, N. D. Lawrence, and K. Q. Weinberger (Curran Associates, Inc.), 2672–2680.

Gordon, G. J. (2007). "No-regret algorithms for online convex programs," in *Advances in Neural Information Processing Systems 19*, eds B. Schölkopf, J. C. Platt, and T. Hoffman (MIT Press), 489–496.

Griewank, A., and Walther, A. (2008). "Society for industrial and applied mathematics," in *Evaluating Derivatives: Principles and Techniques of Algorithmic Differentiation*. Philadelphia: SIAM.

Hafner, R., and Riedmiller, M. (2011). Reinforcement learning in feedback control: challenges and benchmarks from technical process control. *Mach. Learn.* 84, 137–169. doi:10.1007/s10994-011-5235-x

Hardt, M., Recht, B., and Singer, Y. (2015). *Train Faster, Generalize Better: Stability of Stochastic Gradient Descent*. arXiv:1509.01240.

Heess, N., Wayne, G., Silver, D., Lillicrap, T., Tassa, Y., and Erez, T. (2015). "Learning continuous control policies by stochastic value gradients," in *Advances in Neural Information Processing Systems 28*, eds A. Cortes, N. D. Lawrence, D. D. Lee, M. Sugiyama, and R. Garnett (Curran Associates, Inc.), 2926–2934.

Hinton, G., Deng, L., Yu, D., Dahl, G. E., Mohamed, A., Jaitly, N., et al. (2012). Deep neural networks for acoustic modeling in speech recognition. *IEEE Signal Process. Mag.* 29, 82–97. doi:10.1109/MSP.2012.2205597

Hopcroft, J. E., and Ullman, J. D. (1979). *Introduction to Automata Theory, Languages, and Computation*. Reading, MA: Addison-Wesley.

Kingma, D. P., and Welling, M. (2014). "Auto-encoding variational Bayes," in *ICLR*.

Klopf, A. H. (1982). *The Hedonistic Neuron: A Theory of Memory, Learning and Intelligence*. Washington, DC: Hemisphere Pub. Corp.

Krizhevsky, A., Sutskever, I., and Hinton, G. E. (2012). "Imagenet classification with deep convolutional neural networks," in *Advances in Neural Information Processing Systems 25*, eds F. Pereira, C. J. C. Burges, L. Bottou, and K. Q. Weinberger (Curran Associates, Inc.), 1097–1105.

Kschischang, F., Frey, B. J., and Loeliger, H.-A. (2001). Factor graphs and the sum-product algorithm. *IEEE Trans. Inf. Theory* 47, 498–519. doi:10.1109/18.910572

Kwee, I., Hutter, M., and Schmidhuber, J. (2001). "Market-based reinforcement learning in partially observable worlds," in *Artificial Neural Networks – ICANN 2001, International Conference Vienna, Austria, August 21–25, 2001 Proceedings*, 865–873. doi:10.1007/3-540-44668-0_120

Lamme, V., and Roelfsema, P. (2000). The distinct modes of vision offered by feedforward and recurrent processing. *Trends Neurosci.* 23, 571–579. doi:10.1016/S0166-2236(00)01657-X

Lay, N., and Barbu, A. (2010). "Supervised aggregation of classifiers using artificial prediction markets," in *Proceedings of the 27th International Conference on Machine Learning (ICML-10), June 21–24, 2010, Haifa, Israel*, 591–598.

LeCun, Y., Bengio, Y., and Hinton, G. (2015). Deep learning. *Nature* 521, 436–444. doi:10.1038/nature14539

LeCun, Y., Bottou, L., Bengio, Y., and Haffner, P. (1998). Gradient-based learning applied to document recognition. *Proc. IEEE* 86, 2278–2324. doi:10.1109/5.726791

Lee, D.-H., Zhang, S., Fischer, A., and Bengio, Y. (2015). "Difference Target Propagation," in *European Conference on Machine Learning and Principles and Practice of Knowledge Discovery in Databases (ECML PKDD)*.

Lewis, D. (1986). *On the Plurality of Worlds*. Oxford: Basil Blackwell.

Lewis, S. N., and Harris, K. D. (2014). *The Neural Market Place: I. General Formalism and Linear Theory*. bioRxiv.

Lillicrap, T. P., Cownden, D., Tweed, D. B., and Ackerman, C. J. (2014). *Random Feedback Weights Support Learning in Deep Neural Networks*. arXiv:1411.0247.

Lillicrap, T. P., Hunt, J. J., Pritzel, A., Heess, N., Erez, T., Tassa, Y., et al. (2015). Continuous control with deep reinforcement learning. *CoRR*.

Maei, H. R., Szepesvári, C., Bhatnagar, S., and Sutton, R. S. (2010). "Toward off-policy learning control with function approximation," in *Proceedings of the 27th International Conference on Machine Learning (ICML-10), June 21–24, 2010, Haifa, Israel*, 719–726.

Minsky, M. (1986). *The Society of Mind*. New York, NY: Simon and Schuster.

Mnih, V., Kavukcuoglu, K., Silver, D., Rusu, A. A., Veness, J., Bellemare, M. G., et al. (2015). Human-level control through deep reinforcement learning. *Nature* 518, 529–533. doi:10.1038/nature14236

Nemirovski, A. (1979). "Efficient methods for large-scale convex optimization problems," in *Ekonomika i Matematicheskie Metody*, 15.

Nemirovski, A., Juditsky, A., Lan, G., and Shapiro, A. (2009). Robust stochastic approximation approach to stochastic programming. *SIAM J. Optim.* 19, 1574–1609. doi:10.1137/070704277

Nemirovski, A., and Yudin, D. B. (1978). On Cezari's convergence of the steepest descent method for approximating saddle point of convex-concave functions. *Sov. Math. Dokl.* 19.

Nemirovski, A. S., and Yudin, D. B. (1983). *Problem Complexity and Method Efficiency in Optimization*. New York, NY: Wiley-Interscience.

Nisan, N., Roughgarden, T., Tardos, É., and Vazirani, V. (eds) (2007). *Algorithmic Game Theory*. Cambridge: Cambridge University Press.

Parkes, D. C., and Wellman, M. P. (2015). Economic reasoning and artificial intelligence. *Science* 349, 267–272. doi:10.1126/science.aaa8403

Pawlak, V., Wickens, J. R., Kirkwood, A., and Kerr, J. N. D. (2010). Timing is not everything: neuromodulation opens the STDP gate. *Front. Syn. Neurosci.* 2:146. doi:10.3389/fnsyn.2010.00146

Pearl, J. (1988). *Probabilistic Reasoning in Intelligent Systems: Networks of Plausible Inference.* Morgan Kaufmann.

Prokhorov, D. V., and Wunsch, D. C. (1997). Adaptive critic designs. *IEEE Trans. Neural. Netw.* 8, 997–1007. doi:10.1109/TNN.1997.641481

Raginsky, M., and Rakhlin, A. (2011). Information-based complexity, feedback and dynamics in convex programming. *IEEE Trans. Inf. Theory* 57, 7036–7056. doi:10.1109/TIT.2011.2154375

Robbins, H., and Monro, S. (1951). A stochastic approximation method. *Ann. Math. Stat.* 22, 400–407. doi:10.1214/aoms/1177729586

Roelfsema, P. R., and van Ooyen, A. (2005). Attention-gated reinforcement learning of internal representations for classification. *Neural Comput.* 17, 2176–2214. doi:10.1162/0899766054615699

Rumelhart, D., Hinton, G., and Williams, R. (1986a). *Parallel Distributed Processing. Vol I: Foundations.* Cambridge, MA: MIT Press.

Rumelhart, D. E., Hinton, G. E., and Williams, R. J. (1986b). Learning representations by back-propagating errors. *Nature* 323, 533–536. doi:10.1038/323533a0

Russell, S., and Norvig, P. (2009). *Artificial Intelligence: A Modern Approach*, 3rd Edn. Upper Saddle River, NJ: Prentice Hall.

Schmidhuber, J. (2015). Deep learning in neural networks: an overview. *Neural Netw.* 61, 85–117. doi:10.1016/j.neunet.2014.09.003

Schulman, J., Heess, N., Weber, T., and Abbeel, P. (2015). "Gradient estimation using stochastic computation graphs," in *Advances in Neural Information Processing Systems 28*, eds C. Cortes, N. D. Lawrence, D. D. Lee, M. Sugiyama, R. Garnett, and R. Garnett (Curran Associates, Inc.), 3510–3522.

Schultz, W., Dayan, P., and Montague, P. (1997). A neural substrate of prediction and reward. *Science* 275, 1593–1599. doi:10.1126/science.275.5306.1593

Selfridge, O. G. (1958). "Pandemonium: a paradigm for learning," in *Mechanisation of Thought Processes: Proc Symposium Held at the National Physics Laboratory.*

Seung, H. S. (2003). Learning in spiking neural networks by reinforcement of stochastic synaptic transmission. *Neuron* 40, 1063–1073. doi:10.1016/S0896-6273(03)00761-X

Shalev-Shwartz, S. (2011). Online learning and online convex optimization. *Found. Trends Mach. Learn.* 4, 107–194. doi:10.1561/2200000018

Silver, D., Lever, G., Heess, N., Degris, T., Wierstra, D., and Riedmiller, M. A. (2014). "Deterministic policy gradient algorithms," in *Proceedings of the 31st International Conference on Machine Learning, ICML 2014, Beijing, China, 21–26 June 2014*, 387–395.

Spivak, D. I. (2013). *The Operad of Wiring Diagrams: Formalizing a Graphical Language for Databases, Recursion, and Plug-and-Play Circuits.* arXiv:1305.0297.

Sra, S., Nowozin, S., and Wright, S. J. (2012). *Optimization for Machine Learning.* Cambridge, MA: MIT Press.

Stoltz, G., and Lugosi, G. (2007). Learning correlated equilibria in games with compact sets of strategies. *Games Econ. Behav.* 59, 187–208. doi:10.1016/j.geb.2006.04.007

Storkey, A. J. (2011). "Machine learning markets," in *Proceedings of the Fourteenth International Conference on Artificial Intelligence and Statistics, AISTATS 2011, Fort Lauderdale, USA, April 11–13, 2011*, 716–724.

Sutskever, I., Vinyals, O., and Le, Q. V. (2014). "Sequence to sequence learning with neural networks," in *Advances in Neural Information Processing Systems 27*, eds Z. Ghahramani, M. Welling, C. Cortes, N. D. Lawrence, and K. Q. Weinberger (Curran Associates, Inc.), 3104–3112.

Sutton, R. (1988). Learning to predict by the method of temporal differences. *Mach. Learn.* 3, 9–44. doi:10.1007/BF00115009

Sutton, R. S., Maei, H. R., Precup, D., Bhatnagar, S., Silver, D., Szepesvári, C., et al. (2009a). "Fast gradient-descent methods for temporal-difference learning with linear function approximation," in *Proceedings of the 26th Annual International Conference on Machine Learning, (ICML) 2009, Montreal, Quebec, Canada, June 14–18, 2009*, 993–1000. doi:10.1145/1553374.1553501

Sutton, R. S., Szepesvári, C., and Maei, H. R. (2009b). "A convergent $O(n)$ temporal-difference algorithm for off-policy learning with linear function approximation," in *Advances in Neural Information Processing Systems 21*, eds D. Koller, D. Schuurmans, Y. Bengio, and L. Bottou (Curran Associates, Inc.), 1609–1616.

Sutton, R. S., McAllester, D. A., Singh, S. P., and Mansour, Y. (2000). "Policy gradient methods for reinforcement learning with function approximation," in *Advances in Neural Information Processing Systems 12*, eds S. A. Solla, T. K. Leen, and K. Müller (MIT Press), 1057–1063.

Sutton, R. S., Modayil, J., Delp, M., Degris, T., Pilarski, P. M., White, A., et al. (2011). "Horde: a scalable real-time architecture for learning knowledge from unsupervised sensorimotor interaction," in *The 10th International Conference on Autonomous Agents and Multiagent Systems – Volume 2 AAMAS '11*, (Taipei: International Foundation for Autonomous Agents and Multiagent Systems), 761–768.

Sutton, R. S., and Barto, A. G. (1998). *Reinforcement Learning: An Introduction.* Cambridge, MA: MIT Press.

Syrgkanis, V., Agarwal, A., Luo, H., and Schapire, R. E. (2015). "Fast convergence of regularized learning in games," in *Advances in Neural Information Processing Systems 28*, eds C. Cortes, N. D. Lawrence, D. D. Lee, M. Sugiyama, R. Garnett, and R. Garnett (Curran Associates, Inc.), 2971–2979.

Tamar, A., Toulis, P., Mannor, S., and Airoldi, E. M. (2014). "Implicit temporal differences," in *NIPS Workshop on Large-Scale Reinforcement Learning and Markov Decision Problems.*

Toulis, P., Airoldi, E. M., and Rennie, J. (2014). "Statistical analysis of stochastic gradient methods for generalized linear models," in *Proceedings of the 31th International Conference on Machine Learning, ICML 2014, Beijing, China, 21–26 June 2014*, 667–675.

van Merriënboer, B., Bahdanau, D., Dumoulin, V., Serdyuk, D., Warde-Farley, D., Chorowski, J., et al. (2015). *Blocks and Fuel: Frameworks for Deep Learning.* arXiv:1506.00619.

Vapnik, V. (1995). *The Nature of Statistical Learning Theory.* New York, NY: Springer.

von Bartheld, C. S., Wang, X., and Butowt, R. (2001). Anterograde axonal transport, transcytosis, and recycling of neurotrophic factors: the concept of trophic currencies in neural networks. *Mol. Neurobiol.* 24, 1–28. doi:10.1385/MN:24:1-3:001

von Neumann, J., and Morgenstern, O. (1944). *Theory of Games and Economic Behavior.* Princeton, NJ: Princeton University Press.

Wainwright, M. J., and Jordan, M. I. (2008). Graphical models, exponential families, and variational inference. *Found. Trends Mach. Learn.* 1, 1–305. doi:10.1561/2200000001

Werbos, P. J. (1974). *Beyond Regression: New Tools for Prediction and Analysis in the Behavioral Sciences.* Ph.D. thesis, Cambridge, MA: Harvard.

Williams, R. J., and Peng, J. (1990). An efficient gradient-based algorithm for on-line training of recurrent network trajectories. *Neural Comput.* 2, 490–501. doi:10.1162/neco.1990.2.4.490

Williams, R. J., and Zipser, D. (1995). "Gradient-based learning algorithms for recurrent networks and their computational complexity," in *Backpropagation: Theory, Architectures, and Applications*, eds Y. Chauvin and D. Rumelhart (Lawrence Erlbaum Associates).

Zinkevich, M. (2003). "Online convex programming and generalized infinitesimal gradient ascent," in *Machine Learning, Proceedings of the Twentieth International Conference (ICML 2003), August 21–24, 2003, Washington, DC, USA*, 928–936.

Conflict of Interest Statement: The author declares that the research was conducted in the absence of any commercial or financial relationships that could be construed as a potential conflict of interest.

13

In Search for the Neural Mechanisms of Individual Development: Behavior-Driven Differential Hebbian Learning

Ralf Der [*]

Max Planck Institute for Mathematics in the Sciences, Leipzig, Germany

When Donald Hebb published his 1949 book "The Organization of Behavior" he opened a new way of thinking in theoretical neuroscience that, in retrospective, is very close to contemporary ideas in self-organization. His metaphor of "wiring" together what "fires together" matches very closely the common paradigm that global organization can derive from simple local rules. While ingenious at his time and inspiring the research over decades, the results still fall short of the expectations. For instance, unsupervised as they are, such neural mechanisms should be able to explain and realize the self-organized acquisition of sensorimotor competencies. This paper proposes a new synaptic law that replaces Hebb's original metaphor by that of "chaining together" what "changes together." Starting from differential Hebbian learning, the new rule grounds the behavior of the agent directly in the internal synaptic dynamics. Therefore, one may call this a behavior-driven synaptic plasticity. Neurorobotics is an ideal testing ground for this new, unsupervised learning rule. This paper focuses on the close coupling between body, control, and environment in challenging physical settings. The examples demonstrate how the new synaptic mechanism induces a self-determined "search and converge" strategy in behavior space, generating spontaneously a variety of sensorimotor competencies. The emerging behavior patterns are qualified by involving body and environment in an irreducible conjunction with the internal mechanism. The results may not only be of immediate interest for the further development of embodied intelligence. They also offer a new view on the role of self-learning processes in natural evolution and in the brain. Videos and further details may be found under http://robot.informatik.uni-leipzig.de/research/supplementary/NeuroAutonomy/.

Keywords: robotics, neural networks, dynamical systems theory, learning, self-organization

Edited by:
Mikhail Prokopenko,
University of Sydney, Australia

Reviewed by:
Ivan Tanev,
Doshihsa University, Japan
Eduardo J. Izquierdo,
Indiana University, USA

*Correspondence:
Ralf Der
ralf.der@t-online.de

1. INTRODUCTION

Autonomy is a puzzling phenomenon both in the evolution of species and in individual development. Translating autonomy by "realizing an independent, self-determined development," the question is how autonomy is grounded in the internal mechanisms of the individual, and even more interesting, what are the conditions for the emergence of this phenomenon. Common explanations postulate specific drives eliciting the emergence and subsistence of autonomous behavior. Examples are the selection pressure in evolution or intrinsic motivation in individual development. In recent years, such drives have been formulated in terms of objective functions ranging from the maximization of predictive information (Ay et al., 2008, 2012; Der et al., 2008; Martius et al., 2013) or empowerment (Klyubin et al., 2005; Salge et al., 2014), to the minimization of free energy (Friston, 2010), to the so-called time-loop error in the homeokinesis approach (Der, 2001; Der and Liebscher, 2002; Der and Martius, 2012), see also Prokopenko (2008, 2009) for more examples. Formulated at the level of behavior, those general principles may be translated into specific rules acting in the internal world of the agent.

Different from such a top-down way of thinking, this paper presents a bottom-up approach, claiming that there exist specific internal mechanisms that, while being unspecific for any task or survival strategy, *per se* have the ability to guide systems to self-determined activity. The hope to base the organization of behavior on local synaptic rules has been a major impact on research in neuroscience ever since the seminal work of Donald Hebb (1949). Hebb's thinking was outstanding not only for its insight into the organization of the brain but also for its anticipation of modern ideas of self-organization, explaining how global order (behavior) can be based on local (synaptic) rules. The progress in neuroscience has provoked many variations and refinements of the original idea. For instance, the divergence problem is possibly counteracted by a synaptic scaling based on some homeostatic self-regulation mechanisms (Song et al., 2000; Turrigiano and Nelson, 2000, 2004; Carlson et al., 2013), and the spike-timing-dependent plasticity (STDP) (Gerstner et al., 1996; Markram et al., 1997; Bi and Poo, 1998) may overcome the problem of spurious associations induced by co-active neurons without causal relations. Also, there is some progress in reward-driven STDP invoking reinforcement learning and explaining several interesting experimental results (Fremaux et al., 2010; Kulvicius et al., 2010; Frémaux, 2013).

Nevertheless, with all the variations of the original law, the conclusion to the question how the neuronal mechanisms organize the behavior is still far from a convincing. In particular, unsupervised as they are, these neural mechanisms should be able to explain and realize the self-organized acquisition of sensorimotor competencies, at a basic level at least. I claim that this requires a more substantial change in the local rules and propose one possible solution in this paper.

Oriented at the original metaphor that synaptic development follows the simple law of "wiring" together what "fires together," the new rule changes the synaptic connectivities driven by two incentives. On the one hand, the "wiring" is not driven by the firing activities of the neurons but by their rates of change. This is reminiscent of differential Hebbian learning studied in earlier work, see Kosko (1986), Klopf (1988), Roberts (1999), and Lowe et al. (2011). The advantage of differential over pure Hebbian learning for the self-organized behavior acquisition has been discussed in a concrete setting close to that of this paper in Der and Martius (2015). On the other hand, different from any Hebbian-like learning, the postsynaptic rate is not that of the neuron itself but is generated by a feedback chain from the external world the neuron is controlling. This link to the external world

is the essential new feature and is what makes the new synaptic mechanism behaviorally relevant in an immediate way.

Neurorobotics is an ideal playground for testing this principle. With robots controlled by a neural network, one may expect not only new impacts for behavior generation in realistic settings but also get some feedback on the possible role of the new synaptic rule for biological systems. I will formulate this rule for the case of a flat sensorimotor loop as introduced in Section "Behavior-Driven Differential Hebbian Learning." This minimalist control paradigm rests on the conviction that control should be less a prescription of what the robot is to do, but consists more in the excitation of specific modes emerging from the irreducible coupling of the mechanical system (robot + environment) with the nervous system. This whole-body paradigm is very close to the idea of morphological computation (Pfeifer and Gómez, 2009; Hauser et al., 2011, 2012; Pfeifer, 2012) but emphasizes more the role of an autonomous dynamics, given the morphology. The new unsupervised learning rule may be helpful in the realization of embodied intelligence that has found enormous interest in the last decade, see Pfeifer and Scheier (1999), Pfeifer and Bongard (2006), and Pfeifer et al. (2007) for excellent surveys and Ritter et al. (2009), Maycock et al. (2010, 2011), Mori and Kuniyoshi (2010), and Yamada et al. (2013) for applications, to name just a few.

This paper focuses on the close coupling between body, control, and environment in challenging physical settings, considering a spherical robot in Section "The Spherical," a snake bot in Section "The Snake," and a hexapod in Section "Discovering New Control Paradigms." The examples demonstrate how the new synaptic mechanism induces a self-determined "search and converge" strategy in behavior space: starting from a dynamics germ these mechanisms elicit behavioral patterns, involving the whole body in tight conjunction with the internal mechanism. Additionally, well aware of the no-free-lunch theorem, I discuss in Section "Spontaneous Symmetry Breaking – the Pattern Behind the Patterns" the general phenomenon of spontaneous symmetry breaking, explaining how low-dimensional behavioral modes may emerge in high-dimensional systems seemingly out of nothing.

2. BEHAVIOR-DRIVEN DIFFERENTIAL HEBBIAN LEARNING

Let us start with formulating the rule in a concrete setting, considering generic robotic systems in physically realistic simulations. The robots are mechanical systems actuated by motors and equipped with a certain set of sensors. The controller, a neural network as described below, translates the sensor values observed in a certain time horizon into commands for the motors.

2.1. Synaptic Plasticity

Let us stipulate that the only information about its body and its interaction with the environment is given to the robot by its vector of sensor values $x_t \in \mathbb{R}^n$. The controller is a neural network, mapping inputs x_t into the controls $y_t \in \mathbb{R}^m$. In the concrete application, the controller is a one-layer net of tanh neurons, described as

$$y = g(Cx + h) \tag{1}$$

where C is the $m \times n$ matrix of synaptic connections and $g: \mathbb{R}^m \to \mathbb{R}^m$ with $g_i(z) = \tanh(z_i)$. When learning this controller with a Hebbian law, the rate of change[1] of synapse C_{ij} would be proportional to the input x_j into the synapse j of neuron i multiplied by its activation y_i, i.e., $\dot{C}_{ij} \propto y_i x_j$. Differential Hebbian learning on its hand would use the rates of change, i.e., $\dot{C}_{ij} \propto \dot{y}_i \dot{x}_j$, see Kosko (1986), Klopf (1988), Roberts (1999), and Lowe et al. (2011). However, this must be modified in order to establish the contact with the external world. For that purpose, we need an internal representation for the relation between motor and sensor values. As we need only the rates of change, we relate the new (after a short time step) velocity vector $\dot{x}'$ to the old velocities $\dot{x}$ and $\dot{y}$ as

$$\dot{x}' = S\dot{x} + A\dot{y} + \eta \qquad (2)$$

with the $n \times n$ matrix S and the $n \times m$ matrix A, η denoting the error in this relation. Let us introduce a new quantity $\dot{\tilde{y}}$ that is implicitly defined as

$$\dot{x}' = S\dot{x} + A\dot{\tilde{y}} \qquad (3)$$

so that it incorporates the effect of the error, see Section "From Directed to Circular Causation" below for a more detailed discussion. The equation can be solved for $\dot{\tilde{y}}$ by using the (generalized) inverse of A. Including a decay term, we define the rate of change of the synaptic strength as

$$\tau \dot{C}_{ij} = \dot{\tilde{y}}_i \dot{x}_j - C_{ij} \qquad (4)$$

where τ is the time scale for the synaptic dynamics. We will for the moment put the bias term $h = 0$, giving a law for its dynamics in equation (11) further below.

As we will see by the experiments, this extremely simple, purely deterministic rule [or its even more reduced counterpart given in equation (10) below] generates most complex behavior patterns as observed in the experiments. Metaphorically speaking, equation (4) is the internal law that enables the agent to realize a self-determined, independent development, establishing in this way its autonomy.

This new unsupervised learning rule is characterized by the metaphor of "chaining together what changes together." As compared to Hebbian learning, this rule not only treats time in a more fundamental way (by considering rates of change instead of the pure firing rates) but includes in addition the chain of cause and effect in an inverse way from the behavior level down to the synaptic dynamics. This step grounds the behavior at the physical level deep in the internal world of the agent (at the level of the synaptic dynamics). Both features together make the system able to self-organize its behavior in close coupling between agent and environment. In view of these arguments, we will call the new rule behavior-driven differential Hebbian learning (BDDHL).

2.2. Empirical Gain Factor

The emergence of modes is contingent on the overall feedback strength of the sensorimotor loop. For controlling the latter, we introduce a gain factor for the neuron so that the action is defined as

$$y = g\left(\kappa \hat{C} x + h\right) \qquad (5)$$

instead of equation (1), where the C matrix is obtained from equation (4) and $\hat{C}$ is normalized i.e., $\hat{C} = C/\|C\|$ with $\|C\|$ the (Frobenius) norm of C. κ is an empirical factor of order one that has to be chosen such that the overall feedback strength is slightly overcritical. In the subcritical region, the dynamics converges toward the resting behavior. In order to avoid numerical problems in this regularization procedure, we may either add a very weak noise to the vectors $\dot{\tilde{y}}$ and $\dot{x}$ in equation (4) or regularize the expression for $\hat{C}$. In applications, an individual gain factor for each neuron is often more appropriate. This is strongly supported by neurophysiological findings on synaptic normalization (Carandini and Heeger, 2011), such as homeostatic synaptic plasticity (Turrigiano and Nelson, 2004), and the balanced state hypothesis (Tsodyks and Sejnowski, 1995; Monteforte and Wolf, 2010).

2.3. A First Discussion

Some obvious properties of that controller can easily be seen from following the signal flow in the closed sensorimotor loop (consider $h = 0$ for this argument). In each time step, the controller receives x and generates at first the postsynaptic potential $z = Cx$. By construction, Cx is a linear combination of velocities so that the controls $y = g(Cx)$ also live in velocity space but are interpreted by the actuators as new target positions. So, the system can have a fixed point (FP) only if spatial positions (like the angle of a hinge joint) and the associated velocities (like the angular velocity of that hinge joint) are compatible over a time horizon set by τ. Trivially, there is a FP matching that condition if the controlled system is at rest so that $\dot{x}_t = 0$ for $t \gg \tau$. In fact, with $\dot{x} = 0$ we have $C = 0$ implying $y = 0$, i.e., all actuators are at their central positions (in the present setting and with $h = 0$). As the experiments show, if the gain factor is subcritical, this FP is even an attractor of the system with a wide basin.

However, the actual life of the system happens outside of that basin. In the behaving system, i.e., with velocities $\dot{x} \neq 0$, the target positions of the actuators are defined as a linear combination of these (past) velocities, which in general will be different from the values of x given by the current pose of the robot. This challenges a strong response of the system that on its hand is changing C what is producing new values for y, pulling x into new directions, and so on. This may lead to a self-amplification of modes if the feedback strength of the sensorimotor loop, as regulated by the gain factor κ, is sufficiently large.

2.4. From Directed to Circular Causation

The trick leading from equations (2) to (3) is the essential step for making the system able to self-organize. The common way of seeing a sensorimotor loop is to consider the actions y as the cause for the reaction of the system as reported by its sensor values. This postulates a signal flow from sensors to actions to new sensor values and so on. In this sense, the actions are seen as the causes for the sensorimotor dynamics. Of course, depending on contexts and given the complexity of the physical world, each action may generate a variety of effects. This is reflected by the η term in equation (2). Hence, the signal flow is still directed but not monocausal as it interweaves a specific cause with several effects.

Still, this corresponds to a directed causality. In this already pretty complex scenario, equation (3) introduces the auxiliary

---

[1] We use the common overdot notation for rates of change or velocities.

quantity $\dot{\bar{y}}$, which incorporates implicitly the effects of the actions y. As $\dot{\bar{y}}$ drives C, which on its hand determines the actions, eventually the actions are defined by their own effects. This is a clear case of circular causation in this complex dynamical system of interconnected causes and effects. With positive feedback strength in the sensorimotor loop, this principle elicits the self-organized behavior modes observed in the experiments.

2.5. Some Technical Details

Before going to the applications, let us note some details of the new rule. For a first reading, this part may be skipped.

2.5.1. Learning of the Response Matrix *A*

Inverse relations between sensor and motor spaces are notoriously difficult to find as they involve the physical response of the system to the applied controls. However, as demonstrated already in earlier work on homeokinesis (Der and Martius, 2012), a very coarse relation between these two worlds is entirely sufficient for the phenomenon of emerging whole-body modes. In the experiments of this paper, we put $S = 0$ so that A is updated in each time step as

$$\dot{A} = \varepsilon_A \left(\dot{x}' - A\dot{y} \right) \dot{y}^{\mathsf{T}} \qquad (6)$$

where $\dot{y}$ is the rate of change of y in the current and $\dot{x}'$ that of x in the next time step. The matrix A can be initialized either by hand, reflecting the known couplings between motor values and the corresponding sensor values, or can be found from a low-frequency and low-load motor babbling in an initiation phase of the development. Also, the inversion of A necessary to obtain the virtual controls $\dot{\bar{y}}$ can be done with pseudoinverse techniques involving inversion only in motor space, given an arbitrary number of sensors. As an alternative, the inverse matrix can also be learned directly. Anyway, in most applications of this paper, the matrix A was simply taken as the unit matrix, hypothesizing a one-to-one sensor to motor coupling. Many technical details about these procedures may be found in Der and Martius (2012).

2.5.2. Explicit Expressions

The update rule generates the matrix C as a weighted average over the projectors on the past velocities in sensor space[2], i.e., switching to discrete times and using matrix notation, C_t being the controller matrix at time t

$$C_t = \langle \dot{\bar{y}} \dot{x}^{\mathsf{T}} \rangle_0^t \approx \frac{1}{\tau} \sum_{t'=0}^{t-1} \dot{\bar{y}}_{t'} \dot{x}_{t'}^{\mathsf{T}} \, e^{-(t-t')/\tau} + e^{-t/\tau} C(0) \qquad (7)$$

As the experiments show, the weighted average can be replaced with a simple windowing operation so that (assuming $t \gg \tau$)

$$C_t = \frac{1}{\tau} \sum_{t'=t-\tau}^{t-1} \dot{\bar{y}}_{t'} \dot{x}_{t'}^{\mathsf{T}} \qquad (8)$$

without losing much. In the experiments, τ ranges from 10 to 100 (roughly 0.2 to 2 s).

These explicit expressions also reveal a basic feature of the approach. Using equation (8) in equation (5) shows that the controller is given by a deterministic, explicit expression over the past sensor states. In fact, as $\dot{\bar{y}}_t$ is a function of $\dot{x}_{t+1}$, we may rewrite equation (5) as

$$y_t = K(x_t, x_{t-1}, \ldots, x_{t-\tau}) \qquad (9)$$

by approximating (with a convenient time scale) any $\dot{x}_{t'}$ by $x_{t'} - x_{t'-1}$.

This demonstrates that the actions of the agent are defined by an explicit, fixed function of its recent sensor history. That this function is fixed, i.e., is not depending explicitly on time, does not mean that the behavior of the agent is fixed. Instead, as seen in the experiments, new experiences may well create new histories, leading to new behaviors creating new experiences and so on. Grounding the current behavior on its recent history is the qualitatively new feature of the approach that produces the observed variety of behaviors, seemingly out of nothing.

3. THE SPHERICAL

The Spherical is ideally suited for showing how BDDHL recognizes and amplifies dynamical structures in sensor space, creating thereby specific behavioral patterns in physical space. The spherical has a ball-shaped body and is driven by moving internal masses[3], shifting thereby the center of gravity, see **Figure 1**. The control y_i defines the target position of the mass on its axis i. The only information the "brain" gets comes from three exteroceptive sensors, measuring the axis-orientation (with respect to the z-axis of the world coordinate system), and three proprioceptive sensors measuring the positions of the masses on their respective axes. Altogether the sensors give the robot only an extremely reduced information on its physical state.

I have chosen this machine because it demonstrates how BDDHL develops definite locomotion patterns starting from a dynamics germ. Without control, the physical behavior is quite simple if the weights are fixed in the center and the sphere is rolling on a 2-d plane. It becomes a little more complicated when, including friction and elasticity effects, opening the third dimension. With the masses outside of the center, the dynamics becomes kind of staggering and with an imposed motion of the masses the trajectory of the Spherical may become highly irregular, even if the controls are harmonic. This is similar to the Barrel case treated in some detail in Der and Martius (2012). So, any control scheme has to make its deal with these specific physical conditions. As the experiments show, the controller of this paper elicits without any knowledge of the physics very well-defined locomotion modes. Note that we use the notion of modes here as in physics, meaning that all degrees of freedom are coherently taking part in a behavior.

In the first set of experiments, the robot is moving on level ground that is elastic and has some friction to be as realistic as possible. In all simulations, we choose $\tau = 10$, $h = 0$, $\kappa = 1$ and start with $C = 0$ so that all masses are in the center (least biased initialization). In the beginning, the robot is kicked by a

---

[2]The solution of the matrix differential equation $\tau \dot{C}(t) = R(t) - C(t)$ is used as $C(t) = \frac{1}{\tau} \int_0^t R(t') e^{-(t-t')/\tau} dt' + e^{-t/\tau} C(0)$. The discrete expression is obtained by taking the Riemannian sum, which is exact if $\tau \to \infty$.

[3]Collisions of those masses are ignored in the simulations.

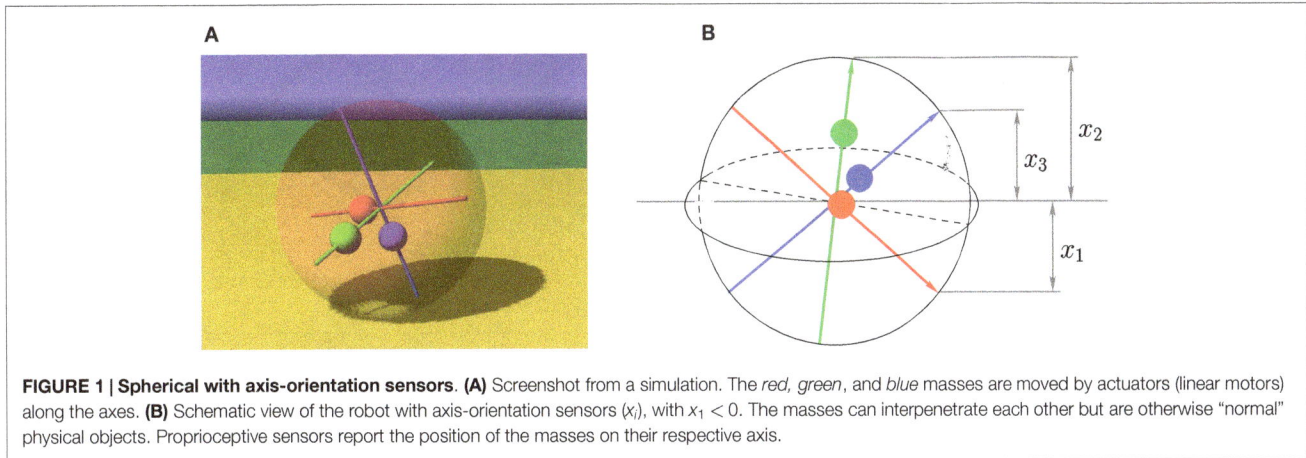

FIGURE 1 | Spherical with axis-orientation sensors. (A) Screenshot from a simulation. The *red, green*, and *blue* masses are moved by actuators (linear motors) along the axes. **(B)** Schematic view of the robot with axis-orientation sensors (x_i), with $x_1 < 0$. The masses can interpenetrate each other but are otherwise "normal" physical objects. Proprioceptive sensors report the position of the masses on their respective axis.

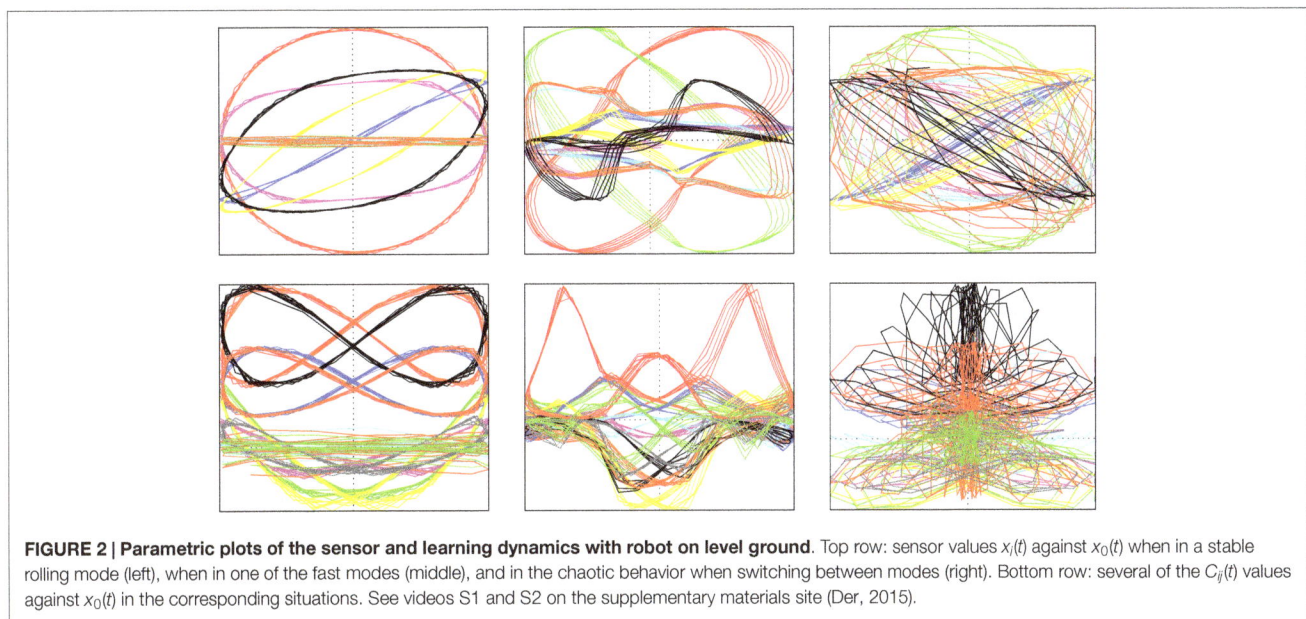

FIGURE 2 | Parametric plots of the sensor and learning dynamics with robot on level ground. Top row: sensor values $x_i(t)$ against $x_0(t)$ when in a stable rolling mode (left), when in one of the fast modes (middle), and in the chaotic behavior when switching between modes (right). Bottom row: several of the $C_{ij}(t)$ values against $x_0(t)$ in the corresponding situations. See videos S1 and S2 on the supplementary materials site (Der, 2015).

mechanical force (an attracting force center marked by a red dot in the simulations) so that it starts rolling. This initial motion is rapidly picked up and amplified by BDDHL[4], the most common mode being the rotation around one of the axes with the masses moving periodically on the other two axes. This mode is very stable against moderate external perturbations but can be switched into another of those modes by a very heavy kick, see video S1 on the supplementary materials site (Der, 2015).

When left alone, the future fate of the robot depends strongly on the learning rate ε_A of the response matrix A, see equation (6). This 6×3 matrix consists of two submatrices A^a and A^w mapping the control vector y to the axis orientation and the weight position sensors, respectively. In the experiments, we choose initially $A^a = \mathbb{I}$ and put all other matrix elements to 0. With ε_A below some critical value ε_{crit}, the emerging mode is stable for a very long time. With faster model learning, the system develops through a sequence of metastable rolling modes with widely differing

characteristics, developing also a kind of lolloping mode and very fast locomotion as demonstrated in video S2 on the supplementary materials site (Der, 2015). In this way, the robot may be said to explore its behavioral spectrum of locomotion. More details are revealed by the parametric plots of the sensorimotor dynamics, showing a high degree of sensorimotor coordination when in the mode and a highly irregular behavior in the transition regions, see **Figure 2**.

In a next series of experiments, the robot is dropped into a large circular basin with an elastic, wavy ground, see video S3 on the supplementary materials site (Der, 2015). Dropped a little outside the center, the robot starts rolling downhill passively but, different from level ground, the robot first has to overcome an initial "orientation" phase with irregular motions, although there is no noise or any external stochasticity. After that, the robot goes into a stable mode running at constant height in the basin. Upon increasing ε_A a little, the sphere goes through several metastable states – running in orbits at a certain height – and leaves the basin after another increase in the learning rate ε_A, see also **Figure 3**.

---

[4]Provided the parameter κ is large enough.

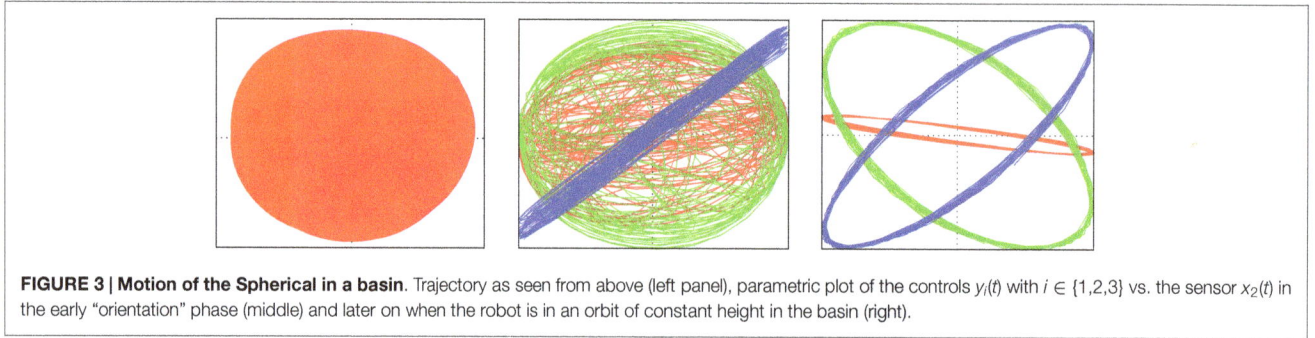

FIGURE 3 | Motion of the Spherical in a basin. Trajectory as seen from above (left panel), parametric plot of the controls $y_i(t)$ with $i \in \{1,2,3\}$ vs. the sensor $x_2(t)$ in the early "orientation" phase (middle) and later on when the robot is in an orbit of constant height in the basin (right).

Video S4 shows the behavior when five robots are started at the same time. Despite strong interactions between the individual robots, all five reach (essentially) the same orbit after some time. Two of the robots are seen to even return to their orbits after colliding. Later, the learning rate ε_A was increased making the robots to spiral higher and higher, eventually leaving the basin.

Interestingly, without knowing anything about the geometric structure of the world the robot is moving in, the emergence of the stable mode demonstrates that BDDHL, in a particular way, is sensitive to the *agent–environment* coupling (ACE). This will even be more obvious from the next example.

4. THE SNAKE

Let us consider another machine – the Snake – in order to demonstrate the emergence of fundamental modes by the self-amplification process. This machine is completely different in physics as compared to the Spherical. Yet, we apply the same controller network, differing only in the number of motor neurons and the nature of the sensors, and apply the DHL rule as defined above, tuning nothing but the overall feed-back strength given by κ and the time scale τ. In this setting, mode generation will be established as a *scalable* phenomenon in the following experiments.

The Snake robot is composed of k capsules connected pairwise by a ball joint with two degrees of freedom (DOF) corresponding to two angles running from $-\phi$ to ϕ coded as $y = -1$ and $y = 1$. In every step, each angle is measured and reported as sensor value $-1 < x_i < 1$. The motors driving each DOF receive a target value for the desired joint angle in the next step. The translation into the physical forces (like the torques of a joint) is done by an embedded PID controller tuned such as to simulate the elasticity of muscles. Driven by the physical forces acting between the individual elements, this elasticity effect makes the true angles to differ vastly from their target values once the robot is in full activity. Let me emphasize that all the information the robot has comes from these proprioceptive sensors giving no information about the physical situation of the robot in its environment.

4.1. Simplifying the Learning Rule

Despite this high physical complexity, the experiments show that the forward model, given by the matrix A, can be kept very simple – it turns out that it is sufficient to learn the model by a simple, low-frequency motor babbling when the joints are without load (as it would be in 0 gravity space without obstacles and/or

ground contact). In the current setting, this reduces the forward model to $A = \mathbb{I}$. As it turns out, the concomitant model learning can be switched off altogether so that equation (4) becomes

$$\tau \dot{C}_{ij} = \dot{x}'_i \dot{x}_j - C_{ij}, \tag{10}$$

where $\dot{x} = \dot{x}(t)$ and $\dot{x}' = \dot{x}(t+1)$ are the rates of change of the joint angles as reported by the sensors. This simplified BDDHL rule is what was used in the experiments discussed in the following, i.e., in both the Snake and in the Hexapod case, as well as in many other applications done so far. Together with the normalization procedure [equation (5)], this extremely simple rule was found in the applications to elicit within minutes (real time) an amazing variety of sensorimotor patterns without any scaffolding from outside.

In the experiments, in all cases, the controller was initialized in the least biased setting, meaning $C = 0$ and $h = 0$ so that all actuators are in their central positions. In the Snake case, this means that the body is completely stretched. This situation corresponds to the trivial attractor, where the system is at rest so that, initially, the system must be started by a mechanical impact on the robot body[5] or by adding some noise to the sensor values.

4.2. Emerging Locomotion Patterns

In a first set of experiments, we put the snake on even ground giving it a kick in the very beginning and whenever it comes to rest. What we observe is that the system develops right from the outset after a very short time (seconds) a collective mode with all degrees of freedom changing coherently. What kind of mode may develop in this initial phase depends on the initial kick and/or the sequence of kicks one is applying in order to get the system going. In the experiments, we see two qualitatively different modes emerging, either a meandering motion like crawling with a certain velocity over ground, or a kind of siderolling, reminiscent of the sidewinding motion known from snakes in sandy deserts, which has also been reproduced by artificial evolution (Prokopenko et al., 2006), see videos S5 and S6 on the supplementary materials site (Der, 2015). **Figure 4** gives a few examples of emerging motion patterns together with the C matrices. Obviously, the latter show distinct structures in close relation to the motion patterns. This will be discussed in Section "Spontaneous Symmetry Breaking – the Pattern Behind the Patterns" below.

These emerging locomotion modes in general are metastable but may last for very long times if there are no external

---

[5] In the simulator, this is realized by a force center marked by a red dot in the videos.

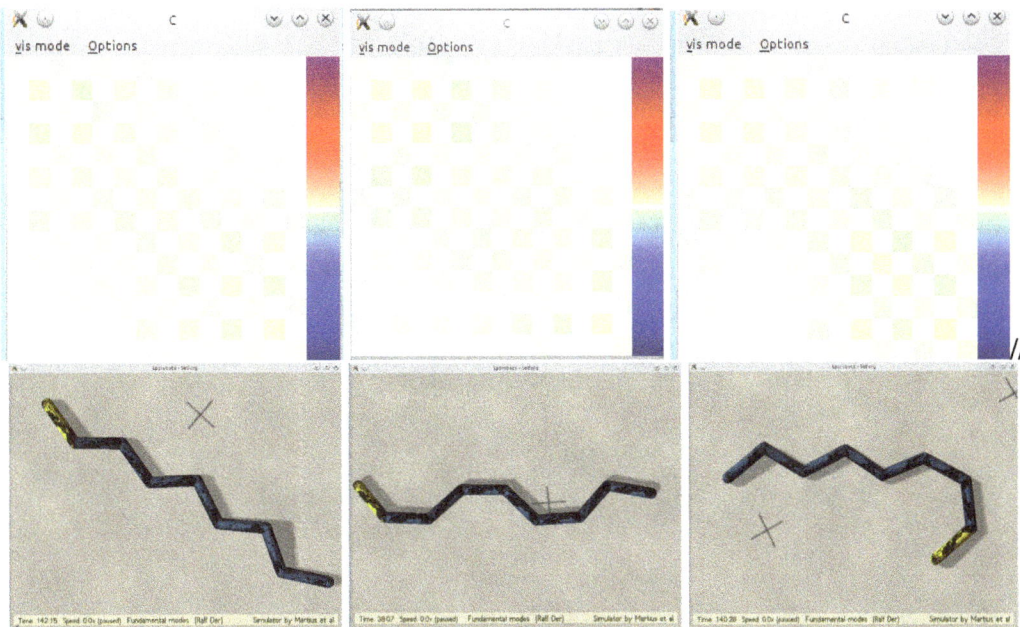

FIGURE 4 | The controller matrix *C* (above) with the Snake in various stable siderolling locomotion modes (below).

perturbations. The system can be forced by mechanical kicks to leave the current mode, but is engaging into another mode after some time, often just a few seconds real time, see video S6 on the supplementary materials site (Der, 2015). In the transitional phase, the system is highly irregular, as is best seen by considering the parametric plots in **Figure 5**.

4.3. The Constitutive Role of the Agent–Environment Coupling

Another point of interest is the active role of the agent–environment coupling in generating the behavior. This is demonstrated by two effects. On the one hand, modes may switch in a definite way by interacting with the environment. When colliding with a wall, or any other obstacle, the Snake will change actively its direction of motion so that it kind of reacts to the collision with the boundary, see videos S8 and S9. Note that there is no contact sensor or the like, the Snake simply reacts to the different coupling with the environment it experiences at the boundary. In the collision, the velocities of the joint angles go to 0 (due to the friction between body and obstacle) giving rise to a mode switching.

On the other hand, the modes are a direct consequence of the agent–environment coupling. This becomes most obvious when this coupling is switched off. Let us consider an experiment with the robot in a stable mode (rolling or crawling). As video S7 on the supplementary materials site (Der, 2015) shows that the modes decay rapidly once the gravity is temporarily being switched off so that the contact with the ground is lost. So, the coupling with the ground is an inexorable ingredient of the mode formation and existence. As an explanation, we note that, given the friction and elasticity of the ground, the different degrees of freedom strongly interact by the forces exerted on each other when moving on the ground. More details on this so-called physical cross-talk

effect may be found in Der (2014). This physical cross-talk has an immediate influence on the velocities of the sensor values, which feeds back to the behavior via the *C* matrix in a synchronizing way.

Videos S8 and S9 demonstrate another effect of the agent–environment coupling: when colliding with a wall or any other obstacle, the Snake experiences a strong physical cross talk by the reactive forces exerted on the joints. This leads, via the induced synaptic dynamics, to a collective reorganization of the system that expresses itself as a reversion of the locomotion velocity. Note that there is nothing like a contact sensor reporting the collision. Instead, the emerging reaction is a pure whole system effect, generating a variety of reactions depending on the circumstances.

4.4. Emergent Dimensionality Reduction

Another important feature is the emergent reduction of dimensionality when converging to the mode. At a more qualitative level, the parametric plots are a convincing indication that the system is confined to a low-dimensional manifold at least in a blurred sense, see **Figure 5**. Interestingly, we also observe that the life time of a mode is directly related to the structure of the parametric plot. The more stable a mode is, the cleaner are the orbits of the sensor values in the parametric plot and the closer is the system to moving on a low-dimensional manifold. Moreover, the life time of the modes is observed to be directly related to the velocity of the Snake over ground. So, the higher the velocity of locomotion, the cleaner the plot. This may be explained by the assumption that the synchronization via the forces across the ground is best if the DOFs cooperate in producing an effective locomotion pattern.

On a quantitative level, there are first results (Martius and Olbrich, 2013) that the dimension of the manifold is a little above two, stipulating an appropriate coarse graining. Note that the phase space of the constrained physical system is of $2 \times (6 + 2K)$

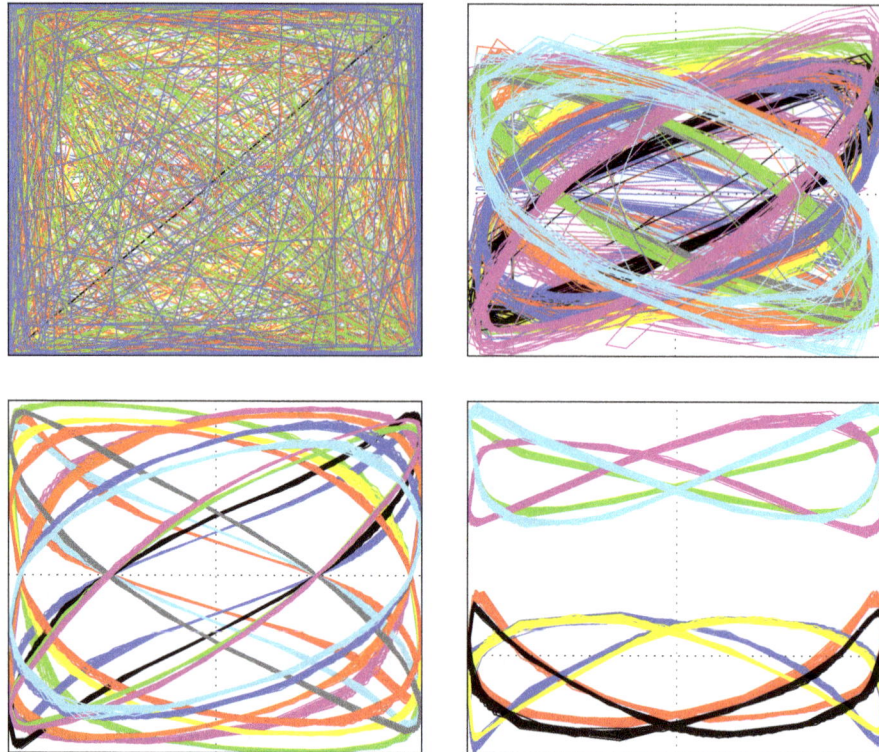

FIGURE 5 | Parametric plots demonstrating the formation of a mode and the emerging dimensionality reduction. The first three panels depict the sensor values $x_i(t)$ with $i = 0. 13$ against the controller output $y_6(t)$ in a time interval of 500 steps (10 s). Top left depicts the behavior in the transition phase between two modes with fully developed chaos, top right depicts the formation of a new mode, and bottom left corresponds to the fully developed mode demonstrating the high degree of sensorimotor coordination. The time between these three phases is about 1 min. The bottom right depicts the behavior of some of the matrix elements of C vs. y_6, demonstrating the tight correlation of the synaptic dynamics with the behavior of the physical system.

dimensions, with $K + 1$ the number of segments and $K = 8$ in the video.

5. DISCOVERING NEW CONTROL PARADIGMS

The last example is to demonstrate the "creative" power of BDDHL to discover new ways of controlling high-dimensional systems. For this purpose, let us modify our system by adding a rule for the dynamics of the bias vector $h \in \mathbb{R}^m$ in equation (1). The idea is that the bias dynamics drives the neurons toward their region of maximal sensitivity. With the bipolar tanh neurons used in this paper a convenient choice is

$$\dot{h} = -\varepsilon_h y, \qquad (11)$$

where y is the output vector of the controller neurons as defined by equation (5) and ε_h is the update rate to be chosen by hand. This dynamics is of particular interest in hysteresis systems as discussed in earlier work (Der and Martius, 2012).

For a discussion, let us consider still another machine, the so-called Hexapod, see **Figure 6**. Apart from its morphology, the robot is constructed in its functionality like the Snake robot, with synaptic dynamics given by equation (10) and the bias dynamics of equation (11). As with the Snake, we may put $A \approx \mathbb{I}$ and consider first the case of a diagonal controller matrix $C = c\mathbb{I}$. Then,

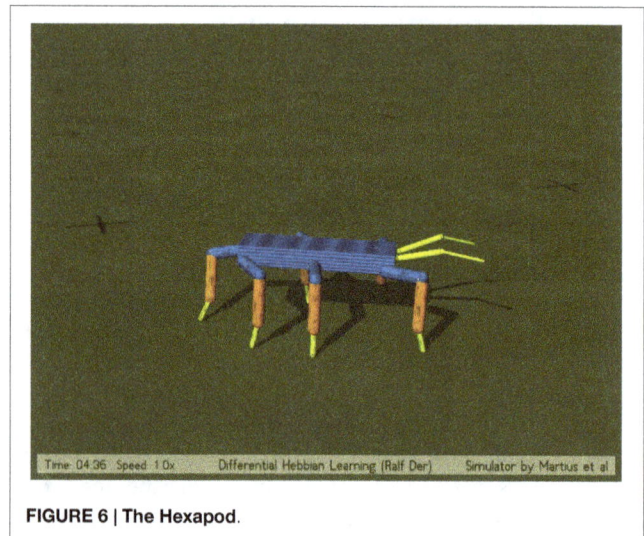

FIGURE 6 | The Hexapod.

the system dynamics is split into individual, decoupled feedback loops. As discussed in Der and Martius (2012) and Der (2014) and others, if $h = 0$ and the coupling strength c is overcritical, each of these loops has two FPs. Considering only the six vertical shoulder joints, the system has 2^6 FPs corresponding to each joint angle either high or low.

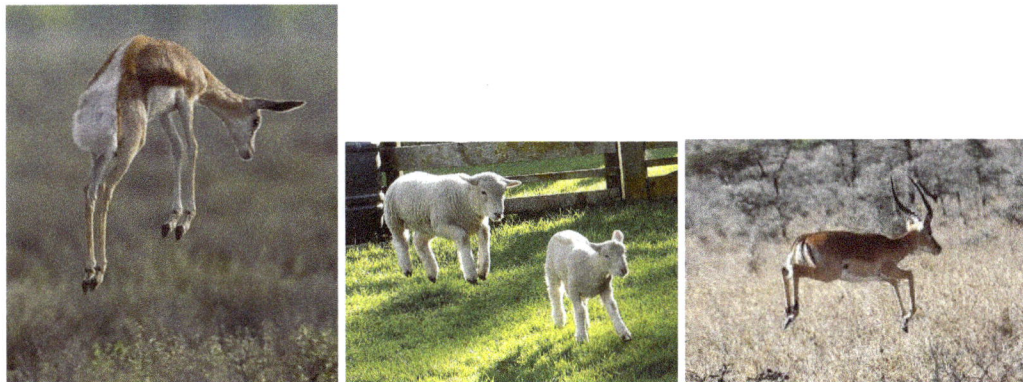

FIGURE 7 | Just for fun? Stotting behavior in animals. The animals spring into the air vertically with all feet lifting simultaneously. The evolutionary advantage seems to be unclear. Authors: Yathin Sk (left), Pam from near Matamata (middle), Rick Wilhelmsen (right), all pictures from Wikimedia commons.

If the h dynamics is switched on, each of those individual circuits becomes a so-called hysteresis oscillator producing a periodic oscillation. Interestingly, these non-linear oscillators have a high tendency to synchronize so that collective locomotion and other dynamical patterns are created, see for instance the so-called ARMBAND robot in Der and Martius (2012) and the videos under playfulmachines.com.

5.1. Experiments – From Motion Germs to Organized Behavior

Now let us drive the parameters of the controller by equation (10) together with the bias dynamics equation (11). Depending on both the starting conditions of the mechanical system and of the controller, we get a vast variety of possible behavior patterns. In all experiments, we start with the least biased controller using $C = 0$ and $h = 0$ so that $y = 0$, corresponding to the central position of all actuators. Jumping patterns emerge in a natural way if we drop the robot from a certain height. With the trunk in a horizontal pose, the robot hits ground with all its legs at the same time. Due to the muscle-like flexibility of the motor-joint system, the robot responds with a damped vertical oscillation. Similar to the Spherical case treated in Section "The Spherical," the BDDHL controller picks up and amplifies this motion germ so that the robot almost immediately executes a more or less stable hopping motion, see video S10 on the supplementary materials site (Der, 2015).

Jumping with all four legs into the air simultaneously is called stotting or pronking and is observed in many quadruped animals, with gazelles in particular, see **Figure 7**. It seems not to be observed in hexapods but we will call the emerging motion patterns also a stotting behavior. With quadrupeds, there must be some evolutionary advantage for the development of this behavior but it seems not to be clear what exactly that is. Nevertheless, it is interesting that BDDHL automatically develops such a behavior without any rewards or evolutionary pressure. I will give an argument in terms of the symmetry group considerations that may explain the preferential emergence of this behavior, see Section "Spontaneous Symmetry Breaking – the Pattern Behind the Patterns" below. Moreover, the frequency of this stotting motion pattern can be regulated by the value of ε_h in a certain range.

FIGURE 8 | Structure of the control matrix when in the jumping motion pattern (left). The upper left 12×12 matrix, referring to the shoulder joints, shows a regular chess board-like pattern. In a later, more complex motion (right) the chess board pattern has been resolved into a more complex geometrical structure, indicating a more intricate cooperation between the various degrees of freedom. Quite generally, different modes correspond to different geometric patterns of the control matrix.

Shaping the pattern is also possible by playing with the time scale parameter τ of equation (4) and the gain factor of the neurons κ. By varying these parameters while the robot is behaving, the robot can also be brought into a forward jumping behavior, sometimes called a bound gait (Yamasaki et al., 2013), see video S11.

5.2. Stotting and Collective Hysteresis

The jumping pattern gets an explanation if we look at the emerged structure of the controller matrix C that defines the behavior together with the bias dynamics of equation (11). As illustrated by **Figure 8**, behaviors are identifiable by the geometrical pattern of their control matrix. It is interesting to see that and how (see below) BDDHL develops control structures that are intimately related to the physical properties, like the dynamics and morphology, of the mechanical system it is controlling.

In the stotting case, it is essentially the synchronous motion of the vertical shoulder joints that generates the jumps. This is directly reflected by the large values of the corresponding sensor to motor coupling elements of the upper left 12×12 matrix. Numerically, we observe that all matrix elements involving the up–down shoulder joints, i.e., $C(0,0)$, $C(0,2)$, ..., $C(10,10)$ are self-regulating toward roughly the same value $c = 0.2$. Ignoring all the other (much smaller elements) for the moment and putting $h = 0$, the system has two stable FPs corresponding to all legs either up or down simultaneously.

This is easily explained by comparing the present situation with that of the *individual* hysteresis loops (see above) generating 2^6 FPs if $c > 1$. In the present situation, a hysteresis loop is generated by the cooperation between the individual feedback loops. In fact, summing the feedback strengths of all loops contributing to one motor output gives just the slightly overcritical feedback $6 \times c \approx 1.2$ for the signal flow through that neuron. So, all the feedback loops are intermingled in a systematic way and when including the h dynamics, a strongly synchronized motion is emerging. Importantly, as the videos show, the repeated contact with the ground helps stabilizing this synchrony by the muscle-like flexibility of the motor-joint system. This is another example of the constitutive role of the agent–environment coupling.

We may call the emerging control scheme a whole-body hysteretic controller and note that this new way of controlling emerged directly from the BDDHL rule under the given physical initialization. With more complex C matrix patterns, see **Figure 8**, more complex motion patterns are readily produced by this new control paradigm. This will be investigated in a later paper. It is also to be noted that the pronounced systematics in the structure of the controller matrix can also be used for editing behaviors *ex post*.

Many other interesting motion patterns can be seen by the videos on the emergence and decay of locomotion patterns at the beginning of the supplementary material page (Der, 2015).

6. SPONTANEOUS SYMMETRY BREAKING – THE PATTERN BEHIND THE PATTERNS

The role of symmetries of the system for the pertinent motion patterns has been extensively studied in the literature, see Schöner et al. (1990), Collins and Stewart (1993), Strogatz and Stewart (1993), Golubitsky et al. (1998, 1999), van der Weele and Banning (2001), Golubitsky (2012), and Tero et al. (2013). In those works, gaits are driven by central pattern generators (CPGs) that are constructed of non-linear oscillators. The CPGs may be considered as open-loop controllers imposing their rhythms on the mechanical system as was the case in the earlier models (Collins and Stewart, 1993; Strogatz and Stewart, 1993; Golubitsky et al., 1998, 1999), or they may respond to the interaction with the body, closing the sensorimotor loop, see Tero et al. (2013).

6.1. Geometric Symmetries
Systems of coupled oscillators are well known from classical mechanics and can be analyzed to some degree. As the investigations show, the various gaits can be associated with the symmetries of the controlled system. In particular, one can study the role of the invariance of the body's geometry against permutations. With quadrupeds, these are the invariance against right–left or back–front permutations, e.g., in the case of the Hexapod, the

corresponding symmetry group is given by the permutations of all legs (assuming complete forward backward symmetry). In the periodically driven systems, there is still the invariance against time shifting by the period duration T.

In those papers, starting from the most general group comprising the maximum number of invariance, hierarchies of symmetry groups were constructed by progressively reducing the set of invariance. Special behaviors, such as different locomotion patterns, can be associated with a definite symmetry group. For instance, stotting may be associated with the group of maximum symmetry. Transitions between gaits can be associated with symmetry breaking bifurcations (Collins and Stewart, 1993; van der Weele and Banning, 2001). So, the approach leads to natural hierarchies of gaits, ordered by symmetry, and to natural sequences of gait bifurcations.

In the bifurcation scenario discussed in those papers, symmetry breaking was induced by changing the controller parameters, such as the amplitude and frequency of the driving force in Collins and Stewart (1993) and van der Weele and Banning (2001), from outside: when crossing the bifurcation point, the system becomes unstable and the system state jumps into one of the emerging alternatives. With BDDHL, we have a self-referential system (Der and Martius, 2012), a dynamical system that changes its parameters by itself, driven by the BDDHL mechanism.

6.2. Symmetries of the Physical Dynamics
The geometric symmetries are only the upper level of the whole spectrum of symmetries associated with the physical dynamics of the mechanical system. For a sketch, let us consider the robot in its least biased initialization as discussed above, corresponding to the central position of all actuators (with corrections due to the load on the joints). When linearizing around that state, the resulting dynamical system is characterized by a bunch of symmetries like the invariance against inverting the sign of joint angles. These symmetries are approximate since they can be perturbed by non-linearities, the actions of the controller, and the interaction with the environment and/or other body parts. Yet, starting from an unspecific dynamics germ, we observe the emergence of motion patterns reflecting the original symmetries of the physical system to a high degree [the principle of parsimonious symmetry breaking, see Der and Martius (2013) and Der (2014)].

When starting with the least biased initialization, i.e., $C = 0$ and $h = 0$, these broken symmetry patterns are emerging in the interaction process of the controller with the body and the environment. This has been demonstrated by the emerging structures of the controller matrix C, see **Figures 4** and **8**, which directly reflect the permutation symmetries of the body. Why is that? The decisive point in this scenario is the fact that the BDDHL mechanism is invariant against just the involved symmetry groups. Considering the explicit expressions for the C matrix given in equations (7) and (8), this is most obvious in the case of sign inversion and permutations of the sensor–motor pair x, y. A more involved symmetry is given by rotations of the sensor–motor space. For a discussion, let us consider the case $A = \mathbb{I}$ as discussed with the Snake and the Hexapod. With $x \rightarrow Ux$ where U is a rotation matrix so that $U^{\mathsf{T}} = U^{-1}$, we obtain

$Cx \rightarrow UCx$ so that, according to equation (1), $y \rightarrow Uy$. This establishes the (approximate) invariance of the system against $(x, y) \rightarrow (Ux, Uy)$ (if the non-linearities can be ignored, as is the case in the starting phase of a mode out of the least biased initialization).

Having stated that, why do the symmetry broken motion patterns emerge? The point now is that BDDHL, with an over-critical value of the global feedback strength (controlled by the gain factor κ), destabilizes the system so that an initial perturbation is amplified. As this amplification process is (approximately) invariant under the operations of the symmetry group, the emerging dynamics stays in the largest symmetry group that is compatible with the initial perturbation. This is like a kind of conservation rule for the symmetry group the system is in. This argument explains why the system goes for the largest group, the stotting in the case of the Hexapod: the larger the group, the larger is the probability that the initial perturbation is consistent with the group, in the sense that it is approximately invariant under the operations of the group.

The initial perturbation can also be realized by some noise that is invariant under the operations of the group. Then, the perturbation is fully unspecific so that the symmetry breaking is truly spontaneous. Otherwise, as already mentioned, when kicking the system by an external impact, the perturbation is not fully unspecific, one even can usher the system into a specific symmetry group, like in the case of the Snake with its emerging crawling or siderolling modes. This is another interesting feature of the BDDHL approach.

7. CONCLUSION

This paper proposes a new synaptic law for the organization of behavior that replaces Hebb's original metaphor of "wiring" together what "fires together" by "chaining together" what "changes" together, linking the motor neurons by a feedback chain to the behavior in the physical world consisting of body and environment. This feedback chain is the essential new feature of the proposed synaptic rule and is what makes the synaptic dynamics behaviorally relevant in an immediate way.

In applications to a number of complex robots, it was demonstrated that neurocontrollers with that new rule elicit an amazing variety of complex behavior patterns, contingent on the specific embodiment and the agent–environment coupling. The patterns have been shown to directly reflect the symmetries of the body and the agent–environment coupling. In particular, a physical system with many degrees of freedom like the Snake is seen to self-organize into definite locomotion patterns without rewarding or prestructuring the system in any way. This example also showed the tremendous reduction of dimensionality emerging in the controlled systems. For instance, in the examples with the Snake, the *constrained* physical systems live in a phase space of up to 40 dimensions. Yet, the controlled system converges within seconds toward a (blurred) low-dimensional manifold, hosting the definite locomotion pattern.

The results suggest new ways for robotics. With the self-organization ability realized by the BDDHL rule, robots can be driven into different behaviors by external influences and can switch between behaviors just by interaction with the environment or a human trainer. This also opens new ways for a kinesthetic teaching as will be detailed in a later paper. BDDHL realizes a "search and converge" strategy in behavior space that can be guided by just two metaparameters – the time scale τ of the synaptic dynamics and the gain factor κ of the controller neurons. Different from most search strategies, BDDHL realizes a self-determined search, a deterministic process with actions being defined as a plain function of the sensor values (over the recent past) so that all behaviors are repeatable and utilizable as building blocks in behavioral architectures. This is an advantage over other approaches, such as homeokinesis (Der and Liebscher, 2002; Der and Martius, 2012), which also produce interesting behaviors but are more inclined to search and less to a convergence toward definite (though metastable) behaviors.

Neural networks often are blamed for the opacity of the solutions found by a learning procedure. Interestingly, this is different with the present approach. In the experiments with both the Snake and the Hexapod, definite structures of the synaptic matrix were observed, revealing transparent relations between emerging behavior and control structure. So, given the pronounced, behavior-related structure in the controller matrix, the emerging behaviors can be understood and even be edited to shape behaviors into desired directions.

The presented results may also have some impact on biology. It was demonstrated that an extremely simple synaptic rule can, contingent on the morphology and the agent–environment coupling, elicit a vast variety of behavioral patterns that may have an immediate evolutionary advantage. On a speculative level, this may be considered as a new factor in natural evolution. It is commonly assumed in natural evolution that new behaviors are the result of a mutation in morphology accompanied by an appropriate mutation of the controller so that the probability of selection is the product of two (very small) probabilities. Had nature discovered the BDDHL rule, new species with new behaviors could emerge just by mutations of the morphology, trusting that BDDHL will drive the modified system to new, fitness relevant modes of behavior. As proposed by Baldwin (Baldwin, 1896; Weber and Depew, 2003), such new behaviors, reoccurring in every generation, could be made permanent eventually by another mutation freezing the behavior. It would be interesting to look for indications of the new synaptic plasticity in living systems.

Let me conclude with a few words on the question in what sense are the emerging behaviors autonomous? The nature of autonomy is still widely debated, see, for instance Di Paolo (2005) and Bertschinger et al. (2008). My attitude is a very modest one, just stating that autonomy is large if the agent unfolds a rich spectrum of different behavioral patterns, driven by an internal law that is as free as possible on the wishes and intentions of its designer. This freedom postulate is fulfilled almost ideally with the rules of equation (4) or (10) as the behavior of the agent is entirely determined by a universal law that is formulated at the synaptic level, two levels deeper than that of the behavior, and it is formulated entirely in terms of the sensor values the robot has generated by its actions in its recent past. So, in this setting, there is no room for the designer to sneak its intentions in. However, quantifying the richness of the behavior spectrum is an open question so that, in this sense, autonomy remains in the eyes of the beholder.

AUTHOR CONTRIBUTIONS

All research is done by the author.

REFERENCES

Ay, N., Bernigau, H., Der, R., and Prokopenko, M. (2012). Information-driven self-organization: the dynamical system approach to autonomous robot behavior. *Theory Biosci.* 131, 161–179. doi:10.1007/s12064-011-0137-9

Ay, N., Bertschinger, N., Der, R., Güttler, F., and Olbrich, E. (2008). Predictive information and explorative behavior of autonomous robots. *Eur. Phys. J. B* 63, 329–339. doi:10.1140/epjb/e2008-00175-0

Baldwin, J. M. (1896). A new factor in evolution. *Am. Nat.* 30, 441–451, 536–553. doi:10.1086/276428

Bertschinger, N., Olbrich, E., Ay, N., and Jost, J. (2008). Autonomy: an information theoretic perspective. *BioSystems* 91, 331–345. doi:10.1016/j.biosystems.2007.05.018

Bi, G. Q., and Poo, M. M. (1998). Synaptic modifications in cultured hippocampal neurons: dependence on spike timing, synaptic strength, and postsynaptic cell type. *J. Neurosci.* 18, 10464–10472.

Carandini, M., and Heeger, D. J. (2011). Normalization as a canonical neural computation. *Nat. Rev. Neurosci.* 13, 51–62. doi:10.1038/nrn3136

Carlson, K. D., Richert, M., Dutt, N., and Krichmar, J. L. (2013). "Biologically plausible models of homeostasis and stdp: stability and learning in spiking neural networks," in *Neural Networks (IJCNN), The 2013 International Joint Conference on* (Dallas, TX: IEEE), 1–8. doi:10.1109/IJCNN.2013.6706961

Collins, J. J., and Stewart, I. N. (1993). Coupled nonlinear oscillators and the symmetries of animal gaits. *J. Nonlinear Sci.* 3, 349–392. doi:10.1007/BF02429870

Der, R. (2001). Self-organized acquisition of situated behaviors. *Theory Biosci.* 120, 179–187. doi:10.1078/1431-7613-00039

Der, R. (2014). "On the role of embodiment for self-organizing robots: behavior as broken symmetry," in *Guided Self-Organization: Inception, Volume 9 of Emergence, Complexity and Computation*, ed. M. Prokopenko (Heidelberg: Springer), 193–221.

Der, R. (2015). *Supplementary Materials to Neuroautonomy*. Available at: http://robot.informatik.uni-leipzig.de/research/supplementary/NeuroAutonomy/

Der, R., Güttler, F., and Ay, N. (2008). "Predictive information and emergent cooperativity in a chain of mobile robots," in *Proceedings Eleventh International Conference on the Simulation and Synthesis of Living Systems*. eds S. Bullock, J. Noble, R. A. Watson, and M. A. Bedau (Winchester, UK: MIT Press).

Der, R., and Liebscher, R. (2002). "True autonomy from self-organized adaptivity," in *Proc. Workshop Biologically Inspired Robotics* (Bristol).

Der, R., and Martius, G. (2012). *The Playful Machine – Theoretical Foundation and Practical Realization of Self-Organizing Robots*. Berlin: Springer.

Der, R., and Martius, G. (2013). "Behavior as broken symmetry in embodied self-organizing robots," in *Proceedings of the Twelfth European Conference on the Synthesis and Simulation of Living Systems*. eds P. Liò, O. Miglino, G. Nicosia, S. Nolfi, and M. Pavone (Taormina: MIT Press), 601–608.

Der, R., and Martius, G. (2015). Novel plasticity rule can explain the development of sensorimotor intelligence. *Proc. Natl. Acad. Sci. U.S.A.* 112, E6224–E6232. doi:10.1073/pnas.1508400112

Di Paolo, E. A. (2005). Autopoiesis, adaptivity, teleology, agency. *Phenomenol. Cogn. Sci.* 4, 429–452. doi:10.1007/s11097-005-9002-y

Frémaux, N. (2013). *Models of Reward-Modulated Spike-Timing-Dependent Plasticity*. Ph.D thesis, IC, Lausanne.

Fremaux, N., Sprekeler, H., and Gerstner, W. (2010). Functional requirements for reward-modulated spike timing-dependent plasticity. *J. Neurosci.* 30, 13326–13337. doi:10.1523/JNEUROSCI.6249-09.2010

Friston, K. (2010). The free-energy principle: a unified brain theory? *Nat. Rev. Neurosci.* 11, 127–138. doi:10.1038/nrn2787

Gerstner, W., Kempter, R., Hemmen, J. V., and Wagner, H. (1996). A neuronal learning rule for sub-millisecond temporal coding. *Nature* 383, 76–81. doi:10.1038/383076a0

Golubitsky, M. (2012). Animal gaits and symmetry. *Spring 2012 Meeting of the APS Ohio-Region Section*, Vol. 57 (Columbus, OH). Available at: http://meetings.aps.org/link/BAPS.2012.OSS.D1.3

Golubitsky, M., Stewart, I., Buono, P.-L., and Collins, J. (1998). A modular network for legged locomotion. *Physica D* 115, 56–72. doi:10.1016/S0167-2789(97)00222-4

Golubitsky, M., Stewart, I., Buono, P.-L., and Collins, J. (1999). Symmetry in locomotor central pattern generators and animal gaits. *Nature* 401, 693–695. doi:10.1038/44416

Hauser, H., Ijspeert, A. J., Füchslin, R. M., Pfeifer, R., and Maass, W. (2011). Towards a theoretical foundation for morphological computation with compliant bodies. *Biol. Cybern.* 105, 355–370. doi:10.1007/s00422-012-0471-0

Hauser, H., Ijspeert, A. J., Füchslin, R. M., Pfeifer, R., and Maass, W. (2012). The role of feedback in morphological computation with compliant bodies. *Biol. Cybern.* 106, 595–613. doi:10.1007/s00422-012-0516-4

Hebb, D. O. (1949). *The Organization of Behavior: A Neuropsychological Theory*. New York, NY: Wiley.

Klopf, A. H. (1988). A neuronal model of classical conditioning. *Psychobiology* 16, 85–125.

Klyubin, A. S., Polani, D., and Nehaniv, C. L. (2005). "Empowerment: a universal agent-centric measure of control," in *Congress on Evolutionary Computation* (Edinburgh: IEEE), 128–135. doi:10.1109/CEC.2005.1554676

Kosko, B. (1986). "Differential Hebbian learning," in *AIP Conference Proceedings*, Vol. 151, 277–282.

Kulvicius, T., Kolodziejski, C., Tamosiunaite, M., Porr, B., and Wörgötter, F. (2010). Behavioral analysis of differential Hebbian learning in closed-loop systems. *Biol. Cybern.* 103, 255–271. doi:10.1007/s00422-010-0396-4

Lowe, R., Mannella, F., Ziemke, T., and Baldassarre, G. (2011). "Modelling coordination of learning systems: a reservoir systems approach to dopamine modulated Pavlovian conditioning," in *Advances in Artificial Life. Darwin Meets von Neumann*. Vol. 5778. eds G. Kampis, I. Karsai, and E. Szathmáry (Berlin Heidelberg: Springer), 410–417.

Markram, H., Lübke, J., Frotscher, M., and Sakmann, B. (1997). Regulation of synaptic efficacy by coincidence of postsynaptic APs and EPSPs. *Science* 275, 213–215. doi:10.1126/science.275.5297.213

Martius, G., Der, R., and Ay, N. (2013). Information driven self-organization of complex robotic behaviors. *PLoS ONE* 8:e63400. doi:10.1371/journal.pone.0063400

Martius, G., and Olbrich, E. (2013). "Quantifying emergent behavior of autonomous robots (extended abstract)," in *Workshop on Artificial Life in Massive Data Flows, ECAL 2013* (Taormina).

Maycock, J., Dornbusch, D., Elbrechter, C., Haschke, R., Schack, T., and Ritter, H. (2010). Approaching manual intelligence. *KI Künstliche Intelligenz* 24, 287–294. doi:10.1007/s13218-010-0064-9

ACKNOWLEDGMENTS

The author gratefully acknowledges the hospitality in the group of Nihat Ay at the Max Planck Institute for Mathematics in the Sciences, Leipzig, and many helpful and clarifying discussions with Nihat Ay, Georg Martius, and Keyan Zahedi.

Maycock, J., Essig, K., Haschke, R., Schack, T., and Ritter, H. (2011). "Towards an understanding of grasping using a multi-sensing approach," in *International Conference on Robotics and Automation (ICRA)* (Saint Paul).

Monteforte, M., and Wolf, F. (2010). Dynamical entropy production in spiking neuron networks in the balanced state. *Phys. Rev. Lett.* 105, 268104. doi:10.1103/PhysRevLett.105.268104

Mori, H., and Kuniyoshi, Y. (2010). "A human fetus development simulation: self-organization of behaviors through tactile sensation," in *Development and Learning (ICDL), 2010 IEEE 9th International Conference on* (Ann Arbor: IEEE), 82–87.

Pfeifer, R. (2012). "Morphological computation" – self-organization, embodiment, and biological inspiration," in *International Joint Conference IJCCI 2012.* Vol. 577. eds K. Madani, C. A. Dourado, A. Rosa, and J. Filipe (Barcelona) p. 110.

Pfeifer, R., and Bongard, J. C. (2006). *How the Body Shapes the Way We Think: A New View of Intelligence*. Cambridge, MA: MIT Press.

Pfeifer, R., and Gómez, G. (2009). "Morphological computation – connecting brain, body, and environment," in *Creating Brain-Like Intelligence: From Basic Principles to Complex Intelligent Systems (Lecture Notes in Computer Science)*, eds S. Bernhard, K. Edgar, S. Olaf, R. Helge (Berlin: Springer), 66–83.

Pfeifer, R., Lungarella, M., and Iida, F. (2007). Self-organization, embodiment, and biologically inspired robotics. *Science* 318, 1088–1093. doi:10.1126/science.1145803

Pfeifer, R., and Scheier, C. (1999). *Understanding Intelligence*. Boston, MA: MIT Press.

Prokopenko, M. (2009). Information and self-organization: a macroscopic approach to complex systems. *Artif. Life* 15, 377–383. doi:10.1162/artl.2009.Prokopenko.B4

Prokopenko, M. (ed.) (2008). *Foundations and Formalizations of Self-Organization*. Berlin: Springer.

Prokopenko, M., Gerasimov, V., and Tanev, I. (2006). "Evolving spatiotemporal coordination in a modular robotic system," in *From Animals to Animats 9, Volume 4095 of LNCS*, eds S. Nolfi, G. Baldassarre, R. Calabretta, J. Hallam, D. Marocco, J.-A. Meyer, and D. Parisi (Berlin: Springer), 558–569.

Ritter, H., Haschke, R., and Steil, J. J. (2009). "Trying to grasp a sketch of a brain for grasping," in *Creating Brain-Like Intelligence: From Basic Principles to Complex Intelligent Systems (Lecture Notes in Computer Science)*. eds S. Bernhard, K. Edgar, S. Olaf, and R. Helge (Berlin: Springer), 84–102.

Roberts, P. D. (1999). Computational consequences of temporally asymmetric learning rules: I. Differential Hebbian learning. *J. Comput. Neurosci.* 7, 235–246. doi:10.1023/A:1008910918445

Salge, C., Glackin, C., and Polani, D. (2014). "Empowerment-an introduction," in *Guided Self-Organization: Inception* (Heidelberg: Springer), 67–114.

Schöner, G., Jiang, W. Y., and Kelso, J. S. (1990). A synergetic theory of quadrupedal gaits and gait transitions. *J. Theor. Biol.* 142, 359–391. doi:10.1016/S0022-5193(05)80558-2

Song, S., Miller, K. D., and Abbott, L. F. (2000). Competitive Hebbian learning through spike-timing-dependent synaptic plasticity. *Nat. Neurosci.* 3, 919–926. doi:10.1038/78829

Strogatz, S. H., and Stewart, I. (1993). Coupled oscillators and biological synchronization. *Sci. Am.* 269, 102–109. doi:10.1038/scientificamerican1293-102

Tero, A., Akiyama, M., Owaki, D., Kano, T., Ishiguro, A., and Kobayashi, R. (2013). Interlimb neural connection is not required for gait transition in quadruped locomotion. *arXiv preprint arXiv:1310.7568.*

Tsodyks, M. V., and Sejnowski, T. (1995). Rapid state switching in balanced cortical network models. *Netw. Comput. Neural Syst.* 6, 111–124. doi:10.1088/0954-898X_6_2_001

Turrigiano, G. G., and Nelson, S. (2000). Hebb and homeostasis in neuronal plasticity. *Curr. Opin. Neurobiol.* 10, 358–364. doi:10.1016/S0959-4388(00)00091-X

Turrigiano, G. G., and Nelson, S. B. (2004). Homeostatic plasticity in the developing nervous system. *Nat. Rev. Neurosci.* 5, 97–107. doi:10.1038/nrn1327

van der Weele, J. P., and Banning, E. J. (2001). Mode interaction in horses, tea, and other nonlinear oscillators: the universal role of symmetry. *Am. J. Phys.* 69, 953. doi:10.1119/1.1378014

Weber, B. H., and Depew, D. J. (2003). *Evolution and Learning: The Baldwin Effect Reconsidered*. Cambridge, MA: MIT Press.

Yamada, Y., Fujii, K., and Kuniyoshi, Y. (2013). "Impacts of environment, nervous system and movements of preterms on body map development: fetus simulation with spiking neural network," in *Development and Learning and Epigenetic Robotics (ICDL), 2013 IEEE Third Joint International Conference on* (Osaka: IEEE), 1–7.

Yamasaki, R., Ambe, Y., Aoi, S., and Matsuno, F. (2013). "Quadrupedal bounding with spring-damper body joint," in *Intelligent Robots and Systems (IROS), 2013 IEEE/RSJ International Conference on* (Tokyo: IEEE), 2345–2350.

Conflict of Interest Statement: The author declares that the research was conducted in the absence of any commercial or financial relationships that could be construed as a potential conflict of interest.

14

A Taxonomy of Deep Convolutional Neural Nets for Computer Vision

Suraj Srinivas, Ravi Kiran Sarvadevabhatla, Konda Reddy Mopuri, Nikita Prabhu, Srinivas S. S. Kruthiventi and R. Venkatesh Babu*

Video Analytics Laboratory, Department of Computational and Data Sciences, Indian Institute of Science, Bangalore, India

Traditional architectures for solving computer vision problems and the degree of success they enjoyed have been heavily reliant on hand-crafted features. However, of late, deep learning techniques have offered a compelling alternative – that of automatically learning problem-specific features. With this new paradigm, every problem in computer vision is now being re-examined from a deep learning perspective. Therefore, it has become important to understand what kind of deep networks are suitable for a given problem. Although general surveys of this fast-moving paradigm (i.e., deep-networks) exist, a survey specific to computer vision is missing. We specifically consider one form of deep networks widely used in computer vision – convolutional neural networks (CNNs). We start with "AlexNet" as our base CNN and then examine the broad variations proposed over time to suit different applications. We hope that our recipe-style survey will serve as a guide, particularly for novice practitioners intending to use deep-learning techniques for computer vision.

Keywords: deep learning, convolutional neural networks, object classification, recurrent neural networks, supervised learning

Edited by:
Peter M. Hall,
University of Bath, UK

Reviewed by:
Ignacio Arganda-Carreras,
Basque Country University, Spain
Nicolas Pugeault,
University of Surrey, UK

*Correspondence:
R. Venkatesh Babu
venky@serc.iisc.ernet.in

1. INTRODUCTION

Computer vision problems, such as image classification and object detection, have traditionally been approached using hand-engineered features, such as SIFT by Lowe (2004) and HoG by Dalal and Triggs (2005). Representations based on the Bag-of-visual-words descriptor (see Yang et al., 2007), in particular, enjoyed success in image classification. These were usually followed by learning algorithms like support vector machines (SVMs). As a result, the performance of these algorithms relied crucially on the features used. This meant that progress in computer vision was based on hand-engineering better sets of features. With time, these features started becoming more and more complex – resulting in a difficulty with coming up better, more complex features. From the perspective of the computer vision practitioner, there were two steps to be followed: feature design and learning algorithm design, both of which were largely independent.

Meanwhile, some researchers in the machine learning community had been working on learning models that incorporated learning of features from raw images. These models typically consisted of multiple layers of non-linearity. This property was considered to be very important – and this lead to the development of the first deep learning models. Early examples, such as Restricted Boltzmann Machines (Hinton, 2002), Deep Belief Networks (Hinton et al., 2006) and Stacked Autoencoders (Vincent et al., 2010), showed promise on small datasets. The primary idea behind these works was to leverage the vast amount of unlabeled data to train models. This was called the "unsupervised pre-training" stage. It was believed that these "pre-trained" models would serve as a good initialization for further supervised tasks, such as image classification. Efforts to scale these algorithms on larger datasets culminated in 2012 during the ILSVRC competition (see Russakovsky et al., 2015), which involved, among other things – the task of classifying an image into 1 of 1000 categories. For the first time, a convolutional neural network (CNN)-based deep learnt model by Krizhevsky et al. (2012) brought down the error rate on that task by half, beating traditional hand-engineered approaches. Surprisingly, this could be achieved by performing end-to-end supervised training, without the need for unsupervised pre-training. Over the next couple of years, "Imagenet classification using deep neural networks" by Krizhevsky et al. (2012) became one of the most influential papers in computer vision. Convolutional neural networks, a particular form of deep learning models, have since been widely adopted by the vision community. In particular, the network trained by Alex Krizhevsky, popularly called "AlexNet" has been used and modified for various vision problems. Hence, in this article, we primarily discuss CNNs, as they are more relevant to the vision community. With the plethora of deep convolutional networks that exist for solving different tasks, we feel the time is right to summarize CNNs for a survey. This article can also serve as a guide for beginning practitioners in deep learning/computer vision.

The paper is organized as follows. We first develop the general principles behind CNNs (see Introduction to Convolutional Neural Networks), and then discuss various modifications to suit different problems (see CNN Flavors). Finally, we discuss some open problems (see Open Problems) and directions for further research.

2. INTRODUCTION TO CONVOLUTIONAL NEURAL NETWORKS

The idea of a convolutional neural network (CNN) is not new. This model had been shown to work well for hand-written digit recognition by LeCun et al. (1998). However, due to the inability of these networks to scale to much larger images, they slowly fell out of favor. This was largely due to memory and hardware constraints, and the unavailability of large amounts of training data. With increase in computational power thanks to wide availability of GPUs, and the introduction of large-scale datasets, such as the ImageNet (see Russakovsky et al., 2015) and the MIT Places

dataset (see Zhou et al., 2014), it was possible to train larger, more complex models. This was first shown by the popular *AlexNet* model that was discussed earlier. This largely kick-started the usage of deep networks in computer vision.

2.1. Building Blocks of CNNs

In this section, we shall look at the basic building blocks of CNNs in general. This assumes that the reader is familiar with traditional neural networks, which we shall call "fully connected layers" in this article. **Figure 1** shows a representation of the weights in the *AlexNet* model. While the first five layers are convolutional, the last three are fully connected layers.

2.1.1. Why Convolutions?

Using traditional neural networks for real-world image classification is impractical for the following reason: Consider a 2D image of size 200×200 for which we would have 40,000 input nodes. If the hidden layer has 20,000 nodes, the size of the matrix of input weights would be $40,000 \times 20,000 = 800$ Million. This is just for the first layer – as we increase the number of layers, this number increases even more rapidly. Besides, vectorizing an image completely ignores the complex 2D spatial structure of the image. How do we build a system that overcomes both these disadvantages?

One way is to use 2D convolutions instead of matrix multiplications. Learning a set of convolutional filters (each of 11×11, say) is much more tractable than learning a large matrix (40000×20000). 2D convolutions also naturally take the 2D structure of images into account. Alternately, convolutions can also be thought of as regular neural networks with two constraints (see Bishop, 2006).

- Local connectivity: this comes from the fact that we use a convolutional filter with dimensions much smaller than the image it operates on. This contrasts with the *global* connectivity paradigm typically relevant to vectorized images.
- Weight sharing: this comes from the fact that we perform convolutions, i.e., we apply the same filter across the image. This means that we use the same *local* filters on many locations in the image. In other words, the weights between all these filters are shared.

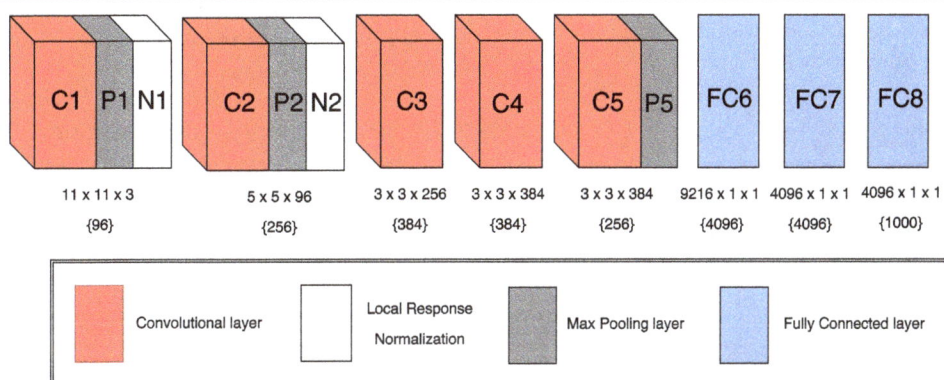

FIGURE 1 | An illustration of the weights in the AlexNet model. Note that after every layer, there is an implicit ReLU non-linearity. The number inside curly braces represents the number of filters with dimensions mentioned above it.

There is also evidence from visual neuroscience for similar computations within the human brain. Hubel and Wiesel (1962) found two types of cells in the primary visual cortex – the simple cells and the complex cells. The simple cell responded primarily to oriented edges and gratings – which are reminiscent of Gabor filters, a special class of convolutional filters. The complex cells were also sensitive to these edges and grating. However, they exhibited spatial invariance as well. This motivated the Neocognitron model by Fukushima (1980), which proposed the learning of convolutional filters in an artificial neural network. This model is said to have inspired convolutional networks, which are analogous to the simple cells mentioned above.

In practical CNNs, however, the convolution operations are not applied in the traditional sense wherein the filter shifts one position to the right after each multiplication. Instead, it is common to use larger shifts (commonly referred to as stride). This is equivalent to performing image down-sampling after regular convolution.

If we wish to train these networks on RGB images, one would need to learn multiple *multi-channel* filters. In the representation in **Figure 1**, the numbers $11 \times 11 \times 3$, along with {96} below **C1** indicates that there are 96 filters in the first layers, each of spatial dimension of 11×11, with one for each of the three RGB channels.

We note that this paradigm of convolution-like operations (location independent feature-detectors) is not entirely suitable for registered images. As an example, images of faces require different feature-detectors at different spatial locations. To account for this, Taigman et al. (2014) consider only locally connected networks with no weight-sharing. Thus, the choice of layer connectivity depends on the underlying type of problem.

2.1.2. Max-Pooling

The Neocognitron model inspired the modeling of simple cells as convolutions. Continuing in the same vein, the complex cells can be modeled as a max-pooling operation. This operation can be thought of as a *max filter*, where each $n \times n$ region is replaced with its max value. This operation serves the following two purposes:

1. It picks out the highest activation in a local region, thereby providing a small degree of spatial invariance. This is analogous to the operation of complex cells.
2. It reduces the size of the activation for the next layer by a factor of n^2. With a smaller activation size, we need a smaller number of parameters to be learnt in the later layers.

2.1.3. Non-Linearity

Deep networks usually consist of convolutions followed by a non-linear operation after each layer. This is necessary because cascading linear systems (like convolutions) is another linear system. Non-linearities between layers ensure that the model is more expressive than a linear model.

In theory, no non-linearity has more expressive power than any other, as long as they are continuous, bounded, and monotonically increasing (see Hornik, 1991). Traditional feedforward neural networks used the sigmoid $\left(\sigma(x) = \frac{1}{1+e^{-x}}\right)$ or the tanh $\left(\tanh(x) = \frac{e^x - e^{-x}}{e^x + e^{-x}}\right)$ non-linearities. However, modern convolutional networks use the ReLU $(\text{ReLU}(x) = max(0,x))$ non-linearity. CNNs with this non-linearity have been found to train faster, as shown by Nair and Hinton (2010).

Recently, Maas et al. (2013) introduced a new kind of non-linearity, called the leaky-ReLU. It was defined as Leaky-ReLU$(x) = max(0,x) + \alpha min(0,x)$, where α is a pre-determined parameter. He et al. (2015) improved on this by suggesting that the α parameter also be learnt, leading to a much richer model.

2.2. Depth

The Universal Approximation theorem by Hornik (1991) states that a neural network with a single hidden layer is sufficient to model any continuous function. However, Bengio (2009) showed that such networks need an exponentially large number of neurons when compared to a neural network with many hidden layers. Recently, Romero et al. (2014) and Ba and Caruana (2014) explicitly showed that a deeper neural network can be trained to perform much better than a comparatively shallow network.

Although the motivation for creating deeper networks was clear, for a long time, researchers did not have an algorithm that could efficiently train neural networks with more than three layers. With the introduction of greedy layerwise pre-training by Hinton et al. (2006), researchers were able to train much deeper networks. This played a major role in bringing the so-called *Deep Learning* systems into mainstream machine learning. Modern deep networks, such as AlexNet, have seven layers. More recent networks, such as VGGnet, by Simonyan and Zisserman (2014b) and GoogleNet by Szegedy et al. (2014) have 19 and 22 layers, respectively, were shown to perform much better than AlexNet.

2.3. Learning Algorithm

A powerful, expressive model is of no use without an algorithm to learn the model's parameters efficiently. The greedy layerwise pre-training approaches in the pre-AlexNet era attempted to create such an efficient algorithm. However, for computer vision tasks, it turned out that a simpler supervised training procedure was enough to learn a powerful model.

Learning is generally performed by minimization of certain loss functions. Tasks based on classification use the softmax loss function or the sigmoid cross entropy function, while those involving regression use the Euclidean error function. In the example of **Figure 1**, the output of the FC8 layer is trained to represent 1 of 1000 classes of the dataset.

2.3.1. Gradient-Based Optimization

Neural networks are generally trained using the backpropagation algorithm (see Rumelhart et al., 1988), which uses the chain rule to speed up the computation of the gradient for the gradient descent (GD) algorithm. However, for datasets with thousands (or more) of data points, using GD is impractical. In such cases, an approximation called the stochastic gradient descent (SGD) is used. It has been found that training using SGD generalizes much better than training using GD. However, one disadvantage is that SGD is very slow to converge. To counteract this, SGD is typically used with a mini-batch, where the mini-batch typically contains a small number of data-points (~100).

Momentum (see Polyak, 1964) belongs to a family of methods that aim to speed the convergence of SGD. This is largely used in practice to train deep networks, and is often considered as an essential component. Other extensions, such as Adagrad by Duchi et al. (2011), Nesterov's accelerated GD by Nesterov (1983), Adadelta by Zeiler and Fergus (2014) and Adam by Kingma and Ba (2014) are known to work equally well, if not better than vanilla momentum in certain cases. For detailed discussion on how these methods work, the reader is encouraged to read Sutskever et al. (2013).

2.3.2. Dropout

When training a network with a large number of parameters, an effective regularization mechanism is essential to combat overfitting. Usual approaches such as λ_1 or λ_2 regularization on the weights of the neural net have been found to be insufficient in this aspect. Dropout is a powerful regularization method introduced by Hinton et al. (2012), which has been shown to work well for large neural nets. To use dropout, we *randomly* drop neurons with a probability p during training. As a result, only a random subset of neurons are trained in a single iteration of SGD. At test time, we use all neurons, however, we simply multiply the activation of each neuron with p to account for the scaling. Hinton et al. (2012) showed that this procedure was equivalent to training a large ensemble of neural nets with shared parameters, and then using their geometric mean to obtain a single prediction.

Many extensions to dropout such as DropConnect by Wan et al. (2013) and Fast Dropout by Wang and Manning (2013) have been shown to work better in certain cases. Maxout by Goodfellow et al. (2013) is a non-linearity that improves performance of a network that uses dropout.

2.4. Tricks to Increase Performance

While the techniques and components described above are theoretically well-grounded, certain *tricks* are crucial to obtaining *state-of-the-art* performance.

It is well known that machine learning models perform better in the presence of more data. Data augmentation is a process by which some geometric transforms are applied to training data to increase their number. Some examples of commonly used geometric transforms include random cropping, RGB jittering, image flipping, and small rotations. It has been found that using augmented data typically boosts performance by about 3% (see Chatfield et al., 2014).

Also well-known is the fact that an ensemble of models perform better than one. Hence, it is the commonplace to train several CNNs and average their predictions at test time. Using ensembles has been found to typically boost accuracy by 1–2% (see Simonyan and Zisserman, 2014b; Szegedy et al., 2014).

2.5. Putting It All Together: AlexNet

The building blocks discussed above largely describe AlexNet as a whole. As shown in **Figure 1**, only layers 1, 2, and 5 contain max-pooling, while dropout is only applied to the last two fully connected layers as they contain the most number of parameters. Layers 1 and 2 also contain local response normalization, which

has not been discussed as Chatfield et al. (2014) showed that its absence does not impact performance.

This network was trained on the ILSVRC 2012 training data, which contained 1.2 million training images belonging to 1000 classes. This was trained on two GPUs over the course of 1 month. The same network can be trained today in little under a week using more powerful GPUs (see Chatfield et al., 2014). The hyperparameters of the learning algorithms, such as learning rate, momentum, dropout, and weight decay, were hand tuned. It is also interesting to note the trends in the nature of features learnt at different layers. The earlier layers tend to learn gabor-like oriented edges and blob-like features, followed by layers that seem to learn more higher order features like shapes. The very last layers seem to learn semantic attributes, such as eyes or wheels, which are crucial parts in several categories. A method to visualize these was provided by Zeiler (2012).

2.6. Using Pre-Trained CNNs

One of the main reasons for the success of the AlexNet model was that it was possible to directly use the pre-trained model to do various other tasks, which it was not originally intended for. It became remarkably easy to download a learnt model, and then tweak it slightly to suit the application at hand. We describe two such ways to use models in this manner.

2.6.1. Fine-Tuning

Given a model trained for image classification, how does one modify it to perform a different (but related) task? The answer is to just use the trained weights as an initialization and run SGD again for this new task. Typically, one uses a learning rate much lower than what was used for learning the original net. If the new task is very similar to the task of image classification (with similar categories), then one need not re-learn a lot of layers. The earlier layers can be fixed and only the later, more semantic layers need to be re-learnt. However, if the new task is very different, one ought to either re-learn all layers, or learn everything from scratch. The number of layers to re-learn also depends on the number of data points available for training the new task. The more the data, the higher is the number of layers that can be re-learnt. The reader is urged to refer to Yosinski et al. (2014) for more thorough guidelines.

2.6.2. CNN Activations as Features

As remarked earlier, the later layers in AlexNet seem to learn visually semantic attributes. These intermediate representations are crucial in performing 1000-way classification. Since these represent a wide variety of classes, one can use the FC7 activation of an image as a generic feature descriptor. These features have been found to be better than hand-crafted features, such as SIFT or HoG for various computer vision tasks.

Donahue et al. (2013) first introduced the idea of using CNN activations as features and performed tests to determine their suitability for various tasks. Babenko et al. (2014) proposed to use the activations of fully connected layers for image retrieval, which they dubbed "Neural Codes". Razavian et al. (2014) used these activations for various tasks and concluded that off-the-shelf CNN features can serve as a hard-to-beat baseline for many tasks. Hariharan et al. (2014) used activations *across* layers as a feature. Specifically, they look at the activations produced by single-image pixels across the network and pool them together. They were found to be useful for fine-grained tasks, such as keypoint localization.

2.7. Improving AlexNet

The performance of AlexNet motivated a number of CNN-based approaches, all aimed at a performance improvement over and above that of AlexNet's. Just as AlexNet was the winner for ILSVRC challenge in 2012, a CNN-based net Overfeat by Sermanet et al. (2013a) was the top-performer at ILSVRC-2013. Their key insight was that training a convolutional network to simultaneously classify, locate, and detect objects in images can boost the classification accuracy and the detection and localization accuracy of all tasks. Given its multi-task learning paradigm, we discuss Overfeat when we discuss hybrid CNNs and multi-task learning in Section "Hybrid Learning Methods."

GoogleNet by Szegedy et al. (2014), the top-performer at ILSVRC-2014, established that very deep networks can translate to significant gains in classification performance. Since naively increasing the number of layers results in a large number of parameters, the authors employ a number of "design tricks." One such trick is to have a trivial 1×1 convolutional layer after a regular convolutional layer. This has the net effect of not only reducing the number of parameters but also results in CNNs with more expressive power. This design trick is laid out in better detail in the work of Szegedy et al. (2014) where the authors show that having one or more 1×1 convolutional layers is akin to having a multi-layer perceptron network processing the outputs of the convolutional layer that precedes it. Another trick that the authors utilize is to involve inner layers of the network in the computation of the objective function instead of the typical final softmax layer (as in AlexNet). The authors attribute scale invariance as the reason behind this design decision.

VGG-19 and its variants by Simonyan and Zisserman (2014b) is another example of a high-performing CNN where the deeper-is-better philosophy is applied in the net design. An interesting feature of VGG design is that it forgoes larger sized convolutional filters for stacks of smaller sized filters. These smaller sized filters tend to be chosen so that they contain approximately the same number of parameters as the larger filters they supposedly replace. The net effect of this design decision is efficiency and regularization-like effect on parameters due to the smaller size of the filters involved.

3. CNN FLAVORS

3.1. Region-Based CNNs

Most CNNs trained for image recognition are trained using a dataset of images containing a single object. At test time, even in case of multiple objects, the CNN may still predict a single class. This inherent problem with the design of the CNNs is not restricted to image classification alone. For example, the problem of object detection and localization requires not only classifying the image but also estimating the class and precise location of the object(s) present in the image. Object detection is challenging since we potentially want to detect multiple objects with varying sizes within a single image. It generally requires processing the image patch-wise, looking for the presence of objects. Neural nets have been employed in this way for detecting specific objects like faces in Vaillant et al. (1994) and Rowley et al. (1998) and for pedestrians by Sermanet et al. (2013c).

Meanwhile, detecting a set of object-like regions in a given image – also called region proposals or object proposals – has gained a lot of attention (see Uijlings et al., 2013). These region proposals are class agnostic and reduce the overhead incurred by the traditional exhaustive sliding window approach. These region proposal algorithms operate at low level and output hundreds of object like image patches at multiple scales. In order to employ a classification net toward the task of object localization, image patches of different scales have to be searched one at a time.

Recent work by Girshick et al. (2014) attempt to solve the object localization problem using a set of region proposals. During test time, the method generates around 2000 category independent region proposals using selective search by Uijlings et al. (2013) from the test image. They employ a simple affine image warping to feed each of these proposals to a CNN trained for classification. The CNN then describes each of these regions with a fixed size high-level semantic feature. Finally, a set of category-specific linear SVMs classify each region, as shown in **Figure 2**. This method achieved the best detection results on the PASCAL VOC 2012 dataset. As this method uses image regions followed by a CNN, it is dubbed R-CNN (Region-based CNN).

A series of works adapted the R-CNN approach to extract richer set of features at patch or region level to solve a wide range of target applications in vision. However, CNN representations lack robustness to geometric transformations restricting their usage. Gong et al. (2014a) show empirical evidence that the global CNN features are sensitive to general transformations, such as translation, rotation, and scaling. In their experiments, they report that this inability of global CNN features translates directly into a loss in the classification accuracy. They proposed a simple

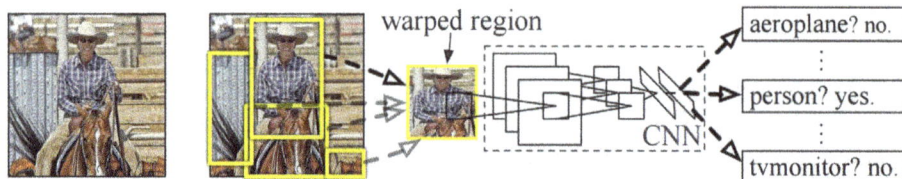

FIGURE 2 | Object detection system of Girshick et al. (2014) using deep features extracted from image regions.

technique to pool the CNN activations extracted from the local image patches. The method extracts image patches in an exhaustive sliding-window manner at different scales and describes each of them using a CNN. The resulting dense CNN features are pooled using VLAD (see Jégou et al., 2011) in order to result in a representation that incorporates spatial as well as semantic information.

Instead of considering the image patches at exhaustive scales and image locations, Mopuri and Babu (2015) utilize the objectness prior to automatically extract the image patches at different scales. They build a more robust image representation by aggregating the individual CNN features from the patches for an image search application.

Wei et al. (2014) extended the capability of a CNN that is trained to output a single label into predicting multiple labels. They consider an arbitrary number of region proposals in an image and share a common CNN across all of them in order to obtain individual predictions. Finally, they employ a simple pooling technique to produce the final multi-label prediction.

3.2. Fully Convolutional Networks

The success of convolutional neural networks in the tasks of image classification (see Krizhevsky et al., 2012; Szegedy et al., 2014) and object detection (see Girshick et al., 2014) has inspired researchers to use deep networks for more challenging recognition problems, such as semantic object segmentation and scene parsing. Unlike image classification, semantic segmentation and scene parsing are problems of structured prediction where every pixel in the image grid needs to be assigned a label of the class to which it belongs (e.g., road, sofa, table, etc.). This problem of per-pixel classification has been traditionally approached by generating region-level (e.g., superpixel) hand crafted features and classifying them using a support vector machine (SVM) into one of the possible classes.

Doing away with these engineered features, Farabet et al. (2013a) used hierarchical learned features from a convolutional neural net for scene parsing. Their approach comprised of densely computing multi-scale CNN features for each pixel and aggregating them over image regions upon which they are classified. However, their method still required the post-processing step of generating over-segmented regions, such as superpixels, for obtaining the final segmentation result. Additionally, the CNNs used for multi-scale feature learning were not very deep with only three convolution layers.

Later, Long et al. (2015) proposed a fully convolutional network architecture for learning per-pixel tasks, such as semantic segmentation, in an end-to-end manner. This is shown in **Figure 3**. Each layer in the fully convolutional net (FullConvNet) performs a location invariant operation i.e., a spatial shift of values in the input to the layer will only result in an equivalent scaled spatial shift in its output while keeping the values nearly intact. This property of translational invariance holds true for the convolutional and max-pool layers that form the major building blocks of a FullConvNet. Furthermore, these layers have an output-centered, fixed-size receptive field on its input blob. These properties of the layers of FullConv Net allow it to retain the spatial structure present in the input image in all of its intermediate and final outputs.

Unlike CNNs used for image classification, a FullConvNet does not contain any densely connected/inner product layers as they are not translation invariant. The restriction on the size of input image to a classification CNN [e.g., 227×227 for AlexNet (Krizhevsky et al., 2012), 224×224 for VGG (Simonyan and Zisserman, 2014b)] is imposed due to the constraint on the input size to its inner product layers. Since a FullConvNet does not have any of these inner product layers, it can essentially operate on input images of any arbitrary size.

During the design of CNN architectures, one has to make a trade-off between the number of channels and the spatial dimensions for the data as it passes through each layer. Generally, the number of channels in the data is made to increase progressively while bringing down its spatial resolution, by introducing stride in the convolution and max-pool layers of the net. This is found to be an effective strategy for generating richer semantic representations in a hierarchical manner. While this method enables the net to recognize complex patterns in the data, it also diminishes the spatial resolution of the data blob progressively after each layer. While this is not a major concern for classification nets that require only a single label for the entire image, this results in per-pixel prediction only at a sub-sampled resolution in case of FullConvNets. For tackling this problem, Long et al. (2015) have proposed a deconvolution layer that brings back the spatial resolution from the sub-sampled output through a learned upsampling operation. This upsampling operation is performed at intermediate layers of various spatial dimensions and are concatenated to obtain pixel-level features at the original resolution.

On the other hand, Chen et al. (2014) adopted a more simplistic approach for maintaining resolution by removing the stride in

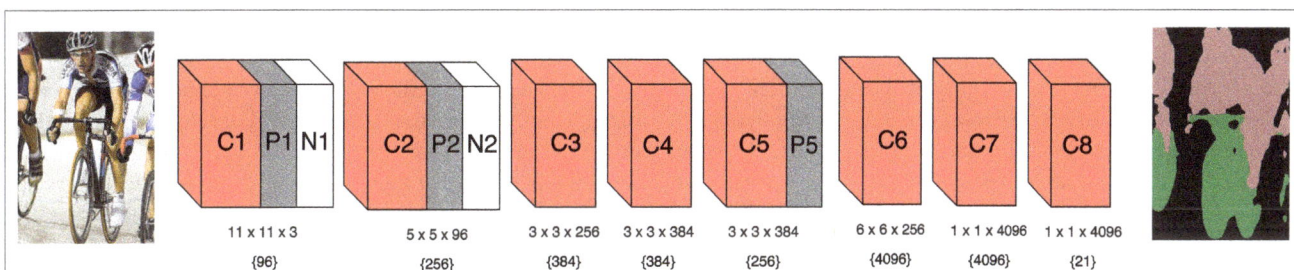

FIGURE 3 | Fully Convolutional Net: AlexNet modified to be fully convolutional for performing semantic object segmentation on PASCAL VOC 2012 dataset with 21 classes.

the layers of FullConvNet, wherever possible. Following this, the FullConvNet predicted output is modeled as a unary term for conditional random field (CRF) constructed over the image grid at its original resolution. With labeling smoothness constraint enforced through pair-wise terms, the per-pixel classification task is modeled as a CRF inference problem. While this post-processing of FullConvNet's coarse labeling using CRF has been shown to be effective for pixel-accurate segmentation, Zheng et al. (2015) have proposed a better approach where the CRF constructed on image is modeled as a recurrent neural network (RNN). By modeling the CRF as an RNN, it can be integrated as a part of any Deep Convolutional Net making the system efficient at both semantic feature extraction and fine-grained structure prediction. This enables the end-to-end training of the entire FullConvNet + RNN system using the stochastic gradient descent (SGD) algorithm to obtain fine pixel-level segmentation.

Visual saliency is another important pixel-level problem considered by researchers. This task involves predicting the salient regions of an image given by human eye fixations. Works by Vig et al. (2014) and Liu et al. (2015) proposed CNN-based approaches for estimating the saliency score for constituent image patches using deep features. By contrast, Kruthiventi et al. (2015) proposed a FullConvNet architecture – DeepFix, which learnt to predict saliency for the entire image in an end-to-end fashion and attained a superior performance. Their network characterized the multi-scale aspects of the image using inception blocks and captured the global context using convolutional layers with large receptive fields. Another work, by Li et al. (2015b), proposed a multi-task FullConvNet architecture – DeepSaliency for joint saliency detection and semantic object segmentation. Their work showed that learning features collaboratively for two related prediction tasks can boost overall performance.

3.3. Multi-Modal Networks

The success of CNNs on standard RGB vision tasks is naturally extended to works on other perception modalities, such as RGB-D and motion information in the videos. Recently, there has been an increasing evidence for the successful adaptation of the CNNs to learn efficient representations from the depth images. Socher et al. (2012) exploited the information from color and depth modalities for addressing the problem of classification. In their approach, a single layer of CNN extracts low-level features from both the RGB and depth images separately. These low-level features from each modality are given to a set of RNNs for embedding into a lower dimension. Concatenation of the resulting features forms the input to the final soft-max layer. The work by Couprie et al. (2013) extended the CNN method of Farabet et al. (2013b) to label the indoor scenes by treating depth information as an additional channel to the existing RGB data. Similarly, Wang et al. (2014a) adapt an unsupervised feature learning approach to scene labeling using RGB-D input with four channels. Gupta et al. (2014) proposed an encoding for the depth images that allows CNNs to learn stronger features than from the depth image alone. They encode depth image into three channels at each pixel: horizontal disparity, height above ground, and the angle the pixel's local surface normal makes with the inferred gravity direction. Their approach for object detection and segmentation

processes RGB and the encoded depth channels separately. The learned features are fused by concatenating and further fed into a SVM.

Similarly, one can think of extending these works for video representation and understanding. When compared to still images, videos provide important additional information in the form of motion. However, majority of the early works that attempted to extend CNNs for video, fed the networks with raw frames. This makes for a much difficult learning problem. Jhuang et al. (2007) proposed a biologically inspired model for action recognition in videos with a predefined set of spatio-temporal filters in the initial layer. Combined with a similar but spatial HMAX (Hierarchical model and X) model, Kuehne et al. (2011) proposed spatial and temporal recognition streams. Ji et al. (2010) addressed an end-to-end learning of the CNNs for videos for the first time using 3-D convolutions over a bunch of consecutive video frames. A more recent work by Karpathy et al. (2014) propose a set of techniques to fuse the appearance information present from a stack of consecutive frames in a video. However, they report that the net that processes individual frames performs on par with the net that operates on a stack of frames. This might suggest that the learnt spatio-temporal filters are not suitable to capture the motion patterns efficiently.

A more suitable CNN model to represent videos is proposed in a contemporaneous work by Simonyan and Zisserman (2014a), which is called two-stream network approach. Though the model in Kuehne et al. (2011) is also a two-stream model, the main difference is that the streams are shallow and implemented with hand-crafted models. The reason for the success of this approach is the natural ability of the videos to be separated into spatial and temporal components. The spatial component in the form of frames captures the appearance information, such as the objects, present in the video. The temporal component in the form of motion (optical flow) across the frames captures the movement of the objects. These optical flow estimates can be obtained either from classical approaches (see Baker and Matthews, 2004) or deep-learnt approaches (see Weinzaepfel et al., 2013).

This approach models the recognition system dividing into two parallel streams as depicted in **Figure 4**. Each is implemented by a dedicated deep CNN, whose predictions are later fused. The net for the spatial stream is similar to the image recognition CNN and processes one frame at a time. However, the temporal stream takes the stacked optical flow of a bunch of consecutive frames as input and predicts the action. Both the nets are trained separately with the corresponding input. An alternative motion representation using the trajectory information similar to Wang and Schmid (2013) is also observed to perform similar to optical flow.

The most recent methods that followed Simonyan and Zisserman (2014a) have similar two-stream architecture. However, their contribution is to find the most active spatio-temporal volume for the efficient video representation. Inspired from the recent progress in the object detection in images, Gkioxari and Malik (2015) built action models from shape and motion cues. They start from the image proposals and select the motion salient subset of them and extract spatio-temporal features to represent the video using the CNNs.

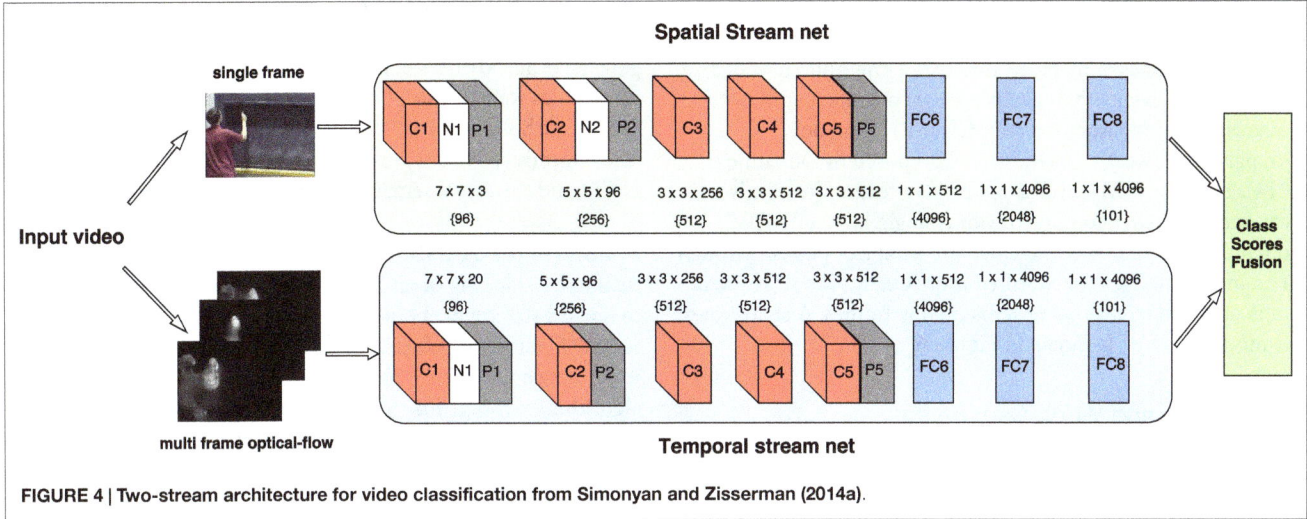

FIGURE 4 | Two-stream architecture for video classification from Simonyan and Zisserman (2014a).

Wang et al. (2015a) employ deep CNNs to learn discriminative feature maps and conduct trajectory constrained pooling to summarize into an effective video descriptor. The two streams operate in parallel extracting local deep features for the volumes centered around the trajectories.

In general, these multi-modal CNNs can be modified and extended to suit any other kind of modality, such as audio, text to complement the image data leading to a better representation of image content.

3.4. CNNs with RNNs

While CNNs have made remarkable progress in various tasks, they are not very suitable for learning sequences. Learning such patterns requires memory of previous states and feedback mechanisms that are not present in CNNs. RNNs are neural nets with at least one feedback connection. This looping structure enables the RNN to have an internal memory and to learn temporal patterns in data.

Figure 5 shows the unrolled version of a simple RNN applied to a toy example of sequence addition. The problem is defined as follows: let a_t be a positive number, corresponding to the input at time t. The output at time t is given by

$$S_t = \sum_{i=1}^{t} a_i.$$

We consider a very simple RNN with just one hidden layer. The RNN can be described by equations below.

$$h_{t+1} = f_h(W_{ih} \times a_t + W_{hh} \times h_t)$$
$$S_{t+1} = f_o(W_{ho} \times h_{t+1})$$

where W_{ih}, W_{hh}, W_{ho} are learned weights and f_h and f_o are non-linearities. For the toy problem considered above, the weights learned would result in $W_{ih} = W_{hh} = W_{ho} = 1$. Let us consider the non-linearity to be ReLu. The equations would then become,

$$h_{t+1} = ReLu(a_t + h_t)$$
$$S_{t+1} = ReLu(h_{t+1}).$$

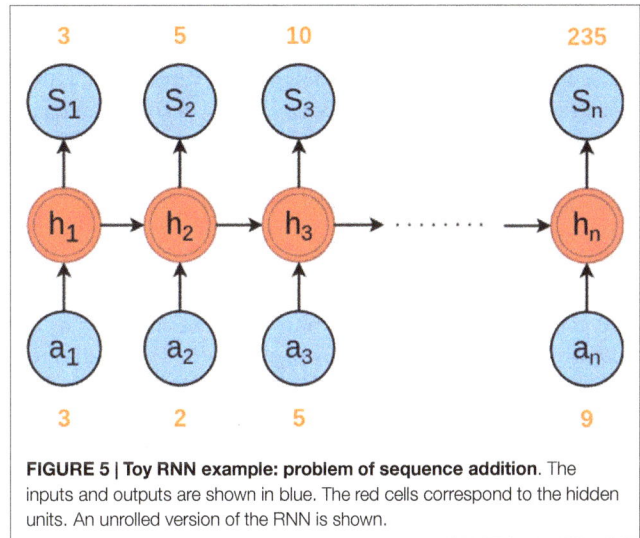

FIGURE 5 | Toy RNN example: problem of sequence addition. The inputs and outputs are shown in blue. The red cells correspond to the hidden units. An unrolled version of the RNN is shown.

Thus, as shown in **Figure 5**, the RNN stores previous inputs in memory and learns to predict the sum of the sequence up to the current timestep t.

As with CNNs, recurrent neural networks have been trained with various back propagation techniques. These conventional methods however, resulted in the *vanishing gradient problem*, i.e., the errors sent backward over the network, either grew very large or vanished leading to problems in convergence. In 1997, Hochreiter and Schmidhuber (1997) introduced LSTM (Long Short Term Memory), which succeeded in overcoming the vanishing gradient problem, by introducing a novel architecture consisting of units called constant error carousels. LSTMs were, thus, able to learn very deep RNNs and successfully remembered important events over long (thousands of steps) durations of time.

Over the next decade, LSTM's became the network of choice for several sequence learning problems, especially in the fields of speech and handwriting recognition (see Graves et al., 2009, 2013). In the sections that follow, we shall discuss applications of RNNs in various computer vision problems.

3.4.1. Action Recognition

Recognizing human actions from videos has long been a pivotal problem in the tasks of video understanding and surveillance. Actions, being events that take place over a finite length of time, are excellent candidates for a joint CNN-RNN model.

In particular, we discuss the model proposed by Donahue et al. (2014). They use *RGB* as well as *optical flow* features to jointly train a variant of Alexnet combined with a single layer of LSTM (256 hidden units). Frames of the video are sampled, passed through the trained network, and classified individually. The final prediction is obtained by averaging across all the frames. A snapshot of this model at time *t* is shown in **Figure 6**.

3.4.2. Image and Video Captioning

Another important component of scene understanding is the textual description of images and videos. Relevant textual description also helps complement image information, as well as form useful queries for retrieval.

RNNs (LSTMs) have long been used for machine translation (see Bahdanau et al., 2014; Cho et al., 2014). This has motivated its use for the purpose of image description. Vinyals et al. (2014) have developed an end-to-end system, by first encoding an image using a CNN and then using the encoded image as an input to a language generating RNN. Karpathy and Fei-Fei (2014) propose a multimodal deep network that aligns various interesting regions of the image, represented using a CNN feature, with associated words. The learned correspondences are then used to train a bidirectional RNN. This model is able not only to generate descriptions for images, but also to localize different segments of the sentence to their corresponding image regions. The multimodal RNN (m-RNN) by Mao et al. (2014) combines the functionalities of the CNN and RNN by introducing a new multimodal layer, after the embedding and recurrent layers of the RNN. Mao et al. (2015) further extend the m-RNN by incorporating a transposed weight sharing strategy, enabling the network to learn novel visual concepts from images.

Venugopalan et al. (2014) move beyond images and obtain a mean-pooled CNN representation for a video. They train an LSTM to use this input to generate a description for the video. They further improve upon this task by developing S2VT (Venugopalan et al., 2015) a stacked LSTM model that accounts for both the RGB as well as flow information available in videos. Pan et al. (2015) use both 2-D and 3-D CNNs to obtain a video embedding. They introduced two types of losses that are used to train both the LSTM and the visual semantic embedding.

3.4.3. Visual Question Answering

Real understanding of an image should enable a system not only to make a statement about it, but also to answer questions related to it. Therefore, answering questions based on visual concepts in an image is the next natural step for machine understanding algorithms. Doing this, however, requires the system to model both the textual question and the image representation, before generating an answer conditioned on both the question and the image.

A combination of CNN and LSTM has proven to be effective in this task too, as evidenced by the work of Malinowski et al. (2015) who train an LSTM layer to accept both the question as well a CNN representation of the image and generate the answer. Gao et al. (2015) use two LSTMs with shared weights along with a CNN for the task. Their experiments are performed on a multilingual dataset containing Chinese questions and answers along with its English translation. Antol et al. (2015) provide a dataset for the task of visual question answering containing both real-world images and abstract scenes.

3.5. Hybrid Learning Methods

3.5.1. Multi-Task Learning

Multi-task learning is essentially a machine learning paradigm wherein the objective is to train the learning system to perform well on multiple tasks. Multi-task learning frameworks tend to exploit shared representations that exist among the tasks to obtain a better generalization performance than counterparts developed for a single task alone.

In CNNs, multi-task learning is realized using different approaches. One class of approaches utilize a multi-task loss

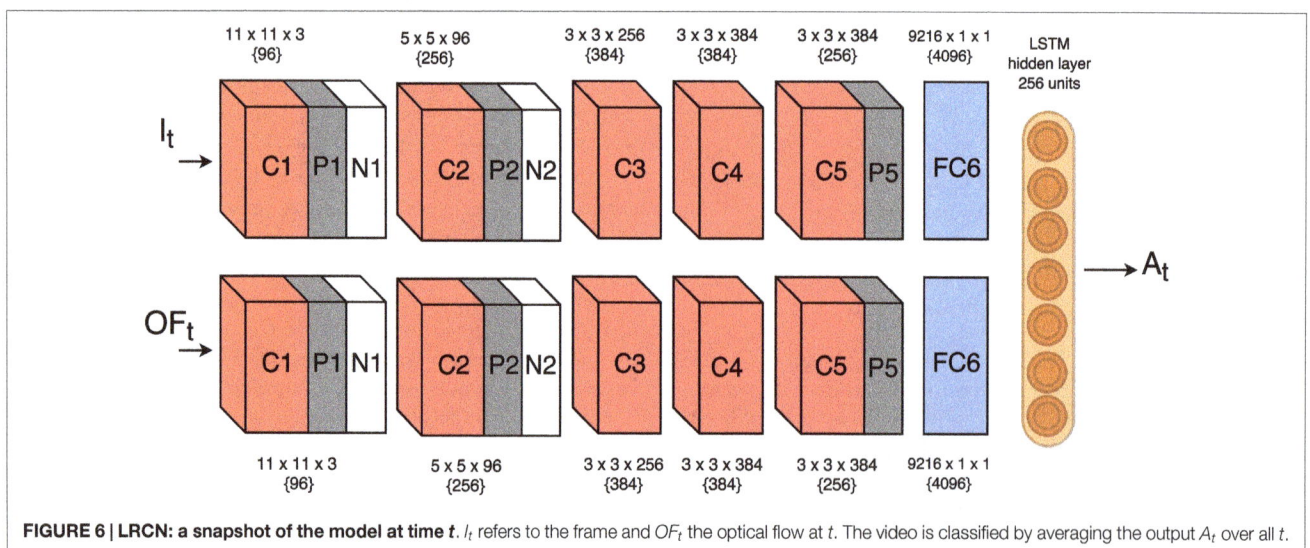

FIGURE 6 | LRCN: a snapshot of the model at time *t*. I_t refers to the frame and OF_t the optical flow at *t*. The video is classified by averaging the output A_t over all *t*.

function with hyper-parameters typically regulating the task losses. For example, Girshick (2015) employ a multi-task loss to train their network jointly for classification and bounding-box regression tasks thereby improving performance for object detection. Zhang et al. (2014) propose a facial landmark detection network that adaptively weights auxiliary tasks (e.g., head pose estimation, gender classification, age estimation) to ensure that a task that is deemed not beneficial to accurate landmark detection is prevented from contributing to the network learning. Devries et al. (2014) demonstrate improved performance for facial expression recognition task by designing a CNN for simultaneous landmark localization and facial expression recognition. A hallmark of these approaches is the division of tasks into primary task and auxiliary task(s) wherein the purpose of the latter is typically to improve the performance of the former (see **Figure 7**).

Some approaches tend to have significant portions of the original network modified for multiple tasks. For instance, Sermanet et al. (2013b) replace pre-trained layers of a net originally designed to provide spatial (per-pixel) classification maps with a regression network and fine-tune the resulting net to achieve simultaneous classification, localization, and detection of scene objects.

Another class of multi-task approaches tend to have task-specific sub-networks as a characteristic feature of CNN design. Li et al. (2015a) utilize separate sub-networks for the joint point regression and body part detection tasks. Wang et al. (2015b) adopt a serially stacked design wherein a localization sub-CNN and the original object image are fed into a segmentation sub-CNN to generate its object bounding box and extract its segmentation mask. To solve an unconstrained word image recognition task, Jaderberg et al. (2014a) propose an architecture consisting of a character sequence CNN and an N-gram encoding CNN that act on an input image in parallel and whose outputs are utilized along with a CRF model to recognize the text content present within the image.

3.5.2. Similarity Learning
Apart from classification, CNNs can also be used for tasks, such as metric learning and rank learning. Rather than asking the CNN to identify objects, we can instead ask it to verify whether two images contain the same object or not. In other words, we ask the CNN to

learn which images are *similar*, and which are not. Image retrieval is one application where such questions are routinely asked.

Structurally, Siamese networks resemble two-stream networks discussed previously. However, the difference here is that both "streams" have identical weights. Siamese networks consist of two separate (but identical) networks, where two images are fed in as input. Their activations are combined at a later layer, and the output of the network consists of a single number, or a *metric*, which is a notion of distance between the images. Training is done so that images that are considered to be similar have a lower output score than images that are considered different. Bromley et al. (1993) separate introduced the idea of Siamese networks and used it for signature verification. Later on, Chopra et al. (2005) extended it for face verification. Zagoruyko and Komodakis (2015) further extended and generalized this to learning similarity between image patches.

Triplet networks are extensions of siamese networks used for rank learning. Wang et al. (2014b) first used this idea for learning fine-grained image similarity learning.

4. OPEN PROBLEMS

In this section, we briefly mention some open research problems in deep learning, particularly of interest to computer vision. Several of these problems are already being tackled in several works.

- Training CNNs requires tuning of a large number of hyper-parameters, including those involving the model architecture. An automated way of tuning such as that by Snoek et al. (2012) is crucial for practitioners. However, that requires multiple models to be trained, which can be both time consuming and impractical for large networks.
- Nguyen et al. (2014) showed that one can generate artificial images that result in CNNs producing a high confidence false prediction. In a related line of work, Szegedy et al. (2013) showed that natural images can be modified in an imperceptible manner to produce a completely different classification label. Although Goodfellow et al. (2014a) attempted to reduce the effects of such adversarial examples, it remains to be seen whether that can be completely eliminated.

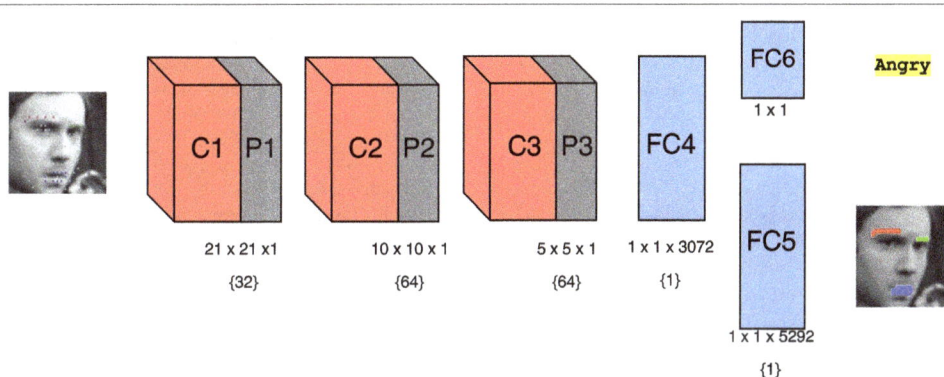

FIGURE 7 | The facial expression recognition system of Devries et al. (2014) that utilizes facial landmark (shown overlaid on the face toward the right of the image) recognition as an auxiliary task that helps improve performance on the main task of expression recognition.

- It is well known (see Gong et al., 2014b) that CNNs are *robust* to small geometric transforms. However, we would like them to be *invariant*. The study of *invariance* for extreme deformation is largely missing.
- Along with using a large number of data points, CNN models are also large and (relatively) slow to evaluate. While there has been a lot of work in reducing number of parameters (see Denil et al., 2013; Collins and Kohli, 2014; Jaderberg et al., 2014b; Hinton et al., 2015; Srinivas and Babu, 2015), it is not clear how to train non-redundant models in the first place.
- CNNs are presently trained in a *one-shot* way. The formulation of an *online* method of training would be desirable for robotics applications.
- Unsupervised learning is one more area where we expect to deploy deep learning models. This would enable us to leverage the massive amounts of unlabeled image data on the web. Classical deep networks, such as autoencoders and restricted Boltzmann machines, were formulated as unsupervised models. While there has been a lot of interesting recent work in the area (see Bengio et al., 2013; Kingma and Welling, 2013; Goodfellow et al., 2014b; Kulkarni et al., 2015), a detailed discussion of these is beyond the scope of this paper.

5. CONCLUDING REMARKS

In this article, we have surveyed the use of deep learning networks – convolutional neural networks in particular – for computer vision. This enabled complicated hand-tuned algorithms being replaced by single monolithic algorithms trained in an end-to-end manner. However, despite our best efforts, it may not be possible to capture the entire gamut of deep learning research – even for computer vision – in this paper. We point the reader to other reviews, specifically those by Bengio (2009), LeCun et al. (2015), and Schmidhuber (2015). These reviews are more geared toward deep learning in general, while ours is more focused on computer vision. We hope that our article will be useful to vision researchers beginning to work in deep learning.

AUTHOR CONTRIBUTIONS

SS contributed to Section 1, part of Section 2, and part of Sections 3 and 4. RS contributed to Section 1, and part of Section 2 and part of Section 3. KM contributed to Sections 3.1 and 3.3. NP contributed to Section 3.4. SK contributed to Section 3.2. VB contributed to overall design and drafting.

REFERENCES

Antol, S., Agrawal, A., Lu, J., Mitchell, M., Batra, D., Zitnick, C. L., et al. (2015). *VQA: Visual Question Answering*. arXiv preprint arXiv:1505.00468.

Ba, J., and Caruana, R. (2014). "Do deep nets really need to be deep?," in *Advances in Neural Information Processing Systems 27*, eds Z. Ghahramani, M. Welling, C. Cortes, N. Lawrence, and K. Weinberger (Curran Associates, Inc.), 2654–2662.

Babenko, A., Slesarev, A., Chigorin, A., and Lempitsky, V. (2014). "Neural codes for image retrieval," in *Computer Vision – ECCV 2014* (Springer), 584–599.

Bahdanau, D., Cho, K., and Bengio, Y. (2014). *Neural Machine Translation by Jointly Learning to Align and Translate*. arXiv preprint arXiv:1409.0473.

Baker, S., and Matthews, I. (2004). Lucas-kanade 20 years on: a unifying framework. *Int. J. Comput. Vis.* 56, 221–255. doi:10.1023/B:VISI.0000011205.11775.fd

Bengio, Y. (2009). Learning deep architectures for AI. *Found. Trends Mach. Learn.* 2, 1–127. doi:10.1561/2200000006

Bengio, Y., Thibodeau-Laufer, E., Alain, G., and Yosinski, J. (2013). *Deep Generative Stochastic Networks Trainable by Backprop*. arXiv preprint arXiv:1306.1091.

Bishop, C. M. (2006). *Pattern Recognition and Machine Learning (Information Science and Statistics)*. Secaucus, NJ: Springer-Verlag New York, Inc.

Bromley, J., Bentz, J. W., Bottou, L., Guyon, I., LeCun, Y., Moore, C., et al. (1993). Signature verification using a siamese time delay neural network. *Int. J. Pattern Recogn. Artif. Intell.* 7, 669–688. doi:10.1142/S0218001493000339

Chatfield, K., Simonyan, K., Vedaldi, A., and Zisserman, A. (2014). "Return of the devil in the details: delving deep into convolutional nets," in *Proceedings of the British Machine Vision Conference* (Nottingham: BMVA Press).

Chen, L.-C., Papandreou, G., Kokkinos, I., Murphy, K., and Yuille, A. L. (2014). *Semantic Image Segmentation with Deep Convolutional Nets and Fully Connected CRFs*. arXiv preprint arXiv:1412.7062.

Cho, K., Van Merriënboer, B., Gulcehre, C., Bahdanau, D., Bougares, F., Schwenk, H., et al. (2014). *Learning Phrase Representations Using RNN Encoder-Decoder for Statistical Machine Translation*. arXiv preprint arXiv:1406.1078.

Chopra, S., Hadsell, R., and LeCun, Y. (2005). "Learning a Similarity metric discriminatively, with application to face verification," in *IEEE Computer Society Conference on Computer Vision and Pattern Recognition, 2005. CVPR 2005*, Vol. 1 (IEEE), 539–546.

Collins, M. D., and Kohli, P. (2014). *Memory Bounded Deep Convolutional Networks*. arXiv preprint arXiv:1412.1442.

Couprie, C., Farabet, C., Najman, L., and LeCun, Y. (2013). *Indoor Semantic Segmentation Using Depth Information*. CoRR, abs/1301.3572.

Dalal, N., and Triggs, B. (2005). "Histograms of oriented gradients for human detection," in *IEEE Computer Society Conference on Computer Vision and Pattern Recognition, 2005. CVPR 2005*, Vol. 1 (IEEE), 886–893.

Denil, M., Shakibi, B., Dinh, L., Ranzato, M. A., and de Freitas, N. (2013). "Predicting parameters in deep learning," in *Advances in Neural Information Processing Systems 26*, eds C. Burges, L. Bottou, M. Welling, Z. Ghahramani, and K. Weinberger (Curran Associates, Inc.), 2148–2156.

Devries, T., Biswaranjan, K., and Taylor, G. W. (2014). "Multi-task learning of facial landmarks and expression," in *Canadian Conference on Computer and Robot Vision (CRV), 2014*, 98–103.

Donahue, J., Hendricks, L. A., Guadarrama, S., Rohrbach, M., Venugopalan, S., Saenko, K., et al. (2014). *Long-Term Recurrent Convolutional Networks for Visual Recognition and Description*. arXiv preprint arXiv:1411.4389.

Donahue, J., Jia, Y., Vinyals, O., Hoffman, J., Zhang, N., Tzeng, E., et al. (2013). *Decaf: A Deep Convolutional Activation Feature for Generic Visual Recognition*. arXiv preprint arXiv:1310.1531.

Duchi, J., Hazan, E., and Singer, Y. (2011). Adaptive subgradient methods for online learning and stochastic optimization. *J. Mach. Learn. Res.* 12, 2121–2159.

Farabet, C., Couprie, C., Najman, L., and LeCun, Y. (2013a). Learning hierarchical features for scene labeling. *IEEE Trans. Pattern Anal. Mach. Intell.* 35, 1915–1929. doi:10.1109/TPAMI.2012.231

Farabet, C., Couprie, C., Najman, L., and LeCun, Y. (2013b). Learning hierarchical features for scene labeling. *IEEE Trans. Pattern Anal. Mach. Intell.* 35, 1915–1929. doi:10.1109/TPAMI.2012.231

Fukushima, K. (1980). Neocognitron: a self-organizing neural network model for a mechanism of pattern recognition unaffected by shift in position. *Biol. Cybern.* 36, 193–202. doi:10.1007/BF00344251

Gao, H., Mao, J., Zhou, J., Huang, Z., Wang, L., and Xu, W. (2015). Are you talking to a machine? Dataset and methods for multilingual image question answering. arXiv preprint arXiv:1505.05612.

Girshick, R., Donahue, J., Darrell, T., and Malik, J. (2014). "Rich feature hierarchies for accurate object detection and semantic segmentation," in *IEEE Conference on Computer Vision and Pattern Recognition (CVPR), 2014* (IEEE), 580–587.

Girshick, R. B. (2015). *Fast R-CNN*. CoRR, abs/1504.08083.

Gkioxari, G., and Malik, J. (2015). "Finding action tubes," in *CVPR*.

Gong, Y., Wang, L., Guo, R., and Lazebnik, S. (2014a). "Multi-scale orderless pooling of deep convolutional activation features," in *Computer Vision–ECCV 2014* (Springer International Publishing), 392–407.

Gong, Y., Wang, L., Guo, R., and Lazebnik, S. (2014b). "Multi-scale orderless pooling of deep convolutional activation features," in *Computer Vision–ECCV 2014* (Springer International Publishing), 392–407.

Goodfellow, I. J., Shlens, J., and Szegedy, C. (2014a). *Explaining and Harnessing Adversarial Examples.* arXiv preprint arXiv:1412.6572.

Goodfellow, I., Pouget-Abadie, J., Mirza, M., Xu, B., Warde-Farley, D., Ozair, S., et al. (2014b). "Generative adversarial nets," in *Advances in Neural Information Processing Systems 27*, eds Z. Ghahramani, M. Welling, C. Cortes, N. Lawrence, and K. Weinberger (Curran Associates, Inc.), 2672–2680.

Goodfellow, I. J., Warde-Farley, D., Mirza, M., Courville, A., and Bengio, Y. (2013). *Maxout Networks.* arXiv preprint arXiv:1302.4389.

Graves, A., Liwicki, M., Fernández, S., Bertolami, R., Bunke, H., and Schmidhuber, J. (2009). A novel connectionist system for unconstrained handwriting recognition. *IEEE Trans. Pattern Anal. Mach. Intell.* 31, 855–868. doi:10.1109/TPAMI.2008.137

Graves, A., Mohamed, A.-R., and Hinton, G. (2013). "Speech recognition with deep recurrent neural networks," in *IEEE International Conference on Acoustics, Speech and Signal Processing (ICASSP), 2013* (IEEE), 6645–6649.

Gupta, S., Girshick, R., Arbelaez, P., and Malik, J. (2014). *Learning Rich Features from RGB-D Images for Object Detection and Segmentation.*

Hariharan, B., Arbeláez, P., Girshick, R., and Malik, J. (2014). *Hypercolumns for Object Segmentation and Fine-Grained Localization.* arXiv preprint arXiv:1411.5752.

He, K., Zhang, X., Ren, S., and Sun, J. (2015). *Delving Deep into Rectifiers: Surpassing Human-Level Performance on Imagenet Classification.* arXiv preprint arXiv:1502.01852.

Hinton, G., Vinyals, O., and Dean, J. (2015). *Distilling the Knowledge in a Neural Network.* arXiv preprint arXiv:1503.02531.

Hinton, G. E. (2002). Training products of experts by minimizing contrastive divergence. *Neural Comput.* 14, 1771–1800. doi:10.1162/089976602760128018

Hinton, G. E., Osindero, S., and Teh, Y.-W. (2006). A fast learning algorithm for deep belief nets. *Neural Comput.* 18, 1527–1554. doi:10.1162/neco.2006.18.7.1527

Hinton, G. E., Srivastava, N., Krizhevsky, A., Sutskever, I., and Salakhutdinov, R. R. (2012). *Improving Neural Networks by Preventing Co-Adaptation of Feature Detectors.* arXiv preprint arXiv:1207.0580.

Hochreiter, S., and Schmidhuber, J. (1997). Long short-term memory. *Neural Comput.* 9, 1735–1780. doi:10.1162/neco.1997.9.8.1735

Hornik, K. (1991). Approximation capabilities of multilayer feedforward networks. *Neural Netw.* 4, 251–257. doi:10.1016/0893-6080(91)90009-T

Hubel, D. H., and Wiesel, T. N. (1962). Receptive fields, binocular interaction and functional architecture in the cat's visual cortex. *J. Physiol.* 160, 106. doi:10.1113/jphysiol.1962.sp006837

Jaderberg, M., Simonyan, K., Vedaldi, A., and Zisserman, A. (2014a). *Deep Structured Output Learning for Unconstrained Text Recognition.* CoRR, abs/1412.5903.

Jaderberg, M., Vedaldi, A., and Zisserman, A. (2014b). *Speeding up Convolutional Neural Networks with Low Rank Expansions.* arXiv preprint arXiv:1405.3866.

Jégou, H., Perronnin, F., Douze, M., Sánchez, J., Pérez, P., and Schmid, C. (2011). Aggregating local image descriptors into compact codes. *IEEE Trans. Pattern Anal. Mach. Intell.*

Jhuang, H., Serre, T., Wolf, L., and Poggio, T. (2007). "A biologically inspired system for action recognition," in *IEEE 11th International Conference on Computer Vision, 2007. ICCV 2007*, 1–8.

Ji, S., Xu, W., Yang, M., and Yu, K. (2010). 3D convolutional neural networks for human action recognition. *IEEE Trans. Pattern Anal. Mach. Intell.* 35, 221–231. doi:10.1109/TPAMI.2012.59

Karpathy, A., and Fei-Fei, L. (2014). *Deep Visual-Semantic Alignments for Generating Image Descriptions.* arXiv preprint arXiv:1412.2306.

Karpathy, A., Toderici, G., Shetty, S., Leung, T., Sukthankar, R., and Fei-Fei, L. (2014). "Large-scale video classification with convolutional neural networks," in *Computer Vision and Pattern Recognition (CVPR)* (Columbus, OH: IEEE), 1725–1732.

Kingma, D., and Ba, J. (2014). *Adam: A Method for Stochastic Optimization.* arXiv preprint arXiv:1412.6980.

Kingma, D. P., and Welling, M. (2013). *Auto-Encoding Variational Bayes.* arXiv preprint arXiv:1312.6114.

Krizhevsky, A., Sutskever, I., and Hinton, G. E. (2012). "Imagenet classification with deep convolutional neural networks," in *Advances in Neural Information Processing Systems 25*, eds F. Pereira, C. Burges, L. Bottou, and K. Weinberger (Curran Associates, Inc.), 1097–1105.

Kruthiventi, S. S., Ayush, K., and Babu, R. V. (2015). *Deepfix: A Fully Convolutional Neural Network for Predicting Human Eye Fixations.* arXiv preprint arXiv:1510.02927.

Kuehne, H., Jhuang, H., Garrote, E., Poggio, T., and Serre, T. (2011). "HMDB: a large video database for human motion recognition," in *IEEE International Conference on Computer Vision (ICCV), 2011* (IEEE), 2556–2563.

Kulkarni, T. D., Whitney, W., Kohli, P., and Tenenbaum, J. B. (2015). *Deep Convolutional Inverse Graphics Network.* arXiv preprint arXiv:1503.03167.

LeCun, Y., Bengio, Y., and Hinton, G. (2015). Deep learning. *Nature* 521, 436–444. doi:10.1038/nature14539

LeCun, Y., Bottou, L., Bengio, Y., and Haffner, P. (1998). Gradient-based learning applied to document recognition. *Proc. IEEE* 86, 2278–2324. doi:10.1109/5.726791

Li, S., Liu, Z.-Q., and Chan, A. (2015a). Heterogeneous multi-task learning for human pose estimation with deep convolutional neural network. *Int. J. Comput. Vis.* 113, 19–36. doi:10.1007/s11263-014-0767-8

Li, X., Zhao, L., Wei, L., Yang, M., Wu, F., Zhuang, Y., et al. (2015b). *Deepsaliency: Multi-Task Deep Neural Network Model for Salient Object Detection.* arXiv preprint arXiv:1510.05484.

Liu, N., Han, J., Zhang, D., Wen, S., and Liu, T. (2015). *Predicting Eye Fixations Using Convolutional Neural Networks.*

Long, J., Shelhamer, E., and Darrell, T. (2015). "Fully convolutional networks for semantic segmentation," in *Proceedings of the IEEE Conference on Computer Vision and Pattern Recognition*, 3431–3440.

Lowe, D. G. (2004). Distinctive image features from scale-invariant keypoints. *Int. J. Comput. Vis.* 60, 91–110. doi:10.1023/B:VISI.0000029664.99615.94

Maas, A. L., Hannun, A. Y., and Ng, A. Y. (2013). "Rectifier nonlinearities improve neural network acoustic models," in *Proc. ICML*, Vol. 30.

Malinowski, M., Rohrbach, M., and Fritz, M. (2015). *Ask Your Neurons: A Neural-Based Approach to Answering Questions about Images.* arXiv preprint arXiv:1505.01121.

Mao, J., Xu, W., Yang, Y., Wang, J., Huang, Z., and Yuille, A. (2015). *Learning like a Child: Fast Novel Visual Concept Learning from Sentence Descriptions of Images.* arXiv preprint arXiv:1504.06692.

Mao, J., Xu, W., Yang, Y., Wang, J., and Yuille, A. (2014). *Deep Captioning with Multimodal Recurrent Neural Networks (m-RNN).* arXiv preprint arXiv:1412.6632.

Mopuri, K. R., and Babu, R. V. (2015). "Object level deep feature pooling for compact image representation," in *Proceedings of the IEEE Conference on Computer Vision and Pattern Recognition*, 62–70.

Nair, V., and Hinton, G. E. (2010). "Rectified linear units improve restricted Boltzmann machines," in *Proceedings of the 27th International Conference on Machine Learning (ICML-10)*, 807–814.

Nesterov, Y. (1983). "A method of solving a convex programming problem with convergence rate o (1/k2)," in *Soviet Mathematics Doklady*, Vol. 27, 372–376.

Nguyen, A., Yosinski, J., and Clune, J. (2014). *Deep neural Networks are Easily Fooled: High Confidence Predictions for Unrecognizable Images.* arXiv preprint arXiv:1412.1897.

Pan, Y., Mei, T., Yao, T., Li, H., and Rui, Y. (2015). *Jointly Modeling Embedding and Translation to Bridge Video and Language.* arXiv preprint arXiv:1505.01861.

Polyak, B. T. (1964). Some methods of speeding up the convergence of iteration methods. *USSR Comput. Math. Math. Phys.* 4, 1–17. doi:10.1016/0041-5553(64)90137-5

Razavian, A. S., Azizpour, H., Sullivan, J., and Carlsson, S. (2014). "CNN features off-the-shelf: an astounding baseline for recognition," in *IEEE Conference on Computer Vision and Pattern Recognition Workshops (CVPRW), 2014* (IEEE), 512–519.

Romero, A., Ballas, N., Kahou, S. E., Chassang, A., Gatta, C., and Bengio, Y. (2014). *Fitnets: Hints for Thin Deep Nets.* arXiv preprint arXiv:1412.6550.

Rowley, H. A., Baluja, S., and Kanade, T. (1998). Neural network-based face detection. *IEEE Trans. Pattern Anal. Mach. Intell.* 20, 23–38. doi:10.1109/34.655647

Rumelhart, D. E., Hinton, G. E., and Williams, R. J. (1988). Learning representations by back-propagating errors. *Cogn. Model.* 5, 3.

Russakovsky, O., Deng, J., Su, H., Krause, J., Satheesh, S., Ma, S., et al. (2015). Imagenet large scale visual recognition challenge. *Int. J. Comput. Vis.* 115, 211–252. doi:10.1007/s11263-015-0816-y

Schmidhuber, J. (2015). Deep learning in neural networks: an overview. *Neural Netw.* 61, 85–117. doi:10.1016/j.neunet.2014.09.003

Sermanet, P., Eigen, D., Zhang, X., Mathieu, M., Fergus, R., and LeCun, Y. (2013a). *Overfeat: Integrated Recognition, Localization and Detection using Convolutional Networks.* arXiv preprint arXiv:1312.6229.

Sermanet, P., Eigen, D., Zhang, X., Mathieu, M., Fergus, R., and LeCun, Y. (2013b). *Overfeat: Integrated Recognition, Localization and Detection using Convolutional Networks.* CoRR, abs/1312.6229.

Sermanet, P., Kavukcuoglu, K., Chintala, S., and LeCun, Y. (2013c). "Pedestrian detection with unsupervised multi-stage feature learning," in *IEEE Conference on Computer Vision and Pattern Recognition (CVPR), 2013* (IEEE), 3626–3633.

Simonyan, K., and Zisserman, A. (2014a). "Two-stream convolutional networks for action recognition in videos," in *Advances in Neural Information Processing Systems 27*, eds Z. Ghahramani, M. Welling, C. Cortes, N. Lawrence, and K. Weinberger 568–576.

Simonyan, K., and Zisserman, A. (2014b). *Very Deep Convolutional Networks for Large-Scale Image Recognition.* arXiv preprint arXiv:1409.1556.

Snoek, J., Larochelle, H., and Adams, R. P. (2012). "Practical bayesian optimization of machine learning algorithms," in *Advances in Neural Information Processing Systems,* 25, eds F. Pereira, C. Burges, L. Bottou, and K. Weinberger (Curran Associates, Inc.), 2951–2959.

Socher, R., Huval, B., Bath, B., Manning, C. D., and Ng, A. Y. (2012). "Convolutional-recursive deep learning for 3d object classification," in *Advances in Neural Information Processing Systems 25*, eds F. Pereira, C. Burges, L. Bottou, and K. Weinberger, 656–664.

Srinivas, S., and Babu, R. V. (2015). "Data-free parameter pruning for deep neural networks," in *Proceedings of the British Machine Vision Conference (BMVC)*, eds M. W. J. Xianghua Xie and G. K. L. Tam (BMVA Press), 31.1–31.12.

Sutskever, I., Martens, J., Dahl, G., and Hinton, G. (2013). "On the importance of initialization and momentum in deep learning," in *Proceedings of the 30th International Conference on Machine Learning (ICML-13).* 1139–1147.

Szegedy, C., Liu, W., Jia, Y., Sermanet, P., Reed, S., Anguelov, D., et al. (2014). *Going Deeper with Convolutions.* arXiv preprint arXiv:1409.4842.

Szegedy, C., Zaremba, W., Sutskever, I., Bruna, J., Erhan, D., Goodfellow, I., et al. (2013). *Intriguing Properties of Neural Networks.* arXiv preprint arXiv:1312.6199.

Taigman, Y., Yang, M., Ranzato, M., and Wolf, L. (2014). "Deepface: closing the gap to human-level performance in face verification," in *IEEE Conference on Computer Vision and Pattern Recognition (CVPR), 2014* (IEEE), 1701–1708.

Uijlings, J., van de Sande, K., Gevers, T., and Smeulders, A. (2013). Selective search for object recognition. *Int. J. of Comput. Vision* doi:10.1007/s11263-013-0620-5

Vaillant, R., Monrocq, C., and Le Cun, Y. (1994). Original approach for the localisation of objects in images. *IEE Proc. Vision Image Signal Process.* 141, 245–250. doi:10.1049/ip-vis:19941301

Venugopalan, S., Rohrbach, M., Donahue, J., Mooney, R., Darrell, T., and Saenko, K. (2015). *Sequence to Sequence – Video to Text.* arXiv preprint arXiv:1505.00487.

Venugopalan, S., Xu, H., Donahue, J., Rohrbach, M., Mooney, R., and Saenko, K. (2014). *Translating Videos to Natural Language Using Deep Recurrent Neural Networks.* arXiv preprint arXiv:1412.4729.

Vig, E., Dorr, M., and Cox, D. (2014). "Large-scale optimization of hierarchical features for saliency prediction in natural images," in *IEEE Conference on Computer Vision and Pattern Recognition (CVPR), 2014* (IEEE), 2798–2805.

Vincent, P., Larochelle, H., Lajoie, I., Bengio, Y., and Manzagol, P.-A. (2010). Stacked denoising autoencoders: learning useful representations in a deep network with a local denoising criterion. *J. Mach. Learn. Res.* 11, 3371–3408.

Vinyals, O., Toshev, A., Bengio, S., and Erhan, D. (2014). *Show and Tell: A Neural Image Caption Generator.* arXiv preprint arXiv:1411.4555.

Wan, L., Zeiler, M., Zhang, S., Cun, Y. L., and Fergus, R. (2013). "Regularization of neural networks using dropconnect," in *Proceedings of the 30th International Conference on Machine Learning (ICML-13),* 1058–1066.

Wang, A., Lu, J., Wang, G., Cai, J., and Cham, T.-J. (2014a). "Multi-modal unsupervised feature learning for Rgb-D scene labeling," in *Computer Vision – ECCV 2014*, volume 8693 of *Lecture Notes in Computer Science*, eds D. Fleet, T. Pajdla, B. Schiele, and T. Tuytelaars (Springer), 453–467.

Wang, J., Song, Y., Leung, T., Rosenberg, C., Wang, J., Philbin, J., et al. (2014b). "Learning fine-grained image similarity with deep ranking," in *IEEE Conference on Computer Vision and Pattern Recognition (CVPR), 2014* (IEEE), 1386–1393.

Wang, H., and Schmid, C. (2013). Action recognition with improved trajectories. In *IEEE International Conference on Computer Vision*, Sydney.

Wang, L., Qiao, Y., and Tang, X. (2015a). "Action recognition with trajectory-pooled deep-convolutional descriptors," in *CVPR*, 4305–4314.

Wang, X., Zhang, L., Lin, L., Liang, Z., and Zuo, W. (2015b). Deep joint task learning for generic object extraction. CoRR, abs/1502.00743.

Wang, S., and Manning, C. (2013). "Fast dropout training," in *Proceedings of the 30th International Conference on Machine Learning (ICML-13)*, 118–126.

Wei, Y., Xia, W., Huang, J., Ni, B., Dong, J., Zhao, Y., et al. (2014). *CNN: Single-Label to Multi-Label.* CoRR, abs/1406.5726.

Weinzaepfel, P., Revaud, J., Harchaoui, Z., and Schmid, C. (2013). "Deepflow: Large displacement optical flow with deep matching," in *IEEE International Conference on Computer Vision (ICCV), 2013* (IEEE), 1385–1392.

Yang, J., Jiang, Y.-G., Hauptmann, A. G., and Ngo, C.-W. (2007). "Evaluating bag-of-visual-words representations in scene classification," in *Proceedings of the International Workshop on Multimedia Information Retrieval* (ACM), 197–206.

Yosinski, J., Clune, J., Bengio, Y., and Lipson, H. (2014). "How transferable are features in deep neural networks?," in *Advances in Neural Information Processing Systems*, 27, eds Z. Ghahramani, M. Welling, C. Cortes, N. Lawrence, and K. Weinberger (Curran Associates, Inc.), 3320–3328.

Zagoruyko, S., and Komodakis, N. (2015). *Learning to Compare Image Patches via Convolutional Neural Networks.* arXiv preprint arXiv:1504.03641.

Zeiler, M., and Fergus, R. (2014). "Visualizing and understanding convolutional networks," in *Computer Vision – ECCV 2014*, volume 8689 of *Lecture Notes in Computer Science*, eds D. Fleet, T. Pajdla, B. Schiele, and T. Tuytelaars (Springer International Publishing), 818–833. doi:10.1007/978-3-319-10590-1_53

Zeiler, M. D. (2012). *Adadelta: An Adaptive Learning Rate Method.* arXiv preprint arXiv:1212.5701.

Zhang, Z., Luo, P., Loy, C. C., and Tang, X. (2014). *Learning and Transferring Multi-Task Deep Representation for Face Alignment.* CoRR, abs/1408.3967.

Zheng, S., Jayasumana, S., Romera-Paredes, B., Vineet, V., Su, Z., Du, D., et al. (2015). *Conditional Random Fields as Recurrent Neural Networks.* arXiv preprint arXiv:1502.03240.

Zhou, B., Lapedriza, A., Xiao, J., Torralba, A., and Oliva, A. (2014). "Learning deep features for scene recognition using places database," in *Advances in Neural Information Processing Systems 27*, eds Z. Ghahramani, M. Welling, C. Cortes, N. Lawrence, and K. Weinberger (Curran Associates, Inc.), 487–495.

Conflict of Interest Statement: The authors declare that the research was conducted in the absence of any commercial or financial relationships that could be construed as a potential conflict of interest.

15

Soft Pneumatic Actuator Skin with Piezoelectric Sensors for Vibrotactile Feedback

Harshal Arun Sonar[†] and Jamie Paik[*][†]

Reconfigurable Robotics Laboratory, Institute of Mechanical Engineering, Ecole Polytechnique Fédérale de Lausanne, Lausanne, Switzerland

The latest wearable technologies demand more intuitive and sophisticated interfaces for communication, sensing, and feedback closer to the body. Evidently, such interfaces require flexibility and conformity without losing their functionality even on rigid surfaces. Although there have been various research efforts in creating tactile feedback to improve various haptic interfaces and master–slave manipulators, we are yet to see a comprehensive device that can both supply vibratory actuation and tactile sensing. This paper describes a soft pneumatic actuator (SPA)-based skin prototype that allows bidirectional tactile information transfer to facilitate simpler and responsive wearable interface. We describe the design and fabrication of a 1.4 mm-thick vibratory SPA – skin that is integrated with piezoelectric sensors. We examine in detail the mechanical performance compared to the SPA model and the sensitivity of the sensors for the application in vibrotactile feedback. Experimental findings show that this ultra-thin SPA and the unique integration process of the discrete lead zirconate titanate (PZT)-based piezoelectric sensors achieve high resolution of soft contact sensing as well as accurate control on vibrotactile feedback by closing the control loop.

Keywords: wearable technology, soft skin, smart actuators, soft sensor, soft actuator, vibrotactile feedback

Edited by:
Carlo Menon,
Simon Fraser University, Canada

Reviewed by:
Gursel Alici,
University of Wollongong, Australia
Robert Shepherd,
Cornell University, USA

*Correspondence:
Jamie Paik
jamie.paik@epfl.ch

[†]Harshal Arun Sonar and Jamie Paik
have contributed equally to this work.

1. INTRODUCTION

Over the past decade, research on the use of robotic haptic devices in neuro rehabilitation has accelerated (Maciejasz et al., 2014). The use of haptic feedback has proven effective in aided rehabilitation in subjects following stroke or paralysis (Viau et al., 2004; Takahashi et al., 2008; Alahakone and Senanayake, 2009; Kapur et al., 2009; Wall, 2010; Carmeli et al., 2011). Haptic feedback is generally divided into two classes – namely tactile and kinesthetic. The devices required for providing kinesthetic feedback are comparatively large and heavy (Hirose et al., 2001; Shahoian et al., 2004; Viau et al., 2004; Takahashi et al., 2008) and thus less suitable for a wearable scenario. In wearable devices, vibrotactile feedback is considered as one of the safest and most popular ways to interact with the human body (Lindeman et al., 2006; Choi and Kuchenbecker, 2013).

The human tactile sensory response is limited in frequency range, and both temporal and spatial resolution making the determination of the vibrotactile actuator's specifications difficult. In general, tactile sensation is perceived through four different types of mechanoreceptors inside human skin. The mechanoreceptors responsible for vibrotactile sensation are the rapidly adapting (RA) and Pacinian corpuscle (PC) receptors with perceptible frequencies ranging from 3 to 100 and 100 to 400 Hz, respectively (Choi and Kuchenbecker, 2013). Therefore, for effective tactile sensing, the operational frequency range is selected between one of the two along with a minimum spatial resolution of 1–2 mm for human fingers (Kaczmarek et al., 1991; Asamura et al., 2001). The majority of vibrotactile systems developed so far have used electromagnetic eccentric motors (Shahoian et al., 2004), electrostatic piezo actuators, or electro-active polymer-based actuators (Koo et al., 2008). These solutions have certain limitations for wearable applications due to bulkiness, rigidity, and complexity. Linear resonant actuators have improved efficiency and a compact size compared to electromagnetic eccentric motors (Mortimer et al., 2007) but still suffer from problems of rigidity, complexity, and narrow bandwidth. An ideal device for wearable tactile interfaces should be light weight, compliant, safe, and incorporate multiple sensing and actuation points over a distributed surface. Koo et al. (2008) and Frediani et al. (2014) demonstrated novel ways to use electro-active polymers (EAPs)-based soft actuators in wearable tactile applications to solve most of the problems associated with conventional vibrotactile devices. Rosset et al. (2013) used the capacitance change for sensing and closed loop control of actuation using EAPs, which can be adopted for tactile sensing. The actuators require special safety considerations to avoid direct high voltage contact with human skin.

We have developed soft pneumatic actuator (SPA) skin to tackle the problem of actuation requirements through its thin, lightweight, easily customizable, and compliant design (Suh et al., 2014). The design uses soft silicone-based 2D monolithic manufacturing to achieve pneumatic vibrotactile actuation with the desired properties. The addition of an extra sensing layer over the SPA skin enables a closed-loop control of the actuation amplitude and also dampens manufacturing defects. This additional sensing layer requires the sensing elements to be ultra-thin, flexible, customizable, and to be distributed over the surface. Recent developments in wearable sensor technology have made it possible to embed piezoelectric element-based sensors (Acer et al., 2015), conductive fabrics, electro-active polymers (Rosset et al., 2013; Maiolino et al., 2015), and other families of stretchable and flexible sensors (Xu et al., 2014; Gerratt et al., 2015) into soft silicone. The sensor selection has a trade-off between cost, design customization, stretchability, sensitivity and non-linearity. We selected piezo ceramic (PZT) sensors due to their high sensitivity, low cost and customizability while losing on stretchability properties for integration into the SPA skin. Multiple PZT sensors can be distributed over the actuation surface to provide multiple points for sensing. To demonstrate the benefit of the SPA skin design we developed a circular shaped multi-actuator SPA skin for application on human finger tips. The similar construction could be adopted for other vibrotactile feedback applications. The main contributions of the presented SPA skin are as follows:

- Design and development of a novel soft, flexible, ultra-thin SPA skin with integrated PZT sensors. We demonstrated its applicability as a wearable system with a single input–output device generating vibrotactile feedback for a variety of application frequencies.
- Prototyping of the highly customizable physical interface and the closed-loop control for various input signals.
- Evaluation of new signal decoupling algorithm that optimized input control signal for the desired actuation amplitude and detected external interactions. The algorithm involves digital

filtering, vibratory peak detection, and separation of external interaction signals from SPA generated signals.

The paper is organized into five sections including the introduction. Section 2 provides description of the design and fabrication of the various components of the SPA skin and PZT sensors. Section 3 considers the embedded sensing and actuation mechanism followed by the feedback control of the SPA skin. The results obtained from the feedback controller and external force interactions are documented in Section 4. The last section presents the summary and conclusions.

2. SPA SKIN DESIGN

The SPA skin is a thin wearable device that contains sensor-embedded pneumatic actuators with multiple actuation points for vibrotactile feedback, as shown in **Figure 1**. Since it is made of silicone, the SPA skin's overall material property is dictated by the pliant silicone. It is lightweight, inherently compliant, and highly customizable, therefore, could cover wide and curvy surfaces (**Figure 1C**). These characteristics make the SPA skin ideal as a wearable device covering various parts of the human body. Further, we used Dragon Skin 30® silicone to make SPA skin for a medically safe (Smooth_On_Inc, 2015) and compliant interface with human skin. In this section, we explain the design and fabrication process of SPA skin (actuation layer), PZT sensing layer, and their integration.

2.1. Actuator Design

The actuation layer of SPA skin consists of two silicone layers and a masking layer sandwiched in between. The first three layers in **Figure 1A** constitutes for the SPA layer. The masking layer fabricated using a polypropylene adhesive tape, avoids bonding between these two soft silicone layers which after curing facilitates the passage of air through the masked layer creating desired shape inflation. The shape and design of actuator is determined by laser cutting of the masking tape according to the application. The requirements for the actuator diameter and distributed actuator density are based on the feedback application location on human body (Kaczmarek et al., 1991). From these requirements, we can then decide the average input pressure and the desired output blocked force needed for manipulation from a single actuator. Our prototype demonstrates distributed sensing and vibrotactile feedback capabilities of SPA skin designed for human fingers. We obtained a sufficient blocked force of 0.3 N with a 3–4 mm diameter bubble shaped actuator with 0.4 mm silicone thickness for this application. The distributed actuator single input channel SPA skin is capable of generating tactile feedback for a variety of actuation frequencies ranging from 5 to 100 Hz.

2.2. Sensor Design

Sensors for the SPA skin need to be distributable over an area, thin and flexible to be wearable. For our application, we focused on achieving a high-level of sensitivity as well. Soft material matrix-based sensors are often considered for wearable devices but display high drift and slow response that are not suitable for

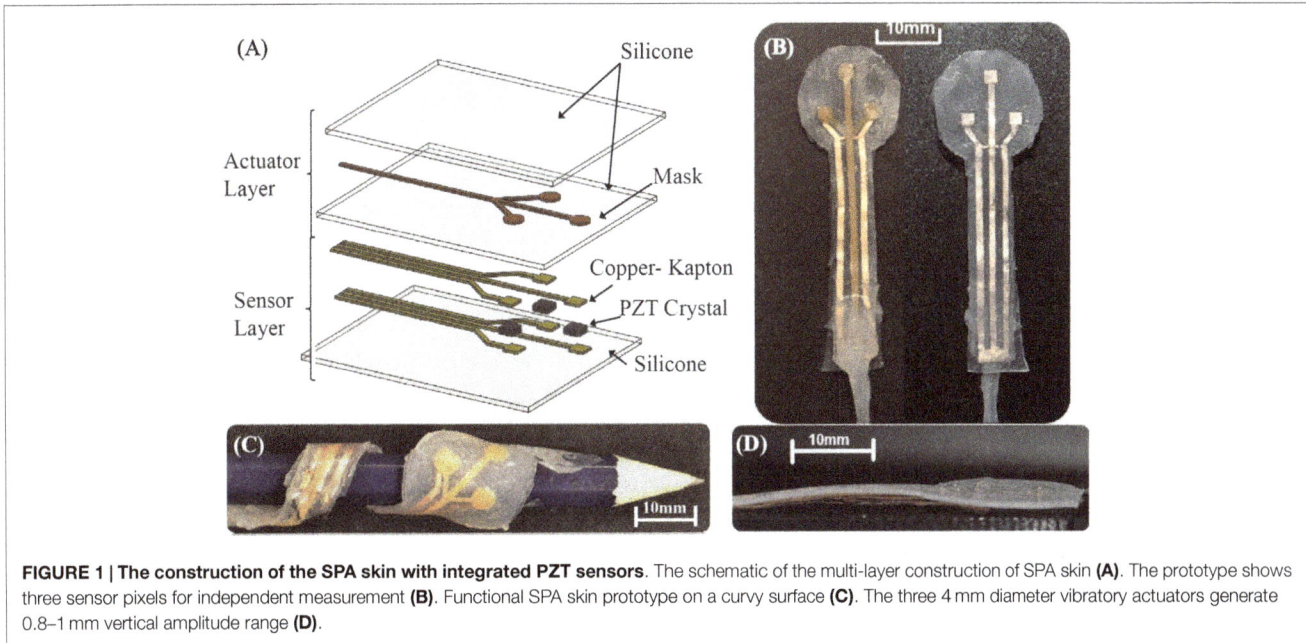

FIGURE 1 | The construction of the SPA skin with integrated PZT sensors. The schematic of the multi-layer construction of SPA skin **(A)**. The prototype shows three sensor pixels for independent measurement **(B)**. Functional SPA skin prototype on a curvy surface **(C)**. The three 4 mm diameter vibratory actuators generate 0.8–1 mm vertical amplitude range **(D)**.

our goal. Piezoelectric ceramics (PZT) are known to be highly sensitive to normal forces applied and can be embedded into silicone substrates (Acer et al., 2015). A PZT element can measure dynamic forces by converting the applied mechanical stress into electrical voltage. Our sensors have multiple pixels of PZT elements discretely distributed over a surface area in the form of grid. The PZT elements are connected using flexible circuit tracks manufactured by laser cutting the copper-plated kapton (polyamide) material. The construction is then embedded inside silicone material for additional support and electrical insulation as shown in **Figure 1A**.

The sensors for SPA skin use 2 mm × 2 mm size PZT element sandwiched between flexible copper–kapton circuit track. Initially the copper–kapton sheet is finely engraved with a low power laser to remove the kapton layer and make the copper layer visible for connection with PZT element as shown in **Figure 2A**. After the engraving process the copper–kapton sheet is cut into the specific shape required to form electrodes. A two-way Z-direction conductive adhesive tape (3M™-9705) is used to bond individual PZT element with the copper–kapton electrodes as shown in **Figure 2B**. This sensor manufacturing procedure enables both signal conducting electrodes to be placed on top of the other without causing a short circuit. The Kapton layer between the electrodes acts as an electrical insulator as depicted in **Figure 2D**. This helps to reduce the area required for electronic tracks and also improves the sensor density for distributed sensing. For our prototype design, we manufactured three sensors, evenly distributed on a 6-mm circular periphery to sense both the internal and external interactions with the SPA skin, as shown in **Figure 2C**.

2.3. Sensor Actuator Integration

The key challenge for multi-layer-based sensors is their integration in parallel. The integration process is simplified by independently manufacturing the sensors and SPA skin. Additionally, this helps to reduce the failure modes as both

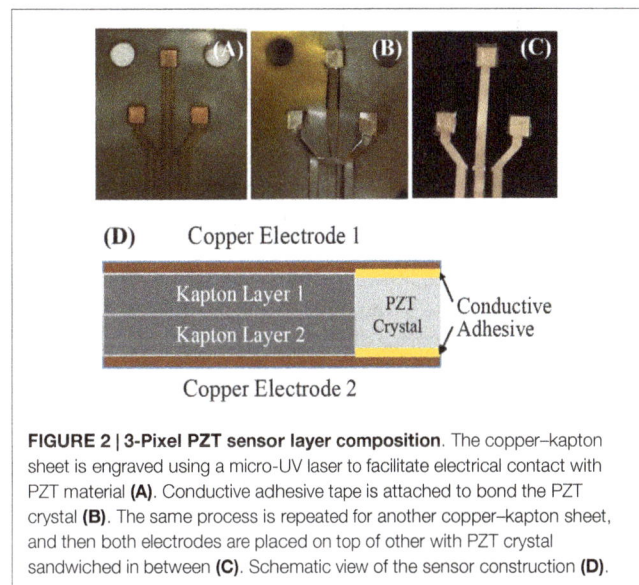

FIGURE 2 | 3-Pixel PZT sensor layer composition. The copper–kapton sheet is engraved using a micro-UV laser to facilitate electrical contact with PZT material **(A)**. Conductive adhesive tape is attached to bond the PZT crystal **(B)**. The same process is repeated for another copper–kapton sheet, and then both electrodes are placed on top of other with PZT crystal sandwiched in between **(C)**. Schematic view of the sensor construction **(D)**.

sensing and actuation components can be tested before final integration. The layerwise construction of SPA skin and the integrated PZT sensors is depicted in **Figure 1A**. The sensitivity of vibration feedback is maximized by placing the sensing elements exactly below the bubble-shaped actuator. The flexible circuit tracks are then designed and laser cut, based on the distributed configuration of the sensor elements. The integration process of the PZT sensors into SPA skin starts by spin coating a thin layer of uncured Dragon Skin 30® onto a cured 400 μm thick layer for bonding purpose. A properly connected working sensor is then placed on this uncured silicone. The SPA skin prototype is then aligned with the sensors and placed on the uncured layer. A small weight is placed on this setup to avoid air bubbles forming inside the bonding layer and the assembly is cured at

TABLE 1 | Geometric and functional specifications of the SPA skin prototype.

SPA skin parameters	Value
Single actuator diameter (mm)	4
Single sensor size (mm)	2×2
Overall thickness (mm)	1.4–1.5
Modulation height (maximum) (mm)	1
PZT sensitivity (Acer et al., 2015)	0.25 V/N
Maximum input pressure (kPa)	90
Actuator bandwidth (Hz)	55

60°C for 1 h to ensure seamless bonding. **Figure 1B** shows the final SPA skin prototype with integrated sensors consisting three sensors and three actuators distributed over the soft surface. Using this design procedure, we obtained following parameters for sensor-integrated SPA skin, as shown in **Table 1**.

3. SPA SKIN: EMBEDDED SENSING AND CONTROL

The sensorized, tactile SPA skin prototype focuses on two main functionalities; it not only detects vibro-tactile feedback amplitude but also measures external interactive forces. Even with high sensitivity PZT sensors, it is challenging to accurately estimate the dynamical forces on the system as the real world interaction forces the human body experiences are small (in the order of 0.1–3 N) (Kaczmarek et al., 1991). Also, the wearable application requires raw sensor signal to be carried along relatively long wires from application area to the electronics unit, which adds extra capacitance and external signal noise. These issues in combination lower the signal to noise ratio, which requires external amplification and active filtering of noise to improve PZT signal quality.

In this section, we discuss the embedded sensing mechanism for SPA skin to obtain the operational specifications for the embedded PZT elements both mathematically and experimentally. These specifications are then used to design the electronics for the distributed sensing system. This is followed by the design of the control system for augmenting the SPA actuation amplitude with sensor data.

3.1. Embedded Sensing Mechanism

A PZT sensor does not provide a static signal as the piezoelectric effect only occurs when external forces cause a change in the PZT crystal's physical dimensions. Due to their high sensitivity, PZT sensors are suitable for measuring dynamically changing signals such as vibration or impact force. The sensors produce electrical field, E, proportional to the stress, σ, generated by an external force, F_{piezo}, specifically in the normal direction. The transfer function for the generated electrical voltage change can be obtained using the direct piezo electric effect given as $E = g_{33} \times \sigma$ where, g_{33} is the piezoelectric voltage coefficient in the normal direction. The relationship in terms of generated open-circuit voltage, V, becomes $V = \left(\frac{g_{33} \times t}{A}\right) \times F_{piezo}$ where t is thickness of PZT and A is the PZT area under stress. For the PZT material (PSI-5H4E), we obtained $V = 0.6 \times F_{piezo}$ with $g_{33} = 19.0 \times 10^{-3}$ V m/N, $t = 0.127$ mm, and $A = 2 \times 2$ mm^2.

Using the equations (124) and (125) from IEEE (1988) for quasi-static applications, the unloaded or free piezo sensitivity can be mapped to a clamped piezo sensitivity value through the electro-mechanical coupling factor, k_{33}, as shown in equations (1) and (2).

$$\epsilon_{r_{free}} = (1 - k_{33}^2)\epsilon_{r_{clamped}} \tag{1}$$

This translates to,

$$V_{clamped} = (1 - k_{33}^2) \times V_{free} = (1 - 0.75^2) \times 0.6 \\ \times F_{piezo} = 0.2625 \ V/N \times F_{piezo} \tag{2}$$

where, ϵ_r is relative permittivity of PZT crystal in normal direction.

Although significantly better than silicone-based sensors, PZTs too suffer from non-linearities in the form of hysteresis and creep as piezo electric constant (d_{33}) and dielectric constant (ϵ_r) changes on application of stress (Damjanovic, 1997; Hall, 2001). In our sensor application, we obtain the maximum stress of 0.75 MPa for 3 N load, which generates the maximum electric field of 0.0145 kV/mm. These values are much lower than the nominal stress (5 MPa) or electric field (0.5 kV/mm) to observe significant hysteresis as presented in (Damjanovic, 1997; Hall, 2001) for PZT-5H (soft PZT). Also, the dielectric constant and piezo electric constants do not change for applied stress of <1 MPa (Zhao and Zhang, 1996). Furthermore, the material dynamic equation for the silicone-embedded PZT sensor is complex to model due to the hyper elastic properties of silicone, which are more dominant than PZT non-linearities. The transfer function is also modulated by the thickness of the silicone layer (Acer et al., 2015).

We experimentally obtained sensitivity value of 0.24–0.26 V/N for a silicone thickness of 0.4 mm from the graph shown in **Figure 3A**. The sensor response is linearly dependent upon the impact force. These results are comparable with the results in equation (2) and results obtained in Acer et al. (2015) for the characterization of silicone-embedded PZT sensors.

3.1.1. Integrated system response

The system response of the 4-mm diameter actuation chamber is measured by applying a 10-Hz on–off input signal to the actuator placed under a *Nano*-17 force sensor and a flat plate. The distance between force sensor and the flat plate is same as the actuator thickness (1.4 mm). **Figure 4** shows the actual blocked force recorded for actuator inflation and deflation. The rise time for the actuator's dynamic response is obtained by first order dynamic system model approximation fitted using the input–output data (**Figure 4A**). The presented model has the first-order transfer function $356/(j\omega + 356)$, which subsequently yields time constant of 28 ms and actuator bandwidth of $356/(2\pi) = 56$ Hz, as plotted in **Figure 4B**. When a similar system response is recorded using the PZT sensor placed just below the actuator, we obtain a bandwidth of 40 Hz (**Figure 4B**) instead of the modeled bandwidth of 56 Hz. This discrepancy is due to the low pass filtering effect introduced by the charge amplification stage in the sensor readings. This amplifier has a cut-off frequency of 86 Hz as discussed in the following section. This limits the overall cut-off frequency to 44 Hz as shown by brown dotted line in **Figure 4B**, which is very close to the actual cut-off frequency observed using PZT sensors.

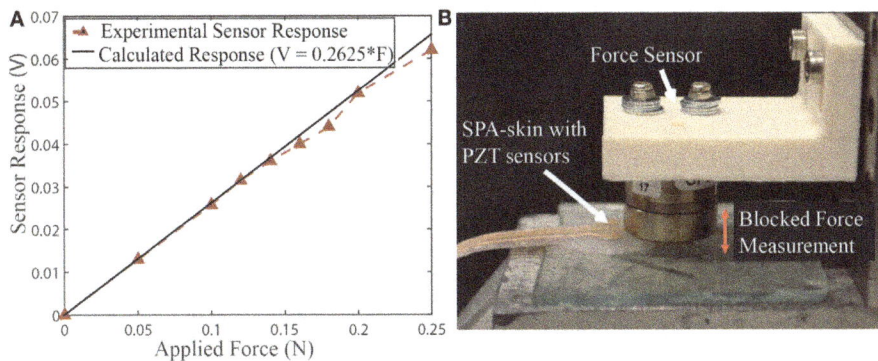

FIGURE 3 | Silicone-embedded PZT sensor characteristics. Silicone-integrated PZT sensor response **(A)** obtained for a 10 Hz stepped force input of different amplitude. Test setup with *Nano*-17 force sensor used for the PZT sensor and soft actuator characterization **(B)**. The *Nano*-17 sensor position can be adjusted precisely in the z-direction to accommodate the exact thickness of SPA skin.

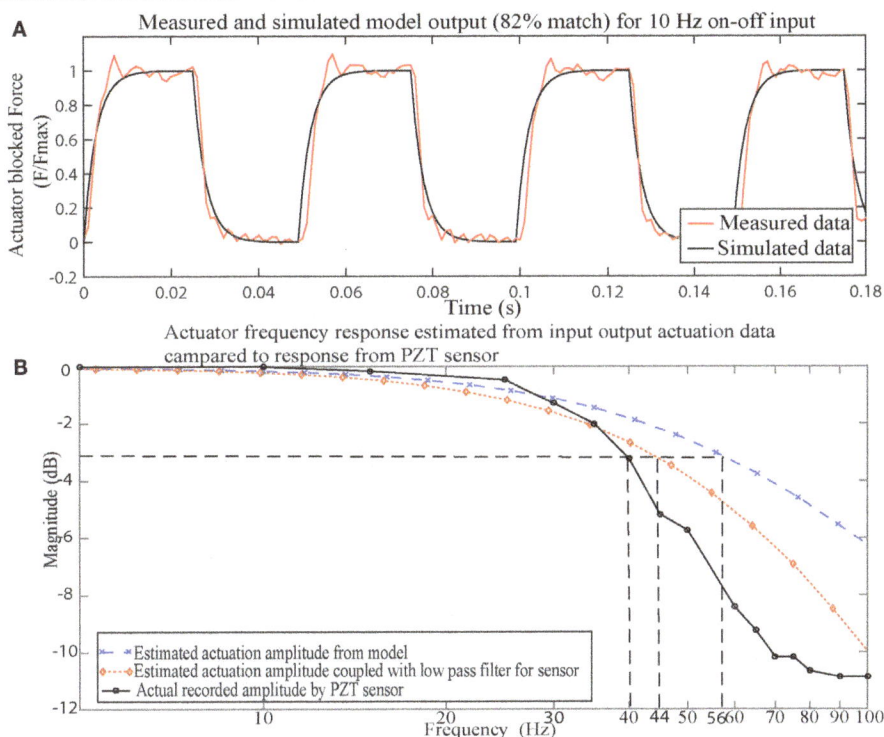

FIGURE 4 | SPA skin system vibration-control response. SPA skin actuated at 10 Hz to experimentally obtain the rise-time and bandwidth of the actuator. First order dynamical system model's output compared to experimentally recorded blocked force output at 10 Hz **(A)**. The rise time of 0.028 ms is obtained with 82% model fit. The corresponding bandwidth of 56 Hz for first order transfer function compared with actual recorded bandwidth of 40 Hz using PZT sensors **(B)**. Low pass filtering at 86 Hz in PZT signal amplification coupled with estimated actuator bandwidth lowers the measured bandwidth to 44 Hz from 56 Hz.

3.2. Distributed Sensing

Distributed sensing is a critical aspect in the design of any wearable platform. A dense grid of distributed-sensing elements makes wearable systems interactive and adaptive toward changes in the physical environment. Furthermore, the sensing layer faces critical challenges including the signal quality in noisy interactions, sensor density over a specific area, bio-compatibility, flexibility, and weight constraints. The PZT sensors are specifically developed for distributed sensing. The novelty in our design includes customizability, high density, parallel information gathering and

the possibility of active feedback control. The system design is optimized for distributed sensing by moving most of the signal processing inside a compact digital platform. The block diagram in **Figure 5** represents the system design for signal acquisition, processing, evaluation, and finally active control of the desired vibrotactile feedback. The PZT signal is processed accordingly to obtain a reliable peak value for the periodic force exerted by the soft actuator. A peak detection algorithm and notch filter are developed for decoupling the line noise, external interaction signal, and SPA-generated interaction signal for

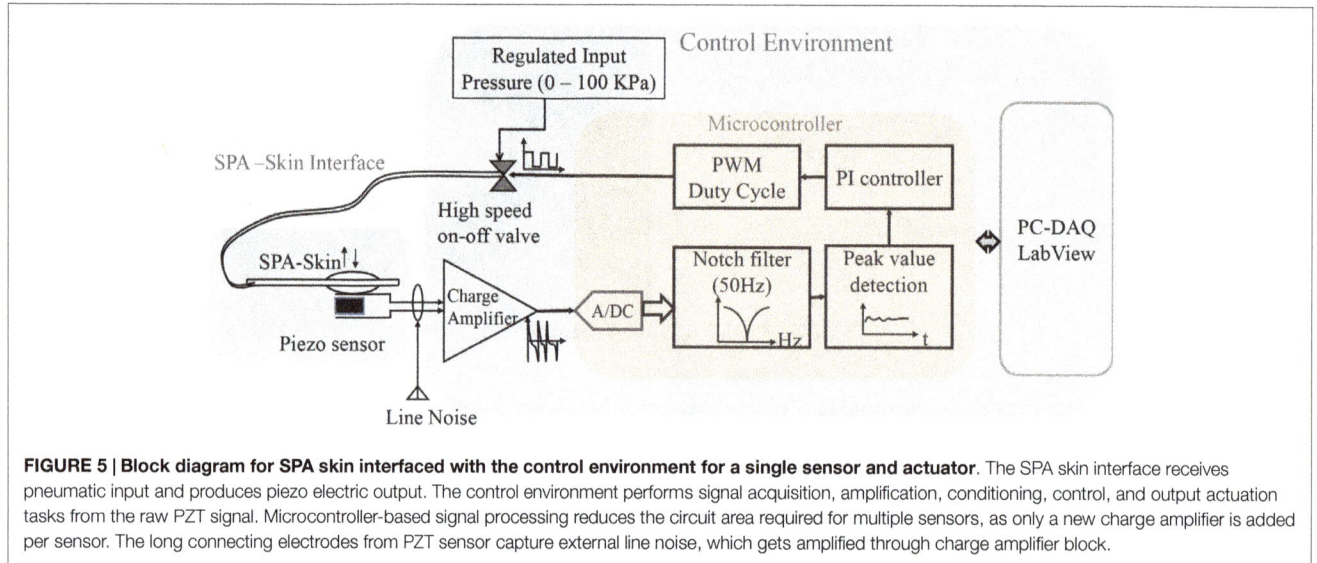

FIGURE 5 | Block diagram for SPA skin interfaced with the control environment for a single sensor and actuator. The SPA skin interface receives pneumatic input and produces piezo electric output. The control environment performs signal acquisition, amplification, conditioning, control, and output actuation tasks from the raw PZT signal. Microcontroller-based signal processing reduces the circuit area required for multiple sensors, as only a new charge amplifier is added per sensor. The long connecting electrodes from PZT sensor capture external line noise, which gets amplified through charge amplifier block.

improving the signal quality to facilitate feedback control as follows.

3.2.1. Input Force Sensing by a Analog Charge Amplification and Filtering

The SPA skin is ultimately aimed for measuring and analyzing the various interaction forces with the human skin in everyday life. The distributed PZT sensors are sensitive to dynamic forces producing electric charge spikes proportional to the impact force. The actual interaction forces exerted on PZT crystals are proportional to the peak value of the electrical voltage generated. Therefore, for the detection of the impact force, the microcontroller tracks the maximum value of the PZT signal over a complete on-period for the given PWM frequency, f. This peak value is then updated for the next timing cycle, and thus, the system has a delay of $1/f$. Also, the forces exerted in this case are expected to be <3 N/finger (Lambercy et al., 2007). The maximum force exerted by the SPA skin upon actuation is around 0.3 N at actuation pressure of 70 kPa. This produces a raw signal peak of 75 mV with the given sensitivity of 0.25 V/N. The signal has low amplitude and is susceptible to electrical loading if driven directly through low impedance circuit. Therefore, a charge amplifier was designed to act both as a buffer circuit and an amplification stage. The PZT signal is amplified and low pass filtered by the charge amplifier. The amplification factor of 3.9 allows a maximum external force detection up to 10 N with a 2.5 V amplifier output swing and a cut off frequency of 86 Hz allows filtering of high frequency noise harmonics. The conditioned signal is then converted into 10-bit digital signal sampled at 1000 Hz for further analysis and recording as shown in **Figure 5**. Every PZT sensor requires an analog signal amplification before it can be connected to a low impedance A/D converter pin.

3.2.2. Distributed Sensor Noise Cancelation

A high density of distributed sensing elements comes with an added cost of increased electrical track and wire lengths. These long wires act like an antenna and capture the noise from a variety of AC sources in their environment. Low amplitude PZT signal and noise signal are amplified to the same extent through charge amplifier; to mitigate this effect, we used a second order digital notch filter. **Figures 6A,B** show the components of 50 Hz noise compared to the PZT sensor generated values at 20 and 70 Hz respectively. A separate second-order digital Butterworth filter is implemented for every sensor to remove the 50-Hz signal component. The dashed lines in **Figures 6A,B** show the signal spectrum after removal of the AC noise and display clear acquisition of the repeated actuator signal. A digital filter is preferred an analog filter as it reduces the size of the physical system considering one physical filter would be required for each sensor. Furthermore, the digital filter is equally reliable due to a relatively high sampling frequency of 1 kHz.

3.3. SPA Skin Control

The SPA skin consists of an array of actuators, which produce the desired inflation for a specific pneumatic input pressure. Any hardware actuator, including DC motors, pneumatic regulators, and pressure valves, requires a feedback mechanism to accurately control the actuation. However, we rarely observe this in vibratory actuators (Precision_Microdrives_Ltd, 2015). When we require closed-loop control, it is not feasible to rely on human sensory perception for actuator tuning or comparison as sensory perception varies drastically from one person to another. The integration of sensing elements in SPA skin provides the necessary feedback to allow for closed-loop control of the vibration amplitude. In this section, we evaluate the soft actuator's dynamics and the response of the integrated PZT sensor to determine the operational bandwidth of SPA skin. This is followed by the design of a feedback controller to accurately control the vibrotactile stimulation with small disturbances.

3.3.1. Operational Bandwidth of SPA Skin

Feedback through the physical interface is dictated by the level of the input frequency. Therefore, the system response of the SPA skin was investigated for a variety of actuation frequencies

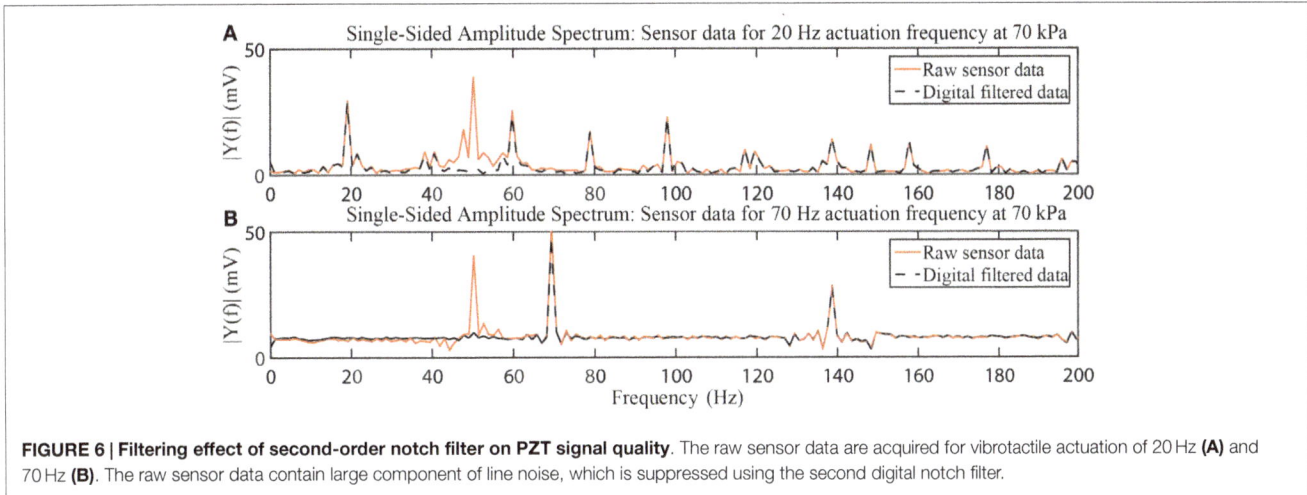

FIGURE 6 | Filtering effect of second-order notch filter on PZT signal quality. The raw sensor data are acquired for vibrotactile actuation of 20 Hz **(A)** and 70 Hz **(B)**. The raw sensor data contain large component of line noise, which is suppressed using the second digital notch filter.

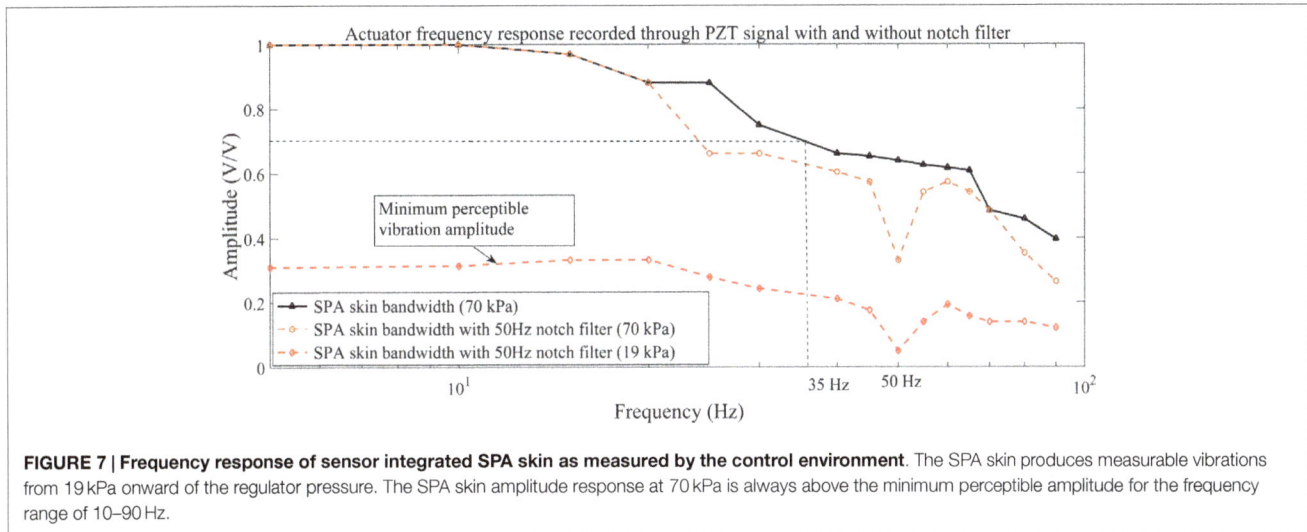

FIGURE 7 | Frequency response of sensor integrated SPA skin as measured by the control environment. The SPA skin produces measurable vibrations from 19 kPa onward of the regulator pressure. The SPA skin amplitude response at 70 kPa is always above the minimum perceptible amplitude for the frequency range of 10–90 Hz.

to determine the actual operational bandwidth and effectiveness of the vibrotactile feedback. The signal conditioning circuit has a cut-off frequency of 86 Hz, so we recorded data for the actuator's response between 10 and 90 Hz. The **Figure 7** shows the raw data recordings for a variety of frequencies at a constant input pressure. The actuator bandwidth obtained for the current setup with a tube length of 50 cm, an inlet diameter of 1 mm and an actuation chamber of 4 mm diameter is approximately 35 Hz. Ideally, this should limit the practical maximum control bandwidth for the vibrotactile feedback to 40 Hz. Human finger skin has sensitivity to vibrotactile stimuli of even lower amplitudes and forces (Mortimer et al., 2007). We experimentally obtained perceivable stimulation and measurable variation in the sensor reading from an input pressure of 19 kPa onward for the SPA skin setup. The response of this minimum perceivable signal is shown in **Figure 7** together with the frequency response at nominal input pressure (70 kPa). We observe that even near the cutoff frequency, the amplitude of the nominal input signal is higher than the minimum perceivable amplitude for tactile sensation. This extends the operational bandwidth of the SPA actuator from 40 to 90 Hz.

3.3.2. Controller for Augmented Vibratory Pneumatic Actuation

As the SPA skin is composed of hyper-elastic material, the relationship between the input pressure and the output amplitude of the actuator is non-linear by nature. Furthermore, minute fabrication defects can cause variations in the properties of the SPA skin. This makes open loop control of the actuation amplitude impractical. We therefore present a closed-loop feedback system for accurate vibrotactile stimulation.

We have approximate model fit for the sensor-actuator system with first order dynamics as discussed in Section 3.1. Assuming this linear range of operation PI(D) controller is implemented to demonstrate the active closed-loop feedback capability. The derivative term is zero to avoid noise amplification on discrete update of input signal (detected peak-value signal). The actuation amplitude is a function of the average pressure inside the inflated chamber. As the pressure regulator used in the setup has a time constant of 1.5 s, it can not be used for high-speed control of the input pressure. Instead, a high-speed on-off valve is added and its duty cycle is used as the control parameter for a desired average input pressure. The relationship between the percentage duty cycle and the maximum average pressure inside the actuation chamber is linear, independent of the frequency of actuation. The slope of this relationship is determined by the input pressure from the regulator. As previously shown in **Figure 3** the sensor response is linearly proportional to the change is the normal applied force. Therefore, the sensor signal can be used to actively compensate for the deviation from the set point (SP) value. The PI control law for this system is as follows:

$$e(t) = SP - y(t) \tag{3}$$

$$u(t) = K_P e(t) + K_I \int_0^\tau e(t) d\tau \tag{4}$$

In the digital domain for sampling time, T_s, equation (4) becomes,

$$U(z) = [K_P + \frac{K_I T_s}{1 - z^{-1}}] E(z) \tag{5}$$

and the corresponding difference equation for the controller is:

$$u[k] = u[k-1] + K_1 e[k] + K_2 e[k-1] \tag{6}$$

where, $K_1 = K_P + K_I T_s$ and $K_2 = -K_P$.

The PI gains are tuned to obtain a stable controller response over a range of operating frequencies between 10 and 90 Hz. For the current SPA skin design, the set point value for a required blocked force ranges from 0 to 0.3 N, which translates to sensor readings of 0–300 mV. The experimental results for this closed-loop system are presented in Section 4 and demonstrate steady response for a given set point.

4. CLOSED-LOOP CONTROL OF THE SPA SKIN PROTOTYPE WITH INTERNAL AND EXTERNAL INPUTS

In the first subsection, we evaluate the controller performance for desired step input for the input signal from SPA skin's actuation at two different frequencies. The second subsection demonstrates the capability of the PZT sensors to detect external interaction forces and differentiate them from the internal vibrotactile actuation.

4.1. Closed-Loop Control of the Vibratory Motors of SPA Skin at 15 and 70 Hz

The haptic feedback research has yet to focus on controlling the vibratory actuation. Not only there is a lack of measurable setup in both actuator and sensor but also it has been more interesting to investigate on the vibrational effect upon the contact. In fact, it has been the human skin that served as the feedback mechanism in the loop. However, to understand and perceive the quantifiable effects of vibrations, we need to close-loop control the actuator. Because we already have the PZT sensors embedded within the actuator layer, we can use these PZT sensors to serve the double duty to control the actuation as well. To do this, we developed a PI-based closed-loop control to achieve active control over the vibration amplitude and consequently control the level of the vibrotactile feedback that will be perceived by the user. The peak

value detected from PZT sensor signal during each on-off cycle of the output actuation is used as the input control signal. The digital controller takes control action as soon as a new peak value is available. The controller gains K_1 and K_2 [in equation (6)] are tuned in order to obtain a stable performance over a range of actuation frequencies without gain re-scheduling. The test results in **Figure 8** show controller response for a set point step at a relatively lower vibration frequency of 15 Hz and at a relatively higher vibration frequency of 70 Hz. The controller output is used to drive the PWM duty cycle value of the on-off control valve, which in turn change the average pressure inside actuation chamber. It can also be observed from **Figure 8** that in both tests the controller settles in <0.5 s. During our experiment, the controller minimized the disturbances originated from minor manufacturing defects, variation in inlet tube length, and external loading during actuator placement on human body.

4.2. Detection of External Interaction Forces

The sensing capability of SPA skin is not only limited to vibrotactile actuation amplitude but can also further be used to detect external interaction forces. Currently, there are very few haptic devices that can embed contact force sensors, which allows monitoring and controlling of the vibration motors in a closed loop. Therefore, the closed-loop controllable SPA skin is even more suitable for the wearable application environments where it not only provides vibrotactile feedback but also measures the external forces the wearable body faces. As we only have a single sensor array for the contact input, we developed decoupling algorithm to process the acquired PZT signal. We categorize the signal into two components such as (A) SPA skin actuation and (B) external interaction forces. External interactions are typically of low frequencies (<20 Hz) as observed from **Figure 9A**. Therefore, we excited SPA skin at a higher frequency (65 Hz) than frequency range of external disturbance signal. The peak value detection algorithm uses the knowledge of actuation frequency to detect the PZT signal envelope. This detected envelope selectively suppresses the high-frequency component due SPA skin actuation and upon low-pass filtering at 25 Hz, the component of the SPA skin actuation frequency is completely decoupled. The frequency spectrum of the peak value envelop signal (**Figure 9A**) shows complete suppression of vibration frequency at 65 Hz and preservation of disturbance signal. This technique is similar to the diode detector or the square law detection algorithm used in recovering the amplitude modulated (AM) signal (Carlson, 1986).

The SPA skin vibrating at 65 Hz is placed on top of a *Nano-17* force sensor to record the external impact force. We tapped on the actuation area multiple times to provide the disturbance input. When finger is tapped on the vibrating SPA skin an impulse is generated as shown in **Figure 9B** recorded by both the *Nano-17* force sensor and the PZT sensors. As PZT sensors are embedded inside the soft silicone structure, they can also sense an impulse in negative direction (**Figure 9B**) on contact release. The rigid force sensor cannot measure this interaction in the opposite direction. The force sensor measured 1 N impact force, which is the coupled force of vibrotactile actuation and impact. The results obtained using force sensor are comparable with the dynamic

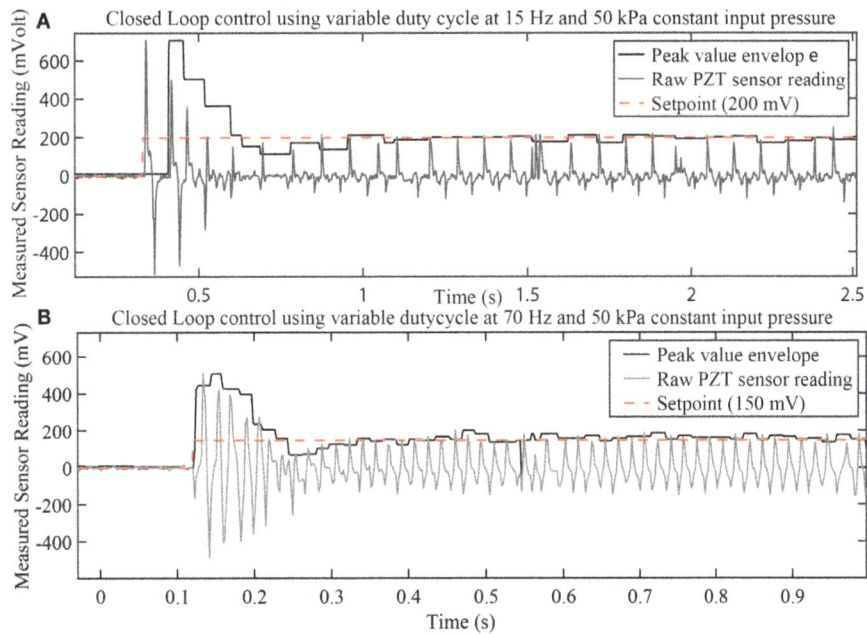

FIGURE 8 | Closed-loop control results for the PI-controller. The SPA skin is actuated at 15 Hz **(A)** and 70 Hz **(B)** at a constant regulator pressure of 50 kPa to validate the controller performance over range of actuation frequencies. The peak value envelope signal derived from the raw PZT sensor reading acts as a control signal to generate the desired duty cycle for the pneumatic valve.

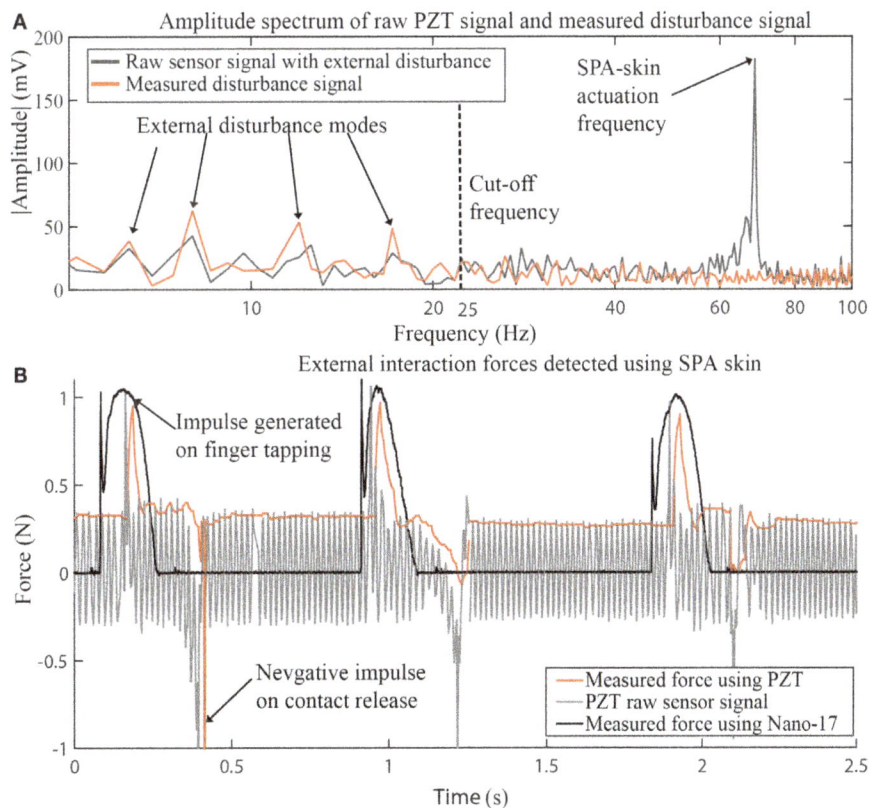

FIGURE 9 | SPA skin external signal detection. Frequency domain representation **(A)** of raw sensor signal and filtered peak value signal clearly demonstrates the capability to filter actuator vibrations and preserve disturbance signal. The external physical disturbances have interaction frequencies <20 Hz. Detection of external interaction due to finger tapping on previously actuated SPA skin at 65 Hz **(B)**. The force sensor data and filtered PZT data in **(B)** read comparable interaction force on positive impact.

interaction forces detected by PZT sensor though this requires initial calibration of the PZT sensor.

5. CONCLUSION

We introduced a unique, compliant, and distributable actuation system with embedded grid of sensing capabilities. The distributed sensing with high customizability makes the SPA skin more versatile to accommodate multiple configurations for wearable applications. Because of its softness and flexibility, it can cover wide and curvy surfaces for actuation and sensing. Furthermore, the presented SPA skin is both an input and output device that produces a modulating vibrotactile feedback over a range of frequencies. The embedded PZT sensors help SPA skin to precisely control vibration and to detect external forces and contacts. The novel method of sensor manufacturing allows placing a grid of sensor electrodes over a surface. However, these capabilities are limited by the tethered wires that create a noisy environment. Digital filtering becomes necessary as the sensed signal amplitude is comparable to the line noise captured by the long running wires acting as an antenna.

We developed an algorithm to decouple the internal and external interaction signals to maximize the detectable range of forces including feedback control of the vibratory actuators. The peak value generated on sensor by step actuation of the SPA contains the information about the blocked force exerted by the single actuation cell on the sensor. The filtered sensor data is therefore recorded over a complete PWM on-time to measure the peak value. This value is used by the closed-loop controller to take the necessary control action. The detected peak value is updated every period of the actuation frequency, limiting the control bandwidth to the frequency of SPA actuation. For the presented prototype, we concentrated on implementation of the controller and external signal detection for a single sensor and actuator. The present hardware and micro-controller platform can simultaneously detect peak values up to 16 sensors and control 16 actuators using independent feedback controllers at 1 kHz. The capabilities can be further extended based on the distributed sensing requirements.

In near future, more experiments will be carried out with variety of sensor and actuator sizes to evaluate the dynamic model and over all transfer function of the sensor integrated SPA skin. Better techniques to minimize the noise through optimal shielding and robust filtering will be investigated. Efforts are being made to integrate the sensor-embedded SPA skin as a plug and play wearable vibrotactile component for a virtual reality system and a feedback mechanism for a rehabilitation device.

FUNDING

This work is funded by the Swiss National Centre of Competence in Research (NCCR) in Robotics.

REFERENCES

Acer, M., Salerno, M., Agbeviade, K., and Paik, J. (2015). Development and characterization of silicone embedded distributed piezoelectric sensors for contact detection. *Smart Mater. Struct.* 24, 075030. doi:10.1088/0964-1726/24/7/075030

Alahakone, A., and Senanayake, S. (2009). "Vibrotactile feedback systems: current trends in rehabilitation, sports and information display," in *IEEE/ASME International Conference on Advanced Intelligent Mechatronics (AIM 2009)* (Singapore: IEEE), 1148–1153.

Asamura, N., Shinohara, T., Tojo, Y., and Shinoda, H. (2001). "Necessary spatial resolution for realistic tactile feeling display," in *ICRA* (Seol: IEEE), 1851–1856.

Carlson, A. B. (1986). *Communication Systems: An Introduction to Signals and Noise in Electrical Communication*, 2nd Edn. New York: McGraw-Hill Education.

Carmeli, E., Peleg, S., Bartur, G., Elbo, E., and Vatine, J.-J. (2011). Handtutortm enhanced hand rehabilitation after stroke a pilot study. *Physiother. Res. Int.* 16, 191–200. doi:10.1002/pri.485

Choi, S., and Kuchenbecker, K. (2013). Vibrotactile display: pe0072ception, technology, and applications. *Proc. IEEE* 101, 2093–2104. doi:10.1109/JPROC.2012.2221071

Damjanovic, D. (1997). Stress and frequency dependence of the direct piezoelectric effect in ferroelectric ceramics. *J. Appl. Phys.* 82, 1788–1797. doi:10.1063/1.365981

Frediani, G., Mazzei, D., De Rossi, D. E., and Carpi, F. (2014). Wearable wireless tactile display for virtual interactions with soft bodies. *Front. Bioeng. Biotechnol.* 2:31. doi:10.3389/fbioe.2014.00031

Gerratt, A. P., Michaud, H. O., and Lacour, S. P. (2015). Elastomeric electronic skin for prosthetic tactile sensation. *Adv. Funct. Mater.* 25, 2287–2295. doi:10.1002/adfm.201404365

Hall, D. (2001). Review nonlinearity in piezoelectric ceramics. *J. Sci. Mater.* 36, 4575–4601. doi:10.1023/A:1012916408297

Hirose, M., Hirota, K., Ogi, T., Yano, H., Kakehi, N., Saito, M., et al. (2001). "Hapticgear: the development of a wearable force display system for immersive projection displays," in *Virtual Reality, 2001. Proceedings IEEE* (Yokohama: IEEE), 123–129.

IEEE. (1988). IEEE standard on piezoelectricity. ANSI/IEEE Std 176-1987. doi:10.1109/IEEESTD.1988.79638

Kaczmarek, K., Webster, J. G., Bach-y Rita, P., and Tompkins, W. J. (1991). Electrotactile and vibrotactile displays for sensory substitution systems. *IEEE Trans. Biomed. Eng.* 38, 1–16. doi:10.1109/10.68204

Kapur, P., Premakumar, S., Jax, S., Buxbaum, L., Dawson, A., and Kuchenbecker, K. (2009). "Vibrotactile feedback system for intuitive upper-limb rehabilitation," in *World Haptics 2009 – Third Joint EuroHaptics conference, 2009 and Symposium on Haptic Interfaces for Virtual Environment and Teleoperator Systems* (Salt Lake City, UT: IEEE), 621–622.

Koo, I. M., Jung, K., Koo, J. C., Nam, J.-D., Lee, Y. K., and Choi, H. R. (2008). Development of soft-actuator-based wearable tactile display. *IEEE Trans. Robot.* 24, 549–558. doi:10.1109/TRO.2008.921561

Lambercy, O., Dovat, L., Gassert, R., Burdet, E., Teo, C. L., and Milner, T. (2007). A haptic knob for rehabilitation of hand function. *IEEE Trans. Neural Syst. Rehabil. Eng.* 15, 356–366. doi:10.1109/TNSRE.2007.903913

Lindeman, R. W., Yanagida, Y., Noma, H., and Hosaka, K. (2006). Wearable vibrotactile systems for virtual contact and information display. *Virtual Real.* 9, 203–213. doi:10.1007/s10055-005-0010-6

Maciejasz, P., Eschweiler, J., Gerlach-Hahn, K., Jansen-Troy, A., and Leonhardt, S. (2014). A survey on robotic devices for upper limb rehabilitation. *J. Neuroeng. Rehabil.* 11, 3. doi:10.1186/1743-0003-11-3

Maiolino, P., Galantini, F., Mastrogiovanni, F., Gallone, G., Cannata, G., and Carpi, F. (2015). Soft dielectrics for capacitive sensing in robot skins: performance of different elastomer types. *Sens. Actuators A Phys.* 226, 37–47. doi:10.1016/j.sna.2015.02.010

Mortimer, B. J., Zets, G. A., and Cholewiak, R. W. (2007). Vibrotactile transduction and transducers. *J. Acoust. Soc. Am.* 121, 2970–2977. doi:10.1121/1.2715669

Precision_Microdrives_Ltd. (2015). *Vibration Motors* [Online]. Available at: https://catalog.precisionmicrodrives.com/order-parts

Rosset, S., OBrien, B. M., Gisby, T., Xu, D., Shea, H. R., and Anderson, I. A. (2013). Self-sensing dielectric elastomer actuators in closed-loop operation. *Smart Mater. Struct.* 22, 104018. doi:10.1088/0964-1726/22/10/104018

Shahoian, E., Martin, K., Schena, B., and Moore, D. (2004). Vibrotactile haptic feedback devices. US Patent 6,693,622.

Smooth_On_Inc. (2015). *Dragon Skin 30* [Online]. Available at: http://www.smooth-on.com/tb/files/DRAGON_SKIN_SERIES_TB.pdf

Suh, C., Margarit, J., Song, Y. S., and Paik, J. (2014). "Soft pneumatic actuator skin with embedded sensors," in *2014 IEEE/RSJ International Conference on Intelligent Robots and Systems (IROS 2014)* (Chicago, IL: IEEE), 2783–2788.

Takahashi, C. D., Der-Yeghiaian, L., Le, V., Motiwala, R. R., and Cramer, S. C. (2008). Robot-based hand motor therapy after stroke. *Brain* 131, 425–437. doi: 10.1093/brain/awm311

Viau, A., Feldman, A. G., McFadyen, B. J., and Levin, M. F. (2004). Journal of neuroengineering and rehabilitation. *J. Neuroeng. Rehabil.* 1, 11. doi:10.1186/1743-0003-1-11

Wall, C. III (2010). Application of vibrotactile feedback of body motion to improve rehabilitation in individuals with imbalance. *J. Neurol. Phys. Ther.* 34, 98. doi:10.1097/NPT.0b013e3181dde6f0

Xu, S., Zhang, Y., Jia, L., Mathewson, K. E., Jang, K.-I., Kim, J., et al. (2014). Soft microfluidic assemblies of sensors, circuits, and radios for the skin. *Science* 344, 70–74. doi:10.1126/science.1250169

Zhao, J., and Zhang, Q. (1996). "Effect of mechanical stress on the electromechanical performance of pzt and pmn-pt ceramics," in *Proceedings of the Tenth IEEE International Symposium on Applications of Ferroelectrics*, Vol. 2 (East Brunswick, NJ: IEEE), 971–974.

Conflict of Interest Statement: The authors declare that the research was conducted in the absence of any commercial or financial relationships that could be construed as a potential conflict of interest.

16

Multi-robot searching with sparse binary cues and limited space perception

Siqi Zhang[1,2], Dominique Martinez[1†]* and Jean-Baptiste Masson[3,4†]

[1] *UMR 7503, Laboratoire Lorrain de Recherche en Informatique et ses Applications, Centre National de la Recherche Scientifique, Vandoeuvre-lès-Nancy, France,* [2] *School of Marine Science and Technology, Northwestern Polytechnical University, Xi'an, China,* [3] *Physics of Biological Systems, Institut Pasteur, Paris, France,* [4] *UMR 3525, Centre National de la Recherche Scientifique, Paris, France*

In this paper, we consider the problem of searching for a source that releases particles in a turbulent medium with searchers having binary sensors and limited space perception. To this aim, we extend an information-theoretic strategy, namely Mapless, to multiple searchers and demonstrate its efficiency both in simulation and robotic experiments. The search time is found to decay as $1/n$ for n cooperative robots as compared to $1/\sqrt{n}$ for independent robots so that significant gains in the search time are obtained with a small number of robots, e.g., $n = 3$. Search efficiency results from pooling sensory information between robots to improve individual decision-making (three detections on average per searcher were sufficient to reach the source) while still maintaining the individual resistivity to various errors during the search. The method is robust to odometry errors and is thus relevant to robots searching in low-visibility conditions, e.g., firefighter robots exploring smoky environments.

Keywords: search and rescue, multi-robot systems, swarm robotics, fire searching, firefighter robot

Edited by:
M. Ani Hsieh,
Drexel University, USA

Reviewed by:
Konstantinos Karydis,
University of Delaware, USA
Roberto Tron,
University of Pennsylvania, USA

*Correspondence:
Dominique Martinez,
UMR 7503, Laboratoire Lorrain de
Recherche en Informatique et ses
Applications, Centre National de la
Recherche Scientifique,
Vandoeuvre-lès-Nancy, France
dominique.martinez@loria.fr
[†] *Senior authors of the paper.*

1. Introduction

Searching for a source releasing particles in the environment (e.g., toxic or explosive materials, pollutants, heat) is particularly challenging given that the chemical transport over long distances is dominated by turbulence (Csanady, 1973; Shraiman and Siggia, 2000). The sensory landscape is thus very heterogeneous in concentration and discontinuous in time, and consists of sporadically located patches traveling with the air flow. The probability of encountering one of these patches decays exponentially with distance from the source. In such turbulent conditions, odor detections become intermittent and no measurement gradient points toward the source (Csanady, 1973; Humphrey and Haj-Hariri, 2012; Celani et al., 2014). Methods based on a measurement gradient like extremum seeking (Zhang et al., 2007; Cochran and Krstic, 2009) are inappropriate in this context because the searcher has to rely on intermittent binary cues (hits with odor patches) rather than continuous sampling of concentration values.

Insects can be very efficient at solving this problem. One example is provided by male moths guided by pheromonal cues and searching for mates located hundreds of meters possibly kilometers away (Baker et al., 1985; Murlis et al., 1992; Mafra-Neto and Carde, 1994; Vickers, 2000). Another exceptional search behavior is the one of *Melanophila* beetles, which detect and track forest fires from infrared (Schmitz et al., 1997) and olfactory (Schutz et al., 1999) cues because their larvae can develop only in freshly burnt wood (Didier, 2010; Schmitz and Bousack, 2012). Artificial robots with searching ability similar to these insects are expected to be very useful in many applications (Gelenbe et al., 1997), e.g., to assist human firefighters in detecting gas leaks and exploring buildings on fire. Models of search processes are therefore important not only to biology but also to applications in robotics. As a search scheme intended to deal with uncertain and dynamic environments, Infotaxis has been shown to produce trajectories similar to those of animals, e.g., moths attracted by a sexual pheromone (Vergassola et al., 2007) or nematodes foraging for food (Calhoun et al., 2014).

Infotaxis is a probabilistic search method based on information theory that relies on a grid map of the environment (Vergassola et al., 2007). The posterior probability for the source position is calculated over the entire map and the searcher moves in the direction that minimizes the entropy of the distribution. Rather than searching for the source position the searcher moves to increase information on the position of the source (Barbieri et al., 2011; Atanasov et al., 2015). Furthermore, Infotaxis, during the Greedy decision process, slightly favors exploration over exploitation of information. Infotaxis has been successfully applied to robotic searches (Martin-Moraud and Martinez, 2010), a prerequisite being that the robot has full access to its position in the environment. Yet, for robots engaged in search missions, space perception can be limited. Think about a firefighter robot searching for fire indoor. As revealed by experiments in this paper, the presence of smoke prevents the use of cameras and laser range finders for localization. In such low-visibility conditions, Infotaxis is not easily applicable as the robot is unable to correct its odometry errors from external cues. Yet, adaptation of Infotaxis to such conditions is not excluded. Another approach, introduced in Masson (2013) as Mapless, allows searching in complex varying environment with limited space perception, possibly corrupted or incomplete information and limited memory. Mapless is based on a standardized projection of the probability map of the source location to remove space perception and on the evaluation of a free energy, whose minimization along the path gives direction to the searcher. Free-energy minimization allows reinforcement of the maximum likelihood decision.

Hereafter, following a similar procedure as the one shown in Masson et al. (2009), we extend Mapless to multiple searchers (swarm Mapless). Whereas decision-making is performed individually by each searcher, the probability of the source location and hence the free energy are jointly estimated by the swarm. The main difference with related works is that the information metric is approximated analytically in swarm Mapless rather than estimated from a grid map or by using (computationally expensive) Monte Carlo sampling techniques in Cortez et al. (2009), Barbieri et al. (2011), Dames and Kumar (2013), and Atanasov et al. (2015). We present here a successful solution with a real robotic system [search for a heat source in a turbulent medium as in Martin-Moraud and Martinez (2010) and Masson (2013)]. This framework is employed as a testbed to assert complete and rigorous evaluations of Mapless and swarm Mapless under real conditions. The paper is organized as follows. Infotaxis, Mapless, and swarm Mapless are detailed in the Section "Materials and Methods." The performance in terms of effectiveness and robustness are assessed, both in simulations and robotic

experiments, in the Section "Results". Our work is discussed in the final section.

2. Materials and Methods

2.1. Infotaxis

Infotaxis was introduced in Vergassola et al. (2007) for searching in complex environments with sparse detections. It is built around two core components: Bayesian inference of the position of the source based on detection history and Greedy decision making based on entropy minimization. The former depends on the modeling of the local environment. An efficient approximation describes the propagation of the cues in the turbulent environment by the advection–diffusion equation (shown in Section "Appendix" for sake of completion). The properties of the medium are encoded by a rate function $R(\vec{r}|\vec{r_0})$ with $\vec{r_0}$ the position of the source and $\vec{r}$ the position of the searcher. Note that a correlation length λ is associated with $R(\vec{r}|\vec{r_0})$ and can be interpreted as the mean distance traveled by the particles before they vanish. The detection process is approximated by a Poisson process, leading to a probability of k detections during time δt $\rho_k = \frac{(R(\vec{r}|\vec{r_0})\delta t)^k \exp(-R(\vec{r}|\vec{r_0})\delta t)}{k!}$. After following a path Θ_t, the posterior distribution of the position of the source at time t reads:

$$P_t(\vec{r_0}|\Theta_t) = \frac{\exp(-\int_0^t R(r_{t'}|r_0)dt') \prod_{i=1}^H R(r_i|r_0)\delta t}{Z_t} \quad (1)$$

with H the total number of detections experienced in Θ_t and $Z_t = \int \exp(-\int_0^t R(\vec{r_{t'}}|\vec{r_0})dt') \prod_{i=1}^H R(\vec{r_i}|\vec{r_0})\delta t d\vec{r_0}$ the normalization constant. The detection process being approximated as Markovian, update of the posterior distribution $P_{t+dt}(\vec{r_0}|\Theta_t)$ is directly obtained from $P_t(\vec{r_0}|\Theta_t)$ by multiplying with the probability of detection or no-detection experienced during δt.

Moving toward the most probable source location, i.e., a maximum likelihood or maximum *a posteriori* strategy, systematically fails far from the source because of the misrepresentation of the environment by $P_t(\vec{r_0}|\Theta_t)$. Infotaxis, searches for information about the position of the source rather than directly trying to reach the source. Upon moving to a neighboring position, $\vec{r}_{t+dt}$, the searcher minimizes the expected variation of entropy of $P_t(\vec{r_0}|\Theta_t)$

$$\Delta S_t(\vec{r}_t \rightarrow \vec{r}_{t+dt}|\Theta_t) = P_t(\vec{r}_{t+dt}|\Theta_t)[0 - S_t]$$
$$+ [1 - P_t(\vec{r}_{t+dt}|\Theta_t)](\sum_{i=0} \rho_i \Delta S_t^i) \quad (2)$$

with $S_t = -\int P_t(\vec{r_0}|\Theta_t)\log P_t(\vec{r_0}|\Theta_t)d\vec{r_0}$ is the entropy of the posterior field computed at time t. The first term encodes the probability of finding the source and promotes maximum likelihood decision and the second, which encodes the probability of not finding the source in $\vec{r}_{t+dt}$, promotes exploration of the environment. In the rest of the paper, the summation will be reduced to zero and one detection as the probability of having more detections during δt is usually extremely low.

2.2. Mapless

A prerequisite for the evaluation of equation (2) based on the probability map [equation (1)] is that the agent perceives space, i.e., the agent is able to (i) build a spatial map of the environment, (ii) locate itself on the map, and (iii) go purposefully to predefined locations. These three tasks have been extensively studied in robotics and are known under the term SLAM for *simultaneously localization and mapping* (Thrun et al., 2005). In the case of fire searching, however, precise localization of the robot and map building operations would be difficult to achieve because infrared light and smoke particles emitted from burning objects prevents the use of cameras and laser range finders. In such low-visibility conditions, it is safer to conduct the search based on a coarse estimation of the robot position.

Mapless was introduced in Masson (2013) as a method for searching with limited space perception, for handling unreliable cues and controlling actively the exploration/exploitation balance. To remove space perception, the posterior distribution $P_t(\vec{r}_0|\Theta_t)$ is projected into a standardized form. The posterior distribution, that later will not directly used by the agent to make direction decision, reads

$$
P_t^M(\vec{r}_0|\Theta_t) = \frac{e^{-\frac{|\vec{r}_0 - \vec{r}_G|^2}{\lambda_G^2}}\left(1 - \frac{1}{N_M}\sum_{j=N_t-N_M+1}^{N_t} e^{-\frac{|\vec{r}_0 - \vec{r}_j|^2}{\lambda_u^2}}\right)}{Z_t^M},
$$

(3)

with $\vec{r}_G$ is the damped center of mass of the detections, $\vec{r}_j$s represent the perceived (by the agent) positions of the agent when there was no detection, λ_G is the scale of the Gaussian approximating the detection term, λ_u is the scale of the Gaussian approximating the non-detection term and Z_t^M is a normalization constant. This projection is based on the separation of the detection and non-detection terms, approximating the former by its main component and the latter by a mean field approximation [Supplementary in Masson (2013)]. This projection allows an essential component of the posterior distribution of the source position to be encoded: the local decrease of the probability around the visited locations where no detections have been experienced. Whereas N_t is the total number of visited positions in Θ_t, only the last N_M positions are recalled to compute the posterior. This prevents storing indefinitely unreliable cues or positioning errors. The parameters λ_u and λ_G are related to the correlation length λ of the source but are not necessarily the same as the non-detection term is made of a larger number of events and is less localized than the detection term.

Instead of the entropy in Infotaxis (Vergassola et al., 2007), a free-energy formulation is used in Mapless. The free energy is written as $F_t = W_t + TS_t$ with T an internal (temperature) parameter that controls the balance between the entropy $S_t = -\int P_t^M(\vec{r}_0|\Theta_t)\log P_t^M(\vec{r}_0|\Theta_t)d\vec{r}_0$ and the "working energy" $W_t = \int_A P_t^M(\vec{r}_0|\Theta_t)d\vec{r}_0$ where the integration domain A is defined as $|\vec{r}_0 - \vec{r}_G| \leq \lambda_G/2$ (Masson, 2013). Note that free energy has been previously used as a principle for linking action to perception (Friston et al., 2010, 2011) and that various functional can be used for the "work term." In Mapless, the free-energy formulation allows an active control between exploration and exploitation through the internal temperature T, see Masson (2013) for the details. When the agent moves from position $\vec{r}_t$ at time t to a

neighboring position, $\vec{r}_{t+dt}$, the expected variation in the free-energy reads

$$
\Delta F_t(\vec{r}_t \to \vec{r}_{t+dt}|\Theta_t) = P_t^M(\vec{r}_{t+dt}|\Theta_t)\,(1 - F_t)
$$
$$
+ ([1 - P_t^M(\vec{r}_{t+dt}|\Theta_t)](\rho_0\Delta F_t^0 + \rho_1\Delta F_t^1)
$$

(4)

where the summation limitation has been applied. The first and second terms on the right-hand side correspond to finding and not finding the source at the new position, respectively. If the source is found at the next step, the free energy F_{t+dt} becomes one. If the source is not found, the agent may or may not detect leading to different variations in the free energy, namely ΔF_t^1 and ΔF_t^0.

An important characteristics of approximating the posterior P_t [equation (1)] by P_t^M [equation (3)] is that the free energy F_t can be computed analytically without the computation of the approximated posterior distribution $P_t^M(\vec{r}_0|\Theta_t)$. All terms involved in the computation of equation (4) are described in Masson (2013). Thus, unlike Infotaxis, Mapless does not require the searcher to build a probability map and locate itself precisely. Efficient searches, far from the source with significant odometry errors are demonstrated in Masson (2013).

2.3. Swarm Mapless

Interest in swarms of agents stems from the expected increase in task efficiency by having multiple agents performing it. As multiple agents can explore an environment more efficiently as a group than as individuals (Berdahl et al., 2013), we propose here an extension of Mapless to collective search. The best performing strategy would be the full collaboration between the agents; that is, the free energy is computed from the shared observations and decision-making is obtained by evaluating the effects of moving the whole swam. Yet, the number of possible actions for n agents on a square 2D grid is 5^n so that performing full collaboration in real time is difficult in practice when $n > 3$. An alternative approach is that the agents share information during their path (i.e., detection and non-detection events) but decision-making is performed individually. Namely, the s-th agent chooses the move $\vec{r}_t^s \to \vec{r}_{t+dt}^s$ that minimizes

$$
\Delta F_t^s(\vec{r}_t^s \to \vec{r}_{t+dt}^s|\Theta_t) = P_t^M(\vec{r}_{t+dt}^s|\Theta_t)\,[1 - F_t]
$$
$$
+ (1 - P_t^M(\vec{r}_{t+dt}^s|\Theta_t))[\rho_0\Delta F_t^{s,0} + \rho_1\Delta F_t^{s,1}]
$$

(5)

where $\Theta_t = \{\Theta_t^1, \Theta_t^2, \ldots, \Theta_t^s, \ldots\}$ denotes the search history for the whole swarm while Θ_t^s is the self-generated path of the s-th agent. It is worth remembering that the robots share their own measurements of their paths and detection history, thus they share paths with odometry errors and possibly anomalous detection, yet as it will be shown Mapless and Swarm Mapless are resistive to these errors. This strategy is referred to swarm Mapless in the following.

3. Results

3.1. Swarm Mapless in Simulation: Three Searchers is Sufficient

Before considering robotic implementations, we first assess the performance of swarm Mapless using numerical simulations in

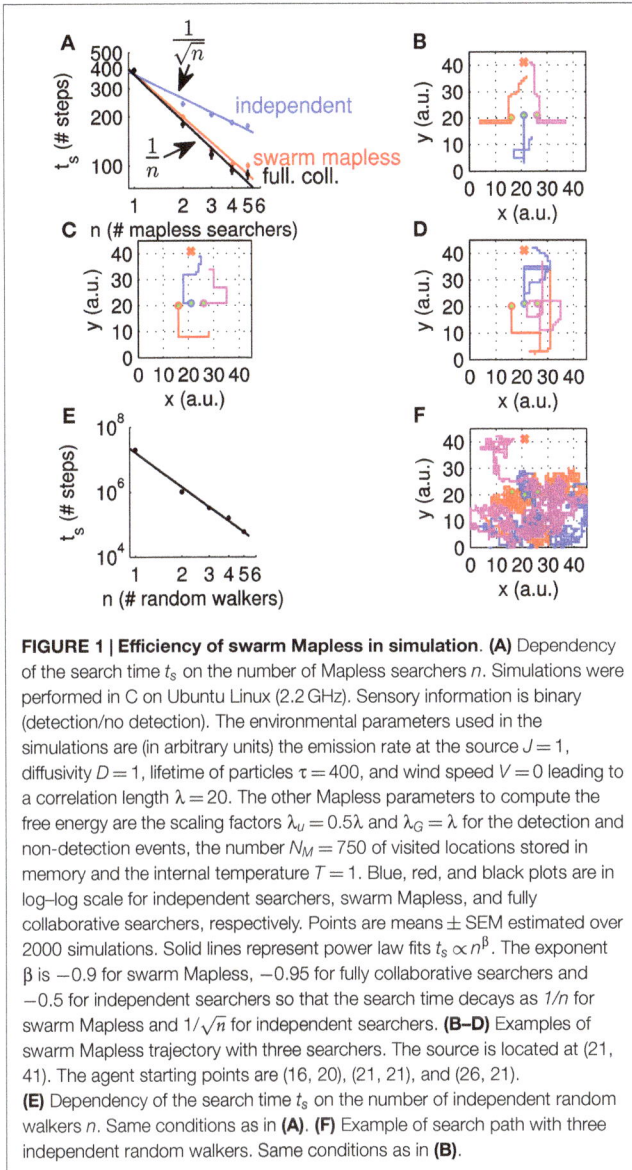

FIGURE 1 | Efficiency of swarm Mapless in simulation. (A) Dependency of the search time t_s on the number of Mapless searchers n. Simulations were performed in C on Ubuntu Linux (2.2 GHz). Sensory information is binary (detection/no detection). The environmental parameters used in the simulations are (in arbitrary units) the emission rate at the source $J = 1$, diffusivity $D = 1$, lifetime of particles $\tau = 400$, and wind speed $V = 0$ leading to a correlation length $\lambda = 20$. The other Mapless parameters to compute the free energy are the scaling factors $\lambda_u = 0.5\lambda$ and $\lambda_G = \lambda$ for the detection and non-detection events, the number $N_M = 750$ of visited locations stored in memory and the internal temperature $T = 1$. Blue, red, and black plots are in log–log scale for independent searchers, swarm Mapless, and fully collaborative searchers, respectively. Points are means $\pm$ SEM estimated over 2000 simulations. Solid lines represent power law fits $t_s \propto n^\beta$. The exponent β is -0.9 for swarm Mapless, -0.95 for fully collaborative searchers and -0.5 for independent searchers so that the search time decays as $1/n$ for swarm Mapless and $1/\sqrt{n}$ for independent searchers. (B–D) Examples of swarm Mapless trajectory with three searchers. The source is located at (21, 41). The agent starting points are (16, 20), (21, 21), and (26, 21). (E) Dependency of the search time t_s on the number of independent random walkers n. Same conditions as in (A). (F) Example of search path with three independent random walkers. Same conditions as in (B).

terms of effectiveness (search time) and robustness (with respect to changes in environmental conditions).

The dependency of the search time t_s on the number of searchers n is shown in **Figure 1A** for swarm Mapless as compared to independent and fully collaborative searchers (see Materials and Methods). The simulations were performed in C with parameters given in figure caption. In all cases, the data are well fitted by a power law $t_s \propto n^\beta$. The exponent β is -0.5 for independent searchers and -0.9 for swarm Mapless so that t_s decays more rapidly ($t_s \propto \frac{1}{n}$) when the searchers cooperate than when they are independent ($t_s \propto \frac{1}{\sqrt{n}}$). Interestingly, both β are consistent with Masson et al. (2009) for swarm of infotactic searchers.

We also note that the gain resulting from full collaboration between the agents is marginal as compared to swarm Mapless with individual decision-making ($\beta = -0.95$ for full collaboration vs. -0.9 for swarm Mapless). The gain with fully collaborative searchers was more significant in Infotaxis (Masson et al.,

2009) than in Mapless. This is the consequence of the reduced representation of the environment in Mapless, the full collaboration between searchers does not improve much Greedy decision processes based on limited information. Yet, the implementation cost of a full collaboration is much higher. Due to the power law, the percentage decrease in the search time for swarm Mapless is 70% from $n = 1$ to 3 searchers and 20% from $n = 3$ to 5, so that swarm Mapless reveals impressive gains in the search time even with a limited number of searchers ($n = 3$).

Some examples of swarm Mapless trajectories obtained with three searchers are depicted in **Figures 1B–D**. We note that some agents originally follow a direction opposite to the one of the source. These incorrect paths are not surprising given the uncertain belief resulting from the lack of detections at the beginning of the search. Yet, the direction toward the source emerges as information from odor detections is gathered over time. On average, we found that only three detections per searcher are sufficient to reach the source, in the search configuration displayed here. It is worth noting that even if the detections are rare (characteristic of searches in dilute or desertic conditions), they are nevertheless crucial to the search process. To assess their importance, we performed complementary simulations with random walkers (see **Figures 1E,F**). The search time of random walkers also exhibits a power law decay with n. Yet, it is many orders of magnitude higher than for swarm Mapless. It is also worth noting that swarm Mapless, Mapless, and obviously Infotaxis exploit the non-detections to explore the search space. Swarms Mapless gains a lot of efficiency from the various parts of the search space where no detections occurred. From the observations above, it is therefore sufficient for experimental purposes to consider a swarm of three searchers to improve effectiveness.

It has been shown that Mapless is resistant to incorrect modeling of the environment (Masson, 2013). Is this robustness preserved when the swarm accumulates information on multiple locations at the same time? To judge it, we tested swarm Mapless under varying conditions, i.e., isotropic diffusivity D in range 0.4–1.6 au in **Figure 2A**, lifetime of particles τ in range 100–800 au in **Figure 2B**, λ_G/λ and λ_u/λ in range 0.1–1 in **Figure 2C**. In each condition, we observe that the variability (as given by the SD of the search time) decreases with the number of searchers (see **Figure 2D**). Moreover, the gain in robustness in more pronounced for $n \leq 3$ than for $n > 3$.

3.2. Swarm Mapless in Robotic Experiments: Resistance to Odometry Errors

Promising results were achieved with Swarm Mapless in simulation (**Figures 1** and **2**). Nevertheless, experimental implementations are necessary to ensure that swarm Mapless can be used in real turbulent environments and can handle the numerous errors encountered. We present hereafter a successful solution for implementing swarm Mapless within a robotic system, and we assess its performance in the real environment. All experiments were performed with Khepera III robots (K-Team SA, Switzerland) and several modules: Korebot II (embedded ARM processor running Linux 2.6 at 600 MHz), KoreIOLE (acquisition board with 12 analog inputs in the 0–5 V range with 5 mV resolution), and KoreWifi (board allowing Wifi communication with the robot).

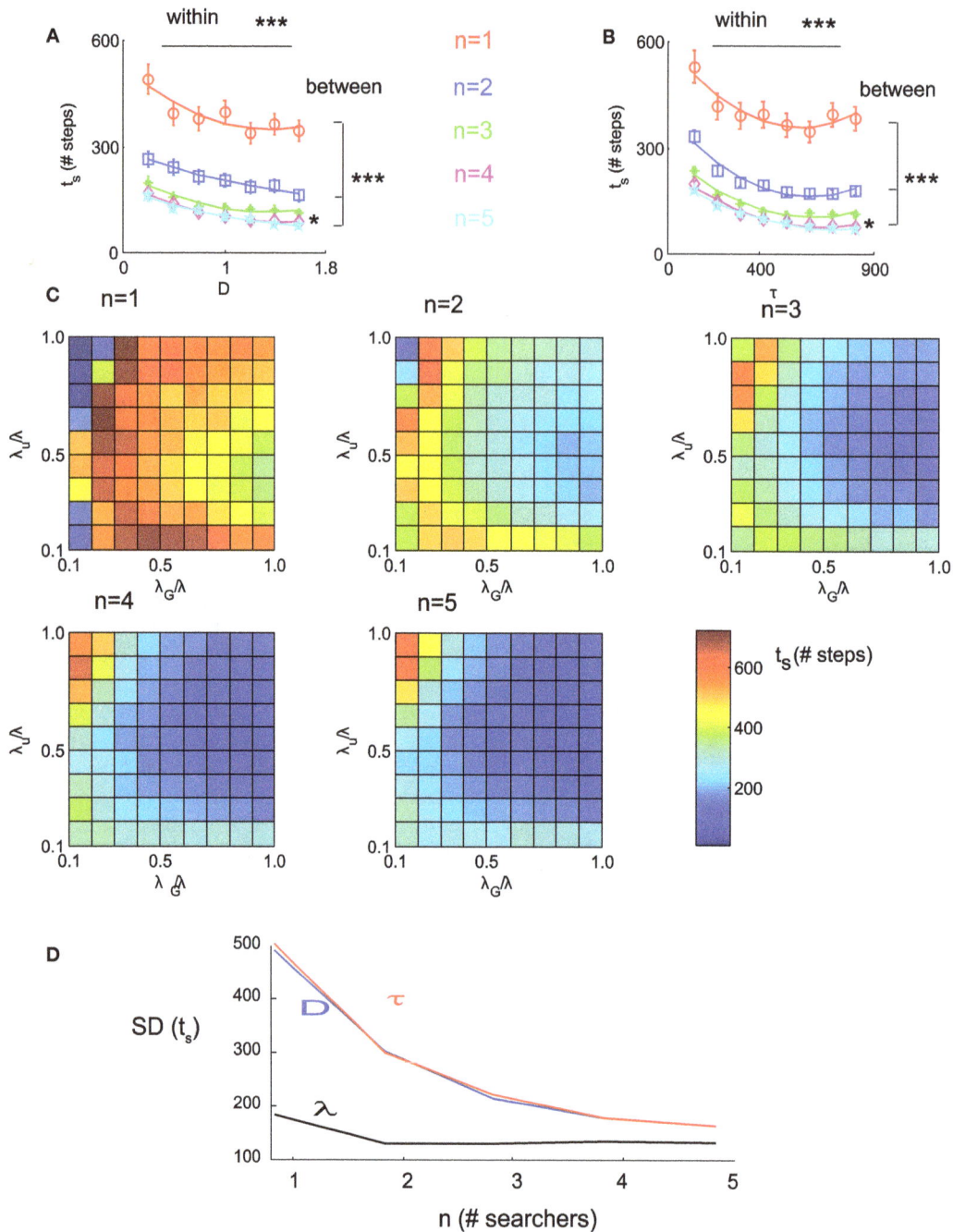

FIGURE 2 | Robustness of swarm Mapless in simulation (n = number of searchers). (A) Swarm Mapless considers isotropic diffusivity D that is different from the one of the source ($D = 1$ au). Points represent the search time t_s as means ± SEM estimated over 2000 simulations. Different colors are associated with the number of searchers as indicated in the figure. Within- and between-group differences are significant (Kruskal–Wallis test). Asterisks indicate significant differences (*$p < 0.05$, ***$p < 0.001$). **(B)** Swarm Mapless considers a lifetime of particles τ that is different from the one of the source ($\tau = 400$ au). Same conditions as in **(A)**. **(C)** Swarm Mapless considers scaling factors λ_u and λ_G for detection and non-detection events that are different from the correlation length λ of the source. **(D)** SD of the search time SD (t_s) versus the number of Mapless searchers n under the three conditions: mismatch in diffusivity D [as in **(A)**], mismatch in lifetime of particles τ [as in **(B)**] and mismatch in correlation length λ [as in **(C)**].

As a proof of concept for fire searching, we consider the search for a heat source with robots equipped with temperature sensors. If the heat source is set to only few degrees above room temperature, the setup is also valid to model olfactory cue searches as in Masson (2013). The environmental conditions inside a building on fire can rapidly deteriorate making visual navigation difficult in the presence of smoke. To assess whether precise localization can still be obtained in low visibility conditions, we equipped a robot with a rangefinder module (**Figure 3A**). The rangefinder sensor is a LIDAR (URG-04LX, Hokuyo) that determines the distance

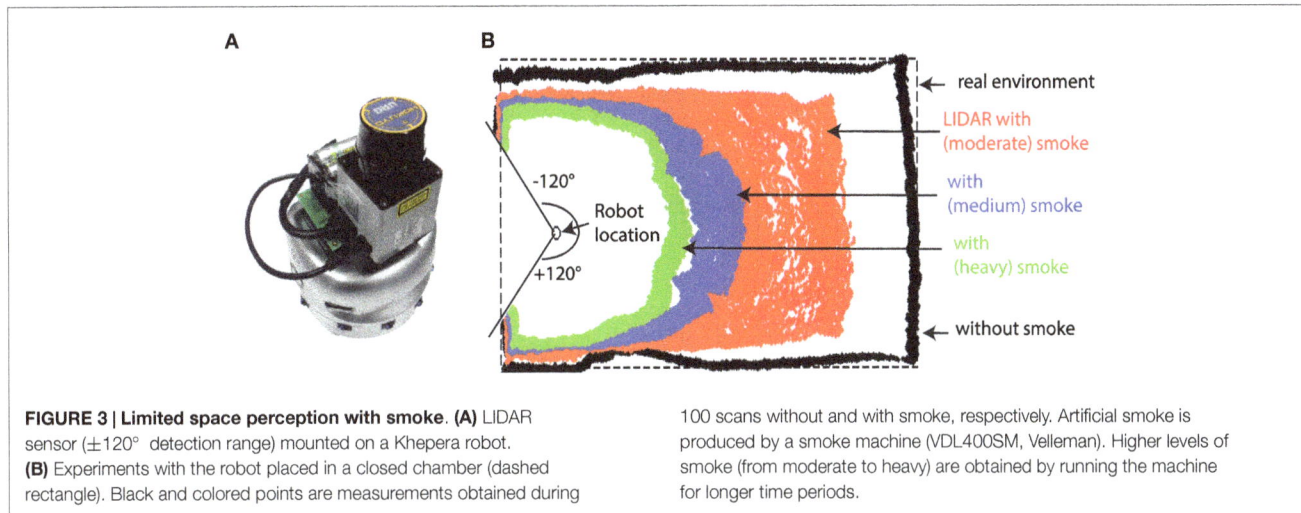

FIGURE 3 | Limited space perception with smoke. (A) LIDAR sensor ($\pm 120°$ detection range) mounted on a Khepera robot. **(B)** Experiments with the robot placed in a closed chamber (dashed rectangle). Black and colored points are measurements obtained during 100 scans without and with smoke, respectively. Artificial smoke is produced by a smoke machine (VDL400SM, Velleman). Higher levels of smoke (from moderate to heavy) are obtained by running the machine for longer time periods.

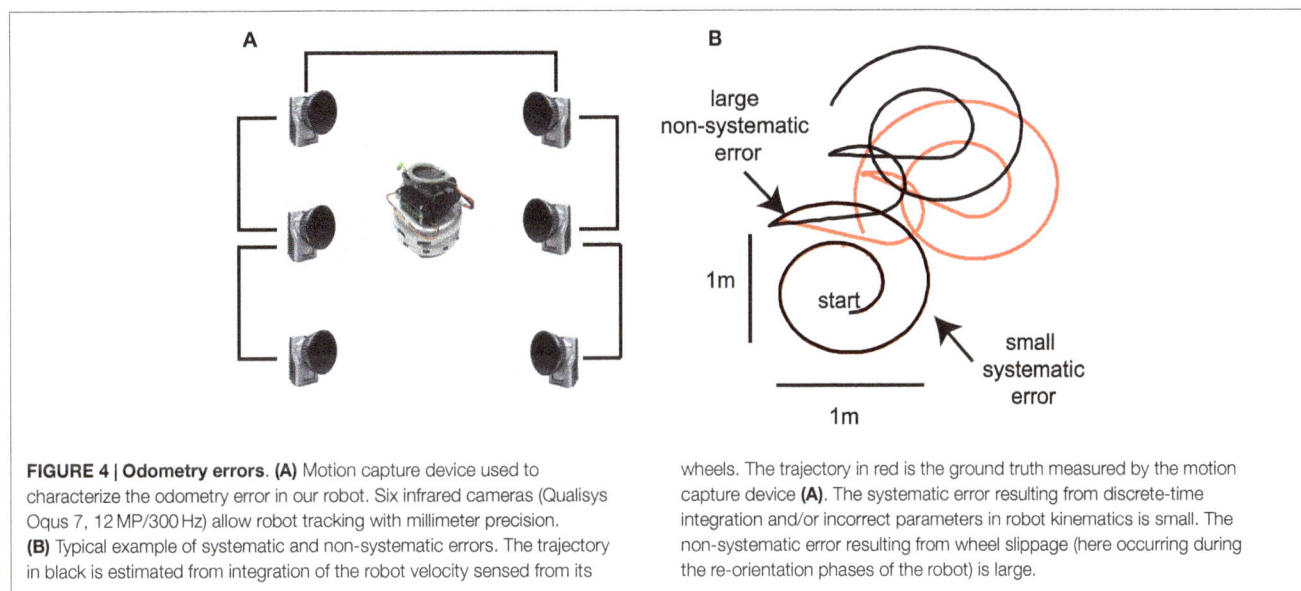

FIGURE 4 | Odometry errors. (A) Motion capture device used to characterize the odometry error in our robot. Six infrared cameras (Qualisys Oqus 7, 12 MP/300 Hz) allow robot tracking with millimeter precision. **(B)** Typical example of systematic and non-systematic errors. The trajectory in black is estimated from integration of the robot velocity sensed from its wheels. The trajectory in red is the ground truth measured by the motion capture device **(A)**. The systematic error resulting from discrete-time integration and/or incorrect parameters in robot kinematics is small. The non-systematic error resulting from wheel slippage (here occurring during the re-orientation phases of the robot) is large.

to objects from the time-of-flight of a rotating laser. In smoky conditions, the LIDAR is not able to detect the boundaries of the test apparatus (**Figure 3B**). Instead, the LIDAR returns the distance to the bottom of the smoke layer so that the measured distance decreases with the smoke density (from moderate to heavy in **Figure 3B**). In agreement with Pascoal et al. (2008) and Starr and Lattimer (2014), this result indicates that a LIDAR would not be capable of providing accurate range finding information in smoky environments.

An alternative method for robot localization is to use odometry, which is path integration of the robot velocity sensed from its wheels. To assess the localization error, we compared robot trajectories obtained from the odometry tracking module of the Khepera III Toolbox (http://en.wikibooks.org/wiki/Khepera_III_Toolbox) to the ground truth provided by a motion capture device (**Figure 4A**). An example of trajectory is shown in **Figure 4B**. The systematic error, resulting from discrete-time integration and/or incorrect parameters in robot kinematics, appeared to be small.

Yet, we noticed the occurrence of large non-systematic errors due to wheel slippage during the re-orientation phases of the robot (**Figure 4B**). Although similar re-orientation phases are used in the experiments below due to step-like movements, we tested swarm Mapless without correcting for odometry errors. Experimental Mapless searches shown in Masson (2013) were very resistant to strong odometry errors, yet accumulating information from multiple searchers is also accumulating errors from all searchers. Thus, it is important to question odometry errors in the context of swarm Mapless experimental searches. In some ways, it allows us to assess the robustness of the algorithm.

In swarm Mapless experiments, we consider the search for a heat source with three robots equipped with temperature sensors (**Figure 5A**). The temperature signal was amplified and filtered with a custom-made board previously designed for biological signals (Martinez et al., 2014). The search was performed in an arena of 6 m long by 4 m large, resulting in a grid-based model of the environment of 30×20 steps (**Figure 5B**). At every step, each

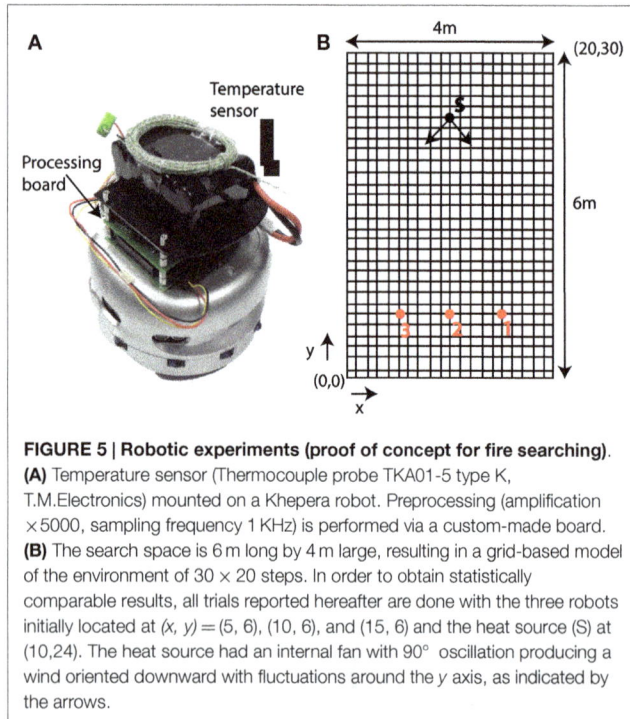

FIGURE 5 | Robotic experiments (proof of concept for fire searching).
(A) Temperature sensor (Thermocouple probe TKA01-5 type K,
T.M.Electronics) mounted on a Khepera robot. Preprocessing (amplification
×5000, sampling frequency 1 KHz) is performed via a custom-made board.
(B) The search space is 6 m long by 4 m large, resulting in a grid-based model
of the environment of 30 × 20 steps. In order to obtain statistically
comparable results, all trials reported hereafter are done with the three robots
initially located at $(x, y) = (5, 6)$, $(10, 6)$, and $(15, 6)$ and the heat source (S) at
(10,24). The heat source had an internal fan with 90° oscillation producing a
wind oriented downward with fluctuations around the y axis, as indicated by
the arrows.

robot chooses the best strategy in terms of free-energy minimization among the five possible actions, i.e., making a move to one of the four neighboring steps or staying still. Linear and angular speeds were set to 10 cm/s and 90°/s as they offer a good compromise between minimizing the errors in the step-like movements and being fast enough (each individual step, including translation and rotation, is performed in ≈3 s). To allow searching with obstacles (e.g., the boundaries delimiting the search space) and prevent the robots running into each other, we added to swarm Mapless a Braitenberg avoidance scheme based on the readings of the Khepera proximity sensors. An example of collective search with three robots is shown in **Figure 6**.

The heat source had an internal fan with 90° oscillation aiming at increasing wind fluctuations and thereby the turbulence level. This dispersion model was also used in Masson (2013). The air conditioning was turned off while other instruments and furniture in the room were placed as usual. This setup led to a complex temperature pattern and the heat source was sufficiently hot for the robots to detect local temperature variations at several meters from the source. **Figure 7** provides two examples of the signal measured by the robot while moving straight toward the heat source (**Figure 7A**) and without the source (**Figure 7B**). Detection events are triggered each time the temperature signal exceeds an adaptive threshold (see figure caption for details). The statistics of detections obtained by repeating the experiment 12 times is shown in **Figure 7** with and without the source. With the heat source (**Figure 7C**), the detection rate decays exponentially with the source distance, in well agreement with the expression of $R(\vec{r}, r_0)$ derived in the Section "Appendix" with a correlation length of $\lambda = 20$ au. Without the source (**Figure 7D**), the false alarm rate is low (≈1 false positive every 12 s) and independent of the source distance.

To test the effectiveness of robot swarm Mapless in the real environment, we repeated experiments in order to obtain 20 successful runs. One successful trial is defined by the fact that one of the robot in the swarm reaches the source within a reasonable search time set at 700, 600, and 500 steps for $n = 1, 2$, and 3 robots, respectively. Above this time limit, the robots are considered to be lost. The total number of trials with (1, 2, 3) robots was (21, 22, 24) and (21, 21, 23) for collaborative and independent robots, respectively. The success rate is high and comparable to what has been previously obtained with one robot (Masson, 2013). More interesting is that the power law dependency of the search time obtained in robotic experiments is similar to the one in simulation (**Figure 8A**). Thus, the search time also decays as $1/n$ for swarm Mapless with n robots as compared to $1/\sqrt{n}$ for independent robots. The $1/n$ decay of swarm Mapless leads to significant gains in the search time. As an example, the mean duration of the search is ≈20 min with one robot as compared to ≈7 min with three robots. An example of swarm Mapless trajectory is shown in **Figure 8B**. It is worth noting that the paths of the 3 robots were reconstructed from an external video camera and not from the odometry of the robots. The reason is that, during the search, the robots have enough time to accumulate odometry errors and their estimated trajectories do not correspond to the reality. Nevertheless, the efficiency of the search confirms that swarm Mapless is resistant to odometry errors. Among the useful properties of swam Mapless, the capability to handle erroneous information is an important one allowing for efficient applications in real environments.

4. Discussion

In the case of diffusion (the signal is maximum at source location and decays with distance from the source), search methods based on a measurement gradient (Ogren et al., 2004; Zhang et al., 2007; Cochran and Krstic, 2009) are guaranteed to converge to the source location. Multiple searchers can similarly be used to locate the plume front in the case of advection (Li et al., 2014). These methods are applicable only in the presence of a homogeneous signal field for which the computation of a measurement gradient is feasible. Here, we addressed the more challenging problem of searching in a turbulent medium. In this context, even if a local gradient could be measured, its direction would not point toward the source, thus the searcher has to rely on intermittent binary cues.

To this aim, we considered an information-theoretic method (Mapless) and its extension to multiple searchers (swarm Mapless). The search strategy is motivated by the fact that the expected search time is bounded by the Shannon's entropy of the probability distribution for the source location (Vergassola et al., 2007). The reduction of entropy in the estimated distribution is thus a necessary (although not sufficient) condition for effective searching. No pure mathematical proof of the algorithmic convergence exists for Mapless and swarm Mapless. Yet, simulations in Masson (2013) show exponentially tailed distributions for the search time ensuring that the average search time is not driven by the tail dynamics. Furthermore, there is a non-nul probability of having no detection during the initial spiraling exploratory behavior, thus

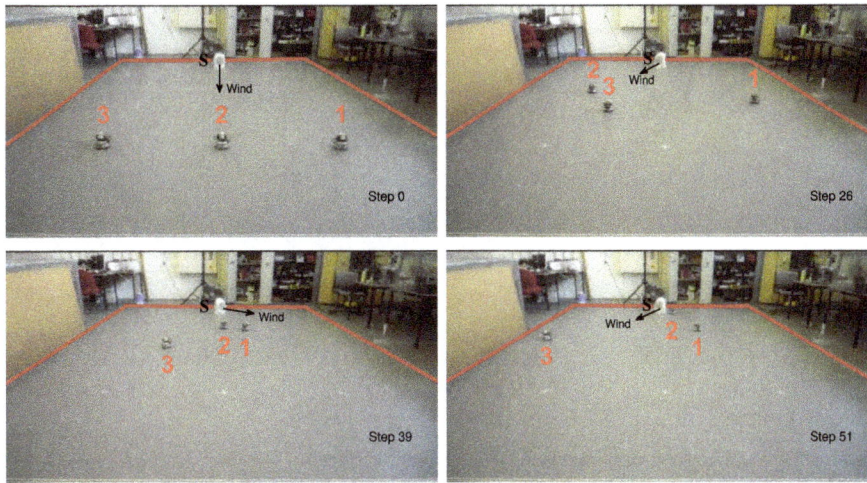

FIGURE 6 | Swarm searching with three robots. Snapshot of the collective search at particular steps. At each step, the source S is at location (10,24) and wind blows as indicated by the arrow. At step 0, the three robots start from locations (5,6), (10,6), and (15,6). At step 51, robot #2 found the source.

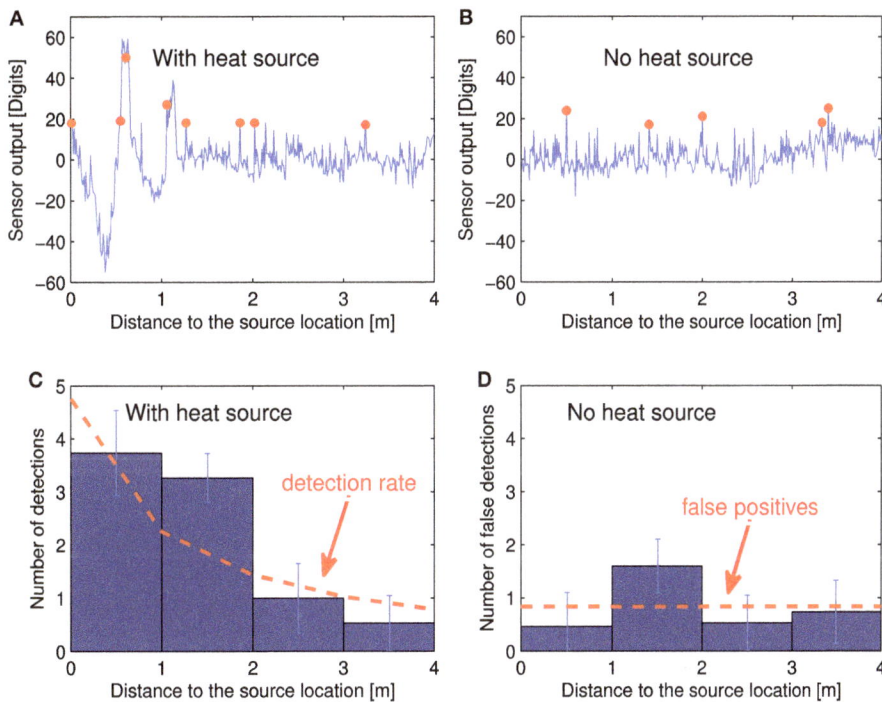

FIGURE 7 | Detection events with and without source. (A,B) Temporal evolution of the measured temperature as a function of the source distance when the robot moves straight toward the source location. The blue curve represents the local variation of the temperature; that is, the difference between the current temperature and a running average calculated over a 10-s sliding window. Red dots correspond to detection events triggered each time the variation in temperature exceeds 15 digits. **(C,D)** Histogram of the number of detections with respect to the source distance d with and without the heat source ($n = 12$ trials in each condition). The dashed curve in C corresponds to a fit with the detection rate $R(\vec{r}, \vec{r_0})$ derived in the Section "Appendix" with a correlation length of $\lambda = 20$ au. The dashed line in D corresponds to a mean false alarm rate of 0.08 detection/s.

not all searches are insured to find the source. Here, we provided statistical measures of the search time based on more than 10^5 simulations (**Figure 1A**) and 10^2 robotic experiments (**Figure 8A**).

The power law dependency of the search time on the number of searchers revealed significant gains even with a small number of robots (e.g., $n = 3$). The search time was found to decay as $1/n$ for swarm Mapless with n robots as compared to $1/\sqrt{n}$ for independent Mapless robots. Search efficiency results from pooling sensory information between robots to improve individual decision-making (three detections

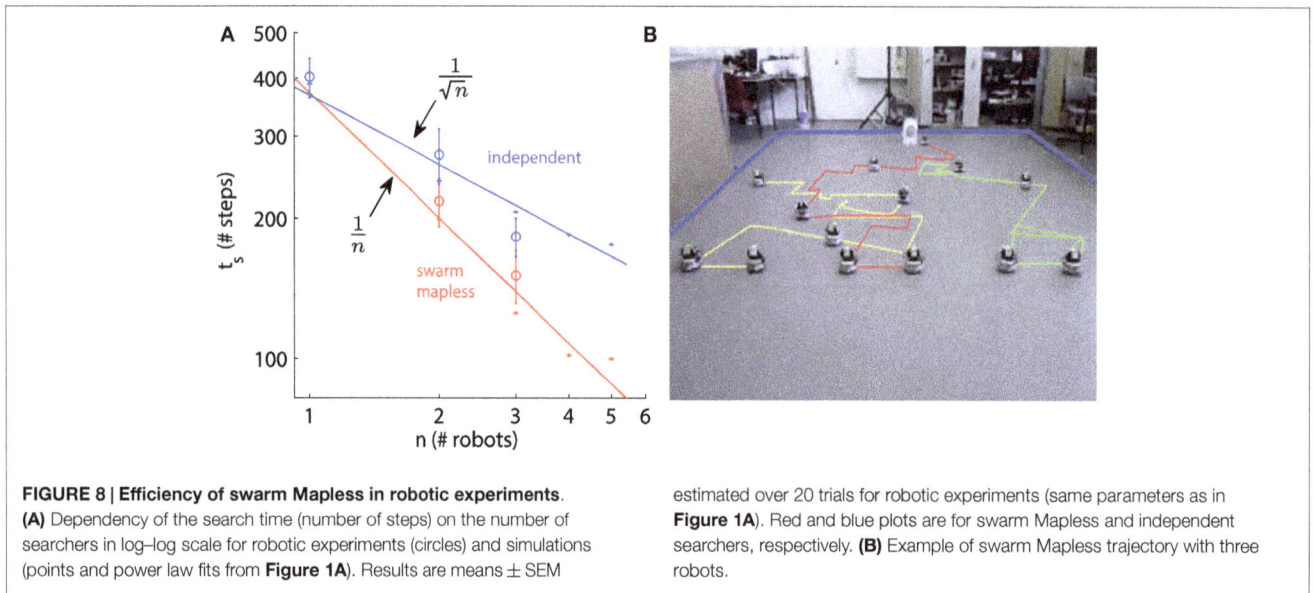

FIGURE 8 | Efficiency of swarm Mapless in robotic experiments.
(A) Dependency of the search time (number of steps) on the number of searchers in log–log scale for robotic experiments (circles) and simulations (points and power law fits from **Figure 1A**). Results are means ± SEM estimated over 20 trials for robotic experiments (same parameters as in **Figure 1A**). Red and blue plots are for swarm Mapless and independent searchers, respectively. **(B)** Example of swarm Mapless trajectory with three robots.

on average per searcher were sufficient to reach the source). In our experiments, loss of efficiency due to collision was not a problem in part because of the small number of robots exploring a relatively large search space and also because the robots tend to repel each other when their distance is inferior to the correlation length of the source, a behavior also observed in infotactic searches (Masson et al., 2009). Yet, it is worth noting that the repellent effect between robots is much weaker than for swarm Infotaxis. It is the consequence of the simplified representation of the environment.

Search methods based on an information gradient, e.g., Atanasov et al. (2015), are related to our work. They are guaranteed to converge to a local maximum of the mutual information (and thereby to the source location provided it corresponds to the local maximum). Yet, it is difficult to judge their efficiency from the literature as no theoretical estimation or upper bound on the search time is given – for example, the searcher may spend a lot of time far from the source where the information gradient is very small – and no evaluation was conducted under real turbulent conditions – mere simulations were performed with a homogeneous signal field in Atanasov et al. (2015). When considering robotic implementation, we also note that calculating an information metric analytically (as in swarm Mapless) is computationally more efficient than estimating it via particle filters.

Future work will then concentrate on comparing the performance of swarm Mapless (in terms of search time and computational complexity) to related approaches on real robotic problems including obstacle-cluttered environments. Another interesting line of research that may prove beneficial and ought to be considered as future work is the generalization of swarm Mapless to cope with multiple sources, as done for example in Masson et al. (2009) for Infotaxis and in Masson (2013) for Mapless.

Author Contributions

JM designed the algorithm, SZ, DM, and JM designed research; SZ, DM, and JM performed research; SZ and DM performed the experiments, DM and JM wrote the paper.

Acknowledgments

This work was funded by the state program investissements d'avenir managed by ANR (grant ANR-10-BINF-05 Pherotaxis). SZ acknowledges support from the National Natural Science Foundation of China under grant 61472325, 51209174, and 51311130137.

References

Atanasov, N., Le Ny, J., and Pappas, G. (2015). Distributed algorithms for stochastic source seeking with mobile robot networks. *J. Dyn. Syst. Meas. Control* 137, 031004. doi:10.1115/1.4027892

Baker, T., Willis, M., Haynes, K., and Phelan, P. (1985). A pulsed cloud of sex pheromone elicits upwind flight in male moths. *Physiol. Entomol.* 10, 257–265. doi:10.1111/j.1365-3032.1985.tb00045.x

Barbieri, C., Cocco, S., and Monasson, R. (2011). On the trajectories and performance of infotaxis, an information based Greedy search algorithm. *Europhys. Lett.* 94, 20005. doi:10.1209/0295-5075/94/20005

Berdahl, A., Torney, C., Ioannou, C., Faria, J., and Couzin, I. (2013). Emergent sensing of complex environments by mobile animal groups. *Science* 339, 574–576. doi:10.1126/science.1225883

Calhoun, A., Chalasani, S., and Sharpee, T. (2014). Maximally informative foraging by *Caenorhabditis elegans*. *eLife* 3, e04220. doi:10.7554/eLife.04220

Csanady, G. T. (1973). *Turbulent Diffusion in the Environment*. Dordrecht: D. Reidel Publishing Company.

Celani, A., Villermaux, E., and Vergassola, M. (2014). Odor landscape in turbulent environments. *Phys. Rev.* 4, 1–17. doi:10.3791/51704

Cochran, J., and Krstic, M. (2009). Nonholonomic source seeking with tuning of angular velocity. *IEEE Trans. Automat. Contr.* 54, 717–731. doi:10.1109/TAC.2009.2014927

Cortez, A., Tanner, H., and Lumia, R. (2009). Distributed robotic radiation mapping. *Exp. Robot.* 54, 147–156. doi:10.1007/978-3-642-00196-3_17

Dames, P., and Kumar, V. (2013). "Cooperative multi-target localization with noisy sensors," in *IEEE International Conference on Robotics and Automation (ICRA)* (Karlsruhe).

Didier, B. (2010). Pyrophiles, ces insectes qui aiment le feu . *Insectes* 156, 13–16.

Friston, K., Daunizeau, J., and Kilner, J. (2010). Action and behavior: a free-energy formulation. *Biol. Cybern.* 102, 227–260. doi:10.1007/s00422-010-0364-z

Friston, K., Mattout, J., and Kilner, J. (2011). Action understanding and active inference. *Biol. Cybern.* 104, 137–160. doi:10.1007/s00422-011-0424-z

Gelenbe, E., Schmajuk, N., Staddon, J., and Rief, J. (1997). Autonomous search by robots and animals: a survey. *Rob. Auton. Syst.* 22, 23–34. doi:10.1016/S0921-8890(97)00014-6

Humphrey, J.-A., and Haj-Hariri, H. (2012). "Stagnation point flow analysis of odorant detection by permeable moth antennae," in *Frontiers in Sensing*. eds F. G. Barth, J. A. C. Humphrey, and M. V. Srinivasan (Wien: Springer-Verlag), 171–192. doi:10.1007/978-3-211-99749-9_12

Li, S., Guo, Y., and Bingham, B. (2014). "Multi-robot cooperative control for monitoring and tracking dynamic plumes," in *IEEE International Conference on Robotics and Automation (ICRA)* (Hong Kong).

Mafra-Neto, A., and Carde, R. T. (1994). Fine-scale structure of pheromone plumes modulated upwind orientation of flying moths. *Nature* 369, 142–144. doi:10.1038/369142a0

Martinez, D., Arhidi, L., Demondion, E., Masson, J.-B., and Lucas, P. (2014). Using insect electroantennogram sensors on autonomous robots for olfactory searches. *J. Vis. Exp.* 90, e51704. doi:10.3791/51704

Martin-Moraud, E., and Martinez, D. (2010). Effectiveness and robustness of robot infotaxis for searching in dilute conditions. *Front. Neurorobot.* 4:1. doi:10.3389/fnbot.2010.00001

Masson, J.-B. (2013). Olfactory searches with limited space perception. *Proc. Natl. Acad. Sci. U.S.A.* 110, 11261–11266. doi:10.1073/pnas.1221091110

Masson, J.-B., Bailly-Bechet, M., and Vergassola, M. (2009). Chasing information to search in random environments. *J. Phys. A Math. Theor.* 42:434009. doi:10.1088/1751-8113/42/43/434009

Murlis, J., Elkinton, J. S., and Card, R. T. (1992). Odor plumes and how insects use them. *Annu. Rev. Entomol.* 37, 479–503. doi:10.1146/annurev.en.37.010192.002445

Ogren, P., Fiorelli, E., and Leonard, N. (2004). Cooperative control of mobile sensor networks: adaptive gradient climbing in a distributed environment. *IEEE Trans. Automat. Contr.* 49, 1292–1302. doi:10.1109/TAC.2004.832203

Pascoal, J., Marques, L., and de Almeida, A. (2008). "Assessment of laser range finders in risky environments," in *IEEE/RSJ International Conference on, Intelligent Robots and Systems, IROS 2008* (IEEE), 3533–3538.

Schmitz, H., Bleckmann, H., and Murtz, M. (1997). Infrared detection in a beetle. *Nature* 386, 773–774. doi:10.1038/386773a0

Schmitz, H., and Bousack, H. (2012). Modelling a historic oil-tank fire allows an estimation of the sensitivity of the infrared receptors in pyrophilous *Melanophila* beetles. *PLoS ONE* 7:e37627. doi:10.1371/journal.pone.0037627

Schutz, S., Weissbecker, B., Hummel, H. E., Apel, K.-H., Schmitz, H., and Bleckmann, H. (1999). Insect antenna as a smoke detector. *Nature* 398, 298–299. doi:10.1038/18585

Shraiman, B. I., and Siggia, E. D. (2000). Scalar turbulence. *Nature* 405, 639–646. doi:10.1038/35015000

Smoluchowski, M. (1917). Versuch einer mathematischen theorie des koagulation-slinetic kolloider losungen. *Z. Phys. Chem.* 92, 129–168.

Starr, J., and Lattimer, B. (2014). Evaluation of navigation sensors in fire smoke environments. *Fire Technol.* 50, 1459–1481. doi:10.3390/s101210953

Thrun, S., Burgard, W., and Dieter, F. (2005). *Probabilistic Robotics*. Cambridge, MA: The MIT Press.

Vergassola, M., Villermaux, E., and Shraiman, B. I. (2007). Infotaxis as a strategy for searching without gradients. *Nature* 445, 406–409. doi:10.1038/nature05464

Vickers, N. (2000). Mechanisms of animal navigation in odor plumes. *Biol. Bull.* 198, 203–212. doi:10.2307/1542524

Zhang, C., Arnold, D., Ghods, N., Siranosian, A., and Krstic, M. (2007). Source seeking with non-holonomic unicycle without position measurement and with tuning of forward velocity. *Syst. Contr. Lett.* 56, 245–252. doi:10.1016/j.sysconle.2006.10.014

Conflict of Interest Statement: The authors declare that the research was conducted in the absence of any commercial or financial relationships that could be construed as a potential conflict of interest.

Appendix

A.1 Detection Rate Function in a Simplified Turbulent Medium

We consider a source located at $\vec{r_0} = (x_0, y_0)$ and emitting "particles" at a rate J. The particles propagate in the environment with diffusivity D, have a mean lifetime τ and are advected by a mean current or wind V (the wind blows in the $-y$ direction). The rate function $R(\vec{r}, \vec{r_0})$ models how particles are detected at location $\vec{r} = (x, y)$ given the source at $\vec{r_0}$. It is obtained by solving the advection–diffusion equation

$$D\nabla^2 C(\vec{r}) + \vec{V} \cdot \nabla C(\vec{r}) - \frac{1}{\tau} C(\vec{r}) - J\delta(\vec{r} - \vec{r_0}) = 0 \quad (A1)$$

where $C(\vec{r})$ is the local concentration of particles at $\vec{r}$ and δ is the Dirac delta function. In the three dimensional case, the solution to equation A1 writes:

$$C(\vec{r}, \vec{r_0}) = \frac{J}{4\pi Dr} e^{\frac{-(y-y_0)V}{2D}} e^{\frac{-r}{\lambda}} \quad (A2)$$

where r is the distance from the source and $\lambda = \sqrt{D\tau/(1 + \frac{V^2\tau}{4D})}$ is the correlation length that can be interpreted as the mean distance traveled by the particles before they vanish. A similar expression is obtained in the 2D case (Vergassola et al., 2007). Considering that particles are detected with a spherical sensor of radius "a," the detection rate follows the Smoluchowski's expression (Smoluchowski, 1917)

$$R(\vec{r}, \vec{r_0}) = 4\pi Da \cdot C(\vec{r}, \vec{r_0}) \quad (A3)$$

17

Prospection in cognition: the case for joint episodic-procedural memory in cognitive robotics

David Vernon[1]*, Michael Beetz[2] and Giulio Sandini[3]

[1] *Interaction Lab, School of Informatics, University of Skövde, Skövde, Sweden,* [2] *Institute for Artificial Intelligence, University of Bremen, Bremen, Germany,* [3] *Department of Robotics, Brain and Cognitive Sciences, Istituto Italiano di Tecnologia, Genova, Italy*

Edited by:
Guido Schillaci,
Humboldt University of Berlin,
Germany

Reviewed by:
John Nassour,
Technische Universität Chemnitz,
Germany
Arnaud Blanchard,
University of Cergy-Pontoise, France
Alessandro Di Nuovo,
Plymouth University, UK

*Correspondence:
David Vernon,
School of Informatics, University of
Skövde, P.O. Box 408, Skövde
SE-54128, Sweden
david.vernon@his.se

Prospection lies at the core of cognition: it is the means by which an agent – a person or a cognitive robot – shifts its perspective from immediate sensory experience to anticipate future events, be they the actions of other agents or the outcome of its own actions. Prospection, accomplished by internal simulation, requires mechanisms for both perceptual imagery and motor imagery. While it is known that these two forms of imagery are tightly entwined in the mirror neuron system, we do not yet have an effective model of the mentalizing network which would provide a framework to integrate declarative episodic and procedural memory systems and to combine experiential knowledge with skillful know-how. Such a framework would be founded on joint perceptuo-motor representations. In this paper, we examine the case for this form of representation, contrasting sensory-motor theory with ideo-motor theory, and we discuss how such a framework could be realized by joint episodic-procedural memory. We argue that such a representation framework has several advantages for cognitive robots. Since episodic memory operates by recombining imperfectly recalled past experience, this allows it to simulate new or unexpected events. Furthermore, by virtue of its associative nature, joint episodic-procedural memory allows the internal simulation to be conditioned by current context, semantic memory, and the agent's value system. Context and semantics constrain the combinatorial explosion of potential perception-action associations and allow effective action selection in the pursuit of goals, while the value system provides the motives that underpin the agent's autonomy and cognitive development. This joint episodic-procedural memory framework is neutral regarding the final implementation of these episodic and procedural memories, which can be configured sub-symbolically as associative networks or symbolically as content-addressable image databases and databases of motor-control scripts.

Keywords: autonomy, cognitive system, development, episodic memory, ideo-motor theory, internal simulation, procedural memory, prospection

Introduction

The goal of this article is to argue the case of the use of joint episodic memory to facilitate prospection and goal-directed action in cognitive robotics. The article begins with insights from the biological sciences regarding the prospective nature of action, leading to a discussion of the role of memory in prospection, and the realization of prospection through internal simulation. This sets the scene for the introduction of ideo-motor theory, *vis-à-vis* sensory-motor theory, and an explanation of the importance of joint perceptuo-motor representations. This is then followed by two examples of how these principles have been applied in cognitive architectures and an argument in favor of explicit perceptuo-motor memory – joint episodic-procedural memory – over perceptuo-motor mappings. We finish with a description of a simple proof-of-principle example implementation of joint episodic-procedural memory for overt attention.

The Goal-Directed and Prospective Nature of Action

Evidence from many different fields of research, including psychology and neuroscience, suggests that the movements of biological organisms are organized as actions and not reactions (von Hofsten, 2004). While reactions are elicited by earlier events, actions are initiated by a motivated subject, they are defined by goals, and they are guided by prospective information (Vernon et al., 2010). For example, when performing manipulation tasks or observing someone else performing them, subjects fixate on the goals and sub-goals of the movements not on the body parts, e.g., the hands or the objects (Johansson et al., 2001; Flanagan and Johansson, 2003). This happens only if a goal-directed action is implied. When showing the same movements without the context of an agent, subjects fixate the moving object instead of the goal.

Evidence from neuroscience also shows that the brain represents movements in terms of actions even at the level of neural processes [see Vernon et al. (2010), Chapter 4]. For example, the primate brain has two areas devoted to controlling movements: the premotor cortex and the motor cortex. The premotor cortex is the area of the brain that is active during motor planning and it influences the motor cortex which then executes the motor program comprising an action. The premotor cortex receives strong visual inputs from a region in the brain known as the inferior parietal lobule. These inputs serve a series visuomotor transformations for reaching (Area F4) and grasping (Area F5). Single neuron studies have shown that most F5 neurons code for specific goal-directed actions, rather than their constituent movements. Furthermore, several F5 neurons, in addition to their motor properties, respond also to visual stimuli. These are referred to as visuomotor neurons. The significance of this is that the premotor cortex of primates encodes actions (including implicit goals and expected states) and not just movements. The goal, therefore, is the fundamental property of the action rather than the specific motoric details of how it is achieved.

In primates, two classes of visuomotor neurons can be distinguished within area F5: canonical neurons and mirror neurons (Rizzolatti and Fadiga, 1998). The activity of both canonical and mirror neurons correlates with two distinct circumstances. In the case of canonical neurons, the same canonical neuron fires when a monkey sees a particular object and also when the monkey actually grasps an object with the same characteristic features. On the other hand, mirror neurons (Gallese et al., 1996; Rizzolatti et al., 1996; Rizzolatti and Craighero, 2004) are activated both when an action is performed and when the same or similar action is observed being performed by another agent. These neurons are specific to the goal of the action and not the mechanics of carrying it out. So, for example, a monkey observing another monkey, or a human, reaching for a nut will cause mirror neurons in the premotor cortex to fire; these are the same neurons that fire when the monkey actually reaches for a nut. However, if the monkey observes another monkey making exactly the same movements but there is no nut present – there is no apparent goal of the reaching action – then the mirror neurons do not fire. Similarly, different motions that comprise the same goal-directed action will cause the same mirror neurons to fire. It is the action that matters: mirror neurons are not active if there is no explicit or implied goal. Since goals focus on the future, not the present, this again demonstrates the importance of prospection in action.

Finally, there is another reason why actions are guided by prospective information as opposed to instantaneous feedback data. Often, events in the agent's world may precede the feedback signals about them because the delays in the control pathways of biological systems may be substantial. If you cannot rely on feedback, the only way to overcome the problem is to anticipate what is going to happen next and to use that information to control one's behavior.

Prospection, then, is central to cognition. The question is how this prospection is achieved. The answer is, somewhat surprisingly, memory.

Memory

Memory facilitates the persistence of knowledge and forms a reservoir of experience. Without it, it would be impossible for the system to learn, develop, adapt, recognize, plan, deliberate, and reason (Vernon, 2014). Memory functions to preserve what has been achieved through learning and development, ensuring that, when a cognitive system adapts to new circumstances, it does not lose its ability to act effectively in situations to which it had adapted previously. But memory has another role in addition to preserving past experience: to anticipate the future. It forms the basis for one of the central pillars of cognitive capacity, i.e., the ability to simulate internally the outcomes of possible actions and select the one that seems most appropriate for the current situation. Viewed in this light, memory can be seen as a mechanism that allows a cognitive agent *to prepare to act*, overcoming through anticipation the inherent "here-and-now" limitations of its perceptual capabilities.

We can distinguish memory in many ways (Squire, 2004; Wood et al., 2012). For example, it can be distinguished on the basis of the nature of what is remembered and the type of access we have to it. Specifically, memory can be categorized as either *declarative* or *procedural*, depending on whether it captures knowledge of things – facts – or actions – skills. Sometimes they are characterized as memory of knowledge and know-how: "knowing that" and "knowing how."[1] This distinction applies mainly to long-term memory but short-term memory too has a declarative aspect. Declarative memory is sometimes referred to as *propositional memory* because it refers to information about the agent's world

---

[1] The distinction between *knowing that* and *knowing how* was made in 1949 by Gilbert Ryle in his book *The Concept of Mind* (Ryle, 1949)

that can be expressed in the form of propositions. This is significant because propositions are either true or false. Thus, declarative memory typically deals with factual information. This is not the case with skill-oriented procedural memory. As a consequence, declarative memories, in the form of knowledge, can be communicated from one agent to another through language, for example, whereas procedural memories can only be demonstrated.

Two different types of declarative memory can be distinguished. These are *episodic memory* and *semantic memory*. Episodic memory (Tulving, 1972, 1984) plays a key role in cognition and in the anticipatory aspect of cognition in particular. It refers to specific instances in the agent's experience while semantic memory refers to general knowledge about the agent's world which may be independent of the agent's specific experiences. In this sense, episodic memory is autobiographical. By its very nature in encapsulating some specific event in the agent's experience, episodic memory has an explicit spatial and temporal context: what happened, where it happened, and when it happened. This temporal sequencing is the only element of structure in episodic memory. Episodic memory is a fundamentally *constructive* process (Seligman et al., 2013). Each time an event is assimilated into episodic memory, past episodes are reconstructed. However, they are reconstructed a little differently each time. This constructive characteristic is related to the role that episodic memory plays in the process of internal simulation that forms the basis of prospection, the key anticipatory function of cognition.

In contrast, semantic memory "is the memory necessary for the use of language. It is a mental thesaurus, organized knowledge a person possesses about words and other verbal symbols, their meaning and referents, about relations among them, and about rules, formulas, and algorithms for the manipulation of the symbols, concepts, and relations."[2]

Episodic memory and semantic memory differ in many ways. In general, semantic memory is associated with how we understand (or model) the world around us, using facts, ideas, and concepts. On the other hand, episodic memory is closely associated with experience: perceptions and sensory stimulus. While episodic memory has no structure other than its temporal sequencing, semantic memory is highly structured to reflect the relationships between constituent concepts, ideas, and facts. Also, the validity (or truth, since semantic memory is a subset of propositional declarative memory) of semantic memories is based on social agreement rather than personal belief, as it is with episodic memory.[3] Semantic memory can be derived from episodic memory through a process of generalization and consolidation. Episodic memory can be both short-term and long-term while semantic memory and procedural memory are long-term.

Memory and Prospection

Memory plays at least four roles in cognition: it allows us to remember past events, anticipate future ones, imagine the

viewpoint of other people, and navigate around our world. All four involve *self-projection*: the ability of an agent to shift perspective from itself in the here-and-now and to take an alternative perspective. It does this by *internal simulation*, i.e., the mental construction of an imagined alternative perspective (Schacter et al., 2008). Thus, there are four forms of internal simulation (Buckner and Carroll, 2007):

1. Episodic memory (remembering the past).
2. Navigation (orienting yourself topographically, i.e., in relation to your surroundings).
3. Theory of mind (taking someone else's perspective on matters).
4. Prospection (anticipating possible future events).

Each form of simulation has a different orientation (past, present, or future) and each refers to the perspective of either the first person, i.e., the agent itself, or another person.

Prospection – the mental simulation of future possibilities – plays a central role in organizing perception, cognition, affect, memory, motivation, and action (Seligman et al., 2013). Prospection is referred to in various ways, e.g., *episodic future thinking*, *memory of the future*, *pre-experiencing*, *proscopic chronesthesia*, *mental time travel*, and just plain *imagination* and it can involve conceptual content and affective – emotional – states (Buckner and Carroll, 2007).

Recent evidence suggests that all four kinds of internal simulation involve a single core brain network and this network overlaps what is known as the *default-mode network*, a set of interconnected regions in the brain that is active when the agent is not occupied with some attentional tasks (Østby et al., 2012).

It is significant that all four forms of simulation are constructive, i.e., they involve a form of imagination. There is a difference between knowing about the future and projecting ourselves into the future. The latter is experiential and the former is not. Thus, episodic memory (memory of experiences) and semantic memory (memory of facts) facilitate different types of prospection. Episodic memory allows you to re-experience your past and pre-experience your future. There is evidence that projecting yourself forward in time is important when you form a goal, creating a mental image of yourself acting out the event and then episodically pre-experiencing the unfolding of a plan to achieve that goal. This use of episodic memory in prospection is referred to as *episodic future thinking*, a term coined by Cristina Atance and Daniela O'Neill to refer to the ability to project oneself forward in time to pre-experience an event (Atance and O'Neill, 2001; Szpunar, 2010).

The constructive aspect of episodic memory, whereby old episodic memories are reconstructed slightly differently every time a new episodic memory is assimilated or remembered, is particularly important in the context of internal simulation of events that have not been previously experienced. While episodic memory certainly needs some constructive capacity to assemble individual details into a coherent memory of a given episode, the *constructive episodic simulation hypothesis* (Schacter and Addis, 2007a,b; Schacter et al., 2008; Szpunar, 2010) suggests that its role in prospection involving the simulation of multiple possible futures imposes an even greater need for a constructive capacity because of the need to extrapolate beyond past experiences. In

---

[2] This quotation explaining the characteristics of semantic memory appears in Endel Tulving's 1972 article (Tulving, 1972), p. 386 and is quoted in his Précis (Tulving, 1984). While this definition of semantic memory dates from 1972, it is still valid today. It also explains the linguistic origins of the term.

[3] Semantic memory and episodic memory can be contrasted in many other ways: twenty-seven differences are listed in Tulving (1983), p. 35.

other words, simulating multiple yet-to-be-experienced futures requires flexibility in episodic memory. This flexibility is possible because episodic memory is not an exact and perfect record of experience but one that conveys the essence of an event and is open to re-combination.

It is also significant that when humans imagine the future, they not only anticipate an event, but they also anticipate how they feel about that event. These are referred to as *hedonic* consequences of the event: whether we feel good about it or bad about it, whether it is associated with pleasure or pain, and lack of concern or fear. Thus, the pre-experience of prospection also involves "pre-feeling." The brain accomplishes prospection by simulating the event and the associated hedonic experience (Gilbert and Wilson, 2007). While pre-feeling is not always reliable because contextual factors also play a part in the hedonic experience, this hedonic aspect of episodic memory is important because it reflects the affective nature of cognition and opens up a plausible way to factor emotional drives and value systems into the operation of memory, prospection, and action selection.

Internal Simulation and Action

In the preceding section, we considered internal simulation entirely in terms of memory-based self-projection, using re-assembled combinations of episodic memory to pre-experience possible futures, re-experience (and possibly adjust) past experiences, and project ourselves into the experiences of others. However, we know that action plays a significant role in our perceptions so the question then is: does action play a role in internal simulation? The answer is a clear "yes" (Hesslow, 2002, 2012; Svensson et al., 2007). Internal simulation extends beyond episodic memory and includes simulated interaction, particularly embodied interaction. Although the terms simulation, internal simulation, and mental simulation are widely used, you will also see references being made to *emulation*, very often when the approach endeavors to model the exact mechanism by which the simulation is produced (Grush, 2004).

The Simulation Hypothesis

There are a number of simulation theories, but perhaps the most influential is what is known as the *simulation hypothesis* (Hesslow, 2002, 2012). It makes three core assumptions:

1. The regions in the brain which are responsible for motor control can be activated *without* causing bodily movement.
2. Perceptions can be caused by internal brain activity and not just by external stimuli.
3. The brain has associative mechanisms that allow motor behavior or perceptual activity to evoke other perceptual activity.

The first assumption allows for simulation of actions and is often referred to as *covert action* or *covert behavior*. The second allows for simulation of perceptions. The third assumption allows simulated actions to elicit perceptions that are like those that would have arisen if the actions had actually been performed. There is an increasing amount of neurophysiological evidence in support of all three assumptions (Svensson et al., 2013). If we link these assumptions together, we see that the simulation hypothesis

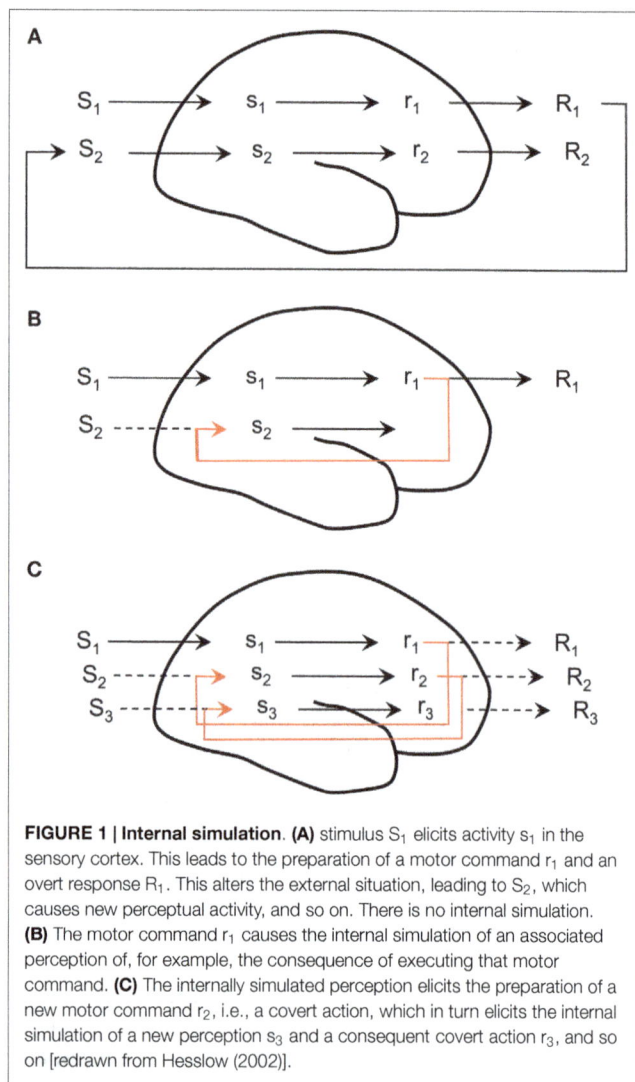

FIGURE 1 | **Internal simulation.** **(A)** stimulus S_1 elicits activity s_1 in the sensory cortex. This leads to the preparation of a motor command r_1 and an overt response R_1. This alters the external situation, leading to S_2, which causes new perceptual activity, and so on. There is no internal simulation. **(B)** The motor command r_1 causes the internal simulation of an associated perception of, for example, the consequence of executing that motor command. **(C)** The internally simulated perception elicits the preparation of a new motor command r_2, i.e., a covert action, which in turn elicits the internal simulation of a new perception s_3 and a consequent covert action r_3, and so on [redrawn from Hesslow (2002)].

shows how the brain can simulate extended perception-action-perception sequences by having the simulated perceptions elicit simulated action which in turn elicits simulated perceptions, and so on. **Figure 1** summarizes the simulation hypothesis, showing three situations, one where there is no internal simulation, one where a motor response to an input stimulus causes the internal simulation of an associated perception, and one where this internally simulated perception then elicits a covert action which in turn elicits a simulated perception and a consequent covert action, and so on.

Motor, Visual, and Mental Imagery

Action-directed internal simulation involves three different types of anticipation: implicit, internal, and external (Svensson et al., 2009). *Implicit anticipation* concerns the prediction of motor commands from perceptions (which may have been simulated in a previous phase of internal simulation). *Internal anticipation* concerns the prediction of the proprioceptive consequences of carrying out an action, i.e., the effect of an action on the agent's own body. *External anticipation* concerns the prediction of the consequences for external objects and other agents of carrying out an action.[4] Implicit anticipation selects some motor activity (possibly covert, i.e., simulated) to be carried out based on an association between stimulus and actions; internal anticipation and external anticipation then predict the consequences of that action. Collectively, they simulate actions and the effects of actions.

Covert action involves what is referred to as *motor imagery* and simulation of perception is often referred to as *visual imagery*. Perceptual imagery would perhaps be a better term since there is evidence that humans use imagery from all the senses. In a way, motor imagery is also a form of perceptual imagery, in the sense that it involves the proprioceptive and kinesthetic sensations associated with bodily movement. However, reflecting the interdependence of perception and action, covert action often has elements of both motor imagery and visual imagery and, *vice versa*, the simulation of perception often has elements of motor imagery. Visual imagery and motor imagery are sometimes referred to collectively as *mental imagery* (Wintermute, 2012). Moulton and Kosslyn (2009) identify several different types of perceptual imagery and distinguish between two different types of simulation: *instrumental simulation* and *emulative simulation*. The former concerns itself only with the content of the simulation while the latter also replicates the process by which that content is created in the simulated event itself. They refer to this as *second-order simulation*.

Joint Perceptuo-Motor Representations

In the foregoing, we remarked on the fact that mental imagery, viewed as another way of expressing the process of internal simulation, comprises both visual imagery (or perceptual imagery) and motor imagery. More importantly, though, we noted that these two forms of imagery are tightly entwined: they complement each other and the simulation of perception and covert action both involve elements of visual and motor imagery.

Classical treatments of memory usually maintain a clear distinction between declarative memory and procedural memory, in general, and between episodic memory and procedural memory, in particular. However, contemporary research takes a slightly different perspective, binding the two more closely, e.g., the mirror neuron system, in particular. While it is still a major challenge to understand how these two memory systems are combined, this coupling is the basic idea underpinning joint perceptuo-motor representations: representations that bring together the motoric and sensory aspects of experience in one framework, such as that anticipated in the simulation hypothesis.

In this section, we look at four approaches that have been developed to address joint perceptuo-motor representations. First, we look at two approaches to implementing ideo-motor theory in cognitive robotics: Shanahan's Global Workspace Theory architecture and Demiris's HAMMER architecture. We follow this by highlighting two additional approaches that endeavor to integrate perceptuo-motor representations more tightly: the Theory of Event Coding (TEC) and Object-Action Complexes. Since none of these explicitly incorporate episodic or procedural memory, we then suggest a way of drawing the principal ideas of each together

in a form of explicit joint episodic-procedural memory. We then argue that this joint episodic-procedural memory allows several of the challenges of cognitive robotics to be addressed.

Before discussing these, to provide the necessary context for prospective perceptuo-motor representations, we first address the difference between sensory-motor theory and ideo-motor theory.

Sensory-Motor Theory and Ideo-Motor Theory

Broadly speaking, there are the two distinct approaches for planning actions: *sensory-motor* action planning and *ideo-motor* action planning (Stock and Stock, 2004). Sensory-motor action planning treats actions as reactive responses to sensory stimuli and assumes that perception and action use distinct and separate representational frameworks. The sensory-motor view builds on the classic unidirectional data-driven information-processing approach to perception, proceeding stage by stage from stimulus to percept and then to response. It is unidirectional in that it does not allow the results of later processing to influence earlier processing. In particular, it does not allow the resultant (or intended) action to impact on the related sensory perception.

Ideo-motor action planning, on the other hand, treats action as the result of internally generated goals. It is the idea of achieving some action outcome, rather than some external stimulus, that is at the core of how cognitive agents behave. This reflects the view of action described above, with action being initiated by a motivated subject, defined by goals, and guided by prospection. The key point of the ideo-motor principle is that the selection and control of a particular goal-directed movement depends on the anticipation of the sensory consequence of accomplishing the intended action: the agent images (e.g., through internal simulation) the desired outcome and selects the appropriate actions in order to achieve it.

There is an important difference, though, between the concrete movements comprising an action and the higher-order goals of an action. Typically, actors do not voluntarily preselect the exact movements required to achieve a desired goal. Instead, they select prospectively guided intention-directed goal-focused action, with the specific movements being adaptively controlled as the action is executed. Thus, ideo-motor theory should be viewed both as an anticipatory idea-centered way of selecting actions and as a way of bridging the higher-order conceptual representations of intentions and goals[5] with the concrete adaptive control of movements when executing that action (Ondobaka and Bekkering, 2012).

In contrast to sensory-motor models, ideo-motor theory assumes that perception and action share a common representational framework. Because ideo-motor models focus on goals, and because they use a common joint representation that embraces both perception and action, they provide an intuitive explanation of why cognitive agents, humans in particular, are so adept at and predisposed to imitation (Iacoboni, 2009). The essential idea is that when I see somebody else's (goal-directed) actions and

---

[4]The terms *internal anticipation* and *external anticipation* are also referred to as *bodily anticipation* and *environmental anticipation* (Svensson et al., 2013).

[5]Michael Tomasello and colleagues note that the distinction between intentions and goals is not always clearly made. Taking their lead from Michael Bratman (1998), they define an intention as a plan of action an agent chooses and commits itself to in pursuit of a goal. An intention therefore includes both a means (i.e. an action plan) as well as a goal (Tomasello et al., 2005).

the consequences of these actions, the representations of my own actions that would produce the same consequences are activated.

At first glance, ideo-motor theory seems to present a puzzle: how can the goal, achieved through action, cause the action in the first place? In other words, how can the later outcome affect the earlier action? This seems to be a case of *backward causation*. The solution to the puzzle is prospection. It is the anticipated goal state, not the achieved goal state, that impacts on the associated planned action. Goal-directed action, then, is a center-piece of ideo-motor theory, which is also referred to as the *goal trigger hypothesis* (Hommel et al., 2001).

Before proceeding to consider two cognitive architectures that build on ideo-motor theory, we mention *cognitive maps* to highlight the importance of joint perceptuo-motor representations in animal and robot cognition. The idea of a cognitive map was introduced by Tolman as a geometric representation to support navigation in biological agents (Tolman, 1948). While there is a certain lack of consensus on what exactly constitutes a cognitive map (Bennett, 1996; Eichenbaum et al., 1999), most agree that it involves metric information rather than purely topological information to encode spatial relationships in an allocentric framework and that it exploits path integration, at least partially, to effect navigation (Gallistel, 1989, 1990; Stachenfeld et al., 2014); for an alternative perspective, see Gaussier et al. (2002). In any case, a cognitive map combines memories of environmental cues (or perceptual landmarks) with geometrical properties of space that are specified by the remembered landmarks (Metta et al., 2010). Based on the existence of the hippocampus place cells (O'Keefe, 1976), O'Keefe and Nadel suggested that the hippocampal formation provides the neural basis for the cognitive map (O'Keefe and Nadel, 1978).

However, the hippocampus does not just create and store cognitive maps but it also plays a part in episodic memory, e.g., helping to minimize the similarities between new representations and representations that already exist in memory (McNaughton et al., 2006). As with episodic memory, it is also responsible for associating information in ways that allow flexible use of past experiences to guide future actions (*flexible memory expression*) (Eichenbaum et al., 1999; McNamara and Shelton, 2003). Furthermore, it has a role as a prediction mechanism for novelty detection and especially as a way to merge planning and sensory-motor function in a single coherent system (Gaussier et al., 2002). As McNaughton et al. note, ". our current understanding of [the hippocampal formation] underscores the growing paradigm shift in the neurosciences away from thinking about neural coding as being driven primarily by bottom-up, sensory inputs, but rather as a reflection of rich and complex internal dynamics" (McNaughton et al., 2006). Taken together, the characteristics of cognitive maps and the operation of the hippocampal formation echo the arguments being put forward in this paper about the importance of joint perceptuo-motor representations in cognition.

The Global Workspace Cognitive Architecture

Shanahan (Shanahan, 2005a,b, 2006; Shanahan and Baars, 2005) proposes a biologically plausible brain-inspired neural-level cognitive architecture in which cognitive functions such as anticipation and planning are realized through internal simulation of interaction with the environment. Action selection, both actual

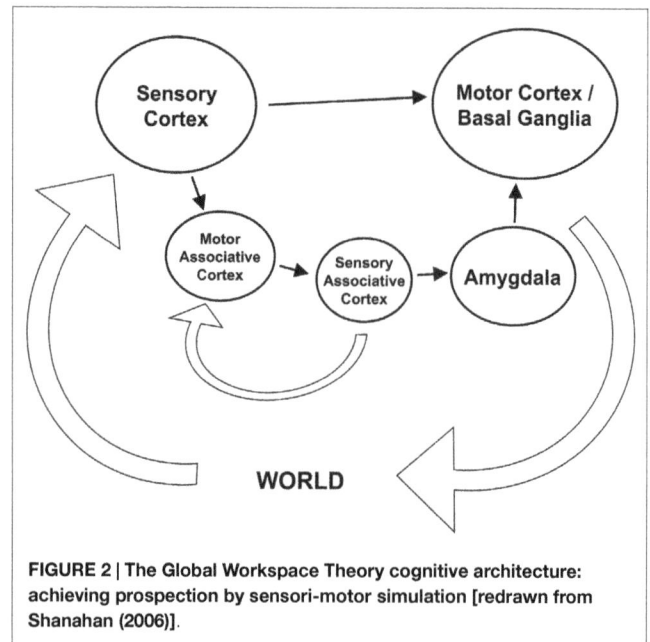

FIGURE 2 | The Global Workspace Theory cognitive architecture: achieving prospection by sensori-motor simulation [redrawn from Shanahan (2006)].

and internally simulated, is mediated by affect. The architecture is based on an external sensori-motor loop and an internal sensori-motor loop in which information passes through multiple competing cortical areas and a global workspace (Baars, 1998, 2002).

Shanahan's cognitive architecture is comprised of the following components: a first-order sensori-motor loop, closed externally through the world, and a higher-order sensori-motor loop, closed internally through associative memories (see **Figure 2**). The first-order loop comprises the sensory cortex and the basal ganglia (controlling the motor cortex), together providing a reactive action-selection sub-system. The second-order loop comprises two associative cortex elements which carry out off-line simulations of the system's sensory and motor behavior, respectively. The first associative cortex simulates a motor output while the second simulates the sensory stimulus expected to follow from a given motor output. The higher-order loop effectively modulates basal ganglia action selection in the first-order loop via an affect-driven amygdala component. Thus, this cognitive architecture is able to anticipate and plan for potential behavior through the exercise of its "imagination" (*i.e.,* its associative internal sensori-motor simulation).

The HAMMER Architecture

While internal simulation is an essential aspect of human cognition, it is also an increasingly important part of artificial cognitive systems. For example, The Hierarchical Attentive Multiple Models for Execution and Recognition (HAMMER) architecture (Demiris and Khadhouri, 2006; Demiris et al., 2014) builds on the simulation hypothesis, accomplishing internal simulation using forward and inverse models which encode internal sensori-motor models that the agent would utilize if it were to execute that action (see **Figure 3**).

HAMMER deploys several inverse-forward pairs to simulate multiple possible futures using a winner-take-all attention process to select the most appropriate action to execute. HAMMER

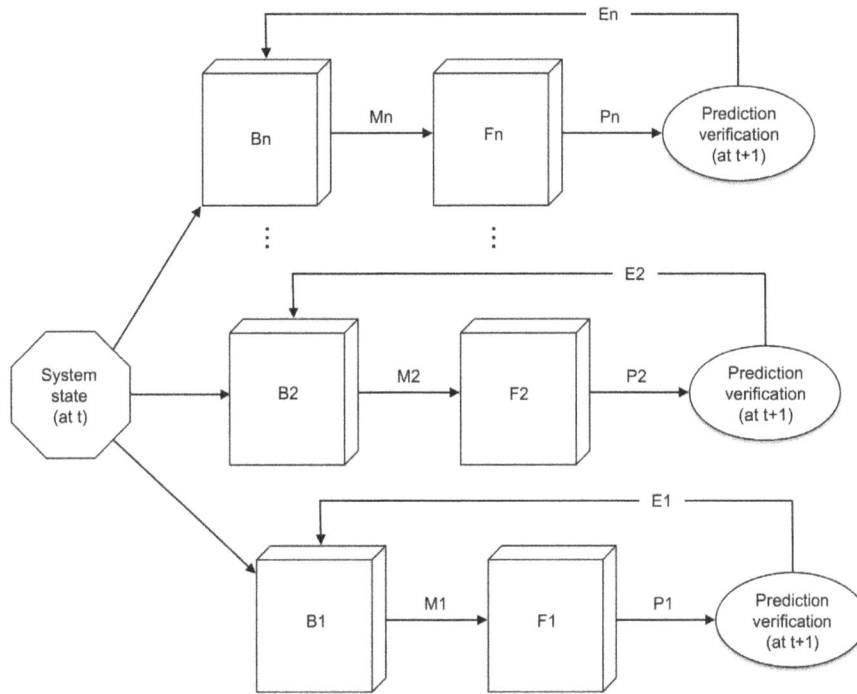

FIGURE 3 | The HAMMER architecture, showing multiple inverse models (B1 to Bn) taking as input the current system state, which includes a desired goal, suggesting motor commands (M1 to Mn), with which the corresponding forward models (F1 to Fn) form predictions of the system's next state (P1 to Pn). These predictions are verified at the next time state, resulting in a set of error signals (E1 to En). Redrawn from Demiris and Khadhouri (2006). See also Demiris et al. (2014) for an alternative rendering of the HAMMER architecture.

includes recurrent connections, thereby allowing multi-stage extended internal simulation and mental rehearsal. This provides the architecture with a way of encapsulating the internal simulation hypothesis proposed by Hesslow (2002, 2012).

The inverse model takes as input information about the current state of the system and the desired goal, and it outputs the motor commands necessary to achieve that goal. The forward model acts as a predictor. It takes as input the motor commands and simulates the perception that would arise if this motor command were to be executed, just as the simulation hypothesis envisages. HAMMER then provides the output of the inverse model as the input to the forward model. This allows a goal state (demonstrated, for example, by another agent or possibly recalled from episodic memory) to elicit the simulated action required to achieve it. This simulated action is then used with the forward model to generate a simulated outcome, i.e., the outcome that would arise if the motor commands were to be executed. The simulated perceived outcome is then compared to the desired goal perception and the results are then fed back to the inverse model to allow it to adjust any parameters of the action.

A distinguishing feature of the HAMMER architecture is that it operates multiple pairs of inverse and forward models in parallel, each one representing a simulation – a hypothesis – of how the goal action can be achieved. The choice of inverse/forward model pair is made by an internal attention process based on how close the predicted outcome is to the desired one. Furthermore, it provides for the hierarchical composition of primitive actions into more complex sequences.

From Perceptuo-Motor Mappings to Perceptuo-Motor Memory

Both Global Workspace Theory and HAMMER are good models of the simulation hypothesis for internal simulation as a vehicle for prospection in cognition. However, they focus on the mapping between perception and motor command, with memory being left implicit (see **Figures 4** and **5**).

Other models, such as the *Theory of Event Coding* (TEC) (Hommel et al., 2001) and *Object Action Complexes* (OACs) (Krüger et al., 2011) attempt to provide a tighter coupling of the perceptual and motor aspect in a joint perceptuo-motor representation.

The Theory of Event Coding (TEC) is a representational framework for combining perception and action planning. It focuses mainly on the later stages of perception and the earlier phases of action. As such, it concerns itself with perceptual features but not with how those features are extracted or computed. Similarly, it concerns itself with preparing actions – action planning – but not with the final execution of those actions and the adaptive control of various parts of the agent's body. The main idea is that perception, attention, intention, and action all work with a common representation and, furthermore, that action depends on both external and internal causes.

TEC provides a basis for combining both sensory-motor and ideo-motor action planning (Stock and Stock, 2004) and to be a joint representation that serves both sensory-stimulated action and prospective goal-directed action. The core concept in TEC is the *event code*. This is effectively a structured aggregation of distal features of an event in the agent's world. These *feature codes* can

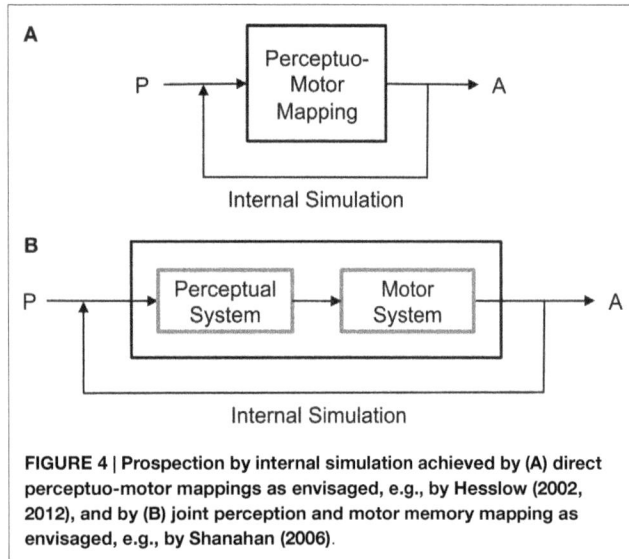

FIGURE 4 | Prospection by internal simulation achieved by (A) direct perceptuo-motor mappings as envisaged, e.g., by Hesslow (2002, 2012), and by (B) joint perception and motor memory mapping as envisaged, e.g., by Shanahan (2006).

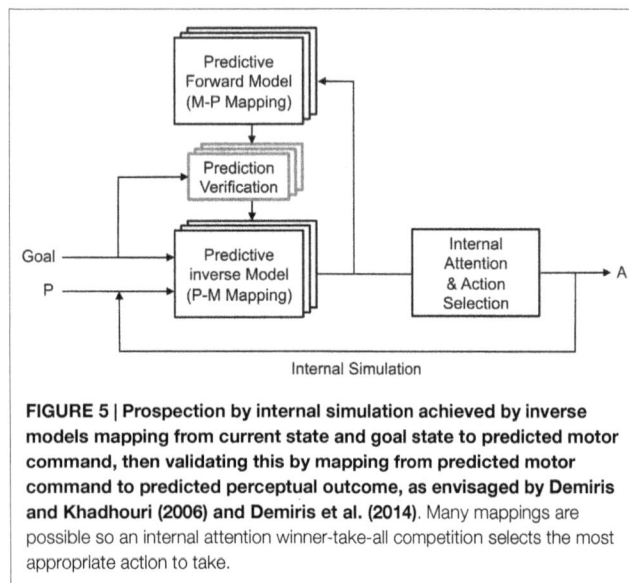

FIGURE 5 | Prospection by internal simulation achieved by inverse models mapping from current state and goal state to predicted motor command, then validating this by mapping from predicted motor command to predicted perceptual outcome, as envisaged by Demiris and Khadhouri (2006) and Demiris et al. (2014). Many mappings are possible so an internal attention winner-take-all competition selects the most appropriate action to take.

be relatively simple (e.g., color, shape, moving to the left, falling) or more complex, such as an affordance. Also, TEC feature codes can emerge through the agent's experience; they do not have to be pre-specified. A given TEC feature code is associated with both the sensory system and the motor system. Typically, a feature code is derived from several proximal sensory sources (sensory codes) and it contributes to several proximal motor actuators (motor codes). Each *event code* comprises several feature codes representing some event, be it a perceived event or a planned event. Feature codes associated with an event are activated both when the event is perceived and when it is planned. Because features can be elements of many event codes, the activation of a given feature effectively primes, i.e., predisposes, all the other events of which this feature is a component.

Inspired by the Theory of Event coding, an Object-Action Complex (OAC) (Krüger et al., 2011) is a triple, i.e., a unit with three components: (E, T, M). E is an "execution specification"

(effectively an action). T is a function that predicts how the attributes that characterize the current state of the agent's world will change if the execution specification is executed. Effectively, of T as a prediction of how the agent's perceptions will change as a result of carrying out the actions given by E. M is a statistical measure of the success of the OAC's past predictions. In this way, an OAC combines the essential elements of a joint representation – perception and action – with a predictor that links current perceived states and future predicted perceived states that would result from carrying out that action. To a large extent, an OAC models an agent's interaction with the world as it executes some motor program (this is referred to as a low-level control program CP in the OAC literature). For example, an OAC might encode how to grasp an object or push an object into a given position and orientation (usually referred to as the object pose). OACs can be learned and executed, and they can be combined into more complex representations of actions and their perceptual consequences.

To date, neither TEC nor OAC has been embedded in the more general internal simulation framework described above. So, it is proposed here that there is a strong case for making memory – episodic and procedural – more explicit and embedding them in an internal simulation framework (such as that envisaged in the simulation hypothesis, the GWT Architecture, and the HAMMER architecture) in a way that makes their links more explicit (such as that envisaged in TEC and OAC). We address such a possible framework on the next section.

A Network-Based Joint Episodic-Procedural Memory for Internal Simulation

The core idea being proposed is to unwind the temporal and causal relationships between specific perceptions and actions that are implicit in the mappings of, e.g., GWT and HAMMER, and make them explicit in a weighted network of associations between perceptions and actions, in the manner of TEC and OAC (see **Figure 6**). In doing so, it makes the input to the joint perceptuo-motor mapping explicit as perceptual episodic memories and motoric procedural memories (see **Figures 7** and **8**). In the case of episodic memory, this provides a way to include other modalities including affective or hedonic memories. Procedural memory operates associatively in their own right: such procedural memories are not static but are dynamic and adapt as the action is executed.

Furthermore, such a framework allows one to expose the mapping dynamics explicitly. This may have several advantages in, for example, cognitive development which focuses on extending the timescale of the agent's prospective capacity and expanding the agent's repertoire of actions. Specifically, development might be facilitated by adjusting and adapting the network structure – its topology and strength of connectivity – as a function of experiential learning, intrinsic value systems (Merrick, 2010), including those derived from autonomic self-maintenance (Bickhard, 2000), and affective homeostasis and allostasis (Sterling, 2004, 2012; Morse et al., 2008; Ziemke and Lowe, 2009).

The network model of joint episodic-procedural memory facilitates prospection in three senses: prospection by predicting the

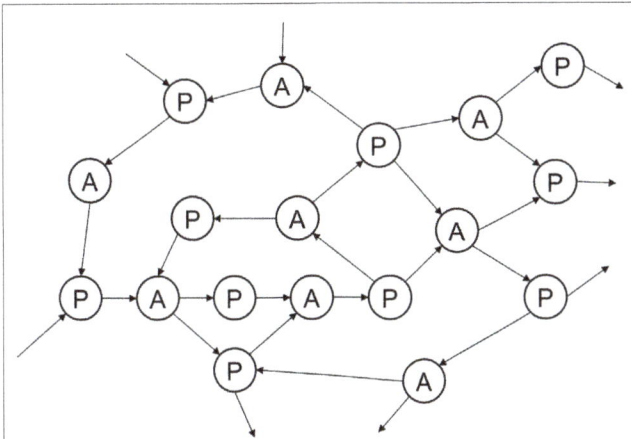

FIGURE 6 | Joint episodic-procedural memory as an explicit network of associations between perceptions and actions, drawn from episodic and procedural memories, unwinding the temporal and causal relationships between specific perceptions and actions that are implicit in the mappings of other perceptuo-motor representations.

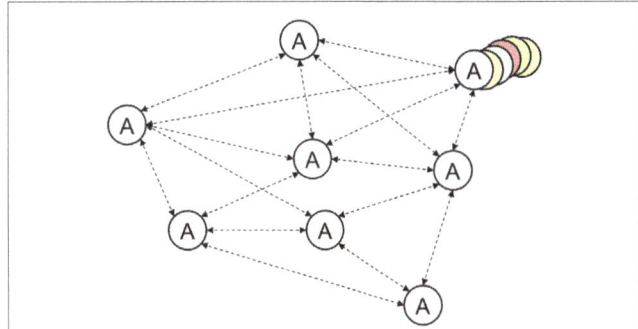

FIGURE 8 | The procedural elements of the joint episodic-procedural memory are drawn from procedural memory and, again, operate associatively in their own right. Such procedural memories are not static but are dynamic and adapt as the action is executed (top right).

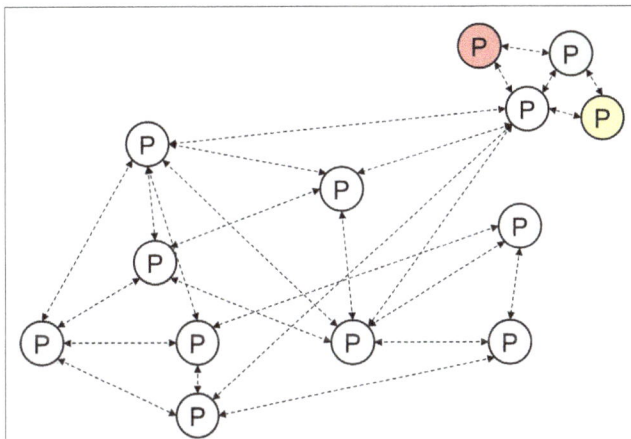

FIGURE 9 | The network model of joint episodic-procedural memory facilitates prospection in three senses: (A) prospection by predicting the outcome of an action carried out in given perceptual circumstances, (B) prospection by predicting the action required to achieve a goal in given perceptual circumstances, and (C) abductive inference of the perceptual states that explains an outcome of a give action.

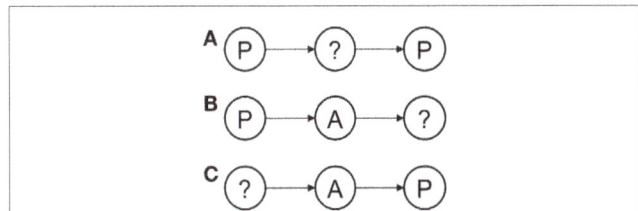

FIGURE 7 | The episodic elements of the joint episodic-procedural memory are drawn from episodic memory and therefore operate associatively in their own right. Furthermore, this provides a way to include other modalities of episodic memory (top right) including affective or hedonic memories.

outcome of an action carried out in given perceptual circumstances, prospection by predicting the action required to achieve a goal in given perceptual circumstances, and abductive inference of the perceptual states that explains an outcome of a give action (see **Figure 9**).

Keeping episodic memory explicit in this framework preserves the flexibility for adaptive reconstruction and novel association. Since episodic memory operates by recombining imperfectly recalled past experience, this allows it to simulate new or unexpected events as outlined above.

There is, however, a potential problem in that the scope for exponential growth in association is significant. Something is needed to constrain this potential combinatorial explosion if such a joint episodic-procedural memory system is to be capable of

useful prospection through internal simulation. Because the associative links are exposed explicitly in the network organization, this framework for a joint episodic-procedural memory allows the internal simulation to be conditioned by current context, semantic memory, and the agent's value system by adjusting the associative links. Context and semantics constrain the combinatorial explosion of potential perception-action associations and allow effective action selection in the pursuit of goals, while the value system modulates the memory network to promote the agent's autonomy and cognitive development.

Finally, the approach being suggested here is an abstract schema and is therefore neutral regarding the final implementation of these episodic and procedural memories. These can be effected either as an emergent cognitive system, instantiating them subsymbolically in a biologically inspired manner as associative networks [e.g., Hopfield nets such as in Mohan et al. (2014) or brain-based devices such as in Krichmar and Edelman (2005, 2006)]. Alternatively, they can be effected symbolically as more traditional AI systems. For example, episodic memory might be implemented using content-addressable image databases with traditional image indexing and recall algorithms, while procedural memory could be encapsulated in databases of motor-control scripts derived from experiential learning or from shared resources [e.g., Tenorth and Beetz (2009) and Tenorth et al. (2012, 2013)]. The traditional AI implementation, for the purpose of practical cognitive robotics, has a number of advantages.

Although episodic memory will typically exploit by iconic representations, these representations are often augmented by symbolic tags when derived from on-line repositories. This symbolic tagging makes the integration of semantic knowledge much easier. The fact that both episodic memory and procedural memory are derived from experience, directly or indirectly, also finesses the symbol grounding problem (Harnad, 1990; Sloman). The traditional AI implementation also renders the knowledge contained in the memory inherently transferrable to other agents, provided their sensory systems are compatible and there is a known mapping – direct or indirect – between the embodiments of each agent, as described in Argall et al. (2009).

An Example Joint Episodic-Procedural Memory for Overt Attention

The iCub is a 53 degree-of-freedom humanoid robot (see **Figure 10**) that was designed to be an open-systems platform for research in cognitive development (Sandini et al., 2007; Tsagarakis et al., 2007; Metta et al., 2010). It is approximately 1 m tall, weighs 22 kg, has visual, vestibular, auditory, and haptic sensors, and is capable of dexterous manipulation. To date, iCubs have been delivered to over 20 research laboratories in Europe and one in the U.S.A.[6]

The original iCub cognitive architecture (Sandini et al., 2007; Vernon et al., 2010) focused on gaze-modulated goal-directed reaching and locomotion. Episodic memory and procedural memory were designed to effect internal simulation in order to provide capabilities for prediction and model construction bootstrapped by learned affordances. Motivations encapsulated in the system's affective state addressed curiosity and experimentation, both of which are exploratory motives, triggered by exogenous and endogenous factors, respectively. This distinction between the exogenous and the endogenous was reflected in the overt attention system that could be triggered by both external and internal events. A simple process of homeostatic self-regulation governed by the affective state provided elementary action selection. Finally, all the various components of the cognitive architecture operated concurrently so that a sequence of states representing cognitive behavior emerges from the interaction of many separate parallel processes rather than being dictated by some pre-programed state-machine.

In the variant of the iCub cognitive architecture presented here, the separate episodic and procedural memories have been replaced by a simple proof-of-principle joint episodic-procedural memory (see **Figure 11**). This is the focus of the current article and the specific objective is to investigate how a joint episodic-procedural memory can be used for representation, development, and adaptation of scan-path patterns that result from overt and covert attention. This particular model of attention uses an information-theoretic saliency map (Bruce and Tsotsos, 2009) with an overt attention system comprising (1) the winner-take-all process effected by a selective tuning model to identify a single focus of attention (Tsotsos et al., 1995; Tsotsos, 2006, 2011), (2) an Inhibition-Of-Return (IOR) mechanism to attenuate the attention

FIGURE 10 | The iCub humanoid robot: an open-systems platform for research in cognitive development.

value of previous winning locations so that new regions become the focus of attention, and (3) a habituation process to reduce the salience of the current focus of attention with time thereby ensuring that attention is fixated on a given point for a limited period (Zaharescu et al., 2004). Fixation points are represented using retinotopic images rather than conventional rectangular regularly sampled images. The retinotopic images are constructed using a scale and rotation-invariant log-polar transform (Braccini et al., 1981; Berton, 2006; Berton et al., 2006; Traver and Bernardino, 2010) to map the Cartesian camera image data to a non-uniformly sampled image that reflects the foveated sampling in the primate retina. The resultant scan path patterns are captured in an elementary joint episodic-procedural memory: the episodes are retinotopic log-polar images of the fixation points and the actions are the saccade angles.

The episodic memory in the iCub cognitive architecture is a simple associatively recalled memory of autobiographical events. It is a form on one-shot learning and does not generalize multiple instances of an observed event. In the current implementation, the episodic memory provides a purely visual iconic memory of landmark appearance using scale- and rotation-invariant[7] retinotopic log-polar images as the landmark representation (Braccini et al., 1981; Berton, 2006; Berton et al., 2006; Traver and Bernardino, 2010) with image recognition being effected using color histogram intersection (Swain and Ballard, 1990, 1991). In essence, the iCub episodic memory implements a form of content-addressable memory which is populated by log-polar landmark images acquired under the control of the iCub's covert and overt attention sub-system.

Procedural memory maintains a very simple repository of elementary actions. The current implementation comprises gaze motor commands in a body-centered frame of reference and symbolic tags denoting one of five possible associated actions (reach, push, grasp, locomote, or wait). These are just placeholders for

---

[6]For more information on the iCub robot see http://www.icub.org.

[7]The rotation invariance of log-polar images is restricted to roll: rotation about the camera's principal axis.

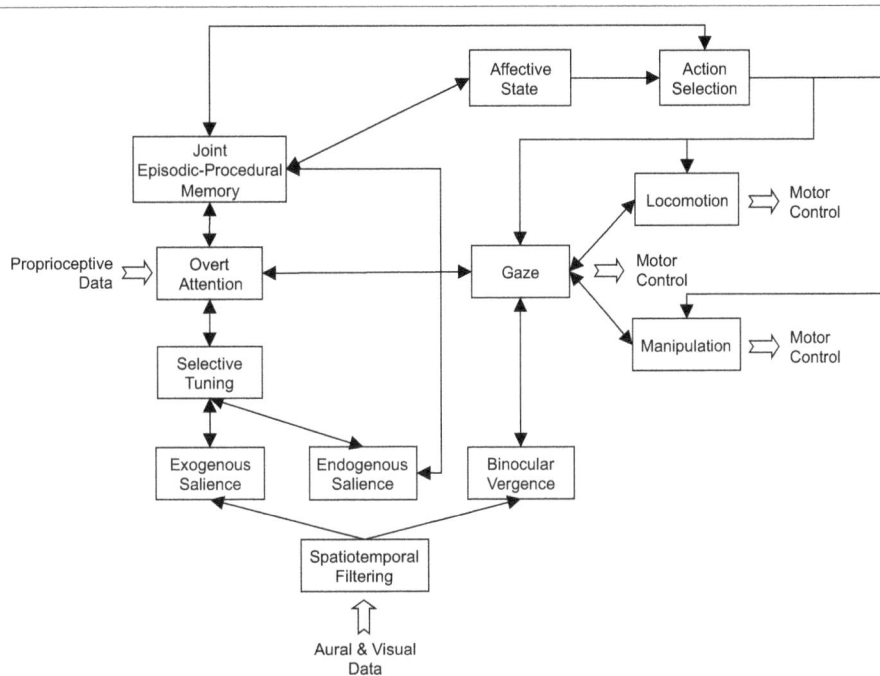

FIGURE 11 | A variant of the iCub cognitive architecture (Vernon et al., 2007, 2010) targeting visual attention with information-theoretic exogenous salience (Bruce and Tsotsos, 2009), the Selective Tuning Model for saccade selection (Tsotsos et al., 1995; Tsotsos, 2006, 2011), overt attention with inhibition of return and habituation modulated scan path dynamics (Zaharescu et al., 2004), and joint episodic-procedural memory.

more flexible and adaptive gaze-directed motor control schemes [e.g., Lukic et al. (2012)] to be implemented later.

The joint episodic-procedural memory itself is a network of associations between motor events and pairs of sensory events. In this variant of the iCub cognitive architecture, a sensory event is a visual landmark which has been acquired by the iCub and stored in the episodic memory. A motor event is a gaze saccade with an optional reaching, grasping, or locomotion movement. Thus, joint episodic-procedural memory can be viewed as a directed network with two types of nodes, one representing sensory patterns – retinotopic log-polar images of the fixation points – and the other representing motor patterns – the saccade motor commands. A path through the network traverses alternately sensory and motor nodes and any clique in this memory network effectively captures a causal relationship between a sensory state, a motor state, and a subsequent sensory state (or a sequence of such associations). An extended path in this memory captures the scan path pattern of the robot as it pays attention to its visual environment (see **Figure 12**).

The key feature of this form of joint episodic-procedural memory representation of the attention pattern of the robot is that it lends itself to development: modulation or dynamically reconfiguration of the connectivity of this network – which is learned from experience – so that its prospective capacity increases as new memories are added as a result of the agent's interaction with its environment. Various forms of adaptive reconfiguration are currently being examined, some based on small world networks (Watts and Strogatz, 1998; Newman, 2000; Bohland and Minai, 2001; Kleinberg, 2006; Telesford et al., 2011) and others based

on information theoretic models that dynamically modulate the pathways in flow networks (Ulanowicz, 2000).

Conclusion

While action and prospection are intimately linked, most research on prospection has tended to focus on the constructive role of episodic memory (Tulving, 1972, 1984; Seligman et al., 2013), i.e., the so-called episodic future thinking (Atance and O'Neill, 2001), often achieved through internal simulation, i.e., the mental construction of an imagined alternative perspectives (Buckner and Carroll, 2007) and simulated embodied interaction (Svensson et al., 2007). Although hedonic affective experience has been addressed to some extent (Gilbert and Wilson, 2007; Lowe and Ziemke, 2011), procedural memory has been neglected in modeling prospective capacities. When it is included, it usually takes the form of distinct forward models that predict the sensory outcome of a given motor command (Hesslow, 2002, 2012; Shanahan, 2006) and inverse models that determine the action required to produce a given goal perception (Demiris and Khadhouri, 2006). Ideo-motor theory (Stock and Stock, 2004; Iacoboni, 2009) is an exception to this. It assumes that perception and action share a common representational framework and that action is the causal result of internally generated goals. Such a joint representation provides greater flexibility in prospection through both inductive inference and abductive inference.

With few exceptions, such as the Theory of Event Coding (Hommel et al., 2001) and object-action complexes

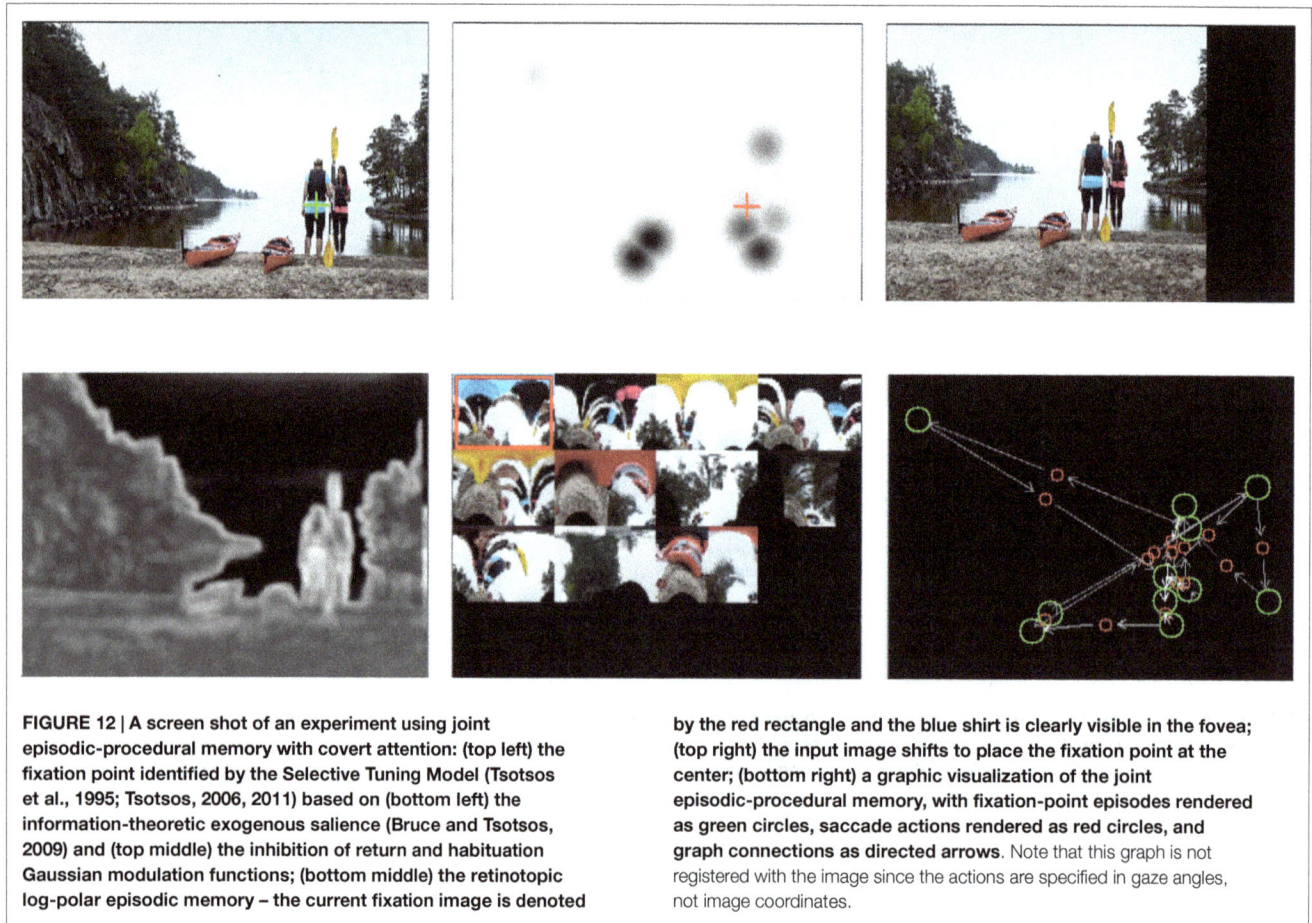

FIGURE 12 | A screen shot of an experiment using joint episodic-procedural memory with covert attention: (top left) the fixation point identified by the Selective Tuning Model (Tsotsos et al., 1995; Tsotsos, 2006, 2011) based on (bottom left) the information-theoretic exogenous salience (Bruce and Tsotsos, 2009) and (top middle) the inhibition of return and habituation Gaussian modulation functions; (bottom middle) the retinotopic log-polar episodic memory – the current fixation image is denoted **by the red rectangle and the blue shirt is clearly visible in the fovea; (top right) the input image shifts to place the fixation point at the center; (bottom right) a graphic visualization of the joint episodic-procedural memory, with fixation-point episodes rendered as green circles, saccade actions rendered as red circles, and graph connections as directed arrows.** Note that this graph is not registered with the image since the actions are specified in gaze angles, not image coordinates.

(Krüger et al., 2011), joint perceptuo-motor representations have received little attention and none have addressed integration of hedonic affective experience. Our conjecture is that an internal simulation capability founded on ideo-motor theory and joint representations, and drawing on recent progress in the modeling-related mirror neuron system (Gallese et al., 1996; Rizzolatti et al., 1996; Rizzolatti and Craighero, 2004; Thill et al., 2013), provides a viable way to approach the integration of procedural and episodic memory as a joint perceptuo-motor system. Our specific contention is that it is helpful to conceive of this joint episodic-procedural memory – for goal-directed internal simulation – as a network of associations between elements of both episodic and procedural memories. This perspective is neutral regarding the final implementation of these episodic and procedural memories and it can facilitate both emergent and cognitivist AI approaches.

We argue that such a framework meets several challenges in cognitive robotics such as the need to accommodate modal and modal episodic data and extended perceptuo-motor sequences, as well as mechanisms for conditioning the association dynamics with external constraints derived from semantic declarative knowledge, current context, and affective value signals. It also addresses the need to integrate the episodic and procedural knowledge gathered by robots as they operate of their physical environment with information extracted from web-based knowledge bases. This is particularly important if the power of indirect knowledge (acquired by interpreting third-party descriptions) is to be harnessed in the development of robot skills.

Acknowledgments

This work was supported in part by the European Commission, Project 611391 DREAM: Development of Robot-enhanced Therapy for Children with Autism Spectrum Disorders (www.dream2020.eu).

References

Argall, B. D., Chernova, S., Veloso, M., and Browning, B. (2009). A survey of robot learning from demonstration. *Rob. Auton. Syst.* 57, 469–483. doi:10.1016/j.robot.2008.10.024

Atance, C. M., and O'Neill, D. K. (2001). Episodic future thinking. *Trends Cogn. Sci.* 5, 533–539. doi:10.1016/S1364-6613(00)01804-0

Baars, B. J. (1998). *A Cognitive Theory of Consciousness*. Cambridge: Cambridge University Press.

Baars, B. J. (2002). The conscious assess hypothesis: origins and recent evidence. *Trends Cogn. Sci.* 6, 47–52. doi:10.1016/S1364-6613(00)01819-2

Bennett, A. (1996). Do animals have cognitive maps. *J. Exp. Biol.* 199, 219–224.

Berton, F. (2006). "A brief introduction to log-polar mapping," in *Technical Report, LIRA-Lab*, (University of Genova).

Berton, F., Sandini, G., and Metta, G. (2006). Anthropomorphic visual sensors. *Encyclo. Sens.* 10, 1–16. doi:10.3389/fnsys.2014.00109

Bickhard, M. H. (2000). *Autonomy, function, and representation.* Available at: http://www.lehigh.edu/~mhb0/autfuncrep.html

Bohland, J. W., and Minai, A. A. (2001). Efficient associative memory using small-world architecture. *Neurocomputing* 3, 489–496. doi:10.1016/S0925-2312(01)00378-2

Braccini, C., Gambardella, G., and Sandini, G. (1981). A signal theory approach to the space and frequency variant filtering performed by the human visual system. *Signal Processing* 3, 231–240. doi:10.1016/0165-1684(81)90038-4

Bratman, M. E. (1998). "Intention and personal policies," in *Philosophical Perspectives*, ed. J. E. Tomberlin (Oxford: Blackwell), 3.

Bruce, N. D. B., and Tsotsos, J. K. (2009). Saliency, attention, and visual search: an information theoretic approach. *J. Vis.* 9, 1–24. doi:10.1167/9.3.5

Buckner, R. L., and Carroll, D. C. (2007). Self-projection and the brain. *Trends Cogn. Sci.* 11, 49–57. doi:10.1016/j.tics.2006.11.004

Demiris, Y., Aziz-Zadeh, L., and Bonaiuto, J. (2014). Information processing in the mirror neuron system in primates and machines. *Neuroinformatics* 12, 63–91. doi:10.1007/s12021-013-9200-7

Demiris, Y., and Khadhouri, B. (2006). Hierarchical attentive multiple models for execution and recognition (HAMMER). *Rob. Auton. Syst.* 54, 361–369. doi:10.1016/j.robot.2006.02.003

Eichenbaum, H., Dudchenko, P., Wood, E., Shapiro, M., and Tanila, H. (1999). The hippocampus, memory, and place cells: is it spatial memory or a memory space? *Neuron* 23, 209–226. doi:10.1016/S0896-6273(00)80773-4

Flanagan, J. R., and Johansson, R. S. (2003). Action plans used in action observation. *Nature* 424, 769–771. doi:10.1038/nature01861

Gallese, V., Fadiga, L., Fogassi, L., and Rizzolatti, G. (1996). Action recognition in the premotor cortex. *Brain* 119, 593–609. doi:10.1093/brain/119.2.593

Gallistel, C. R. (1989). Animal cognition: the representation of space, time and number. *Annu. Rev. Psychol.* 40, 155–189. doi:10.1146/annurev.ps.40.020189.001103

Gallistel, C. R. (1990). *The Organization of Learning.* Cambridge, MA: MIT Press.

Gaussier, P., Revel, A., Banquet, J. P., and Babeau, V. (2002). From view cells and place cells to cognitive map learning: processing stages of the hippocampal system. *Biol. Cybern.* 86, 15–28. doi:10.1007/s004220100269

Gilbert, D. T., and Wilson, T. D. (2007). Prospection: experiencing the future. *Science* 317, 1351–1354. doi:10.1126/science.1144161

Grush, R. (2004). The emulation theory of representation: motor control, imagery, and perception. *Behav. Brain Sci.* 27, 377–442. doi:10.1017/S0140525X04000093

Harnad, S. (1990). The symbol grounding problem. *Physica D* 42, 335–346. doi:10.1016/0167-2789(90)90087-6

Hesslow, G. (2002). Conscious thought as simulation of behaviour and perception. *Trends Cogn. Sci.* 6, 242–247. doi:10.1016/S1364-6613(02)01913-7

Hesslow, G. (2012). The current status of the simulation theory of cognition. *Brain Res.* 1428, 71–79. doi:10.1016/j.brainres.2011.06.026

Hommel, B., Müsseler, J., Aschersleben, G., and Prinz, W. (2001). The theory of event coding (TEC): a framework for perception and action planning. *Behav. Brain Sci.* 24, 849–937. doi:10.1017/S0140525X01000103

Iacoboni, M. (2009). Imitation, empathy, and mirror neurons. *Annu. Rev. Psychol.* 60, 653–670. doi:10.1146/annurev.psych.60.110707.163604

Johansson, R. S., Westling, G., Bäckström, A., and Flanagan, J. R. (2001). Eye-hand coordination in object manipulation. *J. Neurosci.* 21, 6917–6932.

Kleinberg, J. (2006). "Complex networks and decentralized search algorithms," in *Proceedings of the International Congress of Mathematicians (ICM).* Madrid, 1019–1044.

Krichmar, J. L., and Edelman, G. M. (2005). Brain-based devices for the study of nervous systems and the development of intelligent machines. *Artif. Life* 11, 63–77. doi:10.1162/1064546053278946

Krichmar, J. L., and Edelman, G. M. (2006). "Principles underlying the construction of brain-based devices," in *Proceedings of AISB '06 - Adaptation in Artificial and Biological Systems*, Volume 2 of *Symposium on Grand Challenge 5: Architecture of Brain and Mind*, eds T. Kovacs, and J. A. R. Marshall (Bristol: University of Bristol), 37–42.

Krüger, N., Geib, C., Piater, J., Petrickb, R., Steedman, M., Wörgötter, F., et al. (2011). Object-action complexes: grounded abstractions of sensory–motor processes. *Rob. Auton. Syst.* 59, 740–757. doi:10.1016/j.robot.2011.05.009

Lowe, R., and Ziemke, T. (2011). The feeling of action tendencies: on the emotional regulation of goal-directed behaviours. *Front. Psychol.* 2:1–24. doi:10.3389/fpsyg.2011.00346

Lukic, L., Santos-Victor, J., and Billard, A. (2012). " Learning coupled dynamical systems from human demonstration for robotic eye-arm-hand coordination," in *Proceedings of the IEEE-RAS International Conference on Humanoid Robots (Humanoids)*, (Osaka, Japan), 552–559.

McNamara, T. P., and Shelton, A. L. (2003). Cognitive maps and the hippocampus. *Trends Cogn. Sci.* 7, 333–335. doi:10.1016/S1364-6613(03)00167-0

McNaughton, B., Battaglia, F., Jensen, O., Moser, E., and Moser, M. (2006). Path integration and the neural basis of the 'cognitive map'. *Nat. Rev.* 7, 663–678. doi:10.1038/nrn1932

Merrick, K. E. (2010). A comparative study of value systems for self-motivated exploration and learning by robots. *IEEE Trans. Auto. Mental Dev.* 2, 119–131. doi:10.1109/TAMD.2010.2051435

Metta, G., Natale, L., Nori, F., Sandini, G., Vernon, D., Fadiga, L., et al. (2010). The iCub humanoid robot: an open-systems platform for research in cognitive development. *Neural Netw.* 23, 1125–1134. doi:10.1016/j.neunet.2010.08.010

Mohan, V., Sandini, G., and Morasso, P. (2014). A neural framework for organization and flexible utilization of episodic memory in cumulatively learning baby humanoids. *Neural Comput.* 26, 1–43. doi:10.1162/NECO_a_00664

Morse, A., Lowe, R., and Ziemke, T. (2008). "Towards an enactive cognitive architecture," in *Proceedings of the First International Conference on Cognitive Systems*, (Karlsruhe, Germany).

Moulton, S. T., and Kosslyn, S. M. (2009). Imagining predictions: mental imagery as mental emulation. *Philos. Trans. R. Soc. B* 364, 1273–1280. doi:10.1098/rstb.2008.0314

Newman, M. E. J. (2000). Models of the small world: a review. *J. Stat. Phys.* 101, 819–841. doi:10.1023/A:1026485807148

O'Keefe, J. (1976). Place units in the hippocampus of the freely moving rat. *Exp. Neurol.* 51, 78–109. doi:10.1016/0014-4886(76)90055-8

O'Keefe, J., and Nadel, L. (1978). *The Hippocampus as a Cognitive Map.* Oxford: Clarendon.

Ondobaka, S., and Bekkering, H. (2012). Hierarchy of idea-guided action and perception-guided movement. *Front. Cogn.* 3:1–5. doi:10.3389/fpsyg.2012.00579

Østby, Y., Walhovd, K. B., Tamnes, C. K., Grydeland, H., Westlye, L. G., and Fjell, A. M. (2012). Mental time travl and default-mode network functional connectivity in the developing brain. *PNAS* 109, 16800–16804. doi:10.1073/pnas.1210627109

Rizzolatti, G., and Craighero, L. (2004). The mirror neuron system. *Annu. Rev. Physiol.* 27, 169–192. doi:10.1146/annurev.neuro.27.070203.144230

Rizzolatti, G., and Fadiga, L. (1998). "Grasping objects and grasping action meanings: the dual role of monkey rostroventral premotor cortex (area F5)," in *Sensory Guidance of Movement, Novartis Foundation Symposium 218*, eds G. R. Bock, and J. A. Goode (Chichester: John Wiley and Sons), 81–103.

Rizzolatti, G., Fadiga, L., Gallese, V., and Fogassi, L. (1996). Premotor cortex and the recognition of motor actions. *Cogn. Brain Res.* 3, 131–141. doi:10.1016/0926-6410(95)00038-0

Ryle, G. (1949). *The Concept of Mind.* London: Hutchinson's University Library.

Sandini, G., Metta, G., and Vernon, D. (2007). "The iCub cognitive humanoid robot: An open-system research platform for enactive cognition," in *50 Years of AI*, Vol. 4850, eds M. Lungarella, F. Iida, J. C. Bongard, and R. Pfeifer (Heidelberg: Springer), 359–370.

Schacter, D. L., and Addis, D. R. (2007a). The cognitive neuroscience of constructive memory: remembering the past and imagining the future. *Philos. Trans. R. Soc. B* 362, 773–786. doi:10.1098/rstb.2007.2087

Schacter, D. L., and Addis, D. R. (2007b). Constructive memory – the ghosts of past and future: a memory that works by piecing together bits of the past may be better suited to simulating future events than one that is a store of perfect records. *Nature* 445, 27. doi:10.1038/445027a

Schacter, D. L., Addis, D. R., and Buckner, R. L. (2008). Episodic simulation of future events: concepts, data, and applications. *Ann. N. Y. Acad. Sci.* 1124, 39–60. doi:10.1196/annals.1440.001

Seligman, M. E. P., Railton, P., Baumeister, R. F., and Sripada, C. (2013). Navigating into the future or driven by the past? *Perspect. Psychol. Sci.* 8, 119–141. doi:10.1177/1745691612474317

Shanahan, M. P. (2005a). "Cognition, action selection, and inner rehearsal," in *Proceedings IJCAI Workshop on Modelling Natural Action Selection*, 92–99.

Shanahan, M. P. (2005b). "Emotion, and imagination: a brain-inspired architecture for cognitive robotics," in *Proceedings AISB 2005 Symposium on Next Generation Approaches to Machine Consciousness*, 26–35.

Shanahan, M. P. (2006). A cognitive architecture that combines internal simulation with a global workspace. *Conscious. Cogn.* 15, 433–449. doi:10.1016/j.concog. 2005.11.005

Shanahan, M. P., and Baars, B. (2005). Applying global workspace theory to the frame problem. *Cognition* 98, 157–176. doi:10.1016/j.cognition.2004.11.007

Sloman, A. Available at: http://www.cs.bham.ac.uk/research/projects/cogaff/misc/talks/models.pdf.

Squire, L. (2004). Memory systems of the brain: a brief history and current perspective. *Neurobiol. Learn. Mem.* 82, 171–177. doi:10.1016/j. nlm.2004.06.005

Stachenfeld, K., Botvinick, M., and Gershman, S. J. (2014). "Design principles of the hippocampal cognitive map," in *Advances in Neural Information Processing Systems*. eds Z. Ghahramani, M. Welling, C. Cortes, N. D. Lawrence, and K. Q. Weinberger (New York: Curran Associates, Inc.), 2528–2536.

Sterling, P. (2004). "Principles of allostasis," in *Allostasis, Homeostasis, and the Costs of Adaptation*, ed. J. Schulkin (Cambridge: Cambridge University Press), 17–64.

Sterling, P. (2012). Allostasis: a model of predictive regulation. *Physiol. Behav.* 106, 5–15. doi:10.1016/j.physbeh.2011.06.004

Stock, A., and Stock, C. (2004). A short history of ideo-motor action. *Psychol. Res.* 68, 176–188. doi:10.1007/s00426-003-0154-5

Svensson, H., Lindblom, J., and Ziemke, T. (2007). "Making sense of embodied cognition: simulation theories of shared neural mechanisms for sensorimotor and cognitive processes," in *Body, Language and Mind*, Vol. 1, eds T. Ziemke, J. Zlatev, and R. M. Frank (Berlin: Mouton de Gruyter), 241–269.

Svensson, H., Morse, A. F., and Ziemke, T. (2009). "Representation as internal simulation: a minimalistic robotic model," in *Proceedings of the Thirty-first Annual Conference of the Cognitive Science Society*, eds N. Taatgen, and H. van Rijn (Austin, TX: Cognitive Science Society), 2890–2895.

Svensson, H., Thill, S., and Ziemke, T. (2013). Dreaming of electric sheep? Exploring the functions of dream-like mechanisms in the development of mental imagery simulations. *Adapt. Behav.* 21, 222–238. doi:10.1177/1059712313491295

Swain, M., and Ballard, D. (1991). Color indexing. *Int. J. Comput. Vis.* 7, 11–32. doi:10.1007/BF00130487

Swain, M. J., and Ballard, D. H. (1990). "Indexing via colour histograms," in *International Conference on Computer Vision – ICCV90*, 390–393.

Szpunar, K. K. (2010). Episodic future throught: an emerging concept. *Perspect. Psychol. Sci.* 5, 142–162. doi:10.1177/1745691610362350

Telesford, Q. K., Joyce, K. E., Hayasaka, S., Burdette, J. H., and Laurienti, P. J. (2011). The ubiquity of small-world networks. *Brain Connect.* 1, 367–375. doi:10.1089/brain.2011.0038

Tenorth, M., and Beetz, M. (2009). "KnowRob – knowledge processing for autonomous personal robots," in *Proc. IEEE/RSJ International Conference on Intelligent Robots and Systems*, 4261–4266.

Tenorth, M., Perzylo, A. C., Lafrenz, R., and Beetz, M. (2012). "The roboearth language: representing and exchanging knowledge about actions, objects, and environments," in *IEEE International Conference on Robotics and Automation*, 1284–1289.

Tenorth, M., Perzylo, A. C., Lafrenz, R., and Beetz, M. (2013). Representation and exchange of knowledge about actions, objects, and environments in the roboearth framework. *IEEE Trans. Auto. Sci. Eng.* 10, 643–651. doi:10.1109/TASE.2013.2244883

Thill, S., Caligiore, D., Borghi, A. M., Ziemke, T., and Baldassarre, G. (2013). Theories and computational models of affordance and mirror systems: an integrative review. *Neurosci. Biobehav. Rev.* 37, 491–521. doi:10.1016/j.neubiorev. 2013.01.012

Tolman, E. C. (1948). Cognitive maps in rats and men. *Psychol. Rev.* 55, 189–208. doi:10.1037/h0061626

Tomasello, M., Carpenter, M., Call, J., Behne, T., and Moll, H. (2005). Understanding and sharing intentions: the origins of cultural cognition. *Behav. Brain Sci.* 28, 675–735. doi:10.1017/S0140525X05000129

Traver, V. J., and Bernardino, A. (2010). A review of log-polar imaging for visual perception in robotics. *Rob. Auton. Syst.* 58, 378–398. doi:10.1016/j.robot.2009. 10.002

Tsagarakis, N. G., Metta, G., Sandini, G., Vernon, D., Beira, R., Santos-Victor, J., et al. (2007). iCub – the design and realisation of an open humanoid platform for cognitive and neuroscience research. *Int. J. Adv. Robot.* 21, 1151–1175. doi:10.1163/156855307781389419

Tsotsos, J. K. (2006). "Cognitive vision need attention to link sensing with recognition," in *Cognitive Vision Systems: Sampling the Spectrum of Approaches*, Vol. 3948, eds H. I. Christensen, and H.-H. Nagel (Heidelberg: Springer), 25–36.

Tsotsos, J. K. (2011). *A Computational Perspective on Visual Attention*. Cambridge, MA: MIT Press.

Tsotsos, J. K., Culhane, S., Wai, W., Lai, Y., David, N., and Nuflo, F. (1995). Modeling visual attention via selective tuning. *Artif. Intell.* 78, 507–547. doi:10. 1016/0004-3702(95)00025-9

Tulving, E. (1972). "Episodic and semantic memory," in *Organization of Memory*, eds E. Tulving, and W. Donaldson (New York: Academic Press), 381–403.

Tulving, E. (1983). *Elements of Episodic Memory*. Oxford: Oxford University Press.

Tulving, E. (1984). Précis of elements of episodic memory. *Behav. Brain Sci.* 7, 223–268. doi:10.1017/S0140525X0004440X

Ulanowicz, R. E. (2000). *Growth and Development; Ecosystems Phenomenology*. Lincoln: Excel Press.

Vernon, D. (2014). *Artificial Cognitive Systems – A Primer*. Cambridge, MA: MIT Press.

Vernon, D., Metta, G., and Sandini, G. (2007). A survey of artificial cognitive systems: implications for the autonomous development of mental capabilities in computational agents. *IEEE Trans. Evol. Comput.* 11, 151–180. doi:10.1109/TEVC.2006.890274

Vernon, D., von Hofsten, D., and Fadiga, L. (2010). *A Roadmap for Cognitive Development in Humanoid Robots, Cognitive Systems Monographs (COSMOS)*, Vol. 11, (Berlin: Springer).

von Hofsten, C. (2004). An action perspective on motor development. *Trends Cogn. Sci.* 8, 266–272. doi:10.1016/j.tics.2004.04.002

Watts, D. J., and Strogatz, S. H. (1998). Collective dynamics of small world networks. *Nature* 393, 440–442. doi:10.1038/30918

Wintermute, S. (2012). Imagery in cognitive architecture: representation and control at multiple levels of abstraction. *Cogn. Syst. Res.* 1, 1–29. doi:10.1016/j. cogsys.2012.02.001

Wood, R., Baxter, P., and Belpaeme, T. (2012). A review of long-term memory in natural and synthetic systems. *Adapt. Behav.* 20, 81–103. doi:10.1177/1059712311421219

Zaharescu, A., Rothenstein, A. L., and Tsotsos, J. K. (2004). "Towards a biologically plausible active visual search model," in *Proceedings of the Second International Workshop on Attention and Performance in Computational Vision, WAPCV*, Vol. 3368, eds L. Paletta, J. K. Tsotsos, E. Rome, and G. Humphreys (Berlin: Springer), 133–147.

Ziemke, T., and Lowe, R. (2009). On the role of emotion in embodied cognitive architectures: from organisms to robots. *Cogn. Comput.* 1, 104–117. doi:10.1007/s12559-009-9012-0

Conflict of Interest Statement: The authors declare that the research was conducted in the absence of any commercial or financial relationships that could be construed as a potential conflict of interest.

18

Unbiased decoding of biologically motivated visual feature descriptors

Michael Felsberg[1]*, Kristoffer Öfjäll[1] and Reiner Lenz[2]

[1] *Computer Vision Laboratory, Department of Electrical Engineering, Linköping University, Linköping, Sweden,* [2] *Computer Graphics and Image Processing, Department of Science and Technology, Linköping University, Norrköping, Sweden*

Edited by:
Venkatesh Babu Radhakrishnan,
Indian Institute of Science, India

Reviewed by:
George Azzopardi,
University of Groningen, Netherlands
Nicolas Pugeault,
University of Surrey, UK

*Correspondence:
Michael Felsberg,
Computer Vision Laboratory,
Department of Electrical Engineering,
Linköping University,
Linköping SE-58183, Sweden
michael.felsberg@liu.se

Visual feature descriptors are essential elements in most computer and robot vision systems. They typically lead to an abstraction of the input data, images, or video, for further processing, such as clustering and machine learning. In clustering applications, the cluster center represents the prototypical descriptor of the cluster and estimates the corresponding signal value, such as color value or dominating flow orientation, by decoding the prototypical descriptor. Machine learning applications determine the relevance of respective descriptors and a visualization of the corresponding decoded information is very useful for the analysis of the learning algorithm. Thus decoding of feature descriptors is a relevant problem, frequently addressed in recent work. Also, the human brain represents sensorimotor information at a suitable abstraction level through varying activation of neuron populations. In previous work, computational models have been derived that agree with findings of neurophysiological experiments on the representation of visual information by decoding the underlying signals. However, the represented variables have a bias toward centers or boundaries of the tuning curves. Despite the fact that feature descriptors in computer vision are motivated from neuroscience, the respective decoding methods have been derived largely independent. From first principles, we derive unbiased decoding schemes for biologically motivated feature descriptors with a minimum amount of redundancy and suitable invariance properties. These descriptors establish a non-parametric density estimation of the underlying stochastic process with a particular algebraic structure. Based on the resulting algebraic constraints, we show formally how the decoding problem is formulated as an unbiased maximum likelihood estimator and we derive a recurrent inverse diffusion scheme to infer the dominating mode of the distribution. These methods are evaluated in experiments, where stationary points and bias from noisy image data are compared to existing methods.

Keywords: feature descriptors, population codes, channel representations, decoding, estimation, visualization, bias

1. Introduction

We address the problem of decoding visual feature descriptors. Visual feature descriptors, such as SIFT (Lowe, 2004), HOG (Dalal and Triggs, 2005), COSFIRE (Azzopardi and Petkov, 2013), or deep features (LeCun and Bengio, 1995), are essential elements in most computer and robot vision systems. They usually lead to an abstraction of the input data, images, or video, for further processing within a hierarchical system, e.g., for object recognition (Azzopardi and Petkov, 2014). In many cases, *decoding the feature descriptor*, i.e., recovering the essential visual information encoded in the descriptor, is a relevant problem.

For instance, in clustering applications, the cluster center represents the prototypical descriptor of the cluster and the corresponding signal value, such as color value or dominating flow orientation, needs to be estimated. For the analysis of machine learning systems, it is useful to determine the relevance of respective descriptors and to visualize the corresponding decoded information. Thus, the problem of decoding feature descriptors has been addressed frequently in recent work, e.g., for SIFT (Weinzaepfel et al., 2011), HOG (Vondrick et al., 2013), and deep features (Zeiler and Fergus, 2014; Mahendran and Vedaldi, 2015) as well as for quantized descriptors, such as LBP (d'Angelo et al., 2012) and BoW (Kato and Harada, 2014).

In the present work, we address the problem of unbiased decoding from channel representations of visual data. The channel representation (Granlund, 2000) can be used as a visual feature descriptor similar to SIFT and HOG and is closely related to models of population codes used in computational neuroscience.

According to these models, the human brain represents sensorimotor information at a suitable abstraction level through varying activation of neuron populations by means of *tuning curves*. These tuning curves are typically bell-shaped and allow to predict how neurons are activated on an average by a certain stimulus. The brain's task is to invert this relation and to infer an estimate of the stimulus from a *population* of such neurons. Since the neuron activations are noisy, forming a "noisy hill" (Denève et al., 1999) of activations, a robust, stable, and unbiased estimation has to be performed: a good estimator is required. An *unbiased* estimator is supposed to be right on average, i.e., the estimate is not subject to systematic errors caused by factors other than the entity to be estimated. Thus the goal of the present work is to derive an *unbiased* decoding of visual feature descriptors.

One such method is claimed to be the "read out [of] these noisy hills" (Denève et al., 1999), represented by a *population vector*, i.e., orientation is represented by a population coding (Zemel et al., 1998; Pouget et al., 2000). This process is illustrated in **Figure 1**,

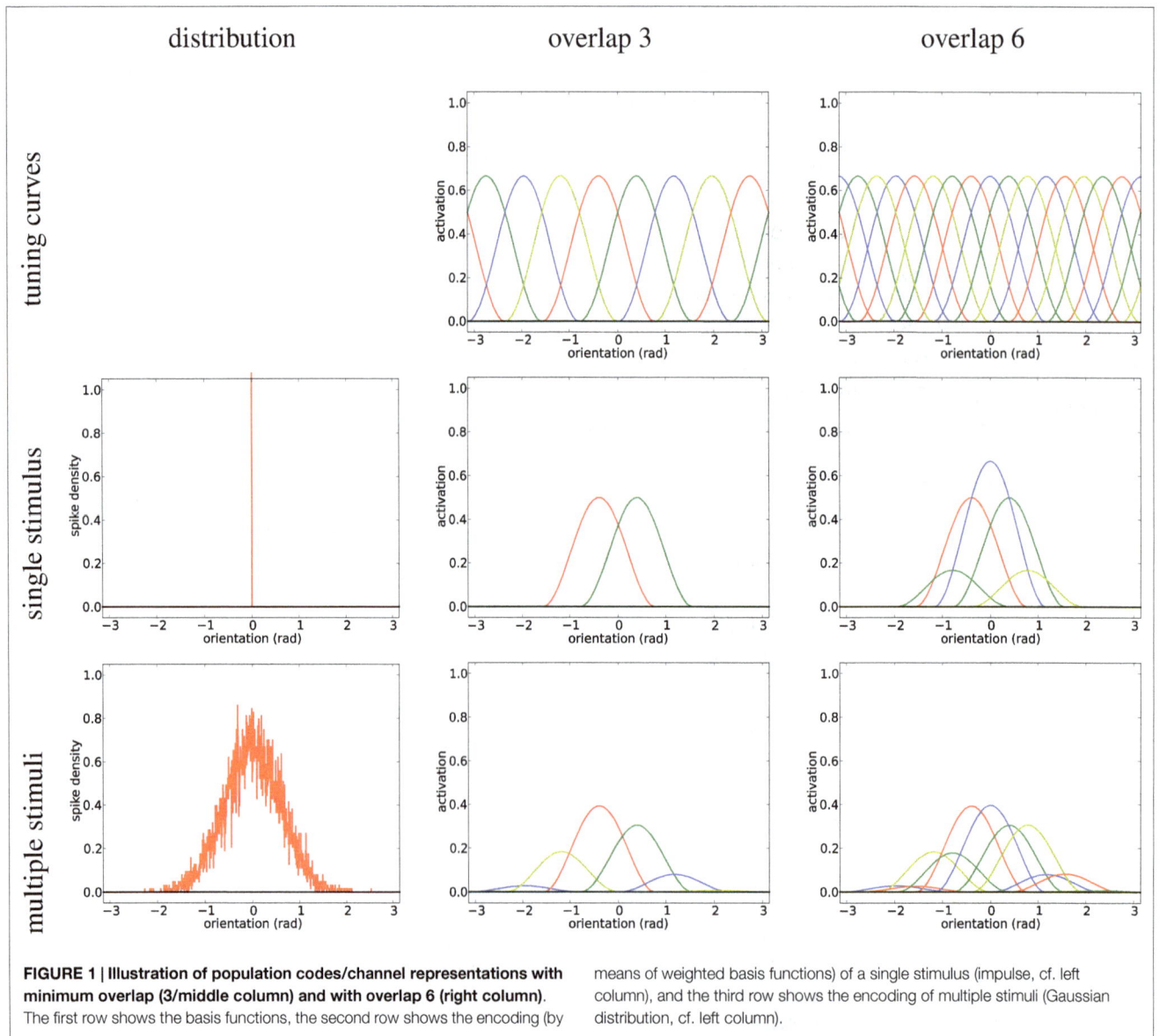

FIGURE 1 | Illustration of population codes/channel representations with minimum overlap (3/middle column) and with overlap 6 (right column). The first row shows the basis functions, the second row shows the encoding (by means of weighted basis functions) of a single stimulus (impulse, cf. left column), and the third row shows the encoding of multiple stimuli (Gaussian distribution, cf. left column).

where 8 orientation tuning curves with overlap 3 and 16 tuning curves with overlap 6 are shown. The activation of the tuning curves is illustrated for a single stimulus at 0 and noisy samples about the preferred orientation at 0 (simulation).

Population codes have earlier also been referred to as channel codes (Snippe and Koenderink, 1992; Howard and Rogers, 1995) and therefore (and in parallel to recent developments in computational neuroscience), the image processing community has developed the concept of *channel representations* (Granlund, 2000). Both coding schemes share many similarities, most importantly approximating kernel density estimators in expectation sense (Scott, 1985; Zemel et al., 1998; Felsberg et al., 2006). The tuning curves are referred to as *channel basis functions* or *kernels* and the coefficient vectors as *channel vectors*. Several analytic models, such as Gaussian kernels, $\cos^2$ kernels, or quadratic B-splines, have been investigated (Forssén, 2004). An overview of channel-based image processing is given in a recent survey (Felsberg, 2012). Other names for population/channel codes are averaged-shifted histograms (Scott, 1985) and distribution fields (Sevilla-Lara and Learned-Miller, 2012; Felsberg, 2013).

The focus in this paper is on decoding channel representations of image information. The purpose of *image representations* is to turn implicitly existing properties of the underlying image signal into a data structure, a *feature descriptor*, which makes the properties explicit. For instance, local orientation information is computed from the image signal and combined with color information in a joint feature descriptor used for classification and regression (Felsberg and Hedborg, 2007).

From a statistical point of view, an image representation consists of estimates of the feature components from a set of measurements in the image. These measurements can differ in number, i.e., in the total *evidence*. The feature component is in most cases assigned the *value* of the maximum likelihood estimate. In many cases, also the variance of the estimate is considered in terms of *coherence* (Jähne, 2005).

A good example for the *three-folded representation* by value/coherence/evidence is the structure tensor (Bigün and Granlund, 1987; Förstner and Gülch, 1987), where the sum of eigenvalues is a measure for the evidence, the difference of eigenvalues (divided by their sum) represents the coherence, and the orientation of the eigenvectors determines the value. It has been pointed out in that the structure tensor results in a 3D cone (Felsberg et al., 2009).[1]

One of the major differences between channel representations and population codings is the decoding: the *reconstruction* from channel codes or the *readout* from population codes. Channel representations make use of signal processing techniques for local reconstruction (Forssén, 2004), the maximum entropy principle to reconstruct a smoothed density function (Jonsson, 2008), or spline-based mode extraction to find the location with maximum likelihood with a minimum of bias (Felsberg et al., 2006). In

contrast, population vector readout uses a *recurrent procedure* to approximate a maximum-likelihood estimation.

Despite the fact that the readout by means of the population vector estimator is unbiased (Denève et al., 1999), the combination of the recurrent procedure and the estimator becomes biased if the recurrent equation is iterated until convergence. This effect is reduced if the tuning functions overlap extensively and if the recurrent equation is only iterated a few (2–3) times. Both approaches are used in previous work (Denève et al., 1999). However, as pointed out recently (Pellionisz et al., 2013), the recursion of covariant ("proprioception") vectors always leads to the eigenvectors of the underlying coefficient matrix and is thus a fundamental obstacle for unbiased estimators if iterated to convergence.

In contrast to mentioned previous works, a truly *unbiased* recurrent network model is derived in the present paper. This is achieved by a theoretical approach and from biologically motivated first principles. The contributions are as follows.

- The analytic form of tuning curves is derived from constancy constraints (Section 2.1).
- The algebraic structure of the resulting minimum-overlap sets is identified. Within this structure, the requirements for unbiased estimates imply Lorentz group constraints on the coefficients (Section 2.2).
- For the decoding of channel vectors with more than three channels, a novel window selection has been derived, establishing a maximum-likelihood estimate (Section 2.3).
- Existing schemes are analyzed with respect to these constraints and a new, unbiased scheme is derived. This scheme is implemented for simulation experiments, where stationary points and bias from noisy image data are compared to existing schemes (Section 2.5).

The presented approaches are based on results from earlier work on channel representations (Sections 2.1 and 2.2) and make use of group theoretic results. The technical details of the work are summarized in the Supplementary Material.

2. Materials and Methods

The current approach to recurrent population codes shows a systematic bias, as we will show further below in this section. This bias can be avoided if the recurrent relations are modified according to certain constraints that we will derive from first principles. These first principles are common practice in many successful image representation techniques.

2.1. Encoding of Population Codes and Minimal Channel Representations

Population codes and channel representations are largely equivalent approaches concerning their encoding properties. In both cases *measurements x* from a *measurements space* $\mathcal{M}$ are encoded using *channel basis functions* $b(x)$ (the tuning curves). These functions are defined on a compact interval, in the case of orientation data, a periodic domain. They are continuous, non-negative, and decay with increasing $|x|$. The *channel support* of b is given by the domain where b is non-zero.

---

[1] Other examples for the three-folded representation are based on local phase estimates, where the coherence (phase-congruency) may be used for simultaneous detection of lines and edges (Reisfeld, 1996; Kovesi, 1999; Felsberg et al., 2005), and color (Lenz and Carmona, 2010). In the latter case, the evidence determines intensity, the coherence is a saturation measure, and the value represents the hue.

The basis function used here is the $\cos^2$ function, defined as (where h denotes the bandwidth parameter)

$$b(x) = \frac{2}{h}\cos^2(\pi x/h) \qquad \text{for } |x| < h/2 \quad \text{and } 0 \text{ otherwise.} \quad (1)$$

Note that we have introduced an additional scaling compared to previous work (Forssén, 2004) to make the basis function *mass preserving*, see below. The channel basis function is shifted n times (w.l.o.g. by integer displacements) to form the *channel basis* $\mathbf{b}(x)$ with components $b_j(x) = b(x - j)$. The number of non-zero components for any x determines the *channel overlap*. Computing the components for K measurements $x_{(k)}$ with weights a_k results in the *channel vector* $\mathbf{c}$ with *channel coefficients*

$$c_j = \sum_{k=0}^{K-1} a_k b(x_{(k)} - j) = \sum_{k=0}^{K-1} a_k b_j(x_{(k)}) \qquad j = 0 \ldots n - 1. \quad (2)$$

Representing elements of $\mathcal{M}$ using channel vectors is called the *channel representation*. If the sum of channel coefficients (l_1-norm) is constant 1 for any x, i.e., the channel vector lies in the $(n-1)$-simplex, the channel representation is mass preserving. We will only consider such channel representations in the sequel and therefore omit "mass-preserving". The simplexes for $n = 3$ and $n = 4$ are illustrated in **Figure 2**.

The channel basis $\mathbf{b}(x)$ determines a 1D curve in nD space, parametrized by x. As illustrated in **Figure 2**, $\cos^2$ functions result in a circle whereas other basis functions produce less regular shapes (A). Also for $n > 3$, the $\cos^2$-generated curve has constant curvature, but is no longer a circle (B). The circle ($n = 3$) respectively constant curvatures are direct consequences from the constant l_1- and l_2-norms of $\mathbf{b}$ for $\cos^2$ functions, restricting the curve to be part of the intersection of a plane and a sphere. Averaging channel representations over a set of measurements $x_{(k)}$ results in a convex combination of points on the curve.

Thus, the $\cos^2$ channel basis $\mathbf{b}$ induces two constancy properties onto the resulting channel vector $\mathbf{c}$: The l_1-norm of $\mathbf{c}$ and the l_2-norm of $\mathbf{c}$ are independent of displacements of x, as long as x is within the range of the tuning functions. The first property

corresponds to the idea of constant probability mass, i.e., each observation provides the same amount of evidence, independently of the specific value. The second property corresponds to the idea of isometry, i.e., scalar products depend only on the relative angle, not on the absolute angle within the reference coordinate system. This property is essential for learning-based approaches that should not be biased toward centers or edges of tuning functions. Thus, both constancy properties are essential for unbiased representations.

A further property of $\cos^2$ kernels as defined above, with the choice $h = 3$, is that they have an overlap of three, i.e., for a single observation x, $\mathbf{b}$ has three non-zero coefficients:

$$\mathbf{b}(x) = (\ldots, 0, \underbrace{b(x - [x] + 1)}_{\text{coefficient}[x]-1}, \underbrace{b(x - [x])}_{\text{coefficient}[x]}, \underbrace{b(x - [x] - 1)}_{\text{coefficient}[x]+1}, 0, \ldots)^{\text{t}},$$

$$(3)$$

where $[x]$ is the closest integer to x. Together with other standard requirements of kernel functions, we obtain that no kernels exist with overlap smaller than three:

Theorem 2.1. *(Minimum-overlap channel basis)* The minimum overlap of a channel basis with constant l_2-norm $\|\mathbf{b}(x)\|_2 = \mu$ for all x is 3. $\mu = \frac{1}{2}$.

Finally, from the two norm constraints, it follows that the $\cos^2$ kernel, which is the simplest non-constant even function, is the only solution (see Supplementary Material for both proofs):

Theorem 2.2. *(Uniqueness of $\cos^2$ kernel)* The unique channel basis function with minimal overlap and integer spacing is given by with equation (1) $h = 3$.

2.2. Decoding Channel Representations

Three-channel systems have some structural similarities with certain computational color models. Due to the existence of three color receptors on the human retina or color sensors in most cameras, one can describe the raw output of a color imaging system by a three-dimensional vector (usually in some form of RGB space), which is structurally similar to the channel vectors introduced above. In this spirit, the color opponent transform (Lenz and Carmona, 2010; van de Sande et al., 2010) and its variant in cylindrical coordinates, i.e., hue–saturation–intensity (van de Weijer et al.,

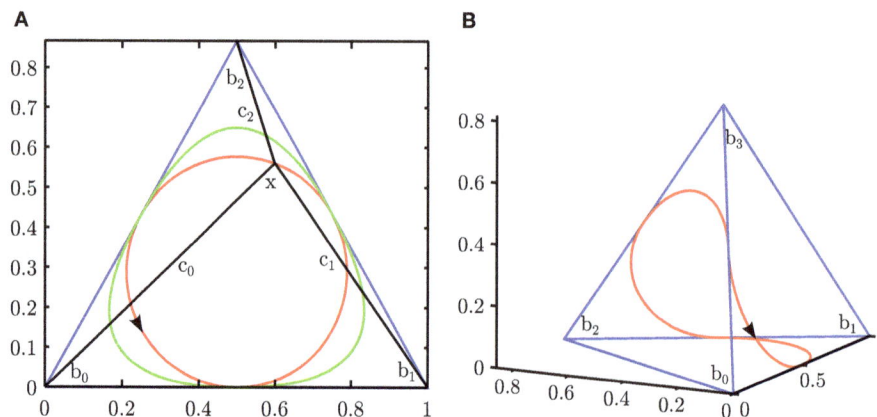

FIGURE 2 | (A) 2-simplex. Blue: edges of a simplex. Red curve: trajectory from $\cos^2$-encoded single measurements. Green: B-spline-encoded single measurements. **(B)** 3-simplex. Blue: edges of simplex. Red: $\cos^2$-encoded single measurements.

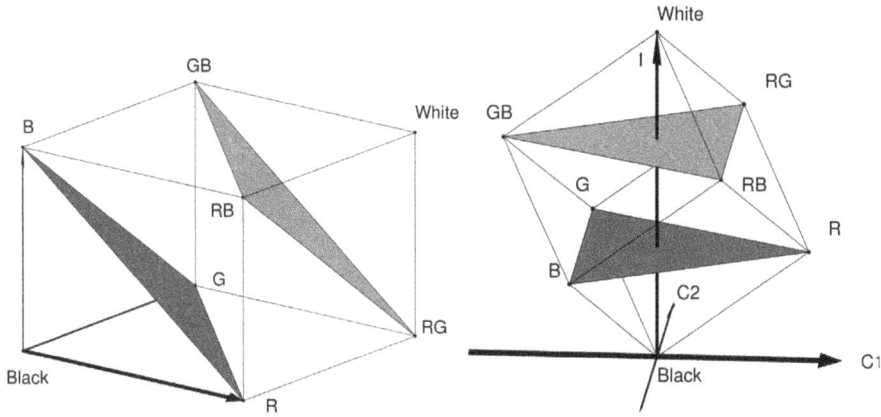

FIGURE 3 | Illustration of the color opponent transform (Åström, 2015). Figure courtesy Freddie Åström.

2005), are three-folded representations, see Section 1. In **Figure 3**, the color opponent transform, an orthogonal transformation of RGB, is illustrated.

The concept of three-folded representations by value/coherence/evidence has been related to channel representations, and thus population codes, in context of decoding $\cos^2$ channels (Forssén, 2004), from which we have adopted the notation in **Figure 4**: value $\hat{x}$, coherence $\hat{r}_1/\hat{r}_2$, and evidence $\hat{r}_2$. Previous work refers to $\hat{r}_1$ and $\hat{r}_2$ as confidence measures and suggests to form the 3-vector $\mathbf{p} = (p_0, p_1, p_2)^t := (\hat{r}_1 \cos 2\pi\hat{x}/3, \hat{r}_1 \sin 2\pi\hat{x}/3, \hat{r}_2)^t$ using a matrix of orthogonal column vectors $\mathbf{W}^t = (\mathbf{w}_0, \mathbf{w}_1, \mathbf{w}_2)$. After normalization[2], we obtain:

$$\mathbf{c} = \mathbf{W}^t \mathbf{p}$$

$$:= \frac{1}{\sqrt{3}} \begin{pmatrix} \sqrt{2} & 0 & 1 \\ \sqrt{2}\cos(2\pi/3) & \sqrt{2}\sin(2\pi/3) & 1 \\ \sqrt{2}\cos(4\pi/3) & \sqrt{2}\sin(4\pi/3) & 1 \end{pmatrix} \begin{pmatrix} r_1\cos(2\pi x/3) \\ r_1\sin(2\pi x/3) \\ r_2 \end{pmatrix}$$

$$\tag{4}$$

Inverting this expression gives $\mathbf{p} = \mathbf{W}\mathbf{c}$, and the estimate for x is obtained by extracting the phase-angle

$$\hat{x} = \frac{3}{2\pi} \arg(p_0 + ip_1) = \frac{3}{2\pi} \arg(\mathbf{c}^t(\mathbf{w}_0 + i\mathbf{w}_1)). \tag{5}$$

Thus, all involved operations (matrix-vector products and extraction of phase) are the same as for the readout of population codes (Denève et al., 1999) and are biologically plausible (Mechler et al., 2002).

Note that the normalization of the matrix columns implies that $r_2 = \sqrt{2}r_1$ in the case of a single encoded value. Thus, the cone becomes acute (70.5° instead of 90°). If a set of measurements is represented using channels, the resulting channel vector $\mathbf{c}$ is a convex combination of points on the surface of this cone, i.e., a point in the interior of the cone, and $r_2 > \sqrt{2}r_1$ in general. Decoding this channel vector is then related to finding the closest point on the cone surface. The exact definition of *closest* in this

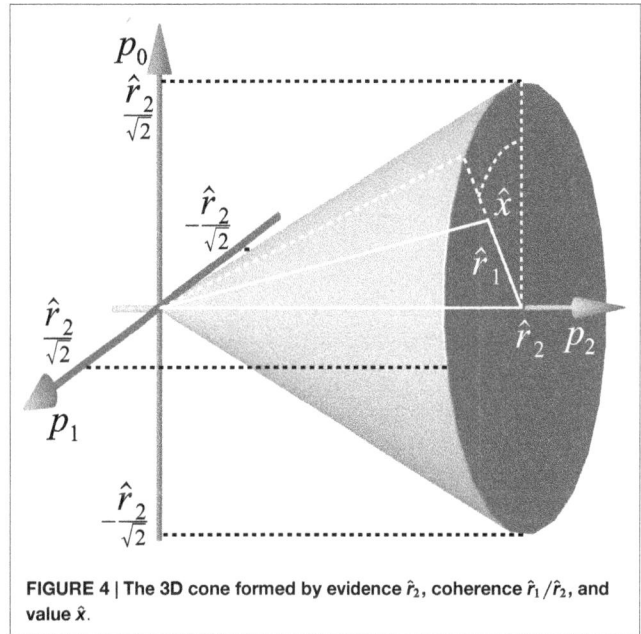

FIGURE 4 | The 3D cone formed by evidence $\hat{r}_2$, coherence $\hat{r}_1/\hat{r}_2$, and value $\hat{x}$.

context is dependent on the transformation group considered. Using just p_0 and p_1 as suggested previously (Forssén, 2004) means to neglect distances along p_2.

The three-fold representations above live in a cone, i.e., if representations are modified or compared, the transformations that leave the cone invariant need to be analyzed and represented. Obviously, we may not use Euclidean transformations for this purpose, because they would produce results outside the cone. Instead, we consider *Lorentz transformations* similar to those we know from relativity theory.[3]

In our case, the Lorentz transformations consist of rotations about the cone axis and Lorentz boosts (contractions). The rotation about the cone axis corresponds to changes of our measurements x (see also **Figure 4**), thus an unbiased scheme must

---

[2]For the sake of convenience, we use the normalized coefficient matrix as common in literature (Fässler and Stiefel, 1992; Lenz and Carmona, 2010)

[3]The cone geometry is easiest defined in $(2+1)$D Minkowski space $\mathbb{R}^{2,1}$. The associated quadratic form has the signature $\{+, +, -1\}$, i.e., $p_0^2 + p_1^2 - p_2^2$. The transformation group that we consider is the special orthogonal group, $SO(2,1)$, more accurately the connected component $SO^+(2,1)$.

be rotation invariant. Using the change of coordinates from the previous section, we derive a constraint on the update of c that must be fulfilled in each step of a recurrent scheme (the proof is given in Supplementary Material):

Theorem 2.3. *(Constraint on channel coefficients)* The decoding of a 3-channel vector $\mathbf{c}$ is invariant under a change of channel coefficients $\nabla_{\mathbf{c}} = (\partial_{c_0}, \partial_{c_1}, \partial_{c_2})^t$ iff

$$(c_0 - c_1)\partial_{c_2} + (c_2 - c_0)\partial_{c_1} + (c_1 - c_2)\partial_{c_0} = 0. \tag{6}$$

2.3. Windowed Decoding by Maximum-Likelihood Estimation

The decoding based on equation (5) is defined for channel representations with three channels. For the general case of more than three channels, recurrent schemes will be discussed below. However, non-iterative methods for more than three channels have been suggested (Forssén, 2004), based on first selecting a window of three channels and then applying equation (4).

The 3-window is selected by various criteria, e.g., maximizing r_2 or having a local maximum at the center channel $c_{j-1} \le c_j \ge c_{j+1}$. From a statistical point of view, however, a maximum-likelihood selection would be preferable and under the assumption of independent Gaussian noise, we arrive at a least squares problem and the corresponding theorem (the proof is given in Supplementary Material):

Theorem 2.4. *(Maximum-likelihood decoding)* Assuming independent Gaussian noise on the channel coefficients, the decoding 3-window with maximum likelihood is given at the index j

$$\hat{j} = \arg\max_j \tilde{r}_1(j) + \sqrt{2}r_2(j), \tag{7}$$

where $r_i(j)$ is computed as r_i of the vector $\mathbf{c}_j := (c_{j-1}, c_j, c_{j+1})^t$,

$$\tilde{r}_1(j) = \begin{cases} r_1(j) & \text{if } |\alpha(\mathbf{c}_j)| \le \pi/3 \\ r_1(j) \cos(|\alpha(\mathbf{c}_j)| - \pi/3) & \text{else,} \end{cases} \tag{8}$$

and $\alpha(\mathbf{c}_j) = \arg(\mathbf{c}_j^t(\mathbf{w}_0 + i\mathbf{w}_1)) - \frac{2\pi}{3}$. The MLE is given as $\hat{x} = \max(\min(\frac{3}{2\pi}\alpha(\mathbf{c}_j), \frac{1}{2}), -\frac{1}{2}) + \hat{j}$.

2.4. Analysis of the Existing Recurrent Scheme for Population Codes

The recurrent procedure suggested for the readout of population codes (Denève et al., 1999) contains the same building blocks as channel decoding and is biologically equally plausible, because such procedures coincide with cortical circuitry. It is suggested to establish two coupled, non-linear equations, the first being a linear mapping of the population vector $\mathbf{o}$ to a slack vector $\mathbf{u} = A\mathbf{o}$, and the second one writing back the squared and normalized $\mathbf{u}$ into $\mathbf{o}$:

$$o_j = u_j^2/(S + \mu|\mathbf{u}|^2). \tag{9}$$

The computational flow of this method is illustrated in **Figure 5**.

The problem is, however, that the scheme given in equation (9) only fulfills the invariance requirement (6) in one trivial case for $\mathbf{A}$. In all other cases, iterating through the scheme leads to successively larger bias toward specific values, e.g., at the channel centers (the proof is given in Supplementary Material):

Theorem 2.5. *(Bias of Denève's scheme)* The scheme (9) is biased unless we choose

$$\mathbf{A} = \sqrt{\frac{S}{3(1-\mu)}} \begin{pmatrix} 1 & 1 & 1 \\ 1 & 1 & 1 \\ 1 & 1 & 1 \end{pmatrix}, \text{ where } \mu < 1. \tag{10}$$

2.5. The New Recurrent Scheme

In this section, we will derive a recurrent scheme that, in contrast to equation (9), fulfills the invariance constraint for the value (6). Moreover, a recurrent scheme should not modify the evidence either, i.e., we require

$$\partial_{r_2} = \mathbf{w}_2^t \nabla_{\mathbf{c}} = 0. \tag{11}$$

The two constraints (6) and (11) imply the existence of a solution in terms of their common null-space. In the case of three channels, this space is spanned by permutations of the vector $(-1, 2, -1)^t$, the (negative) discrete Laplacian kernel. Since the update of $\mathbf{c}$ in the corresponding scheme has to be in a (3-2)-D subspace, only one degree of freedom exists and is uniquely determined by the requirement $\sqrt{2}r_1 = r_2$.

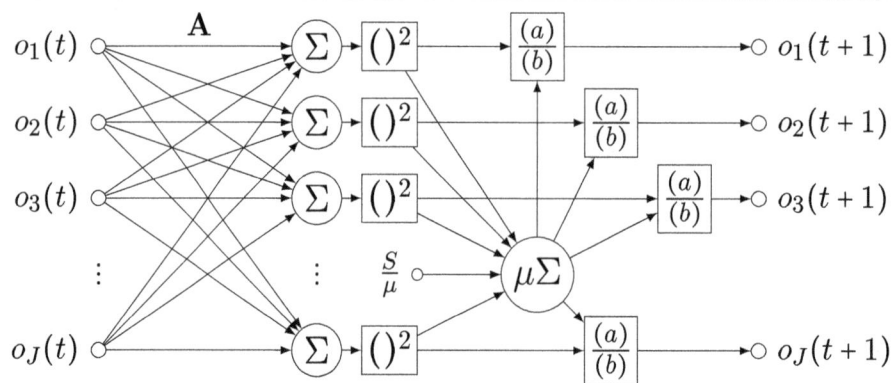

FIGURE 5 | Diagram of the recurrent network for the readout of population codes (Denève et al., 1999).

Thus, the fixpoint solution can be achieved by a single computation and suitable δ as

$$\mathbf{c}_0 = \mathbf{c} + \delta \begin{pmatrix} 2 & -1 & -1 \\ -1 & 2 & -1 \\ -1 & -1 & 2 \end{pmatrix} \mathbf{c}. \tag{12}$$

In the case of more than three channels, the algebraic structure becomes more complicated, because points are no longer projected onto a circle in the plane, but onto a 1-D curve in high-dimensional space. This can no longer be achieved by a linear equation as in equation (12). The proper generalization is obtained by splitting the Laplacian operator according to $\Delta = \partial_j^2$ (j being the index of $\mathbf{c}$, thus ∂_j being the finite difference operator) and defining $\mathbf{d} = \partial_j \mathbf{c}$ as the *flow* between the channels in each incremental step $\mathbf{c}_t$

$$\mathbf{c}_t \propto \partial_j \mathbf{d}, \tag{13}$$

establishing the continuity equation of channel coefficients. The flow coefficient d_j determines how much of the coefficient c_j is moved to c_{j+1}.

In contrast to equation (12), where the flow is computed by the difference $c_{j+1} - c_j$, the case of more than three channels requires a non-linear mapping $f: \mathbb{R}^2 \rightarrow \mathbb{R}$; $(c_j, c_{j+1}) \mapsto d_j$ to compute the flow. The complete computational scheme is illustrated in **Figure 6**.

Assuming that the function f is known and δ being the update step length, we obtain the following iterative algorithm:

while $\sqrt{2}r_1 < r_2$ **do**
  for all j **do**
    $d_j \leftarrow f(c_j, c_{j+1})$
  end for
  for all j **do**
    $c_j \leftarrow c_j + \delta(d_{j-1} - d_j)$
  end for
end while

The implementation of such a scheme and the comparison to the state-of-the-art scheme (Denève et al., 1999) with respect to stationary points using simulation data and with respect to bias using noisy image data is shown in the subsequent section.

3. Results

In this section, we analyze stationary points of the discussed recurrent schemes and we perform experiments on orientation analysis from noisy images. Before looking into the iterative schemes, we first compare the non-iterative decoding according to maximum r_2 (Forssén, 2004) and Theorem 2.4, see **Figure 7**. In the experiment, we compare the respective decoding result for 1000 random measurements encoded into 7 channels and combined with random weights. Obviously, the modification of the window selection leads to exact reproduction of MLE results.

3.1. Stationary Points Analysis

A fully unbiased recurrent scheme must keep the entire measurement domain invariant, i.e., each point in the measurement space is a stationary point under the recurrent scheme. For this subsection, we restrict the discussion to the case of three orientation channels.

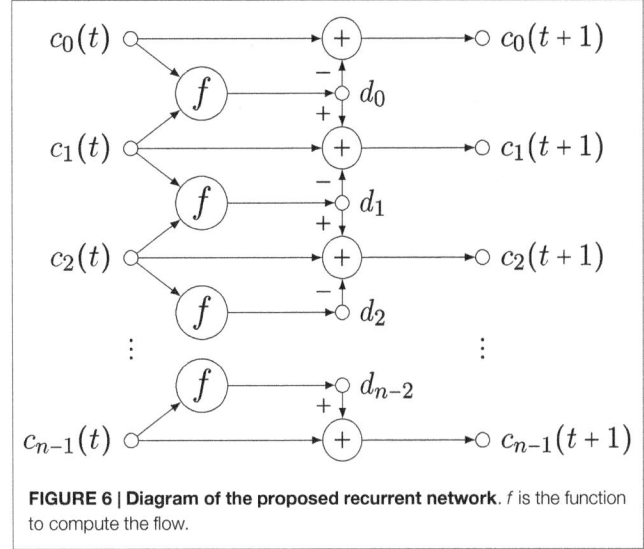

FIGURE 6 | Diagram of the proposed recurrent network. f is the function to compute the flow.

In the case of Denève's scheme, we already know from Theorem 2.5 that only one trivial solution exists. However, we can calculate which *discrete* points are stationary points by requiring

$$\mathbf{o}(t + 1) = \mathbf{o}(t). \tag{14}$$

If $\mathbf{o}$ is a stationary point, $\mathbf{u} = \mathbf{Ao}$ is constant, and thus, also the normalization factor $Q := S + \mu|\mathbf{u}|^2$ is constant. As in the proof of Theorem 2.5, we exploit that $\mathbf{A}$ is circulant, but it is also symmetric (we may reflect all vectors) and thus

$$\mathbf{A} = \begin{pmatrix} \alpha & \beta & \beta \\ \beta & \alpha & \beta \\ \beta & \beta & \alpha \end{pmatrix} \tag{15}$$

for suitable α and β. For a stationary solution, we obtain $o_j = \frac{1}{Q}(\mathbf{Ao})_j^2$ and therefore

$$\begin{aligned} o_0 &= \frac{1}{Q}(\alpha o_0 + \beta o_1 + \beta o_2)^2 \\ o_1 &= \frac{1}{Q}(\beta o_0 + \alpha o_1 + \beta o_2)^2. \\ o_2 &= \frac{1}{Q}(\beta o_0 + \beta o_1 + \alpha o_2)^2 \end{aligned} \tag{16}$$

Q only affects absolute scale of stationary points, as does simultaneously scaling of α and β. Thus, only one degree of freedom remains for changing the stationary points: the quotient of β/α.

Without loss of generality, we now set $S = 0.1$, $\mu = 0.9$, and $\alpha = 1$. For different choices of β, we can now plot the trajectories under iterations of equation (9), see **Figure 8**. For comparison, we also include the result from our proposed method (12).

Obviously, Denève's method has a discrete set of stationary points, which means that iterating until convergence will lead to one out of a set of discrete solutions and thus, the result is strongly biased. Our solution for three channels, however, maps to all points on the unit circle and is thus unbiased.

3.2. Orientation Estimation from Image Data

In this section, we perform a number of experiments using two sets of visual stimuli with different complexity: one contains a

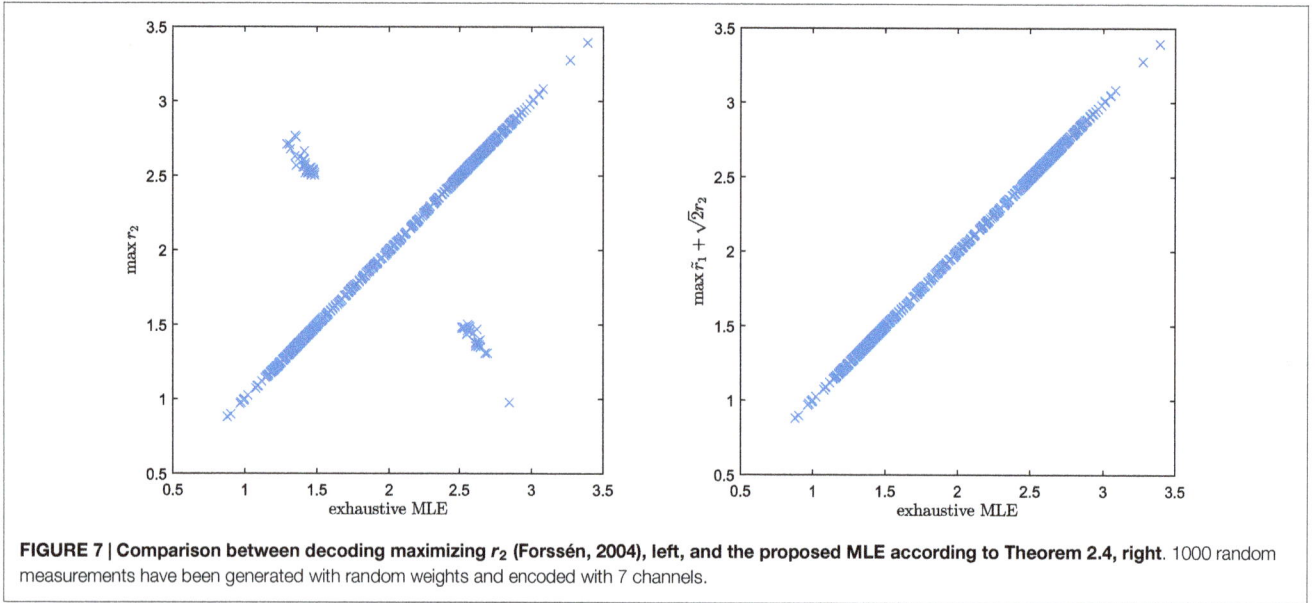

FIGURE 7 | Comparison between decoding maximizing r_2 (Forssén, 2004), left, and the proposed MLE according to Theorem 2.4, right. 1000 random measurements have been generated with random weights and encoded with 7 channels.

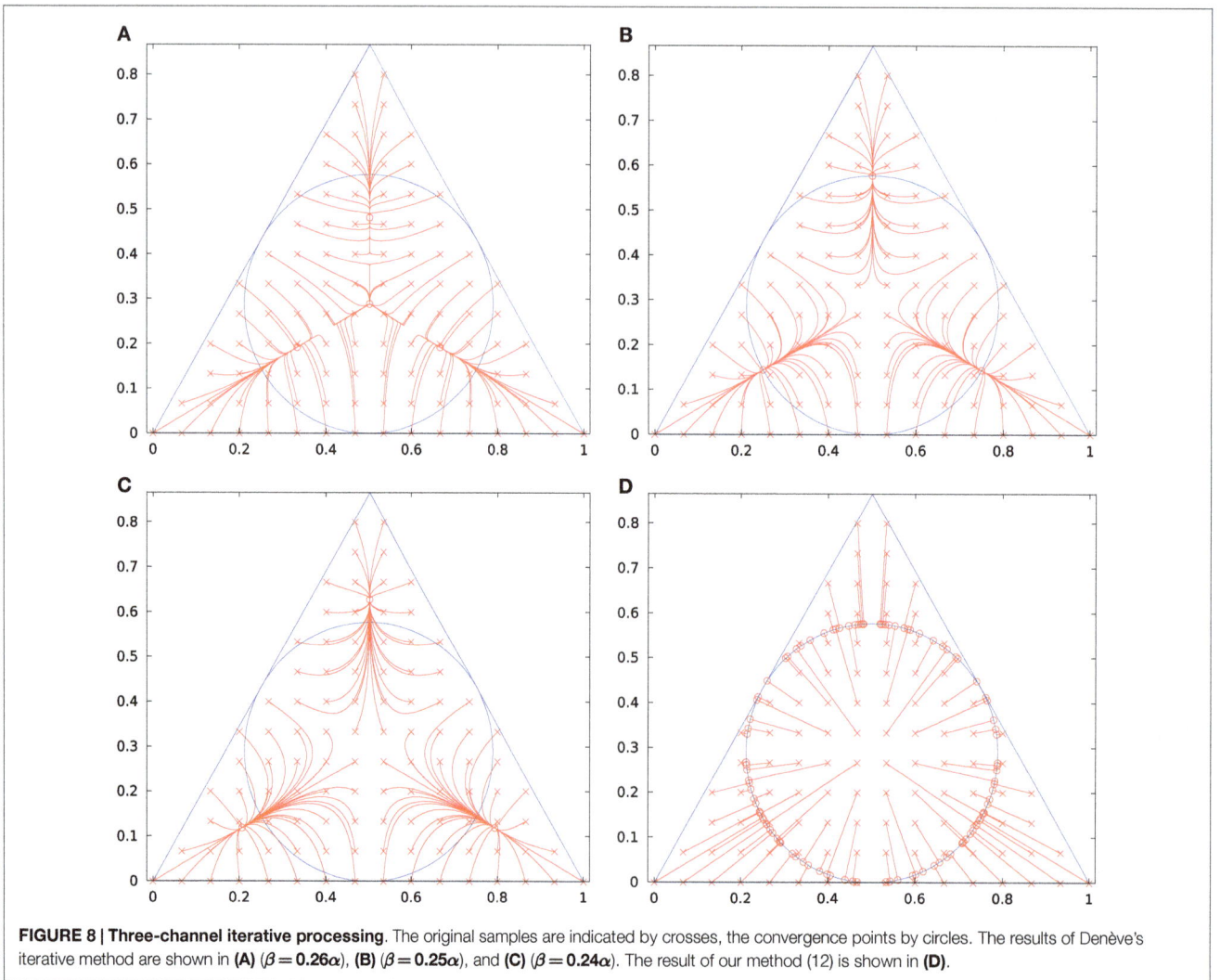

FIGURE 8 | Three-channel iterative processing. The original samples are indicated by crosses, the convergence points by circles. The results of Denève's iterative method are shown in **(A)** ($\beta = 0.26\alpha$), **(B)** ($\beta = 0.25\alpha$), and **(C)** ($\beta = 0.24\alpha$). The result of our method (12) is shown in **(D)**.

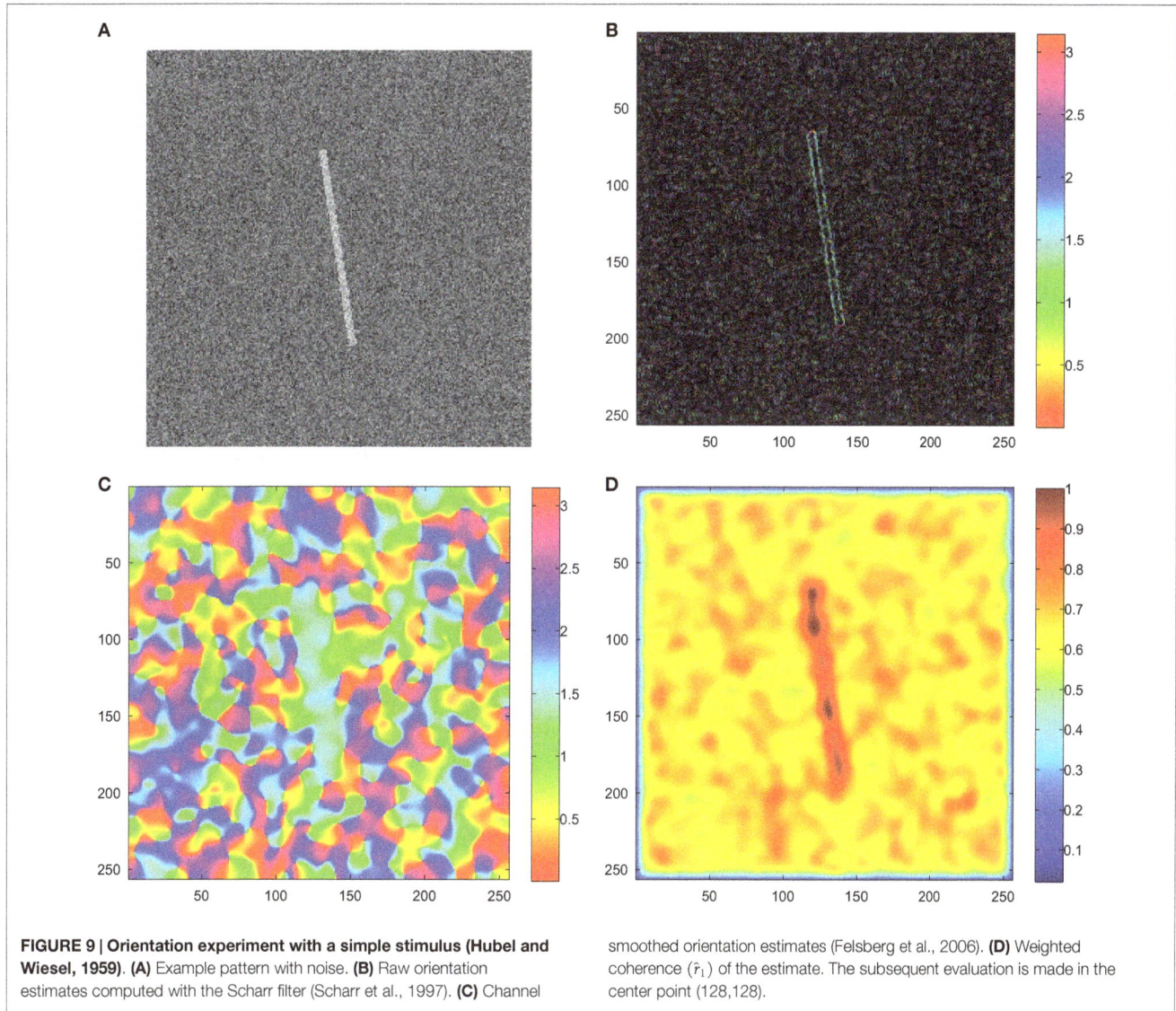

FIGURE 9 | Orientation experiment with a simple stimulus (Hubel and Wiesel, 1959). **(A)** Example pattern with noise. **(B)** Raw orientation estimates computed with the Scharr filter (Scharr et al., 1997). **(C)** Channel smoothed orientation estimates (Felsberg et al., 2006). **(D)** Weighted coherence ($\hat{r}_1$) of the estimate. The subsequent evaluation is made in the center point (128,128).

single orientation without texture (Hubel and Wiesel, 1959), see **Figure 9A**, and the other one multiple orientations with varying textures (Felsberg et al., 2006), see **Figure 10A**. From rotated and noisy instances of these images, local orientation information is extracted with a regularized gradient filter (Scharr et al., 1997). The color coded orientation data is illustrated in **Figures 9 and 10B**.

The extracted orientation estimates are pixel-wise channel coded and the resulting channel images are spatially averaged (Felsberg et al., 2006). The resulting channel representations are then decoded; the result is shown in **Figures 11** and **12A**. If we apply 5, 10, and 100 iterations according to equation (9) before decoding, the results as shown in **Figures 11** and **12B–D**, respectively, are obtained. It is obvious for all cases that the iterative scheme (9) leads to an increasing bias. For three channels, the case of decoding without iterating (9) corresponds to our solution (12) and shows no significant bias.

In **Figure 13**, the corresponding results are shown for 7 channels. The used basis functions are displayed in **Figure 13A**. The

look-up table for the flow function $f(c_j, c_{j+1})$ between the channels is displayed in **Figure 13B**. The look-up table is generated by successively low-pass filtering random samples from a Gaussian distribution, computing the flow, and inverting the sign. The result from the proposed iterative scheme shows no bias effect if iterated to convergence (see **Figures 13D,F**) whereas Denève's method shows clearly visible bias effects (cf. **Figures 13C,E**).

So far, the advantage of applying the proposed scheme compared to direct decoding has not been very explicit. This changes if the noise distribution becomes broader. In **Figure 14**, we show results for simulations with orientation samples (200, respectively, 1800) drawn from a distribution with standard deviation of 0.5 respectively 1.5 times the decoding window width. 10 iterations of Denève's method result already in a clearly visible bias (**Figures 14C,D**), whereas the proposed scheme shows no bias in either case (**Figures 14E,F**). The direct decoding shows no visible bias for the narrower distribution (a) and a clearly visible bias for the wider distribution (b). Note that the bias may not be evaluated if the mean of the distribution is placed at a channel center or

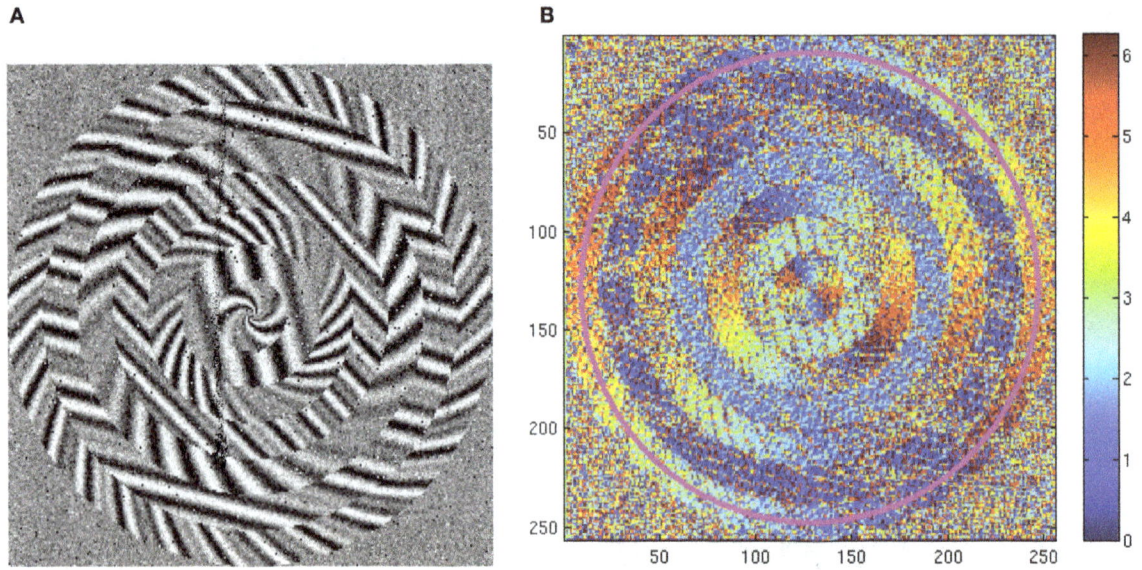

FIGURE 10 | Orientation experiment with a complex pattern (Felsberg et al., 2006). (A) Example pattern with noise. **(B)** Raw orientation estimates computed with the Scharr filter (Scharr et al., 1997). The subsequent evaluation is made along the magenta circle.

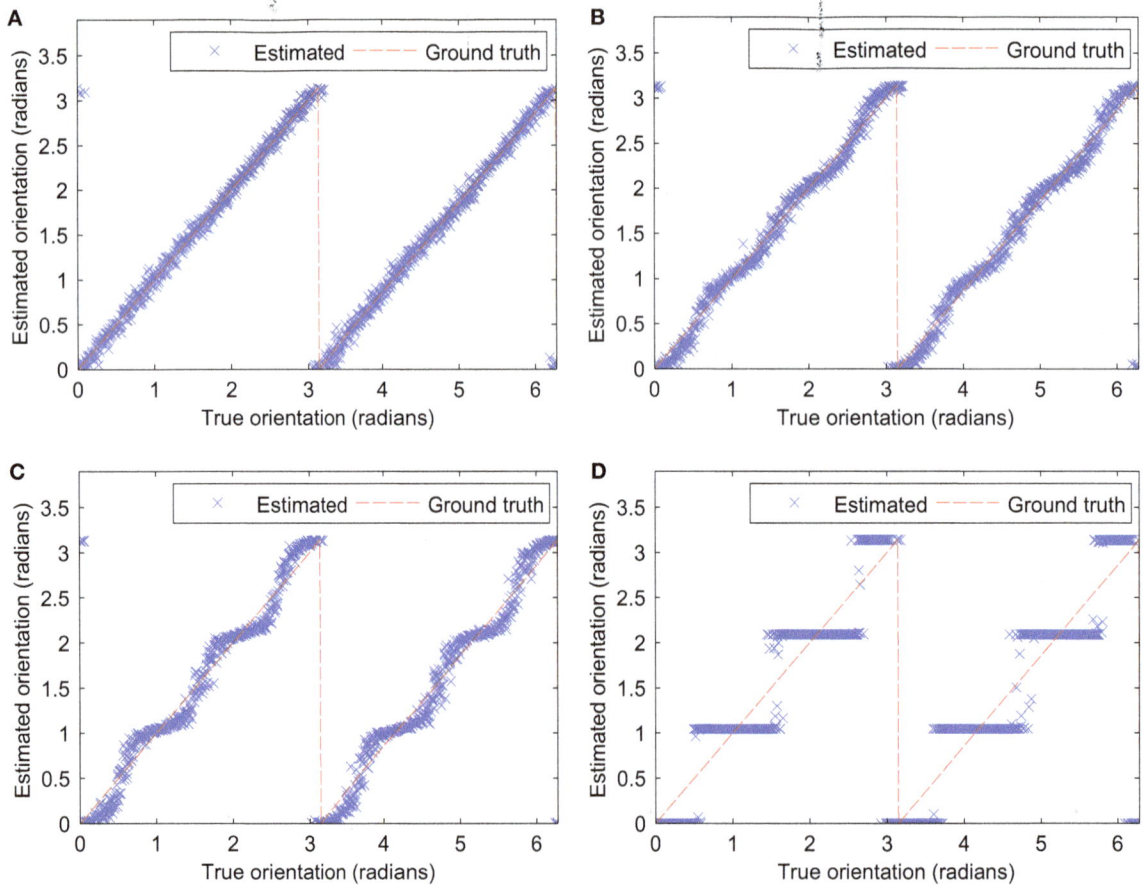

FIGURE 11 | Evaluation results using stimuli from Figure 9 for three channels. (A–D) decoding with respectively 0, 5, 10, and 100 iterations according to (9).

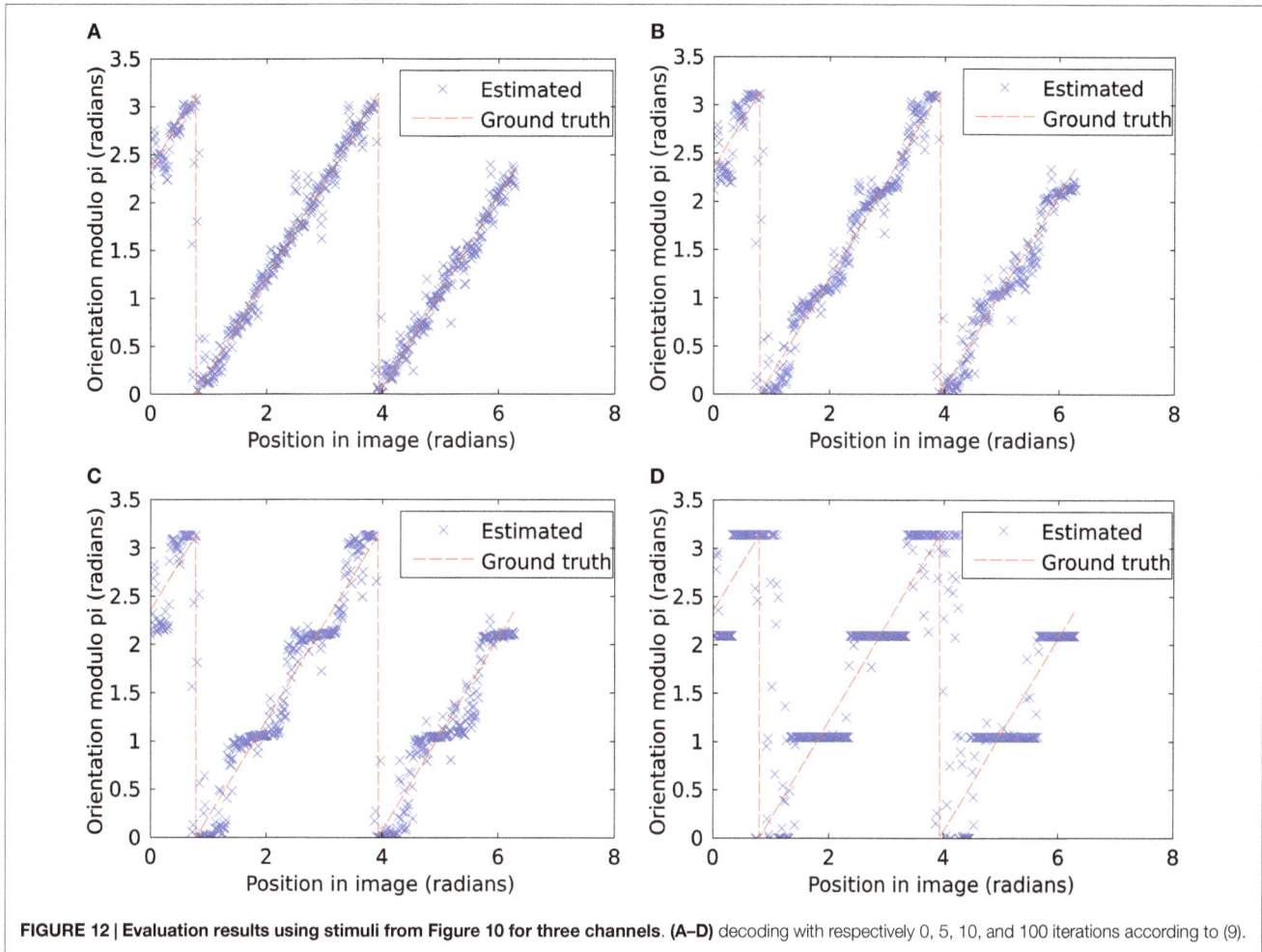

FIGURE 12 | Evaluation results using stimuli from Figure 10 for three channels. (A–D) decoding with respectively 0, 5, 10, and 100 iterations according to (9).

at a boundary. For such setups, all methods show an unbiased behavior.

4. Discussion

We have shown that the recurrent scheme by Denève produces biased output if the mean of the stimuli is not located on a channel center or a channel boundary and if the recurrent scheme is iterated until convergence. The bias of Denève's method has not been visible in the original paper (Denève et al., 1999), because all evaluations have been done either on data centered at a stationary point or as averaged error over all distribution centers.

The suggested extensive overlap of tuning functions and fixed number of iterations (Denève et al., 1999) lead only to a gradual reduction of the bias and have three major practical drawbacks for the design of computational systems:

1. The extensive overlap of tuning functions (oversampling in signal processing terms) increases the redundancy, and thus the computational effort.
2. The number of iterations depends on the noise level of the input: noisier data requires longer processing time as

observed in biological systems. Reducing the number of iterations to a fixed number (in order to limit the bias) will result in useless output if the noise level is above a certain threshold.
3. Dynamic processes with continuous input, i.e., using population codes as an implementation of a Kalman filter (Denève et al., 2007), will make use of the iterative procedure in each time step, and thus the recurrent equation is potentially applied an arbitrary number of times, eventually leading to biased estimates.

All three drawbacks are avoided by using the proposed channel-based approach, established by means of theorems 2.1–2.4 and a simple look-up table. Besides the fact of establishing a new theoretical and computational model for recurrent networks, this possibly leads to two new applications:

• Application 1: denoising for learning – many channel-based learning algorithms (Felsberg et al., 2013) require clean inputs in order to avoid responses to insignificant inputs. If the recurrent scheme is applied before the training, cleaner responses might be the consequence.

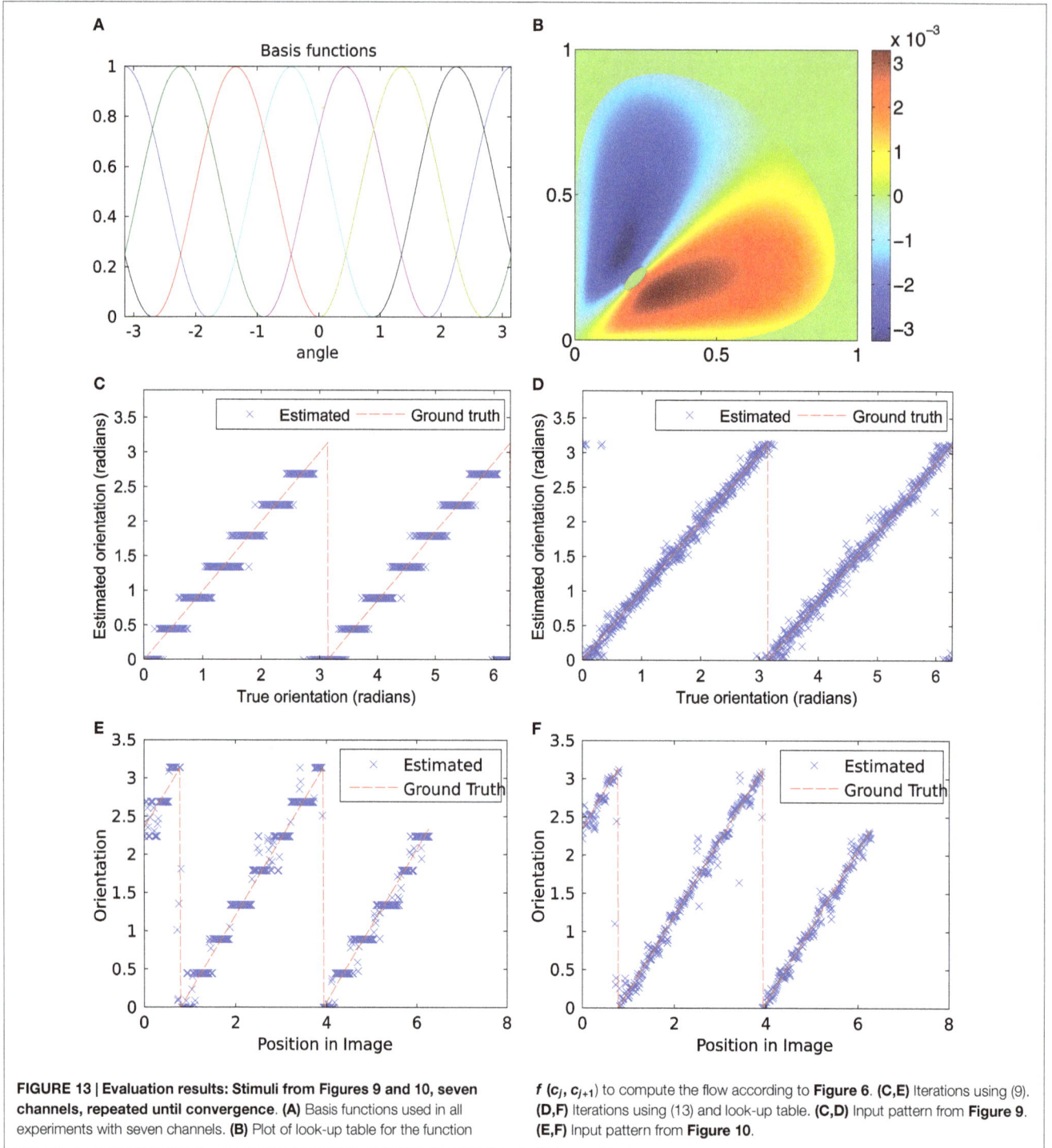

FIGURE 13 | Evaluation results: Stimuli from Figures 9 and 10, seven channels, repeated until convergence. (A) Basis functions used in all experiments with seven channels. **(B)** Plot of look-up table for the function $f(c_j, c_{j+1})$ to compute the flow according to **Figure 6**. **(C,E)** Iterations using (9). **(D,F)** Iterations using (13) and look-up table. **(C,D)** Input pattern from **Figure 9**. **(E,F)** Input pattern from **Figure 10**.

• Application 2: adaptive channel resolutions – in certain cases, it would be useful to increase resolution of the channel representation during processing (Felsberg, 2010). This requires a sharpening of the underlying measured distribution, which can be achieved by the proposed scheme.

In conclusion, we believe that recurrent networks are very useful tools, but have to be applied in the appropriate way in order to avoid introducing unwanted bias effects.

Author Contributions

MF has been the lead-researcher in this work, developing most of the conceptual ideas, proving most of the theorems, and writing most of the text. KÖ has been contributing with ideas concerning decoding-invariant processing, deriving computational schemes, running simulations, and contributing to the text. RL has been contributing with derivations and explanations on Lorentz groups and group theory, in general, including the respective sections of the text.

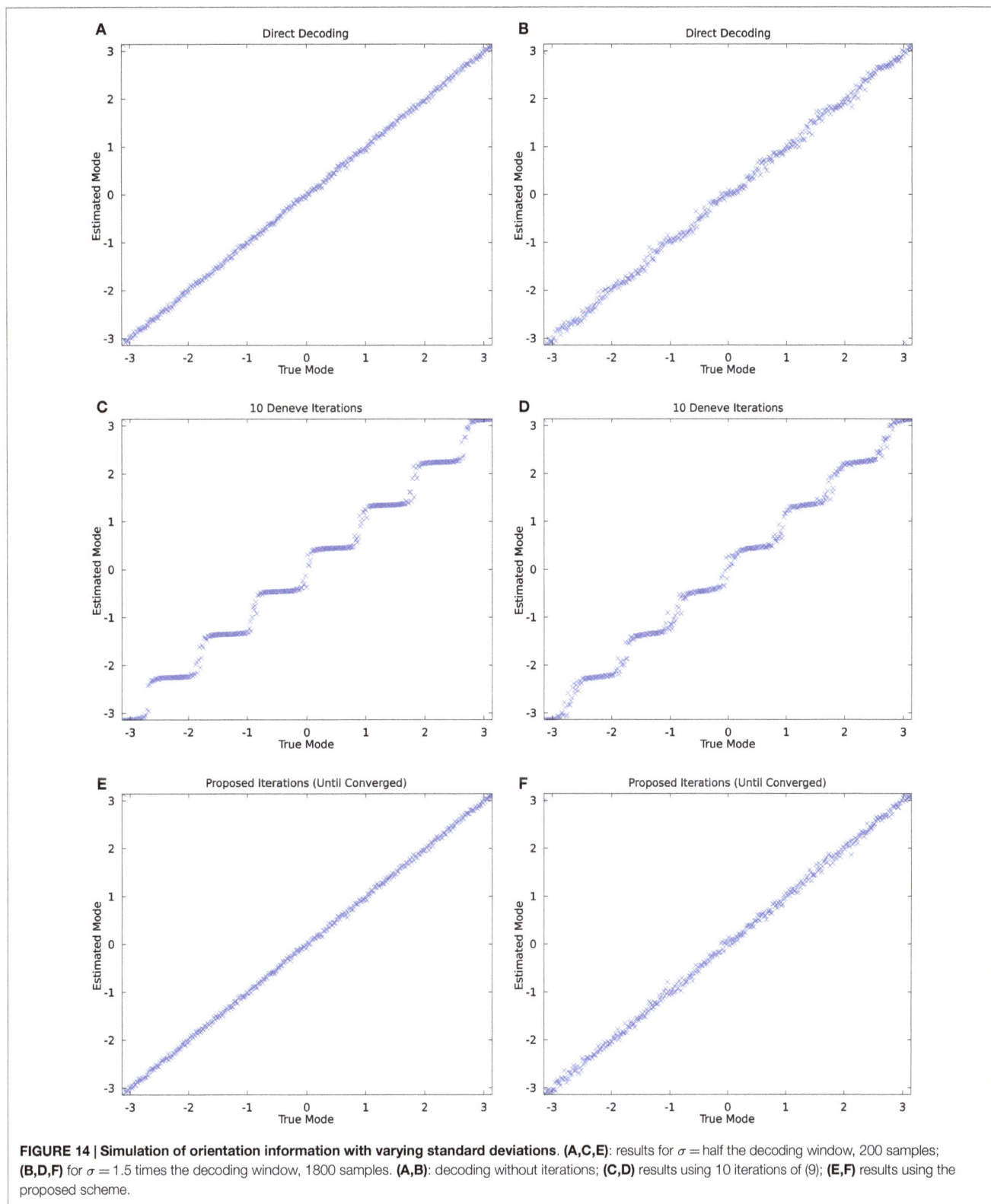

FIGURE 14 | Simulation of orientation information with varying standard deviations. (A,C,E): results for $\sigma =$ half the decoding window, 200 samples; **(B,D,F)** for $\sigma = 1.5$ times the decoding window, 1800 samples. **(A,B)**: decoding without iterations; **(C,D)** results using 10 iterations of (9); **(E,F)** results using the proposed scheme.

Funding

This work has been supported by the VR projects EMC^2 and VIDI, by the SSF projects CUAS and VPS, and by the excellence centers ELLIIT and CADICS.

References

Azzopardi, G., and Petkov, N. (2013). Trainable COSFIRE filters for keypoint detection and pattern recognition. *IEEE Trans. Pattern Anal. Mach. Intell.* 35, 490–503. doi:10.1109/TPAMI.2012.106

Azzopardi, G., and Petkov, N. (2014). Ventral-stream-like shape representation: from pixel intensity values to trainable object-selective cosfire models. *Front. Comput. Neurosci.* 8:80. doi:10.3389/fncom.2014.00080

Åström, F. (2015). *Variational Tensor-Based Models for Image Diffusion in Non-Linear Domains.* Ph.D. thesis, Linköping University, Sweden.

Bigün, J., and Granlund, G. H. (1987). "Optimal orientation detection of linear symmetry," in *Proceedings of the IEEE First International Conference on Computer Vision* (London: IEEE), 433–438.

Dalal, N., and Triggs, B. (2005). "Histograms of oriented gradients for human detection," in *Computer Vision and Pattern Recognition (CVPR) 2005. IEEE Computer Society Conference on,* Vol. 1 (San Diego, CA: IEEE), 886–893. doi:10.1109/CVPR.2005.177

d'Angelo, E., Alahi, A., and Vandergheynst, P. (2012). "Beyond bits: reconstructing images from local binary descriptors," in *Pattern Recognition (ICPR), 2012 21st International Conference on* (Tsukuba: IEEE), 935–938.

Denève, S., Duhamel, J.-R., and Pouget, A. (2007). Optimal sensorimotor integration in recurrent cortical networks: a neural implementation of Kalman filters. *J. Neurosci.* 27, 5744–5756. doi:10.1523/JNEUROSCI.3985-06.2007

Denève, S., Latham, P. E., and Pouget, A. (1999). Reading population codes: a neural implementation of ideal observers. *Nat. Neurosci.* 2, 740–745. doi:10.1038/11205

Fässler, A., and Stiefel, E. (1992). "Preliminaries," in *Group Theoretical Methods and Their Applications.* Boston, MA: Birkhäuser, 1–31.

Felsberg, M. (2010). "Incremental computation of feature hierarchies," in *Pattern Recognition 2010, Proceedings of the 32nd DAGM,* Darmstadt.

Felsberg, M. (2012). "Chap. Adaptive filtering using channel representations," in *Mathematical Methods for Signal and Image Analysis and Representation, volume 41 of Computational Imaging and Vision,* eds L. Florack, R. Duits, G. Jongbloed, M. C. van Lieshout, and L. Davies (London: Springer), 31–48.

Felsberg, M. (2013). "Enhanced distribution field tracking using channel representations," in *IEEE ICCV Workshop on Visual Object Tracking Challenge,* Sydney.

Felsberg, M., Duits, R., and Florack, L. (2005). The monogenic scale space on a rectangular domain and its features. *Int. J. Comput. Vis.* 64, 187–201. doi:10.1007/s11263-005-1843-x

Felsberg, M., Forssén, P.-E., and Scharr, H. (2006). Channel smoothing: efficient robust smoothing of low-level signal features. *IEEE Trans. Pattern Anal. Mach. Intell.* 28, 209–222. doi:10.1109/TPAMI.2006.29

Felsberg, M., and Hedborg, J. (2007). Real-time view-based pose recognition and interpolation for tracking initialization. *J. R. Time Image Process.* 2, 103–116. doi:10.1007/s11554-007-0044-y

Felsberg, M., Kalkan, S., and Krüger, N. (2009). Continuous dimensionality characterization of image structures. *Image Vis. Comput.* 27, 628–636. doi:10.1016/j.imavis.2008.06.018

Felsberg, M., Larsson, F., Wiklund, J., Wadstromer, N., and Ahlberg, J. (2013). Online learning of correspondences between images. *IEEE Trans. Pattern Anal. Mach. Intell.* 35, 118–129. doi:10.1109/TPAMI.2012.65

Forssén, P. E. (2004). *Low and Medium Level Vision using Channel Representations.* PhD thesis, Linköping University, Sweden.

Förstner, W., and Gülch, E. (1987). "A fast operator for detection and precise location of distinct points, corners and centres of circular features," in *ISPRS Intercommission Workshop* (Interlaken: ISPRS), 149–155.

Granlund, G. H. (2000). "An associative perception-action structure using a localized space variant information representation," in *Proc. Int. Workshop on Algebraic Frames for the Perception-Action Cycle* (Heidelberg: Springer).

Howard, I. P., and Rogers, B. J. (1995). *Binocular Vision and Stereopsis.* Oxford: Oxford University Press.

Hubel, D. H., and Wiesel, T. N. (1959). Receptive fields of single neurones in the cat's striate cortex. *J. Physiol.* 148, 574–591. doi:10.1113/jphysiol.1959.sp006308

Jähne, B. (2005). *Digital Image Processing,* 6th Edn. Berlin: Springer.

Jonsson, E. (2008). *Channel-Coded Feature Maps for Computer Vision and Machine Learning.* Ph.D. thesis, Linköping University, Sweden. Dissertation No. 1160, ISBN 978-91-7393-988-1.

Kato, H., and Harada, T. (2014). "Image reconstruction from bag-of-visual-words," in *2014 IEEE Conference on Computer Vision and Pattern Recognition (CVPR)* (Columbus: IEEE), 955–962.

Kovesi, P. (1999). Image features from phase information. *Videre J. Comput. Vis. Res.* 1, 1–26.

LeCun, Y., and Bengio, Y. (1995). Convolutional networks for images, speech, and time series. *Handb. Brain Theory Neural Netw.* 3361, 255–258.

Lenz, R., and Carmona, P. L. (2010). "Hierarchical s(3)-coding of rgb histograms," in *Computer Vision, Imaging and Computer Graphics. Theory and Applications, Vol. 68 of Communications in Computer and Information Science,* eds A. Ranchordas, J. M. Pereira, H. J. Araújo, and J. M. R. S. Tavares (Berlin: Springer), 188–200.

Lowe, D. G. (2004). Distinctive image features from scale-invariant keypoints. *Int. J. Comput. Vis.* 60, 91–110. doi:10.1023/B:VISI.0000029664.99615.94

Mahendran, A., and Vedaldi, A. (2015). Understanding deep image representations by inverting them. *The IEEE Conference on Computer Vision and Pattern Recognition (CVPR),* Boston, MA.

Mechler, F., Reich, D. S., and Victor, J. D. (2002). Detection and discrimination of relative spatial phase by V1 neurons. *Journal of Neuroscience* 22, 6129–6157.

Pellionisz, A., Graham, R., Pellionisz, P., and Perez, J.-C. (2013). "Recursive genome function of the cerebellum: geometric unification of neuroscience and genomics," in *Handbook of the Cerebellum and Cerebellar Disorders,* eds M. Manto, J. Schmahmann, F. Rossi, D. Gruol, and N. Koibuchi (Netherlands: Springer), 1381–1423.

Pouget, A., Dayan, P., and Zemel, R. (2000). Information processing with population codes. *Nat. Rev. Neurosci.* 1, 125–132. doi:10.1038/35039062

Reisfeld, D. (1996). The constrained phase congruency feature detector: simultaneous localization, classification and scale determination. *Pattern Recognit. Lett.* 17, 1161–1169. doi:10.1016/0167-8655(96)00081-5

Scharr, H., Körkel, S., and Jähne, B. (1997). "Numerische isotropieoptimierung von FIR-Filtern mittels Querglättung," in *DAGM Symposium Mustererkennung,* eds E. Paulus and F. M. Wahl (Braunschweig: Springer), 367–374.

Scott, D. W. (1985). Averaged shifted histograms: effective nonparametric density estimators in several dimensions. *Ann. Stat.* 13, 1024–1040. doi:10.1214/aos/1176349654

Sevilla-Lara, L., and Learned-Miller, E. (2012). "Distribution fields for tracking," in *Computer Vision and Pattern Recognition (CVPR), 2012 IEEE Conference on* (Providence, RI: IEEE), 1910–1917. doi:10.1109/CVPR.2012.6247891

Snippe, H. P., and Koenderink, J. J. (1992). Discrimination thresholds for channel-coded systems. *Biol. Cybern.* 66, 543–551. doi:10.1007/BF00204120

van de Sande, K., Gevers, T., and Snoek, C. (2010). Evaluating color descriptors for object and scene recognition. *IEEE Trans. Pattern Anal. Mach. Intell.* 32, 1582–1596. doi:10.1109/TPAMI.2009.154

van de Weijer, J., Gevers, T., and Geusebroek, J.-M. (2005). Edge and corner detection by photometric quasi-invariants. *IEEE Trans. Pattern Anal. Mach. Intell.* 27, 625–630. doi:10.1109/TPAMI.2005.75

Vondrick, C., Khosla, A., Malisiewicz, T., and Torralba, A. (2013). "HOGgles: visualizing object detection features," in *ICCV,* Sydney.

Weinzaepfel, P., Jegou, H., and Perez, P. (2011). "Reconstructing an image from its local descriptors," in *Computer Vision and Pattern Recognition (CVPR), 2011 IEEE Conference on* (IEEE: Colorado Springs), 337–344.

Zeiler, M., and Fergus, R. (2014). "Visualizing and understanding convolutional networks," in *Computer Vision - ECCV 2014, volume 8689 of Lecture Notes in Computer Science,* eds D. Fleet, T. Pajdla, B. Schiele, and T. Tuytelaars (Zürich: Springer International Publishing), 818–833.

Zemel, R. S., Dayan, P., and Pouget, A. (1998). Probabilistic interpretation of population codes. *Neural Comput.* 10, 403–430. doi:10.1162/089976698300017818

Conflict of Interest Statement: The authors declare that the research was conducted in the absence of any commercial or financial relationships that could be construed as a potential conflict of interest.

Permissions

All chapters in this book were first published in FROBT, by Frontiers; hereby published with permission under the Creative Commons Attribution License or equivalent. Every chapter published in this book has been scrutinized by our experts. Their significance has been extensively debated. The topics covered herein carry significant findings which will fuel the growth of the discipline. They may even be implemented as practical applications or may be referred to as a beginning point for another development.

The contributors of this book come from diverse backgrounds, making this book a truly international effort. This book will bring forth new frontiers with its revolutionizing research information and detailed analysis of the nascent developments around the world.

We would like to thank all the contributing authors for lending their expertise to make the book truly unique. They have played a crucial role in the development of this book. Without their invaluable contributions this book wouldn't have been possible. They have made vital efforts to compile up to date information on the varied aspects of this subject to make this book a valuable addition to the collection of many professionals and students.

This book was conceptualized with the vision of imparting up-to-date information and advanced data in this field. To ensure the same, a matchless editorial board was set up. Every individual on the board went through rigorous rounds of assessment to prove their worth. After which they invested a large part of their time researching and compiling the most relevant data for our readers.

The editorial board has been involved in producing this book since its inception. They have spent rigorous hours researching and exploring the diverse topics which have resulted in the successful publishing of this book. They have passed on their knowledge of decades through this book. To expedite this challenging task, the publisher supported the team at every step. A small team of assistant editors was also appointed to further simplify the editing procedure and attain best results for the readers.

Apart from the editorial board, the designing team has also invested a significant amount of their time in understanding the subject and creating the most relevant covers. They scrutinized every image to scout for the most suitable representation of the subject and create an appropriate cover for the book.

The publishing team has been an ardent support to the editorial, designing and production team. Their endless efforts to recruit the best for this project, has resulted in the accomplishment of this book. They are a veteran in the field of academics and their pool of knowledge is as vast as their experience in printing. Their expertise and guidance has proved useful at every step. Their uncompromising quality standards have made this book an exceptional effort. Their encouragement from time to time has been an inspiration for everyone.

The publisher and the editorial board hope that this book will prove to be a valuable piece of knowledge for researchers, students, practitioners and scholars across the globe.

List of Contributors

David Wilkie, Weizi Li and Ming C. Lin
Department of Computer Science, University of North Carolina at Chapel Hill, Chapel Hill, NC, USA

Jason Sewall
Parallel Computing Laboratory, Intel Corporation, Santa Clara, CA, USA

Takuya Otani
Graduate School of Advanced Science and Engineering, Waseda University, Tokyo, Japan
Japan Society for the Promotion of Science, Tokyo, Japan

Kenji Hashimoto
Waseda Institute for Advanced Study, Tokyo, Japan
Humanoid Robotics Institute (HRI), Waseda University, Tokyo, Japan

Masaaki Yahara, Shunsuke Miyamae and Takaya Isomichi
Graduate School of Creative Science and Engineering, Waseda University, Tokyo, Japan

Shintaro Hanawa and Yasuo Kawakami
Faculty of Sport Sciences, Waseda University, Tokyo, Japan

Masanori Sakaguchi
Faculty of Sport Sciences, Waseda University, Tokyo, Japan
Faculty of Kinesiology, University of Calgary, Calgary, AB, Canada

Hun-ok Lim
Humanoid Robotics Institute (HRI), Waseda University, Tokyo, Japan
Faculty of Engineering, Kanagawa University, Yokohama, Japan

Atsuo Takanishi
Humanoid Robotics Institute (HRI), Waseda University, Tokyo, Japan
Department of Modern Mechanical Engineering, Waseda University, Tokyo, Japan

Shlomo Berkovsky and Jill Freyne
Digital Productivity Flagship, CSIRO, Sydney, NSW, Australia

Liyi Dai
Computing Sciences Division, U.S. Army Research Office, Research Triangle Park, NC, USA

Michał Joachimczak, Reiji Suzuki and Takaya Arita
Graduate School of Information Science, Nagoya University, Nagoya, Japan

Rodrigo Verschae
Advanced Mining Technology Center, Universidad de Chile, Santiago, Chile

Javier Ruiz-del-Solar
Advanced Mining Technology Center, Universidad de Chile, Santiago, Chile
Department of Electrical Engineering, Universidad de Chile, Santiago, Chile

Tim Genewein, Felix Leibfried and Jordi Grau-Moya
Max Planck Institute for Intelligent Systems, Tübingen, Germany
Max Planck Institute for Biological Cybernetics, Tübingen, Germany
Graduate Training Centre of Neuroscience, Tübingen, Germany

Daniel Alexander Braun
Max Planck Institute for Intelligent Systems, Tübingen, Germany
Max Planck Institute for Biological Cybernetics, Tübingen, Germany

Simon Gallo, Laura Santos-Carreras, Tristan Vouga and Hannes Bleuler
Laboratory of Robotic Systems, School of Engineering, Ecole Polytechnique Fédérale de Lausanne, Lausanne, Switzerland

Giulio Rognini
Laboratory of Robotic Systems, School of Engineering, Ecole Polytechnique Fédérale de Lausanne, Lausanne, Switzerland
Center for Neuroprosthetics, Ecole Polytechnique Fédérale de Lausanne, Geneva, Switzerland
Laboratory of Cognitive Neuroscience, Brain Mind Institute, Ecole Polytechnique Fédérale de Lausanne, Geneva, Switzerland

Olaf Blanke
Center for Neuroprosthetics, Ecole Polytechnique Fédérale de Lausanne, Geneva, Switzerland
Laboratory of Cognitive Neuroscience, Brain Mind Institute, Ecole Polytechnique Fédérale de Lausanne, Geneva, Switzerland
Department of Neurology, University Hospital of Geneva, Geneva, Switzerland

Yiannis Karayiannidis
Department of Signals and Systems, Chalmers University of Technology, Gothenburg, Sweden
Center for Autonomous Systems, Royal Institute of Technology (KTH), Stockholm, Sweden

Leonidas Droukas, Dimitrios Papageorgiou and Zoe Doulgeri
Department of Electrical and Computer Engineering, Aristotle University of Thessaloniki, Thessaloniki, Greece

Bulcsú Sándor, Tim Jahn, Laura Martin and Claudius Gros
Institute for Theoretical Physics, Goethe University Frankfurt, Frankfurt am Main, Germany

Santiago Morante, Juan G. Victores and Carlos Balaguer
Robotics Lab, Automation and Engineering Systems Department, Universidad Carlos III de Madrid, Madrid, Spain

David Balduzzi
School of Mathematics and Statistics, Victoria University Wellington, Wellington, New Zealand

Ralf Der
Max Planck Institute for Mathematics in the Sciences, Leipzig, Germany

Suraj Srinivas, Ravi Kiran Sarvadevabhatla, Konda Reddy Mopuri, Nikita Prabhu, Srinivas S. S. Kruthiventi and R. Venkatesh Babu
Video Analytics Laboratory, Department of Computational and Data Sciences, Indian Institute of Science, Bangalore, India

Harshal Arun Sonar and Jamie Paik
Reconfigurable Robotics Laboratory, Institute of Mechanical Engineering, Ecole Polytechnique Fédérale de Lausanne, Lausanne, Switzerland

Siqi Zhang
UMR 7503, Laboratoire Lorrain de Recherche en Informatique et ses Applications, Centre National de la RechercheScientifique, Vandoeuvre-lès-Nancy, France
School of Marine Science and Technology, Northwestern Polytechnical University, Xi'an, China

Dominique Martinez
UMR 7503, Laboratoire Lorrain de Recherche en Informatique et ses Applications, Centre National de la RechercheScientifique, Vandoeuvre-lès-Nancy, France

Jean-Baptiste Masson
Physics of Biological Systems, Institut Pasteur, Paris, France
UMR 3525, Centre National de la Recherche Scientifique, Paris, France

David Vernon
Interaction Lab, School of Informatics, University of Skövde, Skövde, Sweden

Michael Beetz
Institute for Artificial Intelligence, University of Bremen, Bremen, Germany

Giulio Sandini
Department of Robotics, Brain and Cognitive Sciences, Istituto Italiano di Tecnologia, Genova, Italy

Michael Felsberg and Kristoffer Öfjäll
Computer Vision Laboratory, Department of Electrical Engineering, Linköping University, Linköping, Sweden

Reiner Lenz
Computer Graphics and Image Processing, Department of Science and Technology, Linköping University, Norrköping, Sweden

www.ingramcontent.com/pod-product-compliance
Lightning Source LLC
Chambersburg PA
CBHW080629200326
41458CB00013B/4568